CANE SUGAR TECHNOLOGY

THE COMPLETE REFERENCE

By Dilip Jambhale

B. Sc. A. V. S. I. (sugar tech,)
Consulting sugar technologist, former technical adviser
And in charge head of sugar technology dept. VSI Pune

2023

Author: Dilip Bhanudas Jambhale

ISBN: 9798859360673

Dedicated to
My wife Varsha,
Daughter Gauri, son in law Amit,
Son Harish, daughter in law Mrunal,
And my grandchildren Mihir, Kaiwalya and Vihang

MODERN ISGEC SUGAR PLANT

CONTENTS

Abbreviations of References of the proceedings and journals

ASSCT	Australian Society of Sugar Cane Technologists
BSTA	Barbados Sugar Technologist Association
DSTA	Deccan Sugar Technologists Association (India)
IS	Indian Sugar
ISJ	International Sugar Journal
ISSCT	International Society of Sugar Cane Technologists
SASTA	South African Sugar Technologist Association
STAI	Sugar Technologist Association of India
SISTA	South Indian Sugar Technologist Association

PREFACE

In last 30 years considerable progress has been made in equipment and process of cane sugar manufacture to improve productivity, efficiencies and quality of sugars. For example steam consumption which was 44 to 48 % on cane was considered as normal. Now in modern plants it has been brought down to 27-28 % on cane. It finds a need of a new book on sugar technology that covers all the modern changes in equipment and process. In writing this book efforts have been made to incorporate all the new concepts in the manufacture of sugar.

This book covers theoretical as well as practical aspects of each and every operation involved in the process of sugar manufacture from sugar cane. Almost all topics from harvesting of sugar cane to production and storage of sugar and molasses have been incorporated in this book.

Different technological options have also been discussed, for instance, in juice extraction from cane apart from milling, diffusion has also been described in details.

In last twenty years many old factories have modernised the plants. New high boilers with backpressure / condensing turbines have been installed. Thus the factories have gone for export of electricity and are getting monetary gain. The single stage drive turbines of mills have been replaced by higher efficient electric drives.

In juice clarification the short retention clarifiers and decanter centrifuge are now used in modern plants have been discussed in details.

Efficient utilisation of steam is very much essential to save the bagasse. This book includes different evaporator configurations and steam economy.

To conserve the steam and heat energy raw juice heating from 45 to 69°C by condensate has been common in modern plants.

Now in last 10 years in modern plants FFE quadruple set with all last effect vapours for pan boiling is being adopted.

The book covers different massecuite boiling schemes, material balance at pan station, diversification of boiling house products, effluent treatment, computerization of the plants etc.

It also includes following chapters that gives basic knowledge of the units –

1) Condensing and cooling system, 2) steam generation, 3) boiler feed water and boiler water treatment, 4) steam turbines, 5) export of power, 6) effluent water treatment.

   The basic concepts of technical control has been included.

It also includes basics of automation of the process followed at each unit and also include some information about computerisation of the plant.

Thus the book contains considerable material, which has not appeared previously in book form.

Hope this book will be useful for process managers, works managers, process superintendents / sugar engineers' working in factories as well as in sugar machinery fabrication workshops as reference book. This book will also be useful for students of sugar technology and sugar engineering.

I must remember late D S Lande, Mangal Singh and A R Patil. Thanks are also due to my colleagues from the industry. I must also mention the inspiration that my family gave me in writing this book.

Dilip B. Jambhale
Pune, India
2nd September 2023

-OOOOO-

1 HISTORY OF SUGARCANE AND SUGAR

Researchers have found that sugarcane is originated in south pacific islands. The people of New Guinea were probably the first to domesticate sugarcane, sometime around 8,000 BC. Wild sugarcane also found in eastern and northern part of Africa. However the centre of origin of sugarcane may be India as the wild sugarcane with smallest number of chromosomes found in India.

Fig. 1.1 Sugarcane

Sugarcane and sugar are mentioned in oldest sacred Sanskrit book of Hindu religion Ayurveda (1500 BC). This indicates that the history of sugarcane cultivation and sugar manufacture has started some 3500 years back in India. In "Ayurveda" sugarcane was known as "Ikshu". The word "Ikshu" has no parallel word in any other Indo-Aryan language. This indicates that the Aryans come to know about the plant only after entering India. Proto-Australoids or Austrics, who came in India before Dravidians, Nordic Aryans and Mongoloids laid the foundation of Indian civilization.

They cultivated rice, vegetables and sugarcane and made sugar from sugarcane. In the "Institution of Manu" numerous passages clearly deal with Ikshu (sugarcane) and sharkara (sugar). According to Hindu mythology the sugarcane plant was originally created by the renowned hermit Vishwamitra to take the place of heavenly sweets in the temporary paradise organised by him for the sake of King Trishanku.

George Watt (in 1889) says, *"There is little room for doubt that the world is indebted India for sugarcane. Botanical evidence is by no means backward in lending confirmation to the idea of India having been the original home of the sugarcane".*

The process of manufacture of cane sugar granules from sugarcane juice had also been developed in India in about 400 to 700 BC. In Sanskrit language Sharkara means crystallized sugar. All the words for sugar in Arabic, Persian, Greek, and Latin and in modern languages of India and Europe are originated from the Sanskrit word 'Sharkara'.

Fig. 1.2 Cane crushing 200 years back

The word for sugar in Arabic 'Kand' is derived from Sanskrit khand (candy sugar). The English word candy is also derived from Kand or khand and in both languages it means sugar crystallised into large pieces. Charaka and Sushruta mentioned the word Guda in Sanskrit. Cane juice boiled to thick mass was called as Guda, Gula or Gur – a solid mass and these words are being used at present in most of the south Asian countries.

Koutilya - the guru and minister of Emperor Chandragupta, the founder of Maurya dynasty (300 BC), notes down the sugarcane in the list of principal crops cultivated. There are lot of references in Tamil sangam literatures like Purananuru, Ainkurunuru, Perumpaanaatruppadai, Paṭṭiṉappālai and Akananuru about cultivation of sugarcane and making sugar. It is mentioned in Purananuru that the sugar cane was brought to Tamil land from an unknown place during the Sangam period. In Purananuru and Ainkurunuru, sugarcane juice extraction with use of huge machineries was compared with the sound made by elephants and the smoke produced during the process of making of sugar was like clouds.

In 510 BC Darius, the Persian Emperor invaded the Indian sub-continent. He found that the people used a substance from a plant to sweeten their food. Until then the Persian people had used honey to sweeten food, and so they called sugar cane 'the reed which gives honey without bees'.

In the fourth century AD, Alexander the Great conquered parts of Indian subcontinent and carried with him the 'sacred reed' sugarcane while returning from Punjab and the plant travelled to west.

Indian sailors carried sugar by various trade routes. During the reign of Harsha (606–647 AD) in North India travelling Buddhist monks brought sugar crystallization methods to China. Chinese documents confirm at least two missions to India, initiated in 647 AD, for obtaining technology for sugar-refining. China established its first sugarcane cultivation in the seventh century. In South Asia, the Middle East and China use of sugar in royal families was common.

Fig. 1.3 traditional gur manufacturing in Punjab

In the Ain-e-Akbari (1590 AD) method of sugarcane cultivation and manufacture of all forms of sugar is described in details. The refined sugar in Hindi language is also called as 'Chini' as the technology of sugar refining seems to be developed in China and then diffused to India.

This clearly indicates that the knowledge of sugarcane cultivation and sugar manufacturing was conveyed from India. The culture of the sugarcane had been spread by the Arabs to northern Africa and south Europe. At the same time, that was being carried into Java and the Philippines by Chinese. The Crusaders brought sugar to France in the eleventh century. In 1420, the Portuguese had conveyed it from Sicily to Madeira and to St. Thomas Islands. The Spanish conveyed the cultivation to the Canary Islands in the fifteenth century. Columbus introduced sugarcane into Santo Domingo on his second voyage in 1494 and from there; it was carried to Cuba, West Indian and central and south America. The Dutch first established sugar works in Brazil in 1580 but on being expelled from that country by Portuguese, they carried the art of sugar manufacture in 1655 to the West Indies. Sugar was manufactured by

English in Barbados in 1643 and in Jamaica in 1664. The cultivation in Louisiana started from 1750.

Fig. 1.4 old sugar refinery in Europe

By the end of the medieval period, sugar was known worldwide but was very expensive and was considered a "fine spice". In the year 1792, sugar prices in Great Britain were rose to very high level. The East India Company was then called upon to lend their assistance to help in the lowering of the price of sugar. On 15 March 1792, his Majesty's Ministers to the British Parliament presented a report related to the production of refined sugar in British India. Lieutenant J. Paterson, of the Bengal establishment, reported that refined sugar could be produced in India more cheaply than in the West Indies. From 1800, technological improvements and New World sources began turning it into a cheaper bulk commodity.

Sugar refineries were built in Germany France and England in the sixteenth century but modern cane sugar refining dates from the early part of the 19th century in England. Until first half of the 20th century sugar was so costly that only noble families can use it. Miraculous curative power was supposed to be attributed in it.

References:

1 George Watt “A dictionary of economic products of India” vol. VI part II (published in 1898) p 29-40

2 ibid. P 29-40

3 ibid. p 29 4 ibid. p 320-322

-OOOOO-

Chapter 2 WORLD CANE SUGAR INDUSTRY

In the world approximately 180 million tonnes (raw value) of sugar is produced every year. Around 80% of global sugar is extracted from sugarcane, and remaining 20% from sugar beet. In USA and in some countries high fructose corn syrup (HFCS) is used as sweetener in food industry. The production of which in the world is about 15 million tonnes.

There are some 1700 cane sugar factories all over the world. The factory capacities vary from 800 tcd to 35000 tcd. In many countries the number of sugar factories is decreasing. Old plants of smaller capacities are being closed and selected plants are being expanded for higher capacities with higher efficiencies. However in India the number of factories is still increasing. The highest sugar recovery is of Australia and India.

Table 2.1 World Sugar Production

Sr. No.	Year	Production MMT	Export MMT
1	2019-20	167	56
2	2020-21	180	65
3	2021-22	180	67
4	2022-23	177	67
5	2023-24*	184	72

*expected

Table 2.2 Top 10 sugar producing countries 2022-23

Sr. No.	**Country**	**No of sugar factories installed/ in operation**	**Production MMT**
1	Brazil	330	38.05
2	India	620	32.00
3	E U	---	14.90
4	Thailand	57	11.04
5	China	400	9.00
6	USA	---	8.42
7	Russia	---	7.18
8	Pakistan	80	6.86
9	Mexico	51	5.71
10	Australia	29	4.20

It should be noted that only in 2019-20 Brazil is ranking second in sugar production otherwise Brazil is always the highest sugar producer in the world. Brazil ranks first in sugar exports. Depending on demand and supply in international market Brazil diverts sugar production to ethanol production and controls its sugar production.

In US, EU, sugar is produced from cane as well as beet. In Russia sugar is produced from beet.

The sugar industry in Australia, South Africa, and USA are comparatively advanced in technology in respect of mechanization, instrumentation, automation, and computerization in process control.

Table2.3 Top sugar exporting countries 2023

Sr. No.	Country	% Share of Exporter in world
1	Brazil	66.69
2	India	19.99
3	Mexico	3.30
4	Guatemala	2.45
5	El Salvador	1.36
6	South Africa	1.36

Table 2.4 Top sugar importing countries

Sr. No.	Country	Approx. Import MMT
1	China	4.0
2	Indonesia	4.0
3	USA	2.8
4	Bangladesh	2.1
5	Algeria	1.9
6	Malaysia	1.8
7	Korea Rep. of	1.7
8	Nigeria	1.3
9	Iran	1.3
10	Sudan	1.3

-OOOOO-

3 SUGAR INDUSTRY IN INDIA

Sugar industry in India is an important agro-based industry. India is the second largest producer of sugar in the world after Brazil however India is also the largest consumer of sugar. At present (in 2021) there are 732 sugar factories installed in the country. The total capacity of the factories to produce sugar is around 34.5 MMT per year. The large number of sugar mills is in Maharashtra state (240), in Uttar Pradesh (155). And Karnataka (72). Nearly 80 % of the sugar production in India is produced in these three states. Employment has been generated in rural areas because of sugar factories. Further various additional activities such as servicing of machinery, transport; trade and supply of agriculture inputs etc. are generated.

3.1 Sugar cane production in India

Sugarcane is an important commercial crop. After gaining Independence, India made plans for overall industrial development of sugar industry. In year 1961 Sugar cane was grown in 2413 thousand hectare and production was 110 million tonnes. Now in 2019 it is being grown in 5061 thousand hectare and production is 405 million tonnes. The sugarcane production in 58 years has been increased from 45 tonnes/hectare to 80 tonnes/hectare... About 50 million farmers cultivate sugarcane and around 5 lakh workers employed in sugar mills. In addition around 1.2-1.5 million harvesting and transportation labour is engaged in the season. Uttar Pradesh has largest cane area almost 50 % of the total nation.

Sugarcane is generally planted during rainy season (June to September) also in some area in October- November and in March. About 60-65 % of the sugar cane crop is crushed by sugar factories, 25-30 % crop is used for production of gur (Jaggery) and khandsari sugar and the remaining 7 to 9 % of the sugarcane crop is used for seeding.

3.2 Sugar Industry:

In India at present there are 732 licensed sugar factories out of which 630 have be installed. The capacities of sugar factories vary from 800 tcd to 20,000 tcd. There are some 220 factories with cogeneration plants. There are some 315 distilleries attached to the sugar factories and there are 20 refineries. However, only 500 to 550 sugar factories are mostly in operation. About 250 sugar factories are in co-operative sector and the remaining in private and public sectors. Nearly 5 lakh manpower is employed in the industry. The crushing season of the factories is from October to April, however in south Karnataka and Tamilnadu some factories run for 7 to 8 months. The yearly sugar production in India now is around 300 lakh MT. But it varies from 280 to 320 lakh MT depending upon monsoon rains and climatic conditions.

Bagasse, molasses and press mud are the co-products generated during sugar production. Molasses is used for manufacture of, rectified spirit, alcohol; ENA, ethanol and many other alcohol based downstream products. As mentioned above there are some 315 distilleries attached to sugar factories. Indian sugar industry meets the demand of potable alcohol as well as 10 % blending in petrol.

The bagasse is burned and used for production of steam to produce electricity and run boiling house in the plant. The excess bagasse can be sold to nearby industries as fuel. This bagasse can be utilized for production of extra electricity that can be supplied to state grid. About 220 sugar factories have gone for cogeneration with installation of high-pressure boilers and exporting the surplus power generated to state electricity boards. The industry is able to export around 1,300 MW of power to the grid.

Some factories are having paper plants or particle board / hardboard plants and use saved bagasse as a raw material for the plants. The filer cake /press mud is utilized as manure in the sugar fields. The industry is mainly established in three states – Uttar Pradesh, Maharashtra and Karnataka and 75 to 80 % of country's sugar production is from these three states.

There are four leading sugarcane research institutions, 22 state sugarcane research stations. There are many sugar equipment and machinery manufacturing companies in India. Some the machinery manufacturers give sugar plants on turnkey basis. They fulfil the demand of the industry in the country as well export sugar machinery to various countries in the world.

India has largest technical manpower. The institutions like National Sugar Institute Kanpur, and Sugar Cane Research Institute Coimbatore are run by government of India and Vasantdada Sugar Institute Pune in Maharashtra state and S. Nijalingappa Sugar Institute at Belgawi in Karnataka state is run by cooperative sugar factories from that state. These institutes provide postgraduate level education in sugar technology, sugar engineering and alcohol technology specially designed for sugar industry.

-OOOOO-

4 SUGARCANE QUALITY AND MATURITY

4.1 Sugar in Sugarcane:

Sugar which is an essential ingredient of our foods as well as hot and soft drinks is extracted in crystal form from the plants sugarcane and sugar beet. There are many sugars in different plants. If we go through the chemistry of sugars, these are carbohydrates - a group of organic compounds. The carbohydrates are composed of carbon, hydrogen and oxygen. The number of hydrogen and oxygen atoms in sugar molecules is usually in 2:1 ratio. The sugars glucose; fructose, etc. cannot be further disintegrated into smaller molecules of carbohydrates. Therefore these sugars are known as monosaccharides or simple sugars. Monosaccharides usually contain five (pentoses) or six (hexoses) carbon atoms. The commercial sugar available in market is sucrose. It is a disaccharide.

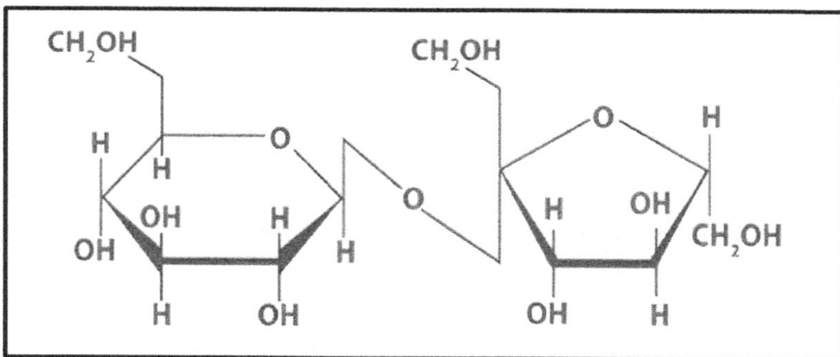

Fig. 4.1 Structural formula of sucrose

Sugarcane and Sugar beet are the two plants from which manufacture of sugar is techno-economically feasible. Now in many countries High Fructose Corn Syrup (HFCS) is being used instead of sugar.

In sugarcane first glucose and fructose are formed from water and atmospheric carbon dioxide taking sunlight as an energy source. The reaction is called as photosynthesis. Chlorophyll a green pigment present in leaves acts as a catalyst in this reaction.

Further the simple sugars glucose and fructose are condensed together to form sucrose. In winter most of the glucose and fructose in the cane are condensed to form sucrose. This phase is called as maturity of cane. When sugarcane is harvested at peak of its maturity it gives higher sugar recovery. For higher sugar recovery the sugarcane is harvested in winter at peak of its maturity when it contains maximum sucrose. Therefore crushing season in India is from November to April.

However, in some tropical areas like Tamilnadu and south Karnataka, where there is not much difference in temperature throughout the year, the maturity of cane comes at the drier period of year when soil moisture is reduced. In these areas factories work round the year with a shutdown of about 3 to 4 months in rainy season only when sugar recovery is low and harvesting and transportation is difficult.

It should be noted that sugar is not really manufactured in a sugar factory but is extracted from the cane which is in dissolved form in juice and made it into crystal form in the factory process.

4.2 Cane Quality:

Growth of sugarcane crop is satisfactory in between temperature 25-38°C. At higher temperature above 38° C and at lower temperature below 21°C the growth rate is reduced. It is almost none below 10°C. The quality of sugar cane is absolutely important in achieving higher sugar recovery throughout the season. The cane quality improves with age, reaches to maximum sucrose content and then declines. Immature cane contains high percentage of water and reducing sugars while less percentage of sucrose and fibre. When plant is fully matured the quantity of simple sugars in stalks is small although there are variations in different varieties and at different region.

After maturity phase if cane is allowed to stand in the fields, it becomes over matured, sucrose content goes down and more undesirable organic impurities are formed at the expense of sucrose. In over matured cane the fibre content increases. This ultimately affects mill extraction. Recoverable sugar is reduced due to higher impurities. Therefore, supply of matured cane is most essential for a sugar factory.

Planning of plantation of early, mid and late maturing varieties and ratoon is helpful in obtaining matured cane supply to a sugar factory with high sucrose content throughout the season.

Exhaustion of applied nitrogen and withdrawal of soil moisture three weeks before harvesting improve the quality of cane. The gradual withdrawal of sheath moisture from 84-85 to 73-74 percent helps in improving the sucrose content. When soil moisture is withdrawn the glucose and fructose molecules present in the plant cells condense together to form sucrose molecules and the water freed is utilized for activity of the plant and ripening is forced. Use of ripening chemicals also can be helpful to bring the maturity earlier.

Preharvest maturity survey improves quality of sugarcane received by the factory giving a rise in recovery by about 0.3 to 0.8 units in the first one/two months of crushing season. After this period most of the cane in the fields is matured and therefore the harvesting is planned according to the date of planting.

There are two methods for testing the maturity of cane crop. One method of testing the maturity of cane is to cut few stalks of cane from the fields and take them to the factory laboratory for analysis for expected recovery. The other method is to extract few drops of juice from the middle internodes of standing cane stalks in the fields with the help of puncturing needle and take brix reading of the juice drops by hand refractometer.

Flowering of sugarcane is observed under certain climatic conditions. The flowers appear on the upper tip of the cane. Some varieties are more liable to flowering than others. Flowering means the vegetative cycle is completed and the stalks further will not grow.

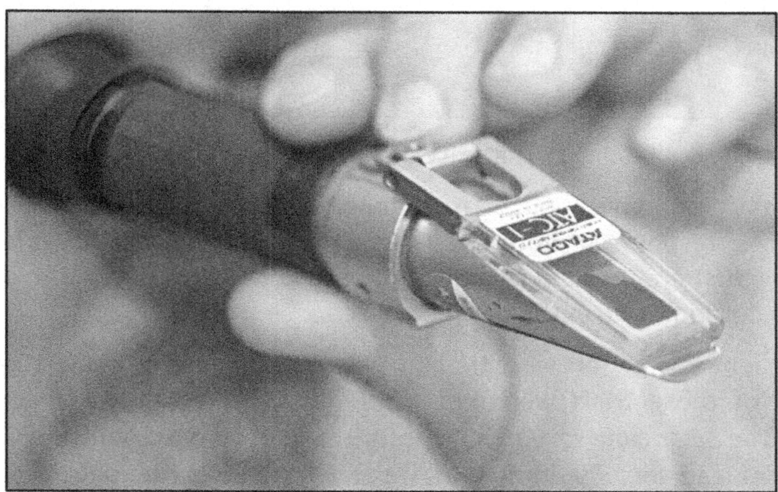

Fig.4.2 Sugarcane maturity testing by refractometric brix measurement

4.3 Forcing maturity of Cane by using Chemical Ripeners:

At the starting of the season the cane is immature and pol% cane is less hence sugar recovery is low. To increase the pol % cane in the early days of crushing season chemical ripeners can be used. So that sugar recovery in factory can be improved. It is followed in some sugarcane growing countries mostly in developed industry. It is used by aerial spray.

Chemicals like Ethrel, Polaris (Glyphosine), Polado (Glyphosates), and sodium metasilicate can be used for forcing the maturity of the cane at the early start of the season. The chemicals are generally used in large scale sugarcane plantations. These are sprayed on cane crop.

With the use of chemical ripeners the pol % cane is increased. Some chemicals have showed good results in improving plant growth and sugar accumulation in sugarcane. The gains in pol % cane have been reported from 0.5-2.0 units, depending upon variety of cane, weather conditions and ripener interaction.

Fig. 4.3 testing of maturity of cane by hand refractometer

Ideal timing for application of ripener with respect to harvest is very much important. The response of different varieties of sugarcane is different. Therefore these usually first are used on trial basis. It is also claimed that the chemicals also prevent the deterioration in the decline phase.

In India chemical ripeners were tried by a number of factories some 50 years back. But results of those trials were not much discussed and then they were not continued. The sugarcane research institutes have developed some early maturing varieties of sugarcane. Plantation of these varieties can also help to improve recovery at the start of the season.

-OOOOO-

5 HARVESTING AND TRANSPORTATION

In developing countries where labour is cheap, harvesting and loading is done mostly manually. However, in developed countries labour is not cheap and is reluctant to work hard in hot sunny atmosphere. Therefore manual harvesting is more costly and therefore almost all cane is harvested mechanically. Manually harvested and loaded cane is comparatively clean and well loaded, as the stalks are being parallel to each other in bundles.

In manual harvesting loss of cane in fields is comparatively less than mechanical harvesting. The extraneous matter with the cane is also less. However now labour is becoming scares and is unwilling for harvesting work. When loading is done manually more time is required and the vehicle utilization and productivity usually adversely affects.

Fig. 5.1 cane uploading in tractor trailer

Harvesting is done in daylight but loading is continued up to late night whereas transportation goes for 24 hours. At Hutatma co-op. sugar factory harvesting in two shifts was tried to reduce the time lag between harvesting and crushing. The average time lag was brought down up to 12 hrs. This resulted in considerable increase recovery in that period. However the system could not be continued as the harvesting labours were reluctant to work late at night.

Fig. 5.2 Cane transportation by carts

Fig. 5.3 Mechanical handling of cane

In India, mechanical harvesting is being followed in some factories by imported mechanical harvesters. But because of small field and different Indian plantation practices the harvesters cannot be used in many cane fields. According to experts even with suitable cane plantation, in India only 30 to 40 % cane can be harvested mechanically.

In Indian Sugar Industry transportation of cane up to 10-30 % is still by bullock carts. The remaining cane transportation is managed by trucks and tractor trolleys. Transporting by 2-ft gauge railway is economic, however because small and scattered sugar cane fields it cannot be effectively used in India and hence transportation by rail is now not in existence in Indian sugar Industry.

-OOOOO-

6 BURNT HARVESTED CANE

In India and in most of the Asian countries green cane harvesting is a common practice. From World War II onwards, burning was become a standard practice in some countries like Australia, U S A, South Africa, Brazil.

In Gujarat state of India, in most of the factories about 40 to 70 % of the cane arrives is burnt cane. However, the burning of the cane is not planned by the factories but farmers burn the cane and then report to the factories for harvesting. The factories have to arrange the harvesting there after. This delays the harvesting and the time lag between burning and harvesting and further crushing in mills is much higher. This causes deterioration of cane. Juices from such cane give trouble in processing.

Green cane harvesting

Advantages: Trash blankets help to suppress weed growth. This provides labour and cost savings in ratoon crop. Trash blankets also help to prevent evaporation of water from the soil surface and allow better water infiltration. This can reduce irrigation requirement and produce higher cane yields in drier areas.

Disadvantages: With green harvested cane the extraneous matter with cane is always on higher side. The cost of harvesting is also more. Some harvesters have difficulty in cutting green cane and therefore burning of cane before harvesting is to be followed for such harvesters.

Burnt harvested cane

Advantages: Burning of standing cane crop removes the trash and to some extent wax on the periphery of the cane stalks. Pre-planned and controlled burning of the standing cane before harvesting reduces trash percentages and extraneous matter arriving at the factory. This makes easier for harvesting labour to cut and load the cane. This also reduces pest and disease problems if the cane is immediately harvested, transported to the factory and crushed in time. While crushing the burnt cane mill capacity is increased by 10-15 %.

Disadvantages: However burnt chopper harvested cane is more susceptible to attack by microorganisms and hence burnt cane deteriorates rapidly. If rain falls upon burnt cane, it is further accelerated.

Intense fire may adversely affect by killing tissues inside the stalks resulting in appreciable loss of sucrose. Dextran and ethanol levels in the green harvested fresh cane are only in traces. However in chopper harvested burnt cane it may increase rapidly. The higher level of dextran gives trouble in the process. In burning the cane, splitting of rinds may occur. It allows bacteria to penetrate in and attack the juice holding tissues of the stalks. This results in rapid deterioration. Delay in delivery further increases this problem. Therefore after burning the cane should be harvested fast and the time lag between harvesting and crushing should be much less. The clarification of juice from burnt cane is usually not so well as green cane. The evaporator heating surface also fouls sooner.

Fig. 6.1 preharvest burning of cane

Now pre-harvest burning of sugarcane is one of the most sensitive environmental issues. Burning of cane before harvesting is being strongly opposed from environmental point of view. Further burning of cane adversely affects health particularly respiratory system of harvesting labours as well as the villagers in the area of burning of cane. Villagers sometimes complain of inconvenience and discomfort caused by smoke and soot from cane fires. But still burnt cane harvesting is being followed in some countries.

-OOOOO-

7 EXTRANEOUS MATTER

The cane received at the factory contains lot of matter like cane tops, green leaves, trash and some soil, which contain very little or no sugar. This material is called extraneous matter of the cane. With Manual harvesting stripping of cane and removal of the tops is done effectively and therefore the percentage of extraneous matter is generally less.

Sethi gives observations made in a factory for percentage of extraneous matter of the cane harvested manually. The average percentage of binding material was 1.48% and the remaining extraneous matter was 2.10% on cane. Thus the total extraneous matter was 3.58%. Some varieties such as Co 671 remain with less adhering to the stalks when those are matured.

With mechanical harvesting removal of extraneous is not done and cane contains generally 7 to 12 % extraneous matter. In South Africa cane crushed contains about 12 % extraneous matter. During rains the extraneous matter increases. More soil is adhered with flood affected cane. In Puerto Rico the amount of extraneous matter may be as high as 25% during rainy season.

Higher extraneous matter, especially soil that causes heavy wear and tear of factory equipment and increases suspended solids in mixed juice. It is observed that the suspended matter percentage in mixed juice from mechanically and loaded cane is always 1.5 –3.0 times higher than manually harvested and loaded cane. Further the bagasse with higher soil content does not burn properly in boilers. Efficiency of mill and boilers is decreased. Smits has described in detail in his paper 'Implication of poor cane quality' the deleterious effects of high extraneous matter with the cane.

Cane tops and green leaves contain little sucrose (about 1%) but pass from mills along with bagasse with more sucrose and sugar loss in bagasse is increased. The little juice extracted from the tops is of very low purity, rich in unwanted nonsugars like starch, amino acids, polyphenols, organic acids, salts etc. These constituents adversely affect the process:

- Starch gives viscosity to intermediate products in the process.
- Amino acids react with reducing sugars to give highly coloured complex organic compounds, which are melassigenic or molassigenic (molasses forming) in nature. Therefore, the recoverable sugar percentage is reduced.
- Polyphenols and soil particles give trouble in settling of mud also polyphenols develop dark colour compounds.

Effect of unwanted nonsugars and soil /clay in the juice

1. Cane Preparation index and extraction is decreased,
2. Mixed juice purity is decreased,
3. Percentage of suspended solids in mixed juice is increased ,
4. Filter cake % cane is increased,
5. Detrimental effect on sugar crystallization,
6. Distortion of crystals - elongated and needle shape crystals may be formed,
7. Increase in massecuite viscosities,
8. Poor exhaustion and hence high final molasses purity,
9. Rapid blinding of centrifugal screens,
10. Pol % cane is decreased,
11. Overall recovery is decreased.

References:

- Clements report HST 1949
- David Eastwood ISJ Nov.1997
- Gly James ISJ Nov.1999
- Sethi IS April. 1996
- Smits SASTA 1976, p 227-230.
- Cabrer - ISSCT 1965, p 1529 - 1538.
- Vignes ISSCT 1980, p 2221-2229
- Smits SASTA 1976, p 227-230

-OOOOO-

8 SUGARCANE CONSTITUENTS

8.1 General:

Clean matured sugarcane contains about 70-75 % water, 13- 17 % sugar and other dissolved solids and the remaining is dry insoluble fibre. The constituents of cane and juice and their percentage depend on maturity and variety of the cane. The type of soil, climatic conditions, agricultural practices, fertilizers used and water availability also effects the composition of cane. The cane tops contains less than 1 percent sucrose and are rich in starch, soluble polysaccharides, and reducing sugars.

The cane juice is opaque and turbid because of presence of colloidal substances such as starch, proteins, gums, waxes, silica etc. The colloids exist in solution in fine dispersion state. The colour of the juice is generally dark green but varies considerably depending on colouring matters present in it. Cane juice is acidic and pH is usually in the range of 5.0 –5.4. The constituents of matured cane and their percentage are given in the following table.

Table 8.1 - Sugarcane Constituents

Sr. No.	Constituent	Percentage
1	water	70--75 %
2	sucrose	11 -- 16 %
3	reducing sugars	0.5 -- 2 %
4	Organic compounds	0.7 -- 2.5 %
5	Inorganic compounds	0.4 -- 0.8 %
6	dry fibre	10 -- 17 %

Table 8.2 The fibre constituents of the Sugarcane

Sr. No.	Constituent	Percentage
1	Cellulose	30 -- 38 %
2	Pentosan	24 -- 30 %
3	Lignin	18 -- 22 %
4	Ash	1 -- 4 %

Immature cane contains more water and reducing sugars percentage and less fibre. When cane is over matured organic acids are increased at the expense of sucrose and therefore sucrose percentage and the pH of the juice is decreased. Therefore more lime is required for clarification of juice from over matured cane. The lime reacts with the excess acids to form soluble salts that pass in clarified juice and form scale in evaporator. Also molasses purity, percentage and consequently sugar loss in molasses is increased.

8.2 Sugars

The sugarcane juice contains sucrose in dissolved form mainly in stalks although it is present in all parts. The juice is stored in parenchyma cells of the stalks. The percentage of sucrose goes on decreasing and that of reducing sugars (glucose and fructose) goes on increasing from bottom to the top of the cane stalks. In early days of season comparatively immature cane arrives at the factory which contains more percentage of reducing sugars. In raw juice the glucose and fructose seldom occurs in equal proportion[1].

8.3 Inversion of Sucrose:

Sucrose solution is optically dextrorotatory and specific optical rotation is + 66.53. When sucrose solution is hydrolysed it gives equimolecular amount of glucose and fructose and after complete hydrolysis, it's the optical rotation becomes – 39.7. This change of dextrorotatory optical activity to levorotatory due to hydrolysis is termed as inversion of sucrose. Inversion can be brought about by heating sucrose solution with dilute acid.

The quantitative elementary composition of sugarcane was determined early in the nineteenth century by Gay Lussac and Thenard[2] who probably have established our present molecular formula of sucrose $C_{12}H_{22}O_{11}$. Sucrose is optically active and its solution rotates a polarised light to right side.

The angle of rotation is proportional to sucrose percentage in solution. This property is used in quantifying the sugar percentage of process materials right from mixed juice to final molasses. This method is quick and more or less exact unless other chemicals of similar property interfere. The reducing sugars do interfere but as they are in small quantities the error is ignored in routine analysis.

As soon as cane is harvested inversion of sucrose starts due to acidity of the juice and enzyme invertase which is present in the cane. Acid inversion may continue throughout process. It depends on temperature, pH and time. Enzymatic activity depends on temperature and variety as well as maturity of cane. The enzymatic inversion is stopped when raw juice is heated and the enzyme invertase is destroyed.

8.4 Destruction of Reducing Sugars:

Glucose contains an aldehyde group and it is therefore an aldohexose. Fructose contains a ketone group and hence is a ketohexose. Both these sugars become acids by oxidation and they become hexavalent alcohols by hydrogenation. The sugars are also called as Reducing Sugar (RS) because they reduce copper in a Fehling solution and Cuprous oxide (Cu_2O) is precipitated. However sucrose does not reduce copper in Fehling solution and therefore this property of RS is used to quantify the RS percentage in sugar solutions or juices. The condensation reaction of glucose and fructose giving sucrose that takes place in sugarcane standing in field is not possible in laboratory.

Table 8.3 Properties of Sucrose, Glucose and Fructose:

Sr. No.	Property	Sucrose	Glucose	Fructose
1	Empirical formula	$C_{12}H_{22}O_{11}$	$C_6H_{12}O_6$	$C_6H_{12}O_6$
2	Molecular weight	342.3	180.2 / 198.2	180.2
3	Crystal structure	Monoclinic	Rhombic (anhydrous}	Orthorhombi c
4	Density	1.588	1.544	1.598
5	Optical activity	Dextrorotary	Dextrorotatory	Laevorotator y
6	Specific rotation at 20°C (At equilibrium for glucose and fructose)	66.53	52.7	-92.4
7	Solubility in water at 20° C(% by weight)	67.09	57.60	81.54
8	Melting point	188° C	146° C	105° C
9	Chemically active group	-	Aldehyde group	Ketone group
10	Effect of alkaline medium soln. and heat	Negligible	Destruction	Destruction
11	Effect of acidic medium solution and heat	Inversion	Negligible	Negligible
12	Refractive index	1.374	-	-

Table 8.4 Sucrose inverted per hour at different temperatures and pH values[3]

Temp°C pH ↓	70	80	90	100	110	120
5.4	0.044	0.13	0.35	0.84	2.15	4.4
5.6	0.026	0.083	0.22	0.53	1.35	2.8
5.8	0.018	0.052	0.14	0.34	0.86	1.8
6.0	0.011	0.033	0.089	0.21	0.54	1.1
6.2	0.007	0.021	0.056	0.13	0.34	0.70
6.4	0.0044	0.013	0.035	0.084	0.22	0.44
6.6	0.0026	0.0083	0.022	0.053	0.14	0.28
6.8	0.0018	0.0052	0.014	0.034	0.086	0.18
7.0	0.0011	0.0033	0.0089	0.021	0.054	0.11
7.2	0.0007	0.0021	0.0056	0.013	0.034	0.07

In clarification process normally 2-4 %, reducing sugars are destroyed[4]. In alkaline medium reducing sugars are decomposed into organic acids at higher temperature. They also react with amino acids to form highly coloured complex organic compound. The latter is known Maillards reaction or browning reactions. In studying the oxidation of glucose in alkaline solution Spoehr and co-workers postulated more than 100 possible products.

8.5 Starch

Cane juice contains little amount of starch. The percentage depends on variety of cane, soil and climatic conditions. Starch is a polymer of glucose. The specific rotation[5] of starch is + 200. It is insoluble in water / juice and remains in fine suspension in juice. Higher starch content in juice gives trouble in clarification. It is partly removed in clarification and remaining passes along with clear juice. It increases viscosity of syrup and rate of crystallization is adversely affected. Therefore, factories prefer to cultivate varieties that are low in starch content. However in South Africa where factories have to process cane of high starch content special techniques in juice clarification have been tried.

8.6 Dextran and Gums:

Dextran and gums are polysaccharides. When growth of microorganisms called Leuconostoc Mesenteroides and Leuconostoc Dextranicum, takes place in cane and juice, an exterior product known as dextran is produced. Dextran is a chain of glucose molecules. In stale cane (particularly chopped harvested cane) the dextran level is considerably increased. Dextran is soluble and is not removed in clarification, but passes along with clear juice and gives difficulties in the process. Viscosity of the molasses is increased, rate of crystallization is reduced and if dextran level is high elongated crystals are formed. The loss in final molasses is increased due to incomplete exhaustion. Dextran is highly dextrorotatory therefore apparent Pol percent and purity is increased therefore unknown loss is increased. Gums are present in cane and are extracted in the juice. Those are removed in clarification to some extent and remaining pass in clear juice. Gums also increase viscosity.

8.7 Organic Acids and Nitrogenous Compound:

The cane juice is acidic with pH 5.0-5.4. Juice from immature cane or stale cane has low pH. There are many organic acids such as - aconitic, citric, oxalic, and many others present in the cane juice. Amongst these aconitic acid is about 1-2% on solids in cane juice. It is not much removed in clarification and passes in final molasses. It can be recovered from final molasses.

There are 16 amino acids like aspargic acid, glutamic acid etc. present in cane juice in free form and as protein constituent. Free amino acids are not removed in clarification and pass into the final molasses. Two amides - aspargine and glutamine are present in cane juice.

Proteins present in the juice (about *0.5% on solids)* are coagulated by lime and heat. They are beneficial for clarification as they adsorb impurities.

Acetic acid and lactic acid are not present in juice / cane but may be formed due to microbial activity when cane deteriorates after harvesting.

Some organic acids may be formed as destruction products of different organic compounds during the process (e.g. Levulinic acid).

8.8 Inorganic Compounds:

The inorganic compounds - phosphates, sulphates, nitrates, chlorides and silicates of sodium, potassium, calcium, magnesium, aluminium and iron are present in the cane juice These are present to the extent of 0.3 to 0.7 percent in the juice.

8.9 Coloured or colouring compounds:

Mainly the following colouring compounds are present in cane juice. These colouring compounds give dark colour to the raw juice.

i) **Chlorophyll:** it is green pigment present in leaves. In juice extraction it passes along with juice and remains in colloidal state. It is removed in clarification and does not affect the process.

ii) **Anthocyanin:** it is of purple colour and present in certain dark varieties only. It is readily soluble in water but is precipitated and removed in clarification process.

iii) **Saccaretin:** It is found associated with fibre. Some quantity of bagacillo always comes along with the juice. In the process of clarification when juice is heated and limed. At high temperature and high pH saccharetin give brewing effect and yellow colour is developed. However it becomes colourless in neutral or acidic medium. Therefore maximum removal of bagacillo from mixed juice by screening is good for process.

iv) **Tannin:** Various soluble green coloured polyphenols derivatives like tannin are present in the tops and buds of the cane. These react with iron and give dark colour to the juice.

8.10 Waxes and Fats:

Raw juice always contains little quantity of waxes and fats. They are of low densities, in suspended and colloidal state. They mostly rise to the top to form a scum at clarifier. Some quantity is carried down with the precipitate. Little part often remains in suspension and pass along with the clarified juice.

8.11 Colloids:

There are two types of colloids present in cane juice, Lyophilic and Lyophobic.

Proteins and gums are main lyophilic colloids in the juice. They are characterised by a strong attraction for liquid. They give viscosity to the juice. Water molecules form a protective wall around the lyophilic colloidal particles. When this water layer over the particles is removed flocculation of lyophilic colloids occurs. This process of dehydration takes place in clarification by heating the juice.

The lyophobic colloids include fats, waxes, soil particles and inorganic compounds such as $Fe(OH)_3$ $Al(OH)_3$. They are less hydrated and are in state of unstable dispersions. They have little effect on the viscosity. The stability of lyophobic colloids is caused by the electrical charge of the particles. The electrical charges prevent them from becoming attached to each other. Neutralization of electric charges results in agglomeration of these colloidal particles. The electric charges on these particles can be neutralized with opposite charges. Variation in pH value in juice clarification influences flocculation of these colloids.

References:

1 Irvine ISSCT 1974 p 1033 – 1039
2 George Watt "A dictionary of economic products of India" 1889 vol. VI part II p 22
3 Staddler G F ISJ 1932 Vol. 34, p. 273
4 Honig P Vol. 1, p 519
5 Imrie and Tilbury STR 1972 p 191 – 361

-OOOOO-

9 DETERIORATION OF CANE

Generally deterioration of cane begins after its cutting. But sometimes in adverse weather conditions or when the cane is diseased, deterioration starts even before harvesting. Cane depending upon climate loses daily about 1-3 % weight due to water evaporation. Water loss may sometimes create an apparent increase in sugar content. The weight loss of burnt harvested cane is negligible in first 24 hours.

Deterioration takes place in three ways

1. Enzymatic (or enzyimic) inversion,
2. Chemical inversion
3. Microbial activity.

Enzymatic inversion:

Enzyme invertase naturally occurs in the cane. This enzyme converts sucrose into invert sugars. The rate of inversion varies with temperature and moisture and is most rapid in hot and dry climate. Sodium metasilicate acts as an invertase inhibitor in cane juice.

Chemical deterioration:

The juice is acidic and due to this acidity inversion of sucrose takes place.

Microbial deterioration:

Microorganisms (mainly leuconostoc mesentoroids and to some extent leuconostoc dextranicum) are responsible for soaring of cane, sucrose loss and dextran formation. These microorganisms consume sucrose and produce long chain of glucose molecules (dextran) and ferment fructose into organic compounds as by-products. Microbial deterioration is rapid in hot and dry weather.

Mechanically harvested, particularly chopper harvested cane deteriorates very fast. Foster[2] has found dextran level as high as 7000 ppm after two day in some chopper harvested cane samples. Some microorganisms convert glucose and fructose to form acids and coloured compounds.

The dextran quantity in juice is relatively small, but it increases viscosity of the juice and is responsible for slow crystallisation, difficulties in purging of massecuites and higher sugar loss in final molasses. When dextran level is abnormally high elongated sugar crystals are formed. Dextran is three times dextrorotatory $[(\alpha)_{D20}$ $^{+}199]$ than sucrose. It gives false higher pol reading which increases apparent purity and the undetermined loss is increased.

Along with dextran the oligosaccharides (low molecular weight polysaccharide) that are produced In between cutting and milling may also be responsible for elongated sugar crystals formation in crystallization process.

Control of deterioration of harvested cane in fields and during transportation is impracticable. The only way to reduce the deterioration losses is to bring down the time lag between harvesting and crushing to a minimum. In case of unexpected crushing stoppage when cane is to be stored for longer time to control the deterioration, it should preferably be stored in shaded place covered with trash. Water should also be spread over it.

References:

1 Smits SASTA 1976 227-230
2 Foster ISSCT 1980 p 2204 – 20
3 Foster ASSCT 1979 p 11-17
4 Lamusse ISJ 1979 p 231-235
5 Alexander et.al. ISSCT 1971 p 794-804
6 Cremata - ISSCT - 1983 p 1157

-OOOOO-

10 SUGAR FACTORY LOCATION AND LAYOUT

Sugar factories are installed where there is large agricultural area under sugarcane or cane cultivation can be developed soon so that sufficient fresh cane is or will be available with minimum transportation cost. The site is generally on a state high way so that transportation of equipment, machinery, cane, raw material like lime sulphur empty bags, further - sugar, molasses or alcohol, filter cake, sold bagasse, can easily be transported. If the factory site is adjacent to railway station it is more advantageous. As the factory is situated in cane growing area mostly abundant water resources are nearby.

The site is usually at slightly a higher level than the surrounding area so that water does not lock in rainy season. Further the disposal of effluent water after treatment becomes easy.

Fig. 10.1 a standard 2500 TCD plant erected in 1995 in India

Layout Plan of a Factory:

In factory layout, adequate area is kept vacant considering the future expansion of the factory and development of ancillary units. The factory requires area for –

- Cane weigh-bridges and cane-yard
- Main factory building
- Workshop
- Laboratory
- Store building
- Sugar Godowns
- Sulphur, lime sheds/ godowns
- Loose and baled bagasse

- Spay pond
- Molasses tanks
- Effluent water treatment plant.

Outside the factory offices, residences, and common service area are always kept.

- Administration, Accounts and Cane Accounts buildings,
- Time office and security office etc.,
- Vehicle parking and garage,
- Staff and workers colony,
- Seasonal and harvesting labour colony,
- Dispensary,
- Petrol pump.

The process in sugar industry in advanced countries mostly automated and hence much less staff and workers are required. However, in countries like India where labour is comparatively cheap, further many factories are in co-operative sector which have to give employment to local labours as social responsibility. Therefore in India the number of workers is always more as compare to other countries. Therefore the area required for the followings is also considered.

Primary, secondary school, junior college etc.

- Playground,
- Cultural hall, library, club
- Market, Bank, post office
- Buses stand etc.

In the layout of a factory the following point are taken into consideration.

Cane weighment platforms are at higher level so that rainwater does pass into basin of the platforms.

- Molasses storage tanks, bagasse storage area, spray pond, effluent water treatment plant etc. are at the backside of the plant.
- Spray pond is away from the godowns. Care is taken that the wind direction is not spray pond to godowns.

-OOOOO-

11 *OUTLINE OF THE PROCESS*

11.1 Cane Preparation:

Sugarcane is received by bullock carts, trucks and trolleys at factory site. After weighment of the cane, it is unloaded directly into the cane carrier or first on the feeder table and then into the cane carrier with controlled feeding. The cane carrier carries the cane to mills. While passing up to the mills the cane is cut into pieces and finally shredded by different combinations of cane preparatory devices like cutter, Fibrizer or shredder etc. In cane preparation no juice is extracted. High cane preparation increases capacity and extraction at mills.

11.2 Juice Extraction:

The prepared cane is crushed in a milling tandem which is a set of 4 to7 three-roller mills. After primary extraction the primary bagasse is imbibed with thin juice to dilute the remaining juice before passing it to the next mill to extract maximum juice. Thus after primary extraction dilution is followed by pressing up to last mill. Before the last mill hot water is used for imbibition purpose.

11.1 milling tandem

Normally 94-97% of the sugar in the cane is extracted in the juice. This percentage is called as mill extraction. The remaining 3-6% sugar is lost in bagasse. The final bagasse from the last mill - generally called as simply bagasse, is woody fibre contains 47-50 % moisture. The juice extracted is screened and send to boiling house for processing.

Diffusers are also used for juice extraction in many cane sugar factories in the world in which dewatering mills are required after diffuser.

The final bagasse goes to boilers and used as fuel for generation of high pressure steam. The high-pressure steam generated is mainly utilised for driving power turbines and mill turbines. The exhaust steam from the turbines is utilised for further process of sugar manufacture. The extra bagasse saved can used as a raw material in the manufacture of particle board or paper or otherwise sold to neighbouring industries as a fuel. Alternatively now many factories are using their total bagasse for higher-pressure steam generation for co-generation of electricity which is supplied to the state electricity board.

A flow diagram of sugar manufacture by double sulphitation process is given herewith.

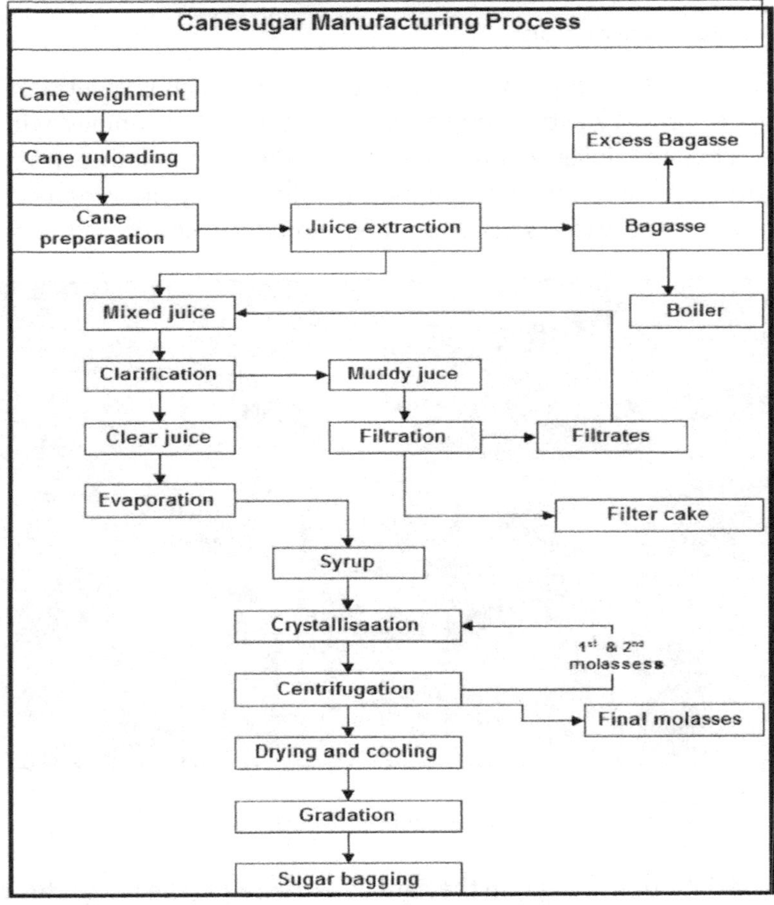

Fig. 11.2 flow chart of plantation sugar manufacturing

11.3 Juice Clarification:

The juice contains sucrose (commercial sugar), reducing sugars (simple sugars) in dissolved form, along with it some organic and inorganic non-sugars are present in dissolved, colloidal and suspended form. The purpose of clarification is to remove the suspended and colloidal non-sugars completely and the dissolved nonsugars as much as possible and to obtain clear transparent light colour 'clarified' or 'clear' juice.

There are three different processes of clarification

1. Defecation
2. Sulphitation
3. Carbonatation or carbonation

Defecation process is used for raw sugar manufacture whereas sulphitation and carbonatation processes are used for the manufacture of direct consumption white sugar which is also called as plantation white sugar or mill white sugar. Addition of milk of lime and then boiling the juice is common in all the processes. By addition of milk of lime acidity of the juice is neutralised. The suspended and colloidal nonsugars and some dissolved non-sugars are precipitated which are separated by settling in clarifier. The clear juice obtained from clarifier is send to evaporator for concentration. The underflow of the clarifier muddy juice is send to rotary vacuum filters for filtration. The filtered juice returns to the mixed juice receiving tank (after weighment). In case of carbonatation settling of the floc is not possible and hence whole the juice is filtered through filter presses.

11.4 Evaporation:

The clarified juice is concentrated to syrup by boiling the juice in multiple effect evaporator set. The exhaust steam from turbines is used to boil the juice in evaporator. The evaporator set may be quadruple, quintuple, vapour cell plus quadruple or any other configuration. In quadruple set there are four evaporator bodies. The exhaust steam is applied to the first body or effect. The next bodies successively work at lower vacuum. Therefore vapours from the first body are used to boil the juice in the second effect. Vapours from the second effect are used to boil the juice in the third effect and so on. So that from 1 kg of exhaust steam in quadruple nearly 4 kg of water is evaporated from the clear juice. The concentrated clear juice is called syrup. In sulphitation process the syrup is bleached by SO_2 gas before it is further boiled for crystallisation.

11.5 Crystallisation:

The sulphited syrup is further boiled and concentrated in vacuum pans in a controlled way so that uniform crystals are developed when fine seed is provided to it. A mass with sugar crystals and mother liquor thus formed is called as massecuite. When the pan is full, the massecuite is discharged in crystallisers from where it is further send to centrifugal machines.

11.6 Centrifugation:

In centrifugal machines the massecuites are cured and the mother liquor called molasses is separated from sugar crystals through screens by high centrifugal force. The separated sugar crystals from high grade massecuite are dried and bagged as commercial white sugar. The molasses obtained from high grade massecuite is again boiled for intermediate massecuite. In reboiling, a stage is reached when further crystallisation of sugar is no longer practically and economically feasible and such molasses is called “final molasses”.

11.7 Different types of sugar factories:

Cane sugar factories can be broadly classified according to the type of sugar they produce.

1. Raw sugar factories
2. Plantation white sugar factories
3. Raw sugar factories with refinery unit attached
4. Independent refineries

Till first half of 20th century, sugar was costly and was a luxury thing to use as food. Only rich people were using sugar. Raw sugar was produced in colonial (tropical and subtropical) countries and was exported to European countries where it was being refined in independent refineries.

In the developing countries like India China, because of techno-economic constrains could not go for production of refined sugar and were producing plantation white sugar. The same process is being followed in most of the factories till today.

-OOOOO-

12 HISTORY OF EQUIPMENT AND PROCESS

12.1 Mills:

Before mechanization vertical three-roller mill was commonly used up to start of 19th century. The mills were driven by animals, wind or water power.

Collinge in 1794 first made a design of mill with three horizontal rollers arranged in triangle. In 1871, Rousselot, a French engineer, of Martinique made some improvement in the design. All the modern mills are based on this design. At the same time hydraulic loading of mill roller was also employed.

Cranes were used for unloading the cane in 1910. Rotating knives to work as equalizer or leveller were introduced in 1870-72. The crushers and shredders were introduced in 1883 – 1887. However before 1920 there were many factories which were not having proper cane preparatory devices like cutters or shredder. Some factories were having crusher in place of cutter or shredder.

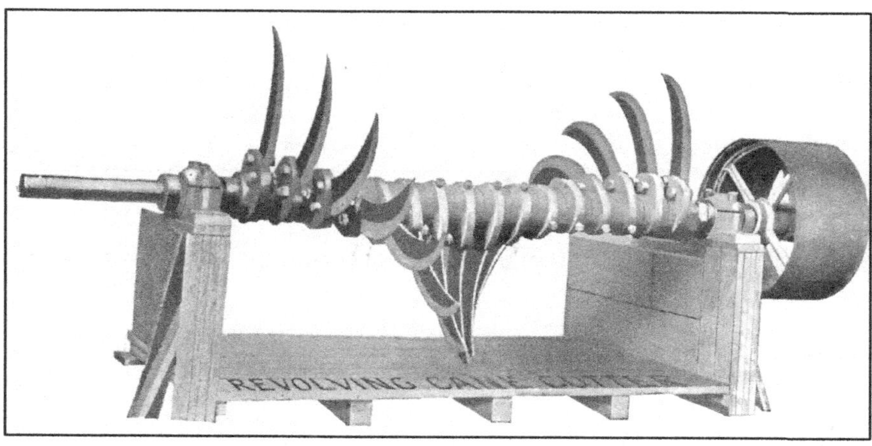

Fig 12.1 Revolving cane equalizer

The introduction of steam as motive power was implemented at the end of the 18th century. Due to this fabrication of large size mills made possible. Three mill trains with use of imbibition water were installed in Louisiana and Australia in1892. Slow speed horizontal steam engine was standard method of driving the mills. Small steam turbines were used to drive the mills from 1947, which is now general in modern plants.

Noel Deerr in his book (1916) gives mill capacities as follows:

A nine-roller mill and crusher of size 30 in. X 60 in. will treat 20 tons of cane per hour, and with mixed juice equal in weight to cane will give an extraction of 93.5, the cane containing 12 per cent. Fibre, a mill 34 in X 78 in will give the same results with 35 tons of cane per hour.

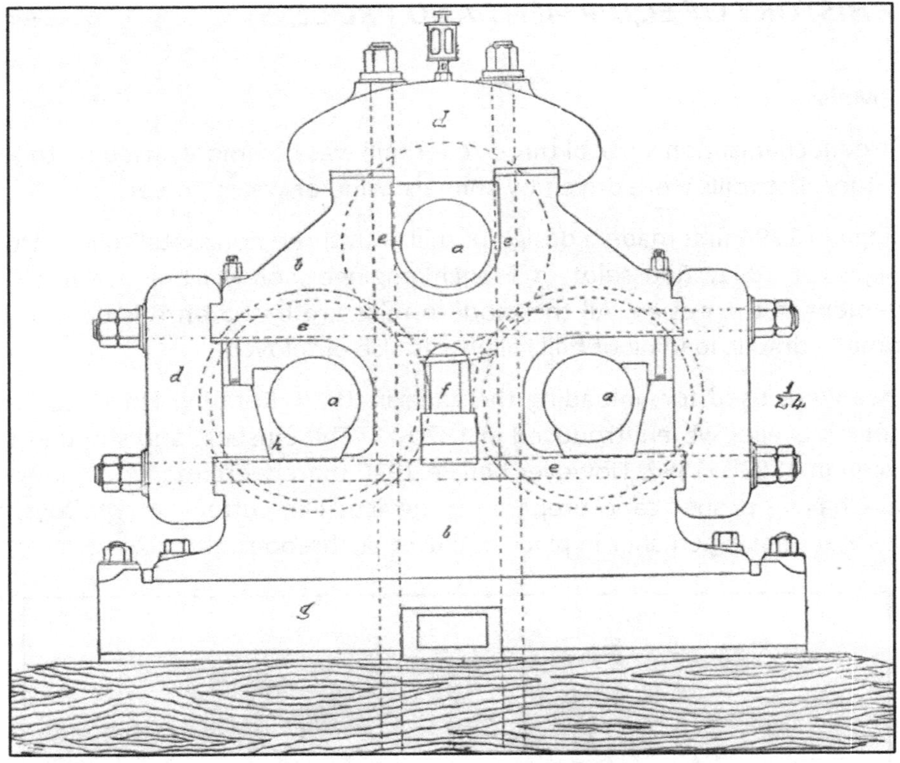

Fig.12.2 Mill design by Rousselot

A twelve-roller train 30 in X 60 in under equal conditions will give an extraction of 95.0.

A fifteen-roller train 30 in. X 60 in. under equal conditions will give an extraction of 96.0.

A twelve-roller train and crusher of 34 in. X 78 in. will treat up to 50 tons of cane per hour and give an extraction of 94.5 under the same conditions as before.

Diffusers:

The diffusion process as applied to the cane sugar industry first came into prominence about 1884, several of plants remain in successful operation. . In 1910 there were many plant working with diffuser. However some factories in Mauritius and Hawaiian Islands have reverted to milling because of faulty design of diffuser, difficulty in constant even supply of cane and excess fuel consumption.

12.2 Juice clarification:

Use of phosphate for juice clarification was started in 1862 in USA. Mr Beans has taken the patent of it. Use of sulphur dioxide gas for cane juice clarification was first tried in the year 1865 in Mauritius. In 1905 PRINSEN GEERLIG describes sulphitation process for clarification of cane juice.

Scientist Horloff first followed the process of hot liming and sulphitation at 70-75°C. It improved the precipitation of calcium sulphite and reduced the scaling problem in juice heaters. Hence this process is called as Horloff's process.

Hundreds of different chemicals/substances have been tried as clarifying agents. Lime is being used for almost 2000 years. The other common substances phosphate/phosphoric acid and sulphur are being used in last 150-200 years.

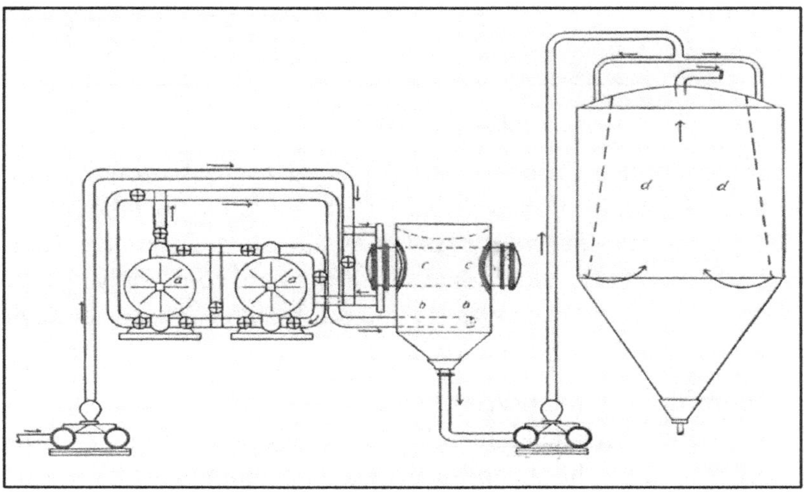

Fig. 12.3 Defecation process

The defecation process has been established in the mid of 19th century. It was batch operation; the juice was being treated with milk of lime, being heated in open tanks to boil and then allowed to settle. The supernatant clear juice was run off to evaporators. Continuous process started first in Australia in 1890. The figure on the next page gives the defecation equipment and process followed. There are two horizontal juice heaters, one liming tank and a continuous clarifier.

The clear juice obtained was send to evaporator and the scum containing suspended solids was send to scum receiving tanks where it was limed and allowed to settle, the clear supernatant liquor being decanted and added to the main juice. The scum then passed through the filter presses, the clear filtered liquor being passed to the evaporator.

Multi-tray subsiders or clarifiers of the Dorr type were introduced in West Indies in 1918. PETER HONIG started the use of pH measurement for better control over the process.

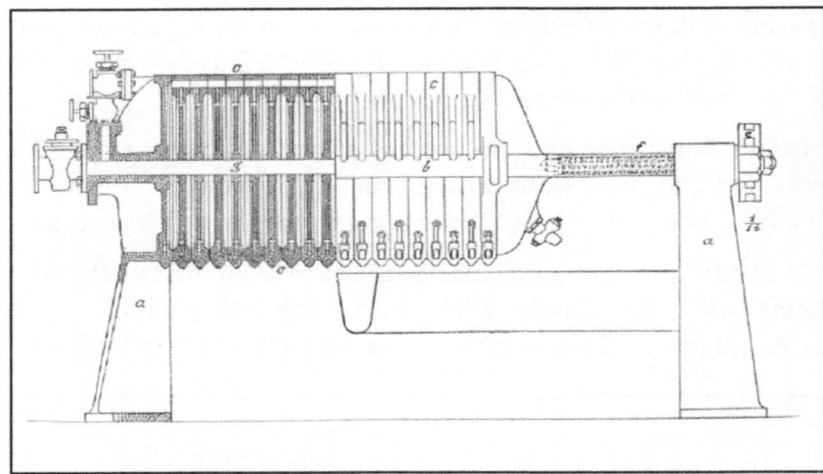

Fig. 12.4 filter presses

The modern continuous process has achieved a great reduction in the total capacity of settlers required as well as a vast reduction in labour requirement. Prototype SRI trayless clarifier was installed at Gin Gin sugar mill in 1969 designed for a crushing rate of 120t/hr. Now trayless clarifiers are commonly installed in mills. Filtration of scum or muddy juice was carried out by filter presses or Taylor filter or sand filtration.

Rotary vacuum drum filter (RVDF), patented in 1872, and is one of the oldest filters used in the industrial liquid-solids separation. It offers a wide range of industrial processing flow sheets and provides a flexible application of dewatering, washing and/or clarification. However, use of rotary vacuum filter in sugar industry has started much late, in the mid of 20th century.

Carbonatation process:

The carbonatation process, which is quite generally adopted in beet sugar factories, has only been applied to the manufacture of cane sugar in few factories in Java and India. This process gave bad results when first introduced into cane sugar factories owing to the presence of glucose in cane juice. However the efforts of Geerligs and Winter made the process practicable. By this process its direct consumption was manufactured. However, Noel Deerr remarks that the best white sugars of Mauritius made by a defecation process combined with the use of sulphur and phosphoric acid are equal to any that he has seen prepared by the carbonation process. The single carbonation process was adapted to factories having inferior quality cane and is not adapted for making white sugars.

12.3 Evaporator:

In the original method of concentrating juice to form crystals was by boiling in shallow open pans with direct heating, either from an open fire or by means of fire enclosed in brickwork (Dutch oven). Use of steam was introduced at about 1800.

Multiple effect evaporation was invented by NORBERT RILLIEUX in Louisiana in 1844. RILLIEUX'S first patent showed two jacketed vacuum pans (HOWARD's original pan) coupled together in double effect, i.e. the vapour from the first acting as heating medium to the second. The jacketed pans were soon replaced by evaporators with horizontal steam tubes. In 1851 the first vertical tube evaporator was introduced. This change in design was made by a German engineer Robert from which modern designs of Robert type evaporator bodies have been developed. The principle was extended in turn from double to triple, quadruple and quintuple effect. In 1880 Rillieux announced the idea of bleeding vapours from the evaporators to replace steam at the vacuum pans. This practice is now extensively adopted. Noel Deerr in 1914 has described seven different evaporator sets. Some of which were of horizontal tubes.

Van Trooyen has given particulars of a Kestner quadruple at 'Pasto-Viejo' Porto Rico. Each body is a vertical cylinder 24 feet high and 3 feet 6 inches in diameter. The separating chamber is 7 feet in diameter and 6 feet high, except in the last body where it is seven feet high. The heating surface of the first three vessels was consisting of 250 tubes, 23 feet long and 1½ inches outside diameter. The fourth vessel there were 130 tubes, also 23 feet long and 2¾ inches diameter.

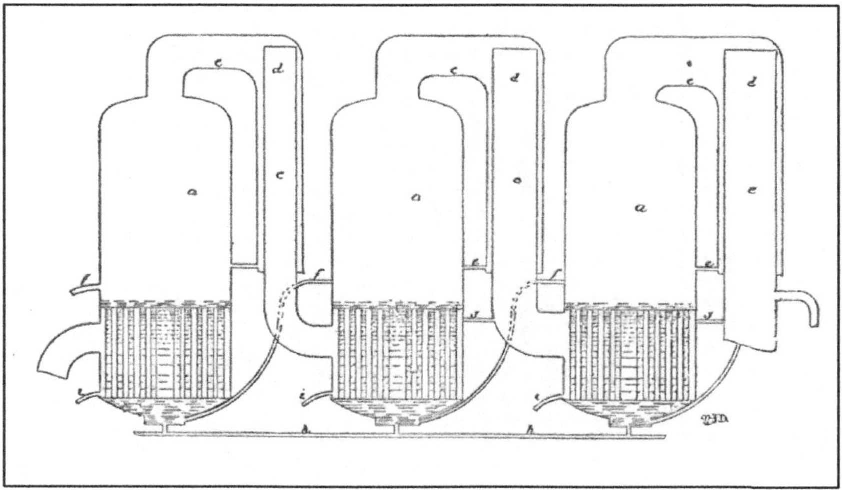

Fig.12.4 Triple effect evaporator

12.4 Vacuum pans

Howard in 1813 has invented vacuum pan. This was really beginning of a new era. His name is also associated with the invention of the filter-press. Howard's original vacuum pan was essentially the shallow pan with a cover that makes the pan airtight. The cover from the top was provided with vapour pipe condenser and vacuum pump. Vacuum was obtained by condensing the vapour given off by a jet of water allowed to gravitate down from an overhead tank. A steam jacket below the lower portion of the pan itself was for heating. Necessary accessories such as proof stick and sight glasses were also provided.

The heating surface provided by the jacket was found insufficient and coil pans with internal heating surface by providing spiral coils working on high pressure steam were developed.

In early years of 20th century mostly were coil pans. The heating surface consists of a number of helical copper coils, reaching from the bottom to a little above the centre of the pan, and so arranged as to divide the heating surface as uniformly as possible. The coils were generally 4½ inches in diameter. They are supported by stay rods fixed to the side of the pan. Now coil pans are obsolete and not used in industry except for manufacture of bold sugar or candy sugar (खडीसाखर) in mini plants.

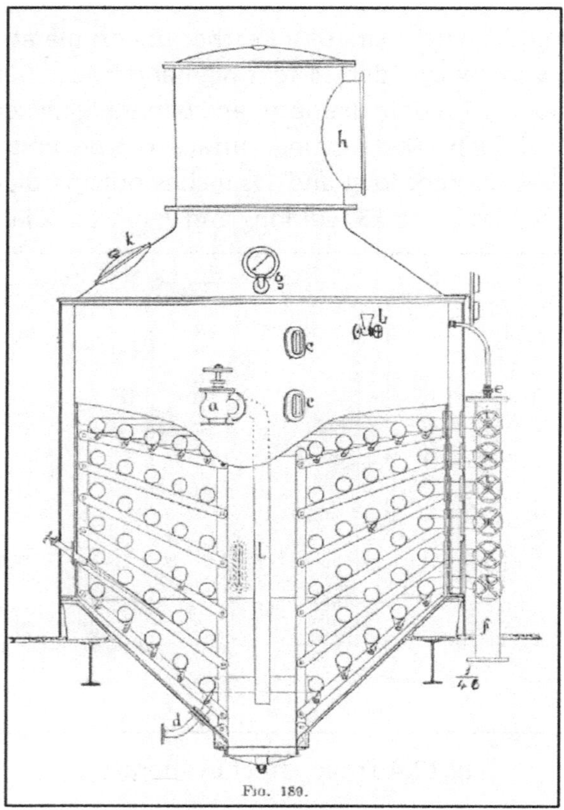

Fig.12.5 vacuum pan

Latter on calandria pans also have been developed with modifications in vertical tube evaporator. Since 1930, systematic study of pan operation and performance has confirmed the importance maximum massecuite height in the pan to avoid excessive hydrostatic head. Attention has also been given on minimizing resistance to circulation of the massecuite, therefore modern low head pans are designed that give good performance with low pressure vapours.

12.5 Crystallizer:

The receivers in which the massecuites were received in order to be cooled in motion are either U shaped or horizontal cylindrical vessels were called as crystallizing tanks. Crystallisation in motion was first described in a patent by WULFF in Germany in 1884. A shaft fitted at both ends passing horizontally through the centre of the vessels is attached the stirring arrangement with worm gears. The crystallizing tanks are made either plain or provided with a jacket, into which hot /cold water can be admitted. So that massecuite cooling rate can be controlled. The use of crystallizers provided with stirring equipment has gradually become general. However the development was slow. With modifications in the old designs various types of crystallizers using water cooling elements have been introduced since 1930.

12.6 Centrifugals:

In old days, the separation of massecuite into sugar crystals and mother liquor was carried out by pouring the massecuite into perforated cast-iron vessels. The molasses dripping conducted with holes while sugar crystals were washed by pouring sugar solution. Mother liquor / molasses was separated by force of gravity. The centrifugal machines that were being used in textile industry was tried in sugar industry.

NOEL DEERR has described the centrifugals as one of the world's great inventions. He mentioned that centrifugals were perhaps introduced into the sugar industry in 1849, by Dubrunfaut. David Mecolley Weston a Scottish engineer working in Hawaii in 1852 patented a design of centrifugal with flexible suspension.

This design then became standard. There were two different type - fixed bearing and suspended pattern ; again they were divided into under and over-driven machines, or according to the method of driving, direct coupled, friction cones, belt, electric, or water drive.

The sugar hoppers and elevators were in use in the industry from the end of 19th century.

Since beginning of the sugar industry bagasse is being used as fuel in sugar industry, both for sugar boiling and power generation. Originally bagasse was burned after air / sun drying. The first boiler furnaces designed for burning bagasse straight from the mills were introduced in Cuba in about 1880.

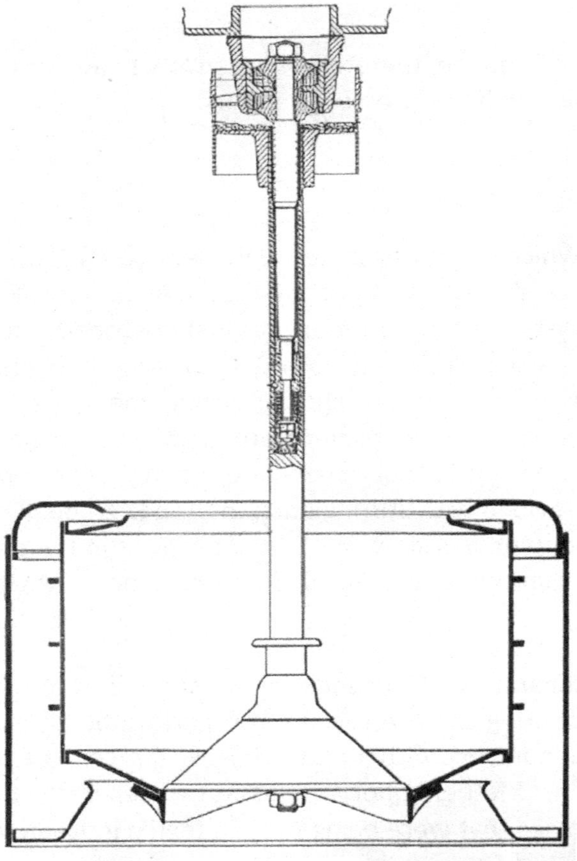

Fig. 12.7 Centrifugal machine

References:

1 Cane sugar – by Noel Deerr, 1911,
2 Cane sugar and its manufacture – by H C Prinsen Geerligs, 1924,
3 On cane sugar and its manufacture - by H C Prinsen Geerligs, 1902,
4 Plantation white sugar manufacture – by W H Th Harloff and H Schmidt,
5 Practical white sugar manufacture – by H C Prinsen Geerligs, 1915,

-OOOOO-

13 CANE WEIGHING AND UNLOADING

13.1 Cane Weighing:

Cane brought to the factory first weighed with loaded vehicles on weighing platforms or weighbridges. Cane weighment is necessary for two reasons –

1. To make payment on the basis of weight of cane that individual cultivator has supplied to the factory,
2. To check and record technical control of the factory.

The cane weight taken on weighbridge is 'Gross Cane Weight', the weight of cane including extraneous matter and binding material. In India 1 % of the gross cane weight is allowed to deduct as binding material to get net cane for making cane payment. Therefore, cane procurement officers should be careful that cane being delivered to factory should be as clean as possible. The tare weights of the vehicles should also be measured accurately. The weighing platforms / bridges should be counter checked regularly. Now electronic weighbridges with cane weighment record have been installed in most of the factories.

Electronic Weighbridges with Computerized Accounting:

A weighbridge is used to weigh vehicles – carts, tractor trolleys, trucks. The weight of cane brought by a vehicle is loaded vehicle weight minus empty vehicle weight.

A weighbridge consists of a large steel platform (the weighbridge deck) resting on, one to six cells (these are the parts that weigh the load). Load cells are a type of sensors which are used to measure force and load. They convert the force into an electrical signal which is proportional to the weight is send to the weight display. Thus sophisticated, accurate and systematic recording by electronic weighing based on single load cell or four-load cells is now done in many factories.

The computerized accounting ensures entry of information of farmer, harvester, transporter, bullock cart without any error. The computerized system also helps in speedy and error free operation of Weigh Bridge. Cane weight is recorded automatically without any intervention.

Token offices and weigh-bridges of sugar factories, comprising of –

1. Complete cane plantation record such as name of cultivator, field area of sugarcane, date of plantation/ previous harvesting, cane variety from each village,
2. Bar code system for the sugar cane transporting vehicles including carts;

3. Harvesting and transportation printed orders are provided to cultivators, harvesters and transporters,
4. Computer in sugarcane token office with bar code/smart card reader is connected to main server.

Fig.13.1 Truck weighing on weighbridge

Now the advanced system is computerized automation system for sugar cane detail plantation record is created online by the agricultural staff of the factory. On maturity of the sugar cane crop, harvesting and transportation orders are generated in the online system by the agriculture assistants working in the fields or by the centralized office.

All vehicles including bullock carts are bar coded. The computers in the token offices and the weigh bridge offices are with bar code readers and/or smart card readers which can be operated by the vehicle drivers themselves. Token number is generated automatically, gross and empty weights of vehicle are accepted automatically and thereby sugar cane receipt /note is generated in the system and printing of the receipt /note is also done automatically on the printer. Thus now there are cost effective, more efficient computerized automation systems for sugarcane token offices and weigh-bridges of sugar factories.

13.2 Cane Unloading:

The level of cane in cane carrier is to be maintained to an optimum. Therefore cane unloading directly in the cane carrier is to be regulated carefully. Hence only small bundles are unloaded directly into the cane carrier when carrier is not running with its full capacity. Most of the cane is unloaded first on feeder tables and then from feeder table into the carrier at controlled rate. Overhead cranes, feeder tables and in some factories hydraulic truck unloaders are used for unloading the cane.

Carts are generally manually unloaded directly into the cane carrier however now in some factories wire ropes are provided and the carts are unloaded by cranes.

Overhead Cranes:

Normally three motion overhead unloading cranes with safe working load capacity (SWL capacity) 5-7.5 MT are in common use. In West and South India generally "Swingle–bar" system is provided to the cane unloading cranes where as in North India "Grab" system is used.

Fig. 13.2 unloading cranes

In factories where swingle –bars are provided, cane is loaded in bundles bound by three slings or wire ropes of which one end hook and the other end ring is provided. At the time of unloading the wire ropes rings are hooked in the swingle-bar of the crane and then the cane bundle is lifted. The swingle bar has three hooks. The cane is unloaded generally on feeder table and then the wire ropes hooks are released. The crane lifts the swingle-bar with the wire ropes hanging. In factories where "grab" is provided to the crane, binding the cane in bundle in wire rope is not required.

According to Indian standard specifications (1987) for a sugar plant of 2500 tcd expandable to 3500 tcd of cane sugar factory, SWL (safe working load) capacity of crane should be 5 tonnes. The crane is electrically operated confirming to class IV of IS specifications. The crane can operate up to 20 lifts per hr. With Grab the crane can operate at normal capacity of 2 MT per lift and with swingle-bar at 4 MT per lift. All the electric motors used for crane operation are TEFC suitable for three hundred operations per hour. Push button type panels are provided. The height of the lift is 8 m.

Fig. 13.2 Cane unloading crane with swingle-bar system

Table 13.1 Overhead crane motors details

Sr. No.	Particulars	Type of Motor	H.P.	Speed m/min.	Rating
1	Hoisting drum drive	Squirrel cage	20	9	1 hr.
2	Long travel drive	Slipring	7.5	15	½ hr.
3	Cross travel drive	Slipring	5	15	½ hr.

Hydraulic truck Unloader:

Fig. 13.3 Hydraulic tilting platform

Hydraulic truck unloader is provided with load handling capacity of 30 tonnes with an angle of tilt 50°, with platform size suitable to truck size. The platform is provided with rear stop and front hooks

The advantages of hydraulic truck unloader are:

- No wire ropes are required.
- Less handling of cane as the cane is directly unloaded in cane carrier and hence less wastage of cane.
- As wire ropes are not used passing of any hooks into the carrier is avoided.
- Time is saved.

In some countries, tipping platforms are installed along the side of the carrier that are operated by mechanical means by electric motor. In such factories truck trolleys and containers / boxes are used. Advantage of this system is that the trucks trolleys can be unloaded as soon as they arrive in the cane yard and that are send immediately for next trip. Therefore, idle hours of the vehicles are minimized. Following are the alternatives of IS specifications for unloading equipment

Table 13.2 alternatives for unloading

Alternative	**2500 TCD Plant**	**3500 TCD Plant**
Alternative no.1	One crane + One hydraulic truck unloader	Two cranes + Two hydraulic truck unloader
Alternative no. 2	Two crane (swingle-bar)	Two cranes (swingle-bar) + One hydraulic truck unloader
Alternative no. 3	Two cranes(grab) + one two motion on auxiliary carrier	Two Cranes + One hydraulic truck unloader

A truck tippler has advantage of minimum maintenance and IS suitable for crushing capacities up to 3500 TCD plants. Grab type unloaders create fluctuating electrical loads and requires very strict maintenance schedule.

13.3 Feeder Table:

Feeder tables are essential equipment of unloading of cane because unloading of cane directly in the cane carrier has to be regulated very carefully and therefore the cane always cannot be unloaded directly into the cane carrier. It is a rectangular table placed alongside the carrier with a fixed platform sloping 5-15° generally towards to the carrier. It is provided with rake to move the cane on it to the main carrier. The movement of the feeder table is interrupted and somewhat jerky.

Fig. 13.4 Feeder table

Standard feeder table recommended is of 5 x 7 m. size with 8 strands of 150 mm. pitch, heavy duty steel drag type chain having braking strength of minimum 40,000 kgs. The chains are driven by a 15 BHP TEFC squirrel cage motor and coupled to a variable speed drive and a reduction gear to control the speed to 1-3 m / min. Approximately 15-20 t of cane can be unloaded at one time on the feeder table. One feeder table of standard size can work well up to 1500 tcd plant.

Auxiliary Cane Carrier:

According to IS specification, an auxiliary cane carrier right angle to the main carrier is recommended. The width of the auxiliary carrier is 3.6 m. and length 20 m. The carrier is inclined towards main carrier with a slope of 15° to the horizontal. It has six strands of chain of 150 mm. pitch. The breaking strength of the chain is 40,000 Kgs. The carrier is driven by a 40 BHP TEFC squirrel cage motor coupled to a variable speed drive and reduction gear to control the speed of the carrier 3-7 m / min. According to IS specifications a cane kicker is recommended over auxiliary cane carrier. The kicker is driven by 50 BHP TEFC squirrel cage motor. Fifty six arms are mounted on a hollow shaft of 0.7 m diameter. Speed of the kicker is 50-70 rpm. The diameter over the tips of the arms is 1200-1400 mm.

-OOOOO-

14 CANE CARRIER

Cane carrier is assembled of endless roller type chains with sprocket wheels at both ends of the carrier. Steel slats are fitted over the chains. Cane unloaded over the upper horizontal portion is conveyed to the mills by cane carrier. The carrier has mainly two parts, the horizontal portion and the inclined portion.

Fig. 14.1 cane carrier

Width of the carrier is always equal to the length of mill rollers. The length of the horizontal portion is decided taking into consideration the length required for feeder table plus length required for manual unloading. Now in most of the factories the carrier is split in two. The first carrier is up to cane leveller with 10 –12 m inclined portion. Installation of Fibrizer at about 2 m from ground level at the end of the first carrier is a general practice in India.

Normally up to 2000 tcd plants two strands and above 2500 tcd plants three strands of roller type chain are provided to carry the slats. The chains are driven at delivery end by sprocket wheels. The upper carrier – loaded with cane, pass over fixed steel joists (frame) that take weight and acts as sliding guides. The lower (return) flight of the carrier is supported at intervals by pulleys. Slip ring AC motors are used as drive for carrier.

According to IS specifications for 2500 tcd plant, two alternatives of cane carrier are recommended.

To suit installation of Fibrizer
To suit installation of shredder

For alternative one, the horizontal length is 30 m whereas for alternative no. 2 it is 40 m. The horizontal portion of the carrier is placed below the ground level. The sloping side of the carrier is about 300 mm above the ground level. Slope of the carrier of the elevated portion varies from 17^0 to 21^0 to the horizontal.

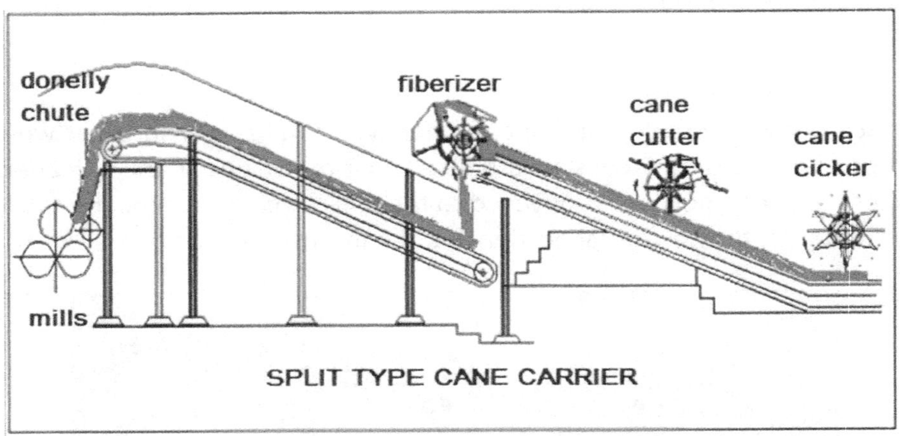

Fig. 14.2 Cane carrier

The highest point of the carrier is determined in such a way that there will be a minimum feeding height of 3 m above top roller centre to feed the Donnelley chute of the first mill. According to IS specification the recommended speed of the cane carrier is 3 to 10 m/min.

Fig, 14.3 inclined cane carrier

Height of Cane over Carrier:

The cane at the kicker is generally partly tangled and partly parallel and therefore bulk density of cane can be considered as 175 kgs / m^3 or 0.175 t / m^3.

Then, tonnes of cane passed over the carrier per hour will be equal to tch = w x l x h x d_b

Where

w = width of the carrier
l = length of the carrier travelled in one hour
= (speed of the carrier m. / min) X 60
= u X 60 where 'u' is the speed of the carrier in m /min.
h = height of the cane over the carrier
d_b = bulk density of the cane

Then,

$$tch = 60u\,whd_b$$

Or

$$h = \frac{tch}{60uwd_b}$$

For a 2500 tcd plant of standard specification

tch = 115 t
u = 8 m. /min.
d_b = 0. 175 t /m^3

Putting the values in the above formula

$$h = \frac{115}{60 \times 8 \times 1.7 \times 0.175}$$

Hence h height of cane over the carrier will be = 0.805 m.

Power Required For Cane Carrier: The recommended drive bhp and speed for cane carriers of 2500 tcd plant are as follows

For first alternative

1) First cane carrier 50 bhp
2) Rake carrier 30 bhp

For second alternative

1) First, cane carrier 40 bhp
2) Second cane carrier 40 bhp
3) Rake carrier 30 bhp

Each carrier drive is directly coupled to electro-dynamic / hydraulic variable speed gear and totally enclosed worm gear box having service factor of 1.7 and suitable open reduction gears / chain and sprocket.

Rake Type Cane Carrier:

When space is limited, the length of the inclined portion of the cane carrier is restricted and slope is to be increased. In such cases rake type cane elevator is used, the slope of which can be as high as 65°. The elevator speed is generally maintained to 8-10 m/min. The slipping of the cane is prevented by providing angle bars. The distance between the two bars is kept about 500- 700 mm.

Fig. 14.4 Rake type cane carrier

According to IS specification for 2500 tcd plant the rake type carrier has a suitable trough to receive the prepared cane from the first / second cane carrier. The width of rake carrier is generally same as that of cane carrier. The inclination of rake is 45°. The rake carrier has two strands of 250 mm. pitch blocked type forged chains (of minimum 60,000 Kg braking strength). The lower end portion receiving the prepared cane is kept below the ground level.

-OOOOO-

15 CANE PREPARATION – EQUALIZER AND LEVELLER

15.1 General:

The extraction of juice by crushing the cane is divided into two steps.

1) Cane Preparation – to cut cane into pieces and further disintegrate it into fine wet fibrous material that is called prepared cane.

2) Cane crushing – The actual crushing of the prepared cane at mills.

In this operation first the cane is made into pieces, the hard rind portion is broken. Further, it is disintegrated into long fibre with minimum dust formation. Hence high quality juice containing parenchyma cells are opened but no juice is extracted. This operation is called as cane preparation. In good cane preparation about 85 % of the juice containing cells from the cane are opened or ruptured. This results in increasing the bulk density of the prepared cane. Therefore higher throughput at the mills becomes possible. Capacity of the milling tandem is increased and at the same time load on mills is also reduced. In addition this, the primary extraction and consequently mill extraction is increased. Five % increase in primary extraction increases the mill extraction by about 0.5 – 0.6%.

The cane preparation is carried out by different combinations of equipment

- **Revolving cane knives** – that cut the cane into chips. Usually the first set is called as leveller and the second as cutter.
- **Revolving hammers** – that further break the cane structure and ruptures the juice containing cells. A Fibrizer or shredder is used for this purpose.

Combination of Cane Preparatory Devices:

Generally one of the following two systems is installed for cane preparation in Indian Sugar Industry

- Leveller + Swing Hammer type Fibrizer
- Leveller + Cutter + Shredder

Usually leveller + leveller system is preferred based on maintenance power consumption etc.

15.2 Cane Kicker, Equaliser or Chopper:

If cane is feed directly to the leveller without using a kicker, it is difficult to avoid jerks and control load on cutting blades of the leveller. There will be frequent jamming of the leveller that will result into tripping of the drive motor. Therefore before feeding the cane to the leveller and leveller or kicker is essential to even out the cane on the carrier. Thus a leveller or kicker is installed to control the height of the cane over the carrier at the starting of the inclined portion of the carrier. It leveller or evenly distributes the cane to a certain height on the cane carrier.

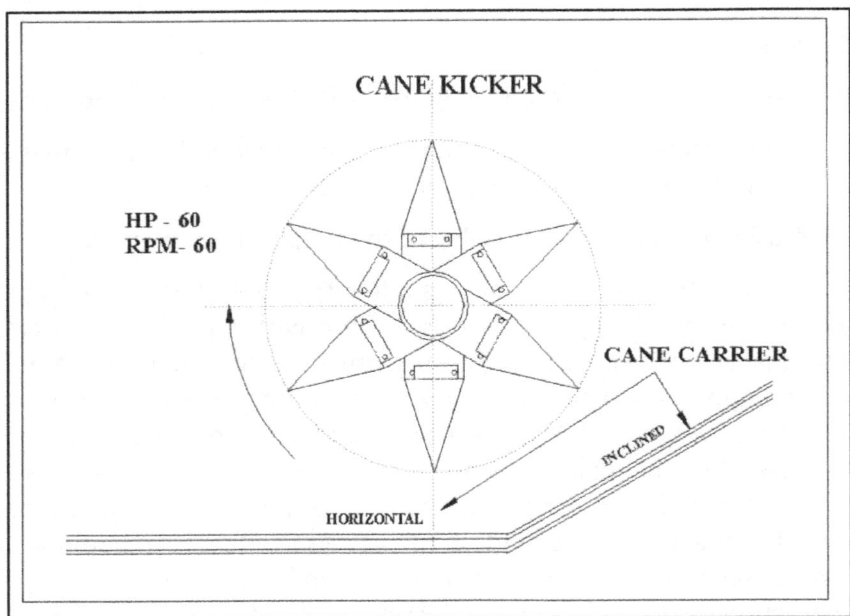

Fig. 15.1 Cane kicker

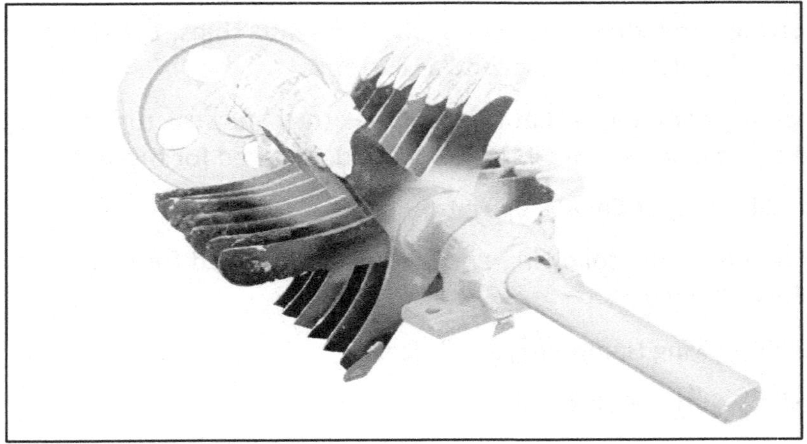

Fig.15.2 Cane kicker

A leveller consists of a horizontal hollow shaft fitted across the carrier having diameter of about 450-500 mm. Straight or curved arms / knives are provided to the shaft. The arms at the lower end rotate in the opposite direction to that of cane. The excess cane level is kicked backward maintaining expected level of cane on the carrier. Therefore cane feeding to the leveller becomes even. The speed of the leveller is generally 40-70 rpm. The height from the carrier slat to the tips of the arm is kept 25-50 mm less than cane height over the carrier. The side plates of the carrier are of 6 mm thick and are raised.

According to IS recommendations for 2500 TCD plant, 24 knives suitably secured to cast steel hubs of IS 1030 grade 26/52 mounted on 180 mm diameter central shaft of 35C8 quality. (Alternatively 24 arms suitably welded to 450 mm diameter central hallow shaft pipe of equivalent strength.) In case of knives the diameter of the tips is 1650 mm and in case of arms it is 1200 mm. In case of arms that are made of angle irons, two angle irons welded pyramidal shape. The shaft is coupled with 40 HP TEFC squirrel cage motor through reduction gear box. Shaft is supported on heavy duty double row self-aligning spherical bearing of minimum 100 mm. bore. The clearance between tips of the kicker knives and the cane carrier slat is maintained at about 750 mm.

15.3 Cane leveller / cutter

Cane leveller / cutter is installed on inclined portion of the carrier. The first set is generally called as leveller operates with normally 30-36 knives running at 500–700 rpm. The clearance between the tips of the knives and the cane carrier is 250-300 mm. The second set of knives called as cutter operates with 50-60 blades also running at speed of 600 rpm, clearance being 30-50 mm.

Fig. 15.3 Leveller

According to IS specification leveller of 32 knives secured to cast steel hubs of IS 1030 grade 26/52 mounted on a forged steel shaft of 200 mm diameter of 40C8 quality. The knives should be of special shock resisting steel having carbide cutting edge of hardness 45-48 hrs. The knives are confirmed to IS 8461, 1977. A suitable flywheel of C.I. (grade 20) duly machined and well balanced is provided at the outer end of the shaft. The leveller is driven by a continuously rated drip proof screen protected 250 bhp slip-ring motor of 600 rpm. The motor is coupled by means of fly wheel transmitting 240 bhp continuously.

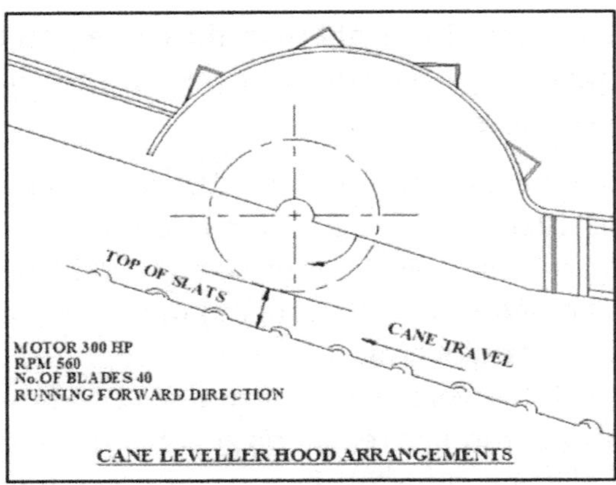

Fig. 15.4 Leveller

Cane knives:

The length of the knives is generally 400-450 mm having hard faces on the surface that is passing through the cane. The hard facing material stands to impact loading and corrosion. Blade material used should be of special steel having tensile strength 21 tonnes/cm^2 hardness 57 Rockwell number or 450-500 BHN number. The knives should be 20-25 per m length of the roller.

- Knives should be of simple designs without any complications
- Knife blade should be individually detachable so that they can quickly and easily be removed for replacing and sharpening.
- The blades should be of superior quality steel,
- A knife set should be well balanced so that the leveller/cutter works well.

There are numerous designs of cane knives.

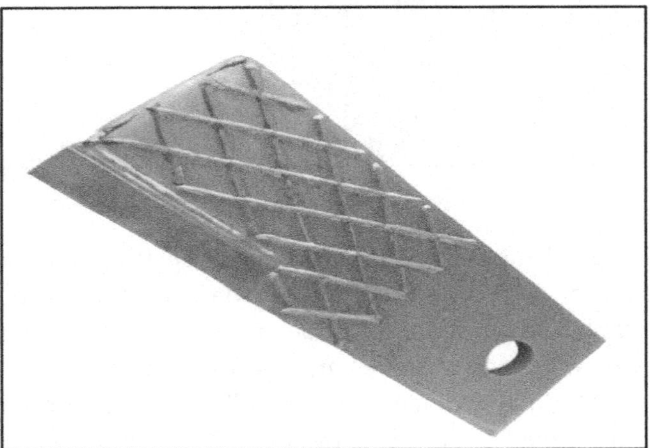

Fig.15.5 Knife

Levelling knives serves primarily to level the tangled volume of cane on the carrier into an even and compact mass. Cutting knives are used essentially for the purpose of cutting and slicing the cane into small bits. Cane knives are used either singly or in double effect i.e. levelling plus cutting.

Axial distance between two successive blades is known as pitch
If 'l' is carrier width and 'p' is pitch then

Number of blades $$n = \frac{l - p}{p}$$

Thus number of blades depends upon the pitch. Pitch distance is generally 25-50mm. For mill size 762 X1525 mm maximum numbers of blades are 60.

$$n = \frac{1525 - p}{p}$$

Where n is the number of blades are 60

Then

$$60 = \frac{1525 - p}{P}$$

Or p = 25 mm.

The cutters can be used by either forward direction or reverse direction. For reverse direction maximum cells are opened and hence preparatory index increases but at the same time HP required is also increased by about 50% than required for forward direction. Power required for cane cutter depends on

- Capacity tch
- Fibre % cane
- Number of blades
- rpm
- Clearance between tip of the blade and top of carrier
- Direction of rotation

The circle described by tips of the blade is called as cutting circle. The clearance is usually 25 to 75mm.

If,

h - The mean height of cane mat on the cane carrier
r- Clearance between tip of the blade and top of carrier

Then, The percentage of uncut cane percentage

$$= \frac{r}{h} \times 100$$

The percentage of uncut portion usually varies from 5 to 9 %.

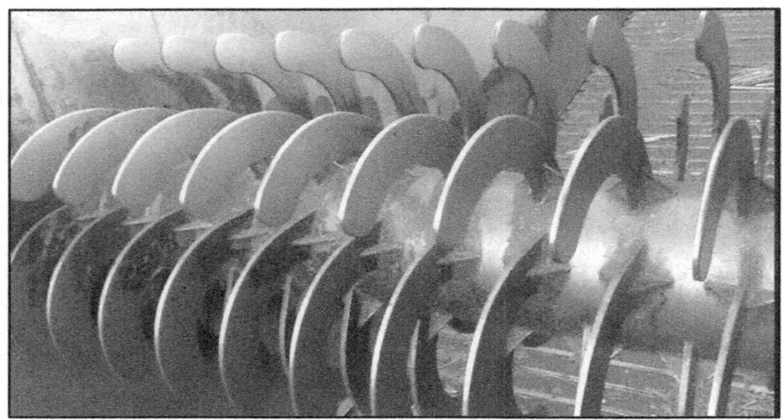

Fig. 15.5 Chopper cutter

-OOOOO-

16 CANE PREPARATION – FIBRIZER, SHREDDER

16.1 Fibrizer:

Cane tissues are very hard and cannot be easily torn out and ruptured. Therefore with level and cutter knives the preparation remains coarse and extraction is not much high even under very high pressure. If the tissues are disintegrated, torn and opened, the juice can be easily extracted in crushing. Hence further fine disintegration of cane is very much essential for higher mill extraction and higher crushing rate. This is achieved by fibrizer or shredder. The pieces of cane are forced by beating by hammers to pass through a very narrow space.

Fig.16.1 Fibrizer

Fibrizer is installed at the head of the cane carrier. The rotor is of 1.6-1.8 m in diameter at the tips and is rotated at 800-1000 rpm in opposite direction to the direction of carrier. The fibrizer is driven by one /two electric motors or steam turbine with reduction gear box. The rotor is completely covered by reinforced mild steel fabricated hood made of 12 mm. thick plate. The cover hood is attached to the cane carrier frame work and is complete with

- deflector plate
- fixed / oscillating type cast steel or mild steel fabricated anvil plate
- anvil suspension system
- front adjustable cover
- rear chute
- Bolted doors on the top of the hood.

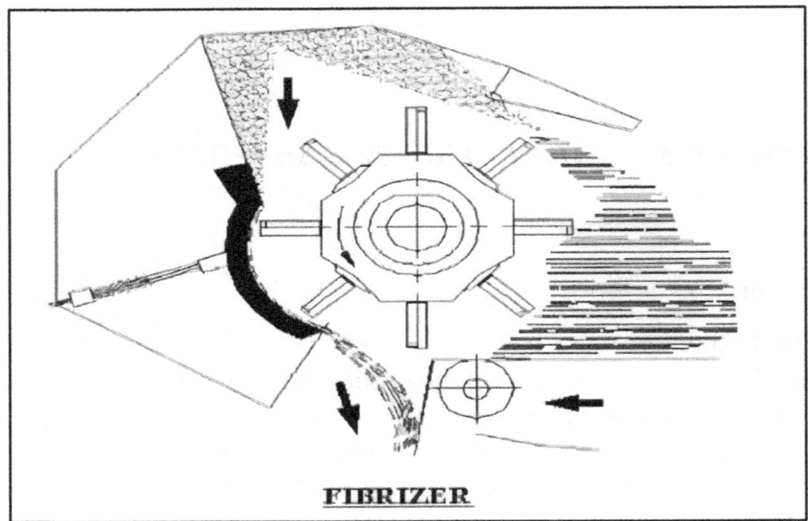

Fig. 16.2 Fibrizer

The partially cut cane pieces pass into the fibrizer through a feeder entrance. Then they are thrown upward by the revolving hammers. The upper part of the casing acts as a deflector and the cane pieces are forced heavily on the way downward against anvil plate. While passing through a narrow slit in between the hammer tips and anvil plate the cane is completely fractured and the cells are mostly ruptured. A provision is made to adjust the clearance.

According to IS specification for 2500 tcd plant the fibrizer should have 44 hammers each weighing 15 Kgs. The rotor shaft should be heavy duty 330 mm. diameter at hubs and 240 mm. diameter at the bearings of 40C8 quality. Hammers are secured to 50 mm. hub mounted on shaft. Special shock resisting steel alloy bolts of EN8 steel are to be used. The deflector plates should be 18 mm thick. An anvil plate 75 mm thick and pocketed serration type plate is machined and hard forged (Hardness min 500 BHN). The rpm of the fibrizer is 690 at 80 % turbine rated speed. Clearance between hammer tips and the anvil plate is about 20-25 mm. Matching flywheels at both ends of the shaft duly machined and well balanced is fitted.

For a 2500 tcd plant, the fibrizer is driven by a steam turbine capable of developing continuously minimum 700 bhp at 80 % of its rated capacity when supplied with steam at normal parameters and all hand operated overload valves in close position. The fibrizer is provided with forced feed lubrication system. The steam turbine has space for grooving and machining in the steam casing for addition of nozzles and throttle valves for increasing turbine power output to minimum 1000 bhp at 80% of its rated speed with a minimum service factor 1.7. No other changes are required for increasing the power. The steam turbine is designed to operate at Inlet steam normal pressure 30 Kgs /cm^2g and maximum exhaust pressure 1.5 Kgs /cm^2.

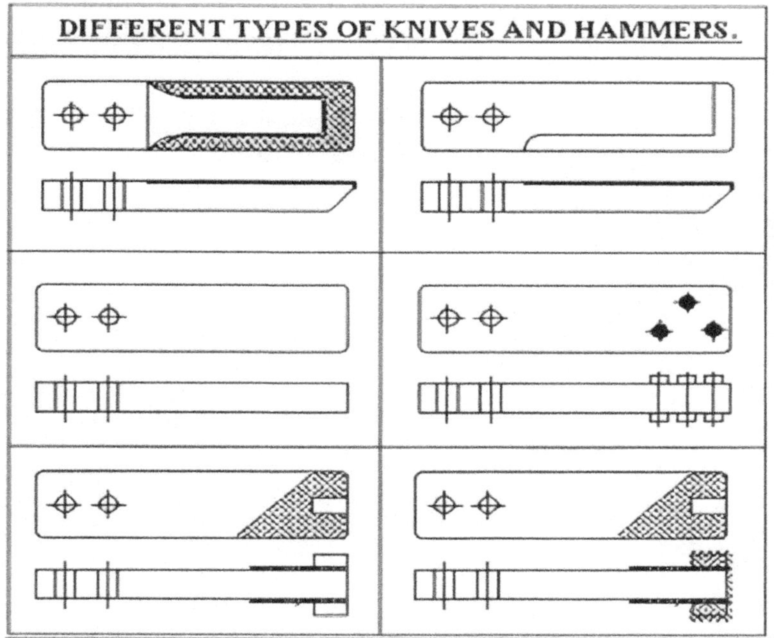

Fig.16.3 knives and hammers

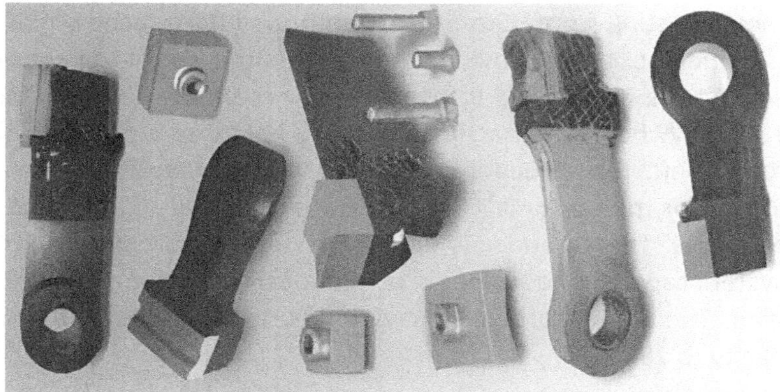

Fig.16.4 Different types of hammers and hammer tips

16.2 Shredder:

The shredder is consists of a rotor carrying hammers. It rotates at 1000-1200 rpm. Diameter at the hammer tips is 1800 mm. The end of hammers passes very close to an anvil plate, which is formed either with toothed profile or of rectangular bars.

Clearance between the anvil bars and rotating hammers is only 10-25 mm. 25 mm at the entry and 10 mm or less at the delivery from the anvil bars. The hammers are straight rectangular bars of silicon-manganese steel. 160 hammers are used. Weight of each hammer is generally 18-22 kg. The size of the hammers varies as per design of shredder. Cane pieces passing through shredder are disintegrated to a fine preparation.

Fig. 16.5 Shredder ready for installation

The hardness of the hammer tips should be minimum 500 BHN. The rotor shaft is of 350 mm. diameter and of 40 C8 quality or equivalent quality.

Shredder housing is fabricated out of 25 mm thick steel plates. Striking plates (with clearance adjusting from 35 mm to 10 mm) and hard faced striking bars are provided. Inspection doors are provided at the discharge end of the housing. Bypass chute of 10 mm. thick is provided to bypass the shredder in case of emergency. A suitably cast steel flywheel duly machined and well balanced is provided at the outer end of the rotor shaft. The shredder is provided with a pressure lubricating system comprising of motor, pump, tank, strainer, valves, pressure gauge and necessary pipings with fittings. The shredder is driven by a steam turbine with forced feed lubrication system capable of developing continuously minimum 500 BHP at 80 % of its rated speed. The steam turbine is having necessary provision to increase the ultimate capacity to 700 BHP.

All types of shredders are using Grid bar assembly for cane shredding. The pocketed grid is designed properly by keeping its width from 200-250 mm for obtaining better results. Defective design of grid bar assembly may lead to choking surging and dusting in the shredder.

The tip speed of hammers varies from 65 m/s to 85 m/s. The power consumption depends on the preparatory index maintained and increases sharply with higher preparatory index.

Power requirement for cane preparation:

The power provided in cane preparation helps in gaining higher capacity and higher extraction. To get desired cane preparation adequate power is required. The power consumed at cane preparatory devices saves some power required at mills. Moreover since power is generated in the factory is very cheap and if adequate power is available it is always beneficial to provide higher power for cane preparation.

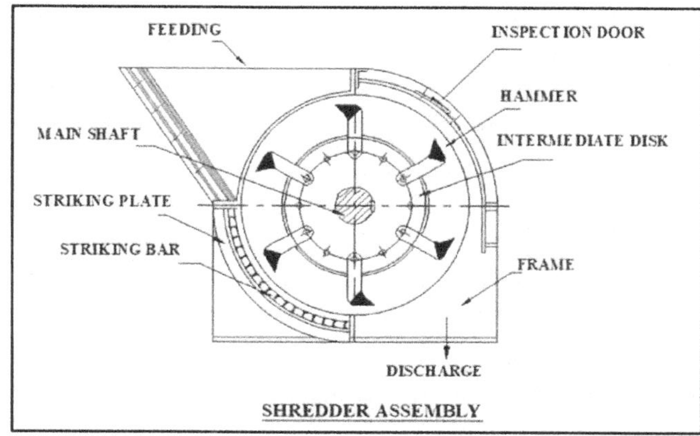

Fig. 16.6 Shredder

Table 16.1 – power installed for cane preparation

Sr. No.	Particulars	Factory A	Factory B
1	Crushing rate tch	150	340
2	Fibre % cane	14.00	14.00
3	Fibre throughput (tfh)	21.0	47.6
4		Installed HP	Installed HP
5	Kicker	40	60
6	Leveller	500	375
7	Cutter	-----	475
8	Fibrizer	1600	4000
9	Total	2140	4910
10	Ihp/tfh	101.9	103.2
11	KW/tfh	76.4	77.4

According to IS Specifications for a 2500 TCD (115 tch) plant total power provided for cane preparation is 990 bhp (40 bhp for kicker, 250 bhp for cutter and 750 bhp for fibrizer). Thus the bhp given is 14.25 bhp/tch or considering 14 % fibre the installed bhp comes out 64.5 bhp/tfh or 48.5 kW/tfh. However, this power is less to achieve the preparatory index above 85%.

The power installed for cane preparation varies considerably from factory to factory. However now the Cane preparatory index 85 + is generally expected and in some factories it is expected up to 90. The installed power for cane preparation of two factories of which preparatory index is above 85 is given here in the following table. Thus it can be seen from the above that to get 85 + preparatory index, the power required should be about 100 hp/tfh or 75 kW /tfh.

Automatic Cane Feed Control system:

Uniform cane feeding is the key of obtaining higher crushing rates and higher extraction. An automatic cane feed control system ensures uniform feed rate to the first mill. Variation in cane feed affects the mill loading, top roller floating condition, imbibition etc. and ultimately it affects the extraction. Automation of cane feeding can control this variation. Therefore automatic cane feed control system is now being adopted in many factories. The control systems are more computable with electric drives as compare to the turbine drives.

Cane Preparation Index:

The prepared cane is analysed for cane preparation index (PI). The cane preparation index reveals the percentage of pol that can be easily extracted because of opening and rupturing of cells in cane preparation. The preparation index (PI) is the percentage of Pol extracted by washing or leaching to the Pol extracted by disintegration in 'Rapi Pol Extractor'.

-OOOOO-

17 TRAMP IRON SEPARATOR

Many times nut-bolts, horse shoes, rope-wire hooks and metal pieces of knives, blades, etc. come along with the cane. Generally when mechanical harvest is followed the quantity of such metal pieces increases.

These metal pieces if not removed before entering the mill they can damage roller grooves to some extent. To avoid damage and losses because of pieces of steel and cast iron that pass through, efforts are made to eliminate the pieces of tramp iron. The amount of tramp iron may be reduced by taking precautions in the loading and unloading of the cane, by insisting on strict tidiness at the cane platform, and by supervising the tightening of bolts at the carrier and the knives.

Further electro-magnets or permanent magnets are installed at the cane carrier to remove tramp iron from the prepared cane. They are commonly called as 'Tramp Iron Separator'. The tramp iron separator is not recommended in Indian standard specifications, however now many factories are installing it.

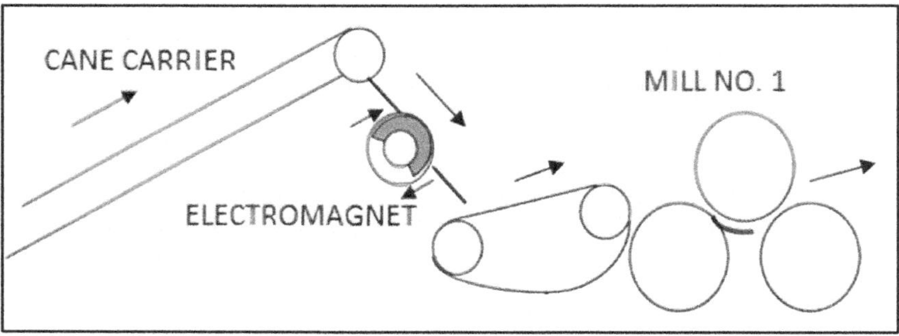

Fig. 17.1 Tramp iron separator before 1st mill

The separator consists of an electromagnet mounted across the whole width of the chute feeding the first mill. It attracts and holds pieces of iron passing through its field. Complete elimination of tramp iron cannot be expected. Some pieces will escape the electromagnet, because they are trapped in the cane.

The electromagnet separator with operating gap up to 500-600 mm is usually used. The selection of proper electromagnet depends on carrier width, operating height of the magnet, speed of the carrier, crushing rate, average weight of tramp iron pieces etc.

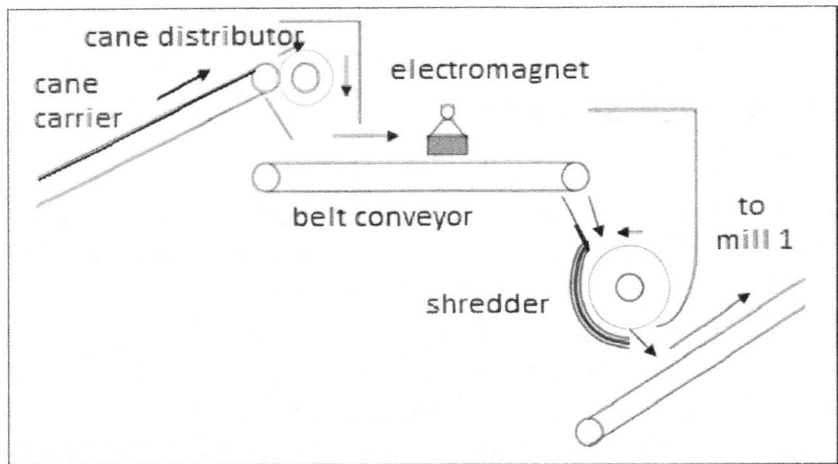

Fig.17.2 Electromagnet fitted before shredder

Magnet is designed for continuous operation and there is loss of power. To avoid this, the present trend is to over excite the magnet only when a tramp iron is encountered, while normally the magnet is in under-excited condition. This leads to saving in energy and at the same time tramp iron pieces are efficiently removed. This is achieved by installation of metal detector prior to the magnet and as and when the tramp iron is detected. It gives signal to the magnetic circuit to increase the magnetic strength temporarily for separation of the iron pieces.

When the iron separator is installed just before fibiriser or shredder, damage of hammers tips of these equiment is avoided. The installation of heavy duty rubber belt is avoided. The magnetic separator gives efficient protection to the rollers.

-OOOOO-

18 MILLS - CONSTRUCTION

The prepared cane is crushed in 4 to 7 number three roller mills in a succession to extract juice. This is also carried by the process called diffusion followed by crushing. Even after number of crushing the juice extraction cannot more than 78-79%. Therefore for higher extraction water / diluted juice is added to the bagasse before crushing it at the next mill. This added water dilutes the remaining juice from bagasse and the diluted juice is extracted in the next mill.

A typical mill constructed of three rollers. The rollers are fitted horizontally in heavy cast steel housing or mill cheeks. The standard housing is symmetrical. The two bottom rollers- feed and discharge, are in fixed position.

Fig. 18.1 a three roller mill with 4th feed roller

A heavy plate called trash plate is fitted in between the feed roller and discharge roller so that the cane / bagasse being crushed can be passed from the feed roller to the discharge roller under the top roller. Thus the trash plate works as a bridge in between the two bottom rollers. It is fitted on heavy steel bracket by high tensile bolts and nuts. The steel bracket is fitted in the headstock. Adequate clearance is provided in between trash plate and delivery roller to drain the extracted juice through it.

The rollers are consisting of a heavy steel shaft or roller journal. A cast iron shell is shrunk fitted on the roller journal. The top roller is driven by prime mover through reduction gear combination. Crown pinions are mounted on all the three rollers. The feed and discharge roller crown pinions are meshing with crown pinion of top roller. Thus the pinions drive the bottom rollers.

The roller shaft is considerably longer for fitting of crown pinion (and coupling in case of top roller). The top roller is hydraulically loaded to squeeze the prepared cane/bagasse at high pressure for maximum extraction.

Scrapers are provided to remove the bagasse jammed in the roller grooves. The scrapers are fitted on shaft and provided with lever or other suitable adjusting device. The scrapers may have renewable cast iron tips.

A mill dimension is specified by the figure for the roller shell diameter followed by its length for e. g. 36" x 72" (915 x 1830 mm). There is a series of roller dimensions that is internationally adopted. Accordingly, the roller size may vary from 24" to 45" (610-1143 mm) in diameter and 48" to 90" (1220-2286 mm). (In India even though metric system is followed the roller dimensions and hence mill sizes are still mentioned in inches)

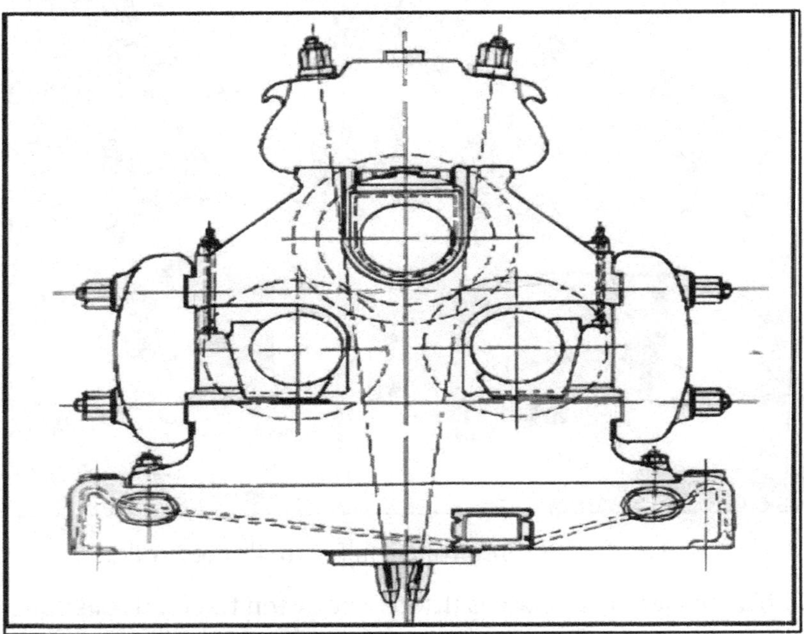

Fig. 18.2 Design of a three roller mill

Juice troughs are provided below the mills to collect the extracted juice. The juice troughs are made up of 8-mm aluminium plates. The troughs are bolted to the headstocks with suitably sealed joints to prevent any juice leakage.

The mills are set on heavy concrete plinth about 2.0 m high above the ground level so that all the juice tanks and pumps can be fixed on ground level and juice can flow in the receiving tanks by gravity. With this the maintenance of the mills also becomes can easy.

The mill rollers rotate in bearings supported in substantial cast steel housings or mill cheeks. The mill cheeks are supported on mill bed or base plate. For alterations in the mill settings a provision is made for adjustment of roller fittings. Trash plate adjustment is also provided.

Fig. 18.3 Mill housing or mill checks

In mills of standard design the top angle varies from 72° to 78°. The top angle should be minimum. This reduces the width of the trash plate and hence friction is reduced. The two half angle 1 and 2 are generally slightly different because of different feed and discharge openings for e.g. A1 = 37° and A2 = 35°.

The top roller is fitted with heavy flanges at the ends. These prevent the cane or bagasse from falling out. However in many factories these side flanges are removed to facilitate juice drainage when mill get over-flooded of juice at the top and feed rollers.

The crown pinions may have 14-21 teeth. Tromp recommends crown pinion 17 teeth. The pinions are fixed to the roller shaft by means of keys. The crown pinions drive the feed and discharge rollers. The pressure at the discharge roller is higher than that at feed roller.

Following are the different mill designs

- Housing with kingbolts or standard design

- A frame house (Squire design)
- Inclined housing (Cali)
- Unsymmetrical housing (fives Lille)
- Self-setting mill (FCB)

Feeding at mills – open chute:

To feed the cane from the carrier to the first mill an open feed chute is provided to descend down the prepared cane by gravity from top end of the carrier to the first mill. The width of the chute is normally equal to the roller length and with slope 60-70° to the horizontal. The open chute is installed in such a way that the line of the feed plate passes through axis of the feed roller of the mill. The feed plate is as close as possible to the feed roller. A clearance in between chute plate and feed roller is kept 5-10 mm. The bottom edge of the plate is serrated to match grooves of the feed roller.

A closed chute called Donnelly chute, which is now used in modern factories, is described later.

Intermediate Carriers:

Mainly there are two types of inter-carriers:

- Slat or Apron type
- Rake or drag type.

Slat type conveyor:

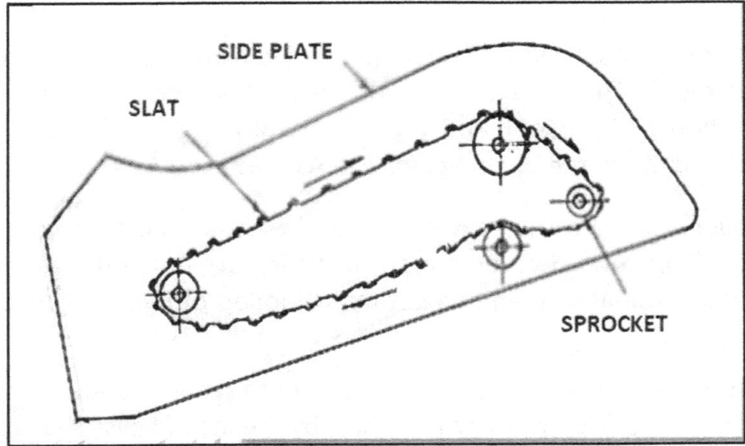

Fig. 18.4 slat apron type inter carrier

An endless slat carrier is provided to convey bagasse from one mill to the next mill. The upper part of the slat carrier is rested on guide and carries the bagasse. It is driven by upper sprocket at the head of the conveyor. The sprocket is driven from the feed roller of the next mill. The inclination of the upper part of the carrier is normally 24-30° to the horizontal. The plane of the feed portion of the conveyor is normal to the axial plane of the top and the feed roller. The horizontal distance between the top roller and the leading sprocket is kept about 0.33 x Roller diameter. The linear speed of the intermediate carrier is kept about 5-10% higher than peripheral speed of the rollers.

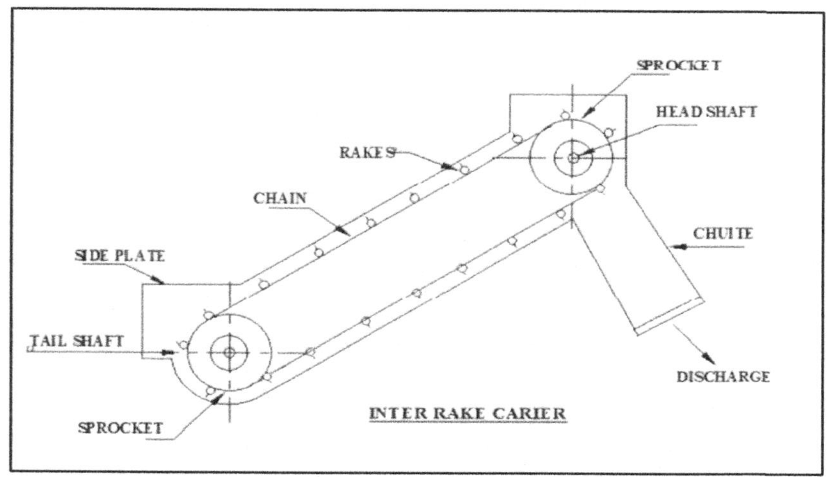

Fig. 18.5 rake type inter carrier

A recent development of using a heavy rubber band conveyor is being followed in some countries. The rubber band is provided with thick ribs to give grip on the bagasse. The rubber conveyor is superior to slat type conveyor in respect of avoiding falling of fine bagasse and in avoiding the mill stoppages for repairs.

When space between two succeeding mills is less, the angle of rising portion of carrier has to be increased to more than 30°. In such case with slat type conveyor the bagasse may descend and feeding at the next mill will be uneven or intermittent. To avoid this rake type inter carriers are used.

Mill Bearings:

The mill roller rotate in bearings supported in cast steel housings, Mill bearings are of cast steel with gunmetal or white metal lined. The bearings are provided with forced feed lubricating system. One to three points are provided for each bearing depending upon size and design of the mill. Heavy oil is used as lubricant. Cooling water circulation arrangement is provided in it.

It is observed that white metal bearings generally do not heat up and wearing is also less in comparison with gun metal bearing, moreover white metal can be re-metaled at the factory site.

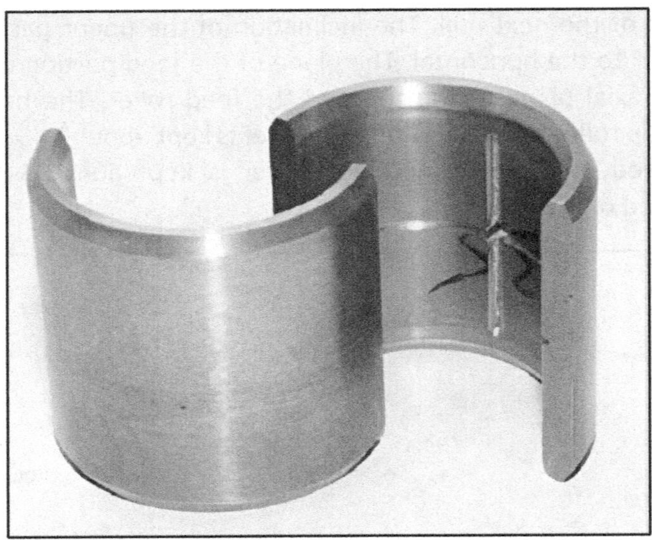

Fig. 18.6 mill bearings

Hydraulic Pressure at Rollers:

18.7 Hydraulic cap

With standard design hydraulic loading system is provided for each journal of the top roller so that the roller can be lifted according to thickness of the prepared cane/ bagasse passing through the mill and still constant pressure is maintained.

The bearings of the top roller are arranged to slide in the gaps of the housing and pressure is maintained by oil acting on hydraulic plunger housed in each top cap. The oil pressure is maintained by an accumulator and transmitted to the plunger by a system of piping. The system is free from leak and generally requires no charging.

The Edward's type accumulator contains a bladder of synthetic rubber containing nitrogen under pressure, which becomes compressed or dilated when the hydraulic plunger rises or falls. Some manufacturers have eliminated the rubber bladder and plastic disc is used to keep minimum contact of the nitrogen and the oil.

In these accumulators a valve is provided at the lower end of the cylinder designed to close immediately in the event of any sudden drop in pressure due to a bad leak in the piping or hydraulic cylinder. This prevents escape of oil with expansion of gas, which could be dangerous. This type of accumulators gives rapid response to changes in load at the mills. With a separate accumulator at each end of the roller pressure can be adjusted individually to compensate for pinion thrust. Indicators of lift of roller are provided at the end of the roller giving a magnified indication of roller lift. Now electronic type roller lifts are provided.

Generally there exists some 1-4 mm play in the hydraulic system. Therefore in mill operation unless the roller is lifted above this play the hydraulic pressure over the bagasse blanket passing through the roller does not become effective and extraction is reduced. To lift the roller above the play minimum bagasse blanket thickness is necessary and hence minimum crush rate is required. Some times with a high degree of compression juice being extracted cannot be escaped out because of improper drainage and reabsorption of juice takes place. Therefore mill engineer has to observe whether proper drainage is taking place or not.

The accumulators, piping and mills are designed for 280 to 320 kg/cm^2 hydraulic pressure. From the point of view of stresses in the roller material, the total pressure exerted on the top roller is generally measured by relating this pressure to the projected area (L X D) of the roller journal.

Total hydraulic load (in tonnes) at the top roller will be

$$= \frac{\text{Hydraulic pressure maintained (kg/cm}^2\text{) X Cross sectional area (cm}^2\text{) of Ram X 2}}{1000}$$

Now this total hydraulic load acts only on the 1/10th roller diameter of the top roller that exerts this pressure on the bagasse blanket.

Thus

$$\text{Specific Hydraulic Pressure (SHP)} = \frac{\text{THP}}{0.1 \text{ X D X L}}$$

The SHP gives a value of pressure exerted on the bagasse blanket per unit area. This is proportional to the hydraulic pressure maintained at the mills. Comparison between the pressures of two different mills can be made by SHP. Extraction depends on specific hydraulic pressure, bagasse blanket thickness, and drainage of juice. Now the bagasse blanket thickness will depend upon crushing rate, fibre % cane and roller rpm. The hydraulic pressure on the top roller should be in the range of 200-250 kg/cm^2.

Reabsorption of juice:

When bagasse (or prepared cane in case of first mill) passes under high pressure through the top roller and discharge roller, juice passes at higher speed than roller and sprays through and jets forward in the bagasse that is just emerged through the rollers which is relatively dry. Thus the juice is immediately reabsorbed in this bagasse. This re-absorption depends upon following factors.

- Fibre per unit area (Fibre index)
- Hydraulic pressure at mills
- Roller speed
- Coarse preparation

With higher hydraulic pressure, extraction increases without much increase in reabsorption. However a point is reached where increase in hydraulic pressure reabsorption overweighed the extraction. This is limiting factor for hydraulic pressure. This depends upon preparation of cane, speed of roller and conditions of roller.

Sugar engineers have given a formula for reabsorption factor. But we do not go in detail about it.

Pinion Reaction:

The power is transmitted from the top roller to the bottom and feed rollers by the crown pinions that are key fitted on the three rollers. The pinions have radial reactions at the contact of two teeth because of the curved surface of the gear teeth. This radial reaction tends to lift the top roller. Hence it is always found that the pinion side of the top roller lifts more.

Different methods are given by sugar engineers to solve this problem. However the simple and practical way to solve this problem with standard mill design is to keep 5% higher hydraulic pressure on the crown than the other side to offset the crown pinion reaction.

Tail Bar coupling;

In standard mill deign the top roller is driven by prime mover. However the top roller is hydraulically loaded and is lifted upward during operation. Therefore the couplings between the final gear shaft and the shaft of the top roller must be flexible enough to allow the up and down movement of the top roller. For this purpose 'tail bar' and couplings (generally of square section) are employed to connect the final stage gearing and the top roller shaft.

While erecting the mills the centre line of the top roller is kept 10-15 mm lower than the centre line of the final gear shaft. The tail bar should be 13-25 mm short in length than length between the roller shaft end and gear shaft end that are to be connected.

Fig. 18.9 Tail bar coupling

Cardan shaft: Some International manufacturers give Cardan shafts with universal joints at each end. This type of connection is more flexible and from mechanical points of view is more practical.

Rope coupling:

When crushing is continued top roller moves upward up to 40 mm depending on thickness of bagasse blanket. The tail bar coupling is not able to overcome with axial and radial force generation.

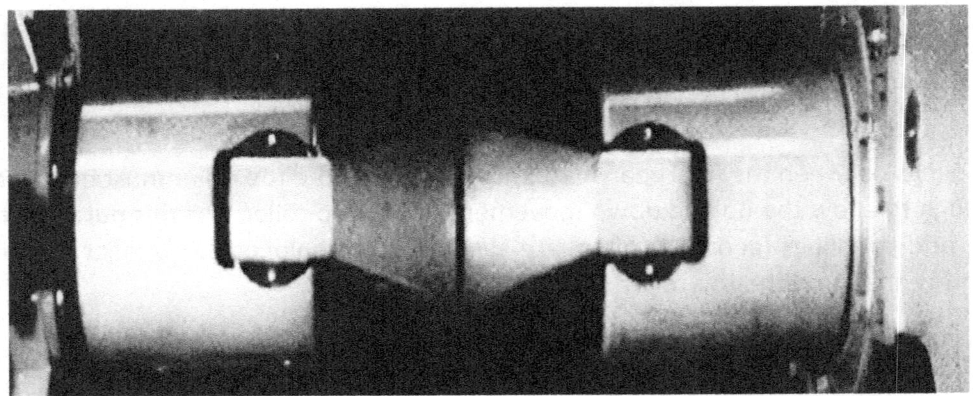

Fig. 18.10 Cardan shaft coupling

Fig.18.11 Rope coupling

On the other hand rope coupling with spherical plain bearings can take care of misalignment up to 60 mm axial displacement and angular misalignment up to 1° and eliminates all axial thrust coming on top roller. The components of rope couplings are bigger in size and weight as compared to tail bar couplings. The cost is always more.

-OOOOO-

19 MILL ROLLERS

The roller shell surface is grooved to increase the roller surface area. This gives better grip over the cane / bagasse entering the mill. Therefore mill capacity is increased. Further extraction is also increased due to effective juice drainage

Circumferential V grooves:

Rollers are cut with V shaped grooves on the periphery of the shell of rollers over the entire length. All grooves are equal in depth for a particular roller and angle of grooves is the same.

The cutting the grooves are such that the top roller grooves mesh with the grooves of bottom rollers. The pitch of grooves i.e. the distance between axes of two successive grooves is gradually reduced from first to last mill. This facilitates better extraction in the following mills. Coarse grooves serve better to disintegrate bagasse. In some countries, uniform pitches are maintained throughout the mill tandem. This practice is followed to reduce maintenance of spare rollers.

Fig. 19.1 mill roller

The angle of the grooves is in between 30-65°, being maximum for the first and minimum for the last mill. For proper drainage of juice the V shaped grooves should not be filled with fine bagasse therefore special scrapers are provided to remove the bagasse from these grooves.

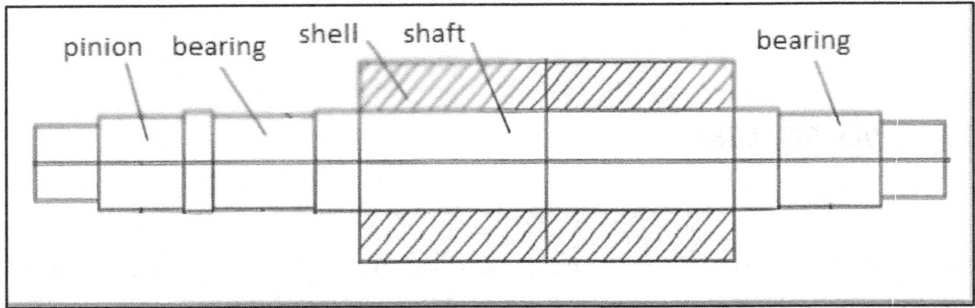

Fig.19.2 Roller details

The grooves can be rationalized as follows:

Table 19.1 Roller pitch and grooves angles

Mill No.	Pitch (mm)	Groove Angle		
		Top Roller	Feed Roller	Discharge Roller
1	65-75	55-60°	30-35°	40-45°
2	65-75	55-60°	30-35°	40-45°
3	30-40	55-60°	35-40°	45°
4	40-45°	55°	40°56+	55°

When setting of mill is to be kept minimum, the top and discharge roller grooves angles should be kept identical. The width of the teeth at the tip and at the bottom is kept 1/10th of the pitch.

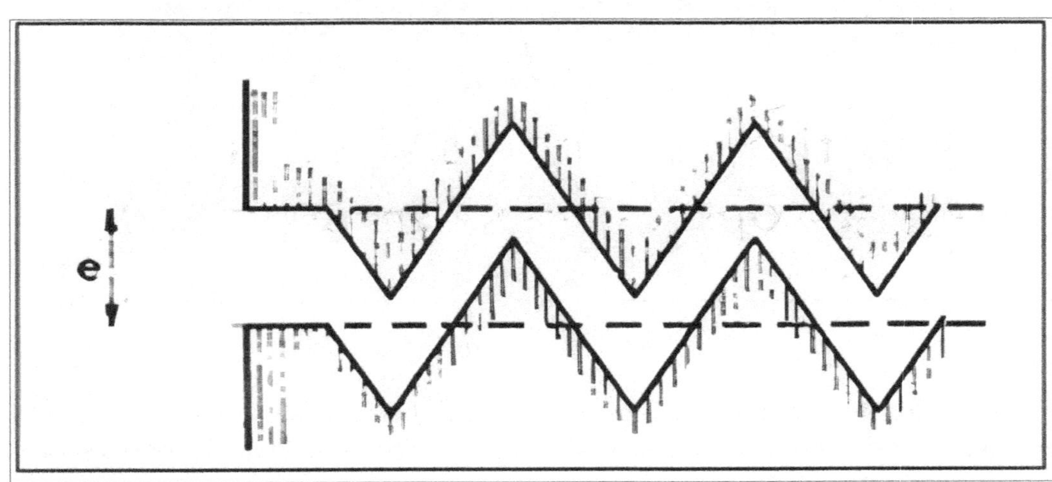

Fig. 19.3 Grooves meshing

The studies made by H. N. Gupta on the roller grooving and juice drainage revealed to prefer differential grooving. Now with TRPF since 30-35% juice is extracted before the cane/bagasse enters the feed roller, in such cases coarse grooving is not necessary.

Messchaert grooves:

Maximum juice drainage is expected at feed roller. Therefore to facilitate juice drainage from the feed roller deep grooves known as juice grooves or MESSCHAERT grooves (MESSCHAERT- inventor of the grooves) are often provided at the base of the circumferential V grooves as shown in the figure. Typical dimensions are 5-8 mm wide and 20-35 mm deep. The Messchaert grooves are provided on the feed roller only.

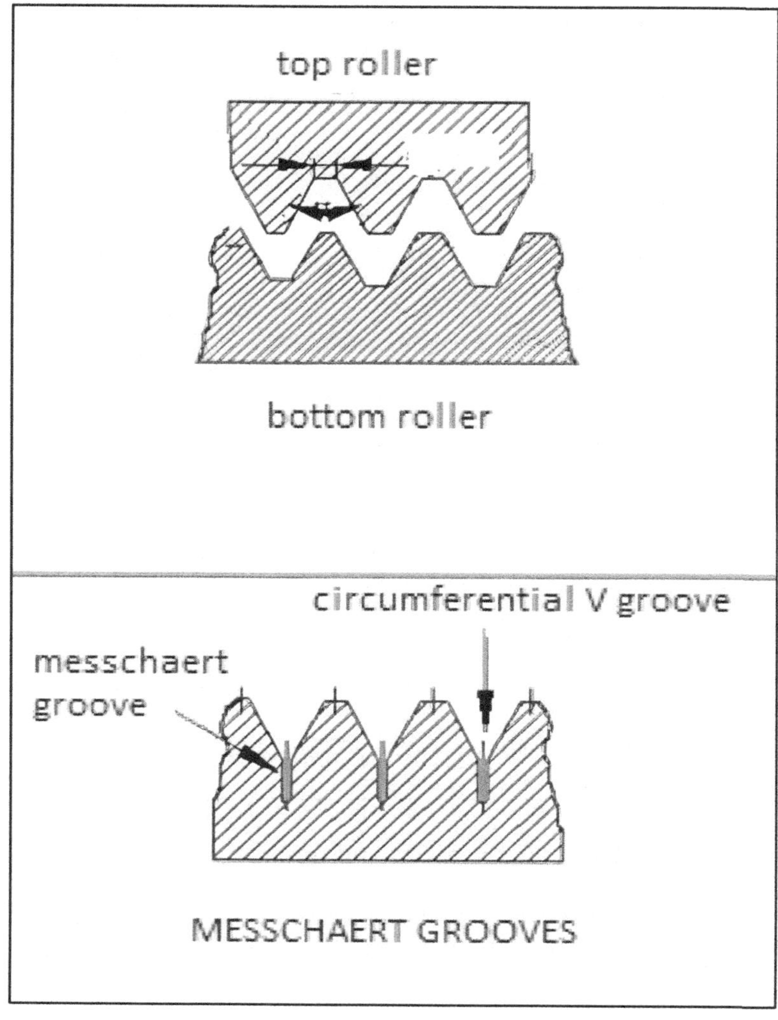

Fig, 19.4 Messchaert grooves and relative position

An alternative arrangement for Messchaert grooves is to use lotus roller with differential grooving at top and feed rollers. This minimizes the re-circulation of fibre. Therefore bagasse feeding and consequently load on the secondary mills is reduced. This helps in increasing the mill extraction.

Chevron grooves:

Now days, there is inclination for higher crush rate with high imbibition. This increases the juice quantity at the mills. Further with the installation of pressure feeder and lotus rollers the mill settings are kept close for higher extraction. In such situation there are chances of slippage of bagasse at the mills. Therefore to have a better grip on the entering bagasse 'Chevron grooves' are recommended for feed and top rollers. Some compression of bagasse is lost at Chevron notches. Therefore the pineapple pattern of cutting of Chevron is preferred. The rollers are coated with surface roughening electrode material.

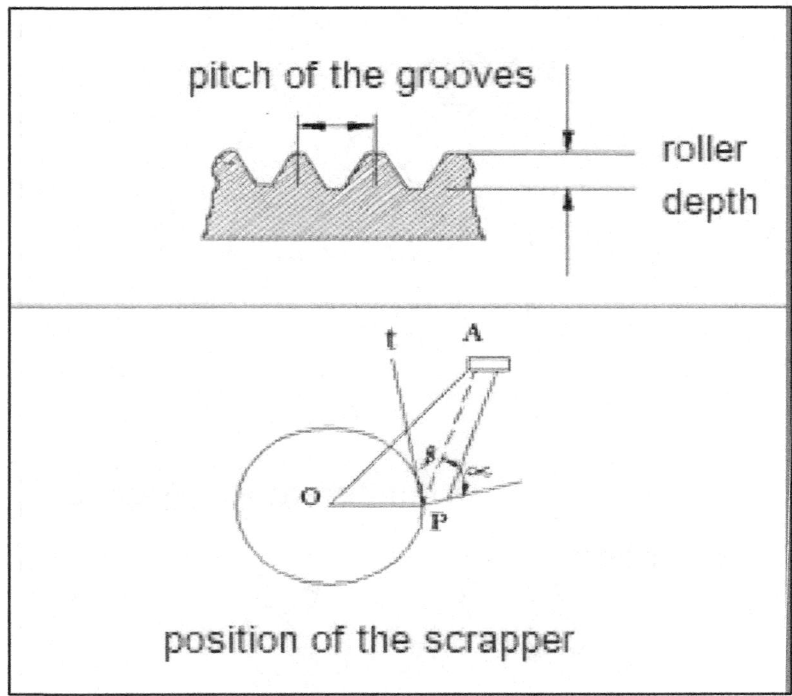

Fig. 19.5 roller grooves and position of scrapper

Now the equal angle grooving in all three rollers is not much followed. Normally a difference of 5° is kept between the grooves of top, discharge and feed roller respectively. Lower angle grooves make the groove sharp and helps not only in feeding but also increase the juice drainage area.

In addition to circumferential V groove Messchaert grooves are also essential to provide the required juice drainage area especially at higher crushing rate. In many factories the Messchaert grooves are provided on alternate or every third groove.

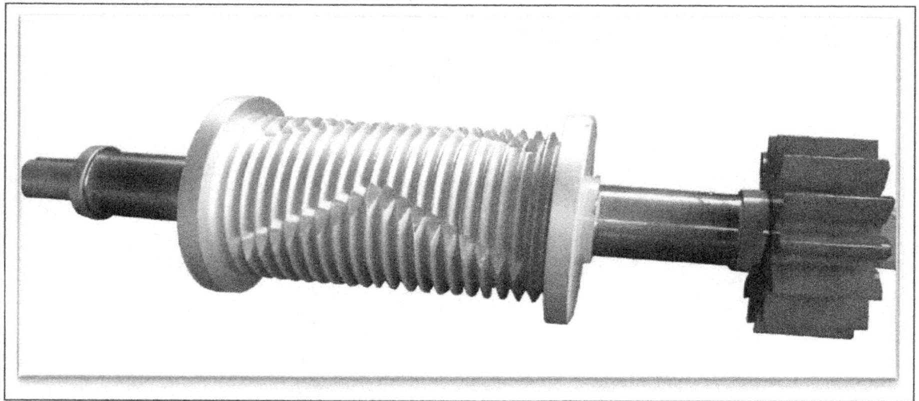

Fig. 19.6 Chevron groves

Main fear in providing Masschaert grooves for every groove is that the shell strength may get reduced and roller may develop circumferential cracks. But Ramkumar says that Messchaert grooves can be provided for each V groove without any adverse effect on the shell strength and this practice helps in improving the mill extraction.

Mill roller surface conditioning:

With higher cane preparation and improvement in feeding devices the feeding at mills have been increased. Now imbibition is also maintained high. In such case to avoid slippage the roller surface are to be maintained. For this purpose roller arcing has become a common practice. For arcing welding machine should be capable of continuous operation with 300 to 350 amperes at approximately 80 volts.

While arcing is attained in running mill, adequate earthing equipment must be fitted to the end of the roller journal prior to arcing. This can prevent the earthing current passing through the shaft journals to the journal bearings. Failure to do so seriously affect the journal and bearing surfaces and lead eventually to surface cracks. The arcing on rollers should be done at low speeds preferably at 3 rpm of roller.

Lotus Roller:

In some factories ordinary top roller is replace by lotus roller. Lotus roller consists of cast steel roller with about 12 to 16 number longitudinal cylindrical passages or pipes located a few centimetres below the roller grooves. At the bottom of each groove or alternative grooves holes are drilled that connect the groove with longitudinal pipes. While crushing, juice can pass through the hole at the bottom of the grooves to the pipes and flows through the pipe to the side ends of the roller shell.

Fig. 19.7 Lotus roller

This reduces flooding and reabsorption and higher imbibition can be taken. The main advantage of lotus roller is to improve the primary extraction at the first mill and to reduce the moisture of the bagasse leaving the last mill. Hence at higher crushing rate, these should be initially installed in first and last mill and latter in other mills.

The lotus roller facilitates ready escape of juice from the zone of maximum compression with consequent reduction of reabsorption and hence is preferred in many factories. There are controversial reports about its performance. The main problem reported is of choking of holes and axial pipes. Compressed air can be used to overcome this problem. However some factories have reported excellent results without any problem of choking.

Couch Roller:

A couch roller is similar to Lotus roller and in place of draining juice to atmosphere, it drains under vacuum. In this system the bottom half of the shell sides are connected to vacuum. The juice drainage through this system is higher.

-OOOOO-

20 MILL DRIVES

General:

The mills are driven by steam turbines in most of the factories. Now with high pressure boilers and cogeneration electric drives are becoming common. Hydraulic drives are also being used in some factories.

For the reasons of economy or simplicity generally in old factories of capacity below 1000 tcd most of the mills are driven by only one turbine. In some old factories of capacity 1000 to 1500 tcd first two mills are driven by one turbine and the next two/ three are driven by one turbine whereas for higher capacities individual drives are provided for each mill. Now in new modern plants generally DC drives are being installed. The different combinations of drive and gears used at milling tandems are as under

Table 20.1 Mill drives and reduction gear system

Sr. No.	Drive	Reduction gear system
1	Steam Turbine	High speed gearbox, low speed gear and helical gear
2	DC drive electrical motors	High speed gearbox and helical gear One helical gearbox, Hydraulic drive
3	AC VFD or DC drive motors	planetary gearbox Shaft mounted planetary gearbox

The foot mounted mill gearbox is resting on civil foundation and is coupled to the mill through tail bar or rope coupling.

Steam Turbine:

The most widely used mill drives are steam turbines. The main advantage of using steam turbine is that it avoids double transformation of energy as in case of electric drive or hydraulic drive. It permits wide range of variation in speed for mills. Steam turbines give good starting torque and speed can be varied up to 25-30% less than designed speed. However specific steam consumption increases and efficiency gown down when turbine is operated at a lower speed. The manufacturers provide mill turbines with supplementary nozzles that can be opened when power demand is increased.

Generally single stage or some time double stage turbines are used as mill drives.

Since the speed of turbine is 5000 to 6000 rpm and the mill roller speed is 3-5 rpm, reduction gear system of 3-4 stages is required. According to Indian Standard Specification for 2500 TCD plant individual turbines of 350 BHP are provided for each mill of 33"x 66" size four mill tandem.

Fig.20.1 Turbine, gear box, open gear and tail bar coupling

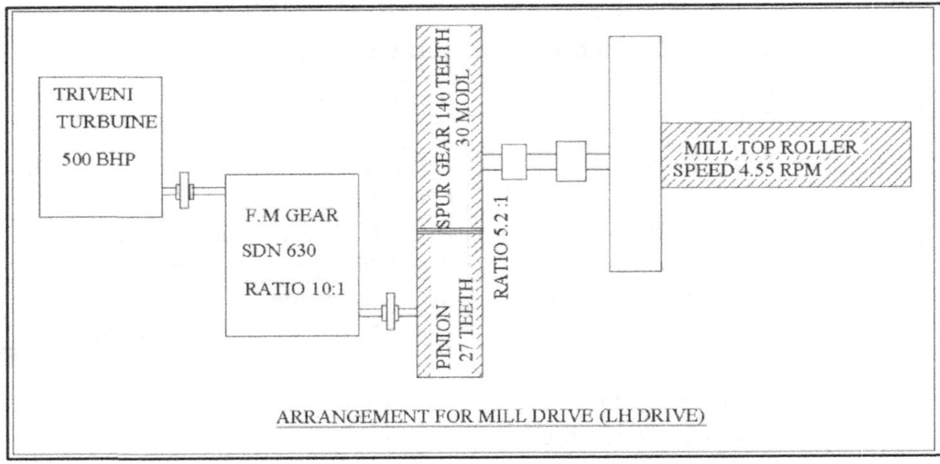

Fig.20.2 Turbine, gear box, open gear and tail bar coupling

While crushing, the juice extracted escapes against the direction of the rotation of the roller. To give sufficient time for the extracted juice to escape and flow down, the speed of the roller should not be more than 5/6 rpm. The most critical unit in the turbine driven mills is the turbine governor. It must be very sensitive and very rapid in action. It must not haunt and must assure perfect regulation. It is the first mill where regulation becomes difficult; however at the second mill the problem in regulation is reduced and becomes negligible for the next mills. Fig. 20.3 Turbine, gear box, open gearing, tail bar, mill roller

The steam turbine is a flexible machine from the point of view of power output.

The power output may be varied over wide range. Steam turbine always works efficiently at its optimum speed. If speed is decreased (or increased) than the optimum speed then the efficiency comes down and steam consumption per kW is increased.

Electric Drive:

Electric drive has been introduced in sugar industry some 90 year ago. It eliminates of steam piping at the milling tandem. It is easy to control. It has lower operating and maintenance cost. The use of electric motor at mills gives ready measurement of power consumption.

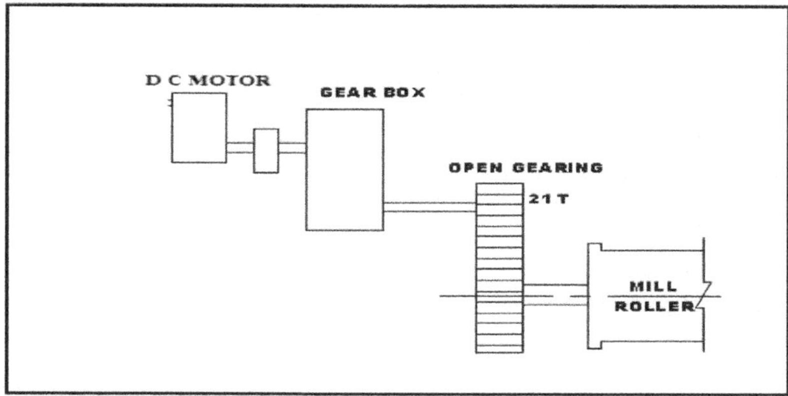

Fig.20.3 DC motor with high speed gear box, open gear

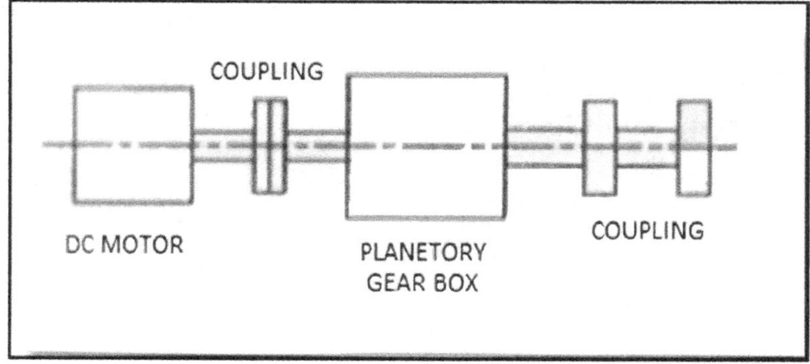

Fig. 20.4 DC motor coupled with planetary gear box

However, it requires specialized personnel familiar with installation and maintenance of big and specialized electrical machinery. Therefore the electric drives were not being widely used. But now new factories are coming with DC electric drives. Further the old plants with installation of high pressure boilers and turbine for cogeneration of electricity are also going for electric drive DC motors.

The electric drives are ideally suitable for cogeneration of surplus electric power.

These drives have wide operating speed range and are suitable for mill automation. These drives require skilled maintenance staff trained in handling staff for electronic control panels used with these drives. The application thyristor controlled drive for mills are more favourable. It offers advantages like

- Continuous variable speed range with step less precise control at high efficiency. (In turbine, the efficiency drops at lower speed).
- Soft start and controlled acceleration feature.
- Desired speed regulation possible.
- Neatness and cleanliness, no old steam piping joint leaking etc.

In many old plants while installing TRPF DC drives are used.

The drive consists of AC/DC converter and DC motor. The AC supply is rectified by thyristor and send to a DC motor which drives the mill through a reduction gearing. The speed can be varied practically zero to rated speed. The only disadvantage is that the DC motors require regular and careful maintenance.

Power consumption of the mill can be observed at any time on ammeter, which is not possible with steam turbine.

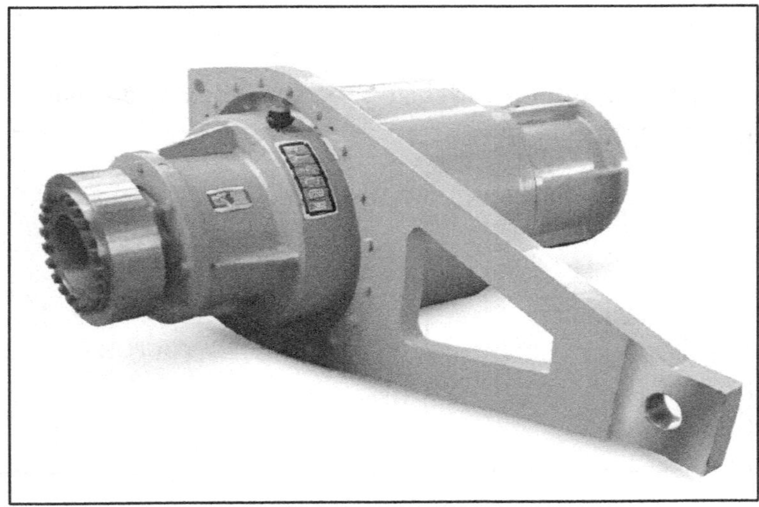

Fig, 20.5 planetary gears

The foot mounted mill gearbox is resting on civil foundation and is coupled to the mill through tail bar or rope coupling. The crown pinion transmission gives stress on the headstocks and is not much efficient. Now shaft mounted planetary gear with motor are being installed. These have following benefits:

Fig.20.6 Shaft mounted electric motor and planetary gear

- Elimination of crown pinion,
- Elimination of tail bar,
- Horizontal forces on roller bearings are reduced,
- Reduction in the mill rollers stress and headstock stress,
- Less wear of mill component.

Hydraulic Drive:

The hydraulic motor excludes gear boxes in between the motor and the mill. The number of hydraulic motors required in a tandem depends on total torque requirement on each mill. A hydraulic pump of variable displacement type is used in the system. Hydraulic pipes are used to connect the pumps with the hydraulic motors and the power is transmitted by the oil flow. AC electric motors can be used on each main pump or a steam turbine to all main pumps. The hydraulic drive can operate up to 200% of the rated torque and from zero speed to the full speed.

The system consists of vane pump, booster pump, hydraulic motor and pipeline with accessories. Vane pump sucks hydraulic oil from tank and delivers it to booster pump at a pressure of about 40 bars. The booster pump delivers the oil to hydraulic motor at 350 bar (Maximum at 450 bar). The high-pressure oil enters to hydraulic motor through pipelines and flexible oil hoses. The hydraulic motors can be fitted individually on each roller of a mill so that no pinion gears are required. The relative speeds of top, feed and discharge rollers also can be kept different. The hydraulic drive can be mounted on a roller shaft at either of the ends. The auxiliary equipment such as filter, cooler and auxiliary pumps are required. With hydraulic drive the mill gearing, tail bar couplings, pinions etc. are not required.

Table 20.2 Efficiency of different mill drives with different reduction gear system

		Efficiency percentage							
Sr. No.	Mill drive	Power turbine	Mill drive	H S Red. Gear box	L S Red. Gear box	Open helical gear	Planetary gear	Tail bar coupling	Overall efficiency
1	Turbine	--	65	98.5	97	88	--	96	52.47
2	Turbine	--	65	98.5	--	--	97	96	59.62
3	DC motor	70	95	--	97	88	--	96	54.49
4	D C motor	70	95	--	--	--	97	96	
5	Hydr. Motor	70	95	92.5 Hydr. Pump	93 Hydr. Motor	--	--	96	54.91
6	Shaft mount	70	95	--	--	--	97	--	64.50

Table 20.3 Installed power for Milling Tandems with conventional turbine and reduction gear system in two factories of India

Sr. No.	Particulars	Factory A	Factory B	
1	Crushing rate tch	150	340	
2	Fibre % cane	14.00	14.00	
3	Fibre throughput	21.0	47.6	
4		HP installed	Mill HP installed	GRPF/TRFP DC drive HP installed
5	Mill no 1	350	1000	------
6	Mill no 2	350	600	250
7	Mill no 3	350	600	250
8	Mill no 4	350	600	150
9	Mill no 5	-----	600	150
10	Total HP installed	1400	4200	
11	Installed hp/tfh	66.7	88.2	
12	KW/tfh	50.0	66.2	

-OOOOO-

21 MILL CAPACITY

Mill capacity:

The capacity of a milling tandem is the crushing rate that can be maintained at standard or normal extraction. The standard extraction is expressed in terms of RME. The normal expected RME from a milling tandem may vary from 94 to 97 depending upon milling tandem, design of the mill, condition of the mill, operational parameters maintained etc. The capacity is generally quoted in terms of tonnes cane per hour (tch) or tonnes cane per day (tcd). Many engineers and technologist have given different formulae for capacity calculations. A formula which is generally followed is given by Hugot and is as follows

$$A = \frac{0.9\,cn\,(1-0.06nD)\,LD^2\sqrt{N}}{f}$$

Where

- A = capacity of the tandem in tch
- f = fibre content per unit cane
- c = coefficient for cane preparation
- l = length of rollers in meter
- D = diameter of rollers in m
- n = speed of rotation of rollers in rpm
- N = number of rollers in the tandem

The coefficient 0.9 may be taken as 0.95 if roller arcing is practised. The coefficient of cane preparation c may be taken as 1.15 to 1.22 depending on cane preparation.

Example:

With the above formula we will calculate capacity of four mill tandem of size 712 X 1525 mm (30" X 60") mill with following normal figures.

- f – Fibre content per unit cane = 0.14
- c – Coefficient for cane preparation = 1.15
- l – Length of rollers in meter = 1,525
- D – Diameter of rollers in m= 0.712
- n – Speed of rotation of rollers in rpm 4.5
- N – Number of rollers in the tandem 12,

Now putting these values in the above equation

$$A = \frac{0.9 \times 1.15 \times 4.5\,(1 - 0.06 \times 4.5 \times 0.712)\,1.525 \times 0.712 \times 0.712 \times \sqrt{12}}{0.14}$$

Solving the above equation we get

A = 72 tcd

Then for 22 hours (as per Indian standard specifications the capacity is given for 22 hrs.) 1584 tcd, or say a 1600 TCD capacity mill. However, as per Indian standard specifications (1972) this milling tandem was recommended for 1250 tcd plant. With UFR to each mill, cane preparation, above 85+ and roller arcing practicing, the capacity can increased up to 1750 tcd with good mill extraction.

Standard mill sizes and their capacities specified by ThyssenKrupp Industries in its technical leaflet are given hereunder.

Table 21.1 Standard Mill Sizes and Capacities

Sr. No.	Mill Size	Three Roller Mills With UF rollers		Three Roller Mills With GRPF / TRPF	
	mm (In)	4 mills	5 mills	4 mills	5 mills
1	610 X 1220 (24 X48)	1000	1400	1300	1900
2	711 X 1422 (28 X 56)	1400	2000	1800	2700
3	762 X 1525 (30 X60)	1700	2500	2200	3300
4	864 X 1728 (34 X 68)	2500	3500	3300	4500
5	915 X 1830 (36 X 72)	3200	4600	4200	6000
6	965 X 1930 (38 X 76)	3500	5000	4500	6500
7	1016 X 2032 (40 X 80)	4400	5700	5300	7000
8	1066 X 2134 (42 X 84)	4400	5700	6600	8800
9	1143 X 2286 (45 X 90)	6400	8200	7500	9800

One should note that these capacities are given for normal or standard reduced mill extraction. However, if extraction is compromised, very high capacity can be attained. We have observed when extraction is overlooked a 30" X 60", four mill tandem with UFR can maintain a regular crushing rate of 2500 + tcd.

-OOOOO-

22 MILL SETTING

The opening between top and feed rollers as well as top and delivery rollers and clearance between top roller and trash plate when mill is working at installed crushing rate with normal parameters is called mill setting. The openings are calculated on certain assumptions in order to give the good results in terms of capacity and extraction. The delivery opening is always smaller than the feed opening, so that the prepared cane/bagasse is subjected to lesser volume and higher pressure at the delivery to get higher extraction.

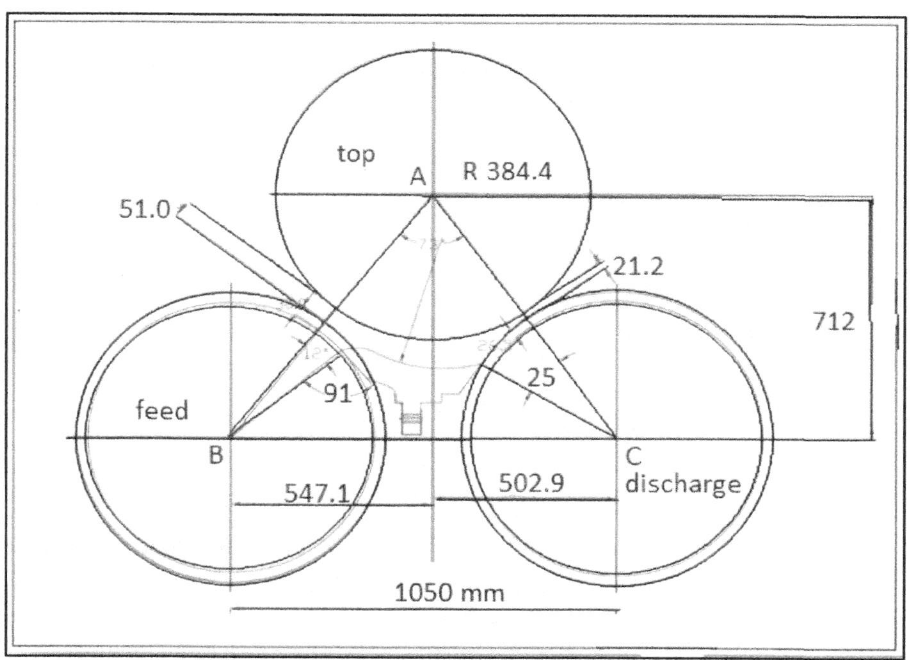

Fig 22.1 Mill setting

There are many methods to calculate the mill setting. In each method certain assumptions are made on the experience of mill engineers. The milling plant design and working conditions change from factory to factory. Therefore by any method we cannot come to any exact and final values of the openings that can be utilized in all factories. Further in mill setting to get good results we have to maintain proper pressure over the feed roller and discharge roller.

The mill setting calculated is an initial guideline. Mill engineers should keenly observe the mill working for followings-

Analysis of feed, discharge, and compound juice of each mill,

- Bagasse analysis of each mill,
- Corresponding crushing rate and roller lift of each mill.

This is called as mill test and taking the brixes of juices a graph is drawn which is called as brix curve. After examining if required the mill should be readjusted. So that correct setting is achieved to give better results.

The discharge work opening (DWO) and Discharge rest opening (DRO) calculated by various methods are more or less same. There is however wide variation in the feed settings. The reason for this variation is due to various milling plant designs and feeding devices. With feeding devices higher crushing rate and higher juice extraction can be achieved at reduced feed setting.

The mill settings do not remain constant. They are subject to change due to wear of mill rollers bearings and mill vibrations. These should therefore be checked periodically and readjusted if required.

For calculating mill setting one of the following methods is generally followed.

- Jawa method
- Fibre method
- Volumetric method
- Ferrell's method

The simpler fibre method widely used is given hereunder

Fibre method:

In this method, setting from the first mill to last mill decreases as the volume of the prepared cane/bagasse from the first mill to the last mill decreases. Feed side setting is always on higher so that prepared cane/bagasse can easily be grabbed squeezed to a limited extent to extract part of the juice. Further a compact blanket of bagasse formed is send to the discharge side by the bridge 'trash plate'.

- The delivery opening is smaller hence prepared cane/bagasse is subjected to lesser volume and high pressure at the delivery to get higher extraction.
- Average mean diameter of top roller and discharge roller is considered while calculating the discharge work opening.
- With experience, logic and mathematics the optimum openings are calculated. First discharge operating (work) opening (DWO) is calculated.
- Then feed operating (work) opening (FWO) is found. The optimum ratios between the different openings are based on empirical consideration.

- The fibre percent bagasse is considered as 33 % to 52 % lowest being for the first mill, increasing for the next mills and highest for the last mill.
- While calculating the discharge work opening (DWO) the density of bagasse passing through the opening is considered as 1200-1270 kg/m^3, lowest being for the first mill and highest being for the last mill.
- The forward slip factor or reabsorption factor depends on speed of the roller and hydraulic pressure applied. It varies from 1.30 to 1.45 in calculating the mill setting it may be taken as 1.32 for the first mill, 1.35 for the mills in between first and last, and 1.40 for the last mill.
- The ratio of discharge rest opening (DRO) to discharge work opening (DWO) i. e. DRO/DWO may be taken as 0.65 to 0.25 highest for the first mill and lowest for the last mill.
- The ratio of feed work opening to discharge work opening [(FWO) /(DWO)] is considered as 2.5 to 1.7, highest being for the first mill and lowest for the last mill.
- The trash plate work opening (TWO) is taken 1.75 times feed work opening of the respective mill for all mills.
- The difference in FWO to FRO will be same as that of DWO to DRO or we can say FWO – FRO = DWO – DRO, Or FRO = FWO – (DWO – DRO)
- The feed and discharge opening will be increased by 80% to the top roller lift for e. g. when in operation the top roller will be lifted by 10 mm the trash plate opening will be increased by 10 mm whereas the feed opening and discharge opening will be increased by 8 mm.
- The difference in trash plate work opening (TWO) and trash plate rest opening (TRO) will be 1.25 times the difference in feed work opening (FWO) and feed rest opening(FRO) Or TWO – TRO = 1.25 (FWO – FRO)

  Hence TRO = TWO – 1.25(FWO – FRO)

Let us take an example of mill setting,

1. mill size – 850 X 1700 mm (34" X 68") four mill tandem,
2. Crushing rate – 120 TCH,
3. Fibre % cane – 14 %,
4. Roller rpm – 4.5

Table 22.1 Mill Setting

Sr. No.	Particulars ↓	Mill number →	1	2	3	4
		Underlined italic numbers are Sr. No. from column no 1				
1	**Crushing rate tch**	**assumed**	**120**			
2	**Fibre % cane**	**assumed**	**14**			
3	**Weight of fibre passing through the mills**	**(_1_X 2X10/60) kg per min**	**280**			
4	**Fibre % bagasse**	**assumed**	**33**	**42**	**47**	**50**
5	**Weight of bagasse passing through the mill.**	**(100 x _3_ /_4_) kg/min**	**848.5**	**666.6**	**595.7**	**560**
6	**Bagasse no void density**	**assumed kg/m³**	**1200**	**1225**	**1250**	**1270**
7	**Volume of bagasse passing per min. through each mill**	**_(5/6)_ m³/min.**	**0.707**	**0.544**	**0.477**	**0.441**
8	**Roller rpm**	**assumed**	**4.5**			
9	**Area escribed by the roller per min.**	**πDLn m²/min**	**20.43**			
10	**Reabsorption Factor**	**assumed**	**1.32**	**1.35**	**1.35**	**1.40**
11	**Discharge Work Opening**	**(_7_ x 1000) / _(9 x 10)_ mm**	**26.2**	**19.7**	**17.3**	**15.4**
12	**DRO/DWO**	**assumed**	**0.6**	**0.5**	**0.4**	**0.3**
13	**Discharge Rest Opening**	**_(11 x 12)_ mm**	**15.7**	**9.9**	**6.9**	**4.6**
14	**FWO/DWO**	**assumed**	**2.4**	**1.9**	**1.9**	**1.8**
15	**Feed Work Opening**	**_(11 x 14)_ mm**	**62.9**	**37.4**	**32.9**	**27.7**
16	**Feed Rest Opening (FWO – (DWO – DRO))**	**_(15 – (11 – 13))_ mm**	**52.4**	**27.6**	**22.5**	**16.9**
17	**T rash plate Work Opening TWO = 1.75 x FWO**	**(1.75 x _15_) mm**	**110.1**	**65.5**	**57.6**	**48.5**
18	**TRO = TWO –1.25(FWO – FRO)**	**_17 –1.25 (15 – 16)_ mm**	**97.0**	**55.7**	**44.6**	**35.0**

Table 22.2 Mill Setting Abstract figures in mm

Sr. No.	Mill number → Particulars ↓	1	2	3	4
1	**Discharge Work Opening**	**26.2**	**19.7**	**17.3**	**15.4**
2	**Discharge Rest Opening**	**15.7**	**9.9**	**6.9**	**4.6**
3	**Feed Work Opening**	**62.9**	**37.4**	**32.9**	**27.7**
4	**Feed Rest Opening (FWO – (DWO – DRO))**	**52.4**	**27.6**	**22.5**	**16.9**
5	**Trash plate Work Opening TWO = 1.75 x FWO**	**110.1**	**65.5**	**57.6**	**48.5**
6	**Trash plate rest opening = TWO –1.25(FWO – FRO)**	**97.0**	**55.7**	**44.6**	**35.0**

A simple formula to calculate delivery work opening (DWO) is given here under

$$DWO = \frac{f \times tch}{DLn} \times \text{mill setting factor}$$

Where

f – Fibre per unit cane
D – Roller diameter
L – Roller length
n – Roller rpm

Table 22.3 Mill setting factor:

Mill no.	4 mill tandem	5 mill tandem	6 mill tandem
1	10.1	10.1	10.1
2	7.7	7.7	7.7
3	6.8	6.9	7.0
4	6.1	6.4	6.5
5	-	6.1	6.2
6	-	-	5.9

The ratio FWO to DWO may vary considerably. Some mills in Australia reports the ratio in the range of 3.0 to 4.2 and achieving excellent results. With pressure feeding arrangements it can be reduced. One Queensland tandem with pressure feeders on all mills reports ratio of 1.48, 1.50, 1.71, and 1.89 for first to fourth mill respectively. Hence Jenkins says *that "any relationship between ratio and feeding qualities must be regarded as a vague generalization. There are other unknown factors which may play important part".*

Trash plate position: The trash plate works as bridge between feed roller and discharge roller under the top roller and conveys the bagasse without extraction of any juice. Many different methods of determining the trash plate setting are used. All are empirical.

- The top angle FTD = 72-78°.
- A – Distance Between top roller and Feed end of the trash plate = 1.5 FOW
- B – Distance Between top roller and trash plate centre = 1.75 FWO
- C – Distance Between top roller and discharge end of trash plate = 1.9 FWO
- D – Distance Between discharge Roller and discharge end of trash plate = 0.8FWO

If the trash plate is on higher side then,

- More hydraulic load is effected on the trash plate
- mill extraction is adversely affected
- Wear of trash plate is more
- Power consumption is increased
- Mill gets frequently choke up

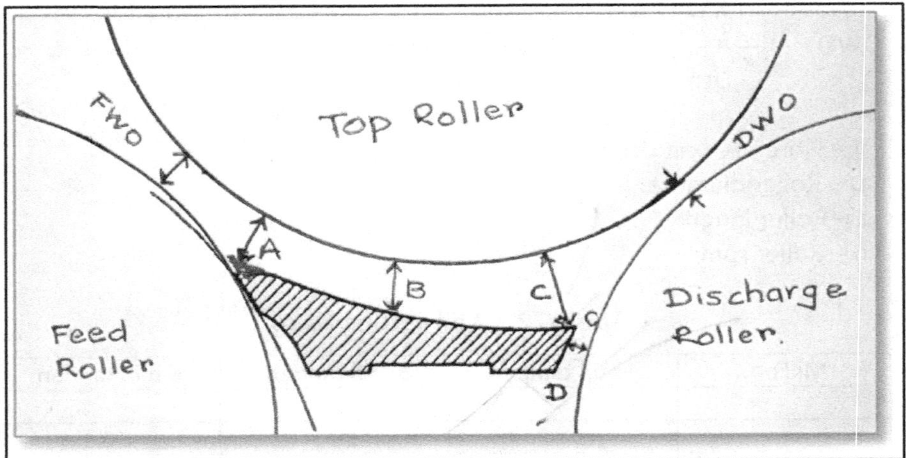

Fig, 22.2 Trash plate position

If the trash plate setting is too low then,

- Angle of contact of bagasse with discharge roller becomes more
- Cush cush falling is increased
- Grumbling noise occurs
- Top roller may slip

Mill Extraction:

In 1960-70 reduced mill extraction of 92-93 was considered a creditable figure but now the reduced mill extraction of 96 with improved capacity is quite common and efforts are being made to achieve 97 RME. The mill extraction mainly depends on the following factors

- Cane preparation
- Specific pressure
- Length of the milling tandem
- Roller speed
- Imbibition
- Mill sanitation.

-OOOOO-

23 MILLS - PRESSURE FEEDERS

Mill Feeding Devices or Pressure Feeders:

In order to improve mill working in respect of crushing rate and extraction, in last 40 year the cane milling has undergone many changes in cane preparation, pressure feeding devices and mill designs. Different auxiliary equipment given here under have been developed.

- Three roller mill with under feed roller
- Three roller mill with under feed roller and Donnelly chute
- Four roller mill with two trash plates
- Three roller mill with two roller pressure feeder (GRPF or TRPF)
- Five roller mill

Six or even seven roller mills have also been installed in few cases for achieving higher crush rate. It is however important to note that basically major portion of the juice is extracted by 3 roller mills.

The additional devices like underfeed with or without Donnelly chute or TRPF /GRPF with Donnelly chute help in feeding of cane / bagasse. They produce required feeding force to push the prepared cane or bagasse into the opening between top roller and feed roller so that slippage of bagasse is avoided. They do not directly contribute much in juice extraction but help in feeding the mills. Therefore they are known as feeding devices. However GRPF / five-roller mill no doubt extract good amount of juice, similar to two-roller crusher used previously before first mill. These have resulted in improving the crushing rates, mill extraction and reducing energy consumption.

Donnelley Chute:

Simple open chutes are often found insufficient to work the milling tandem at a maximum capacity. In such case, an arrangement of an underfeed roller with a closed vertical chute called Donnelly chute is used in many factories. The Donnelly chute is fabricated of 8-mm mild steel plates. The height of the chute is 3 m and inclination above 80° to the horizontal. The location and setting of the chute is important. It is provided stainless steel lining from inside reduces frictional loss on the sides. The weight of bagasse column in the closed chute provides a positive and effective feeding force. The Donnelly chute is provided with electronic / electro-mechanical level sensing device to control the cane feeding.

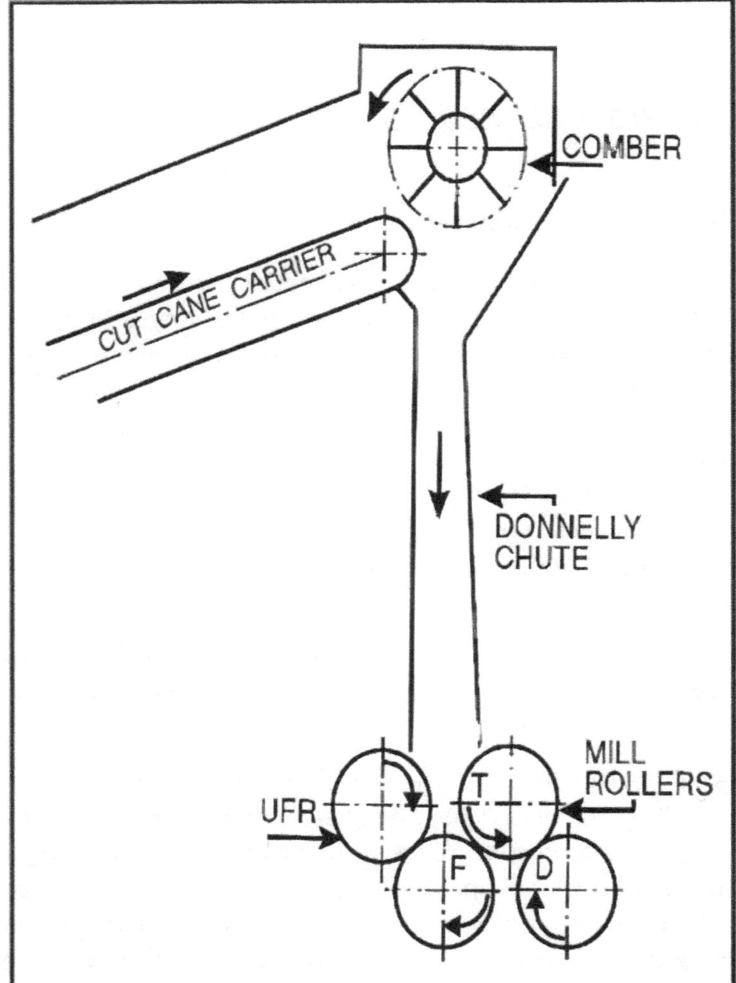

Fig. 23.1 Donnelly chute with comber for the first mill

The Donnelly chute is not suitable for coarse and medium cane preparation even a single uncut cane stalk can create choking problem inside the chute.

Under Feed Roller:

An under-feed roller (UFR) is installed under the bagasse entering the mill close to the feed roller. The diameter of underfeed roller is about 60 to 85 % of the mill roller diameter. It improves feeding at mills. It is a cast iron grooved roller. When UFR are installed to old three roller mills without modification in the intercarrier, the diameter is usually 60-70%. The geometry of the headstock plays an important role in installing the UFR. The diameter of the UFR is selected in such a way that the centre of the UFR should not be above the centre of the top roller shaft and inter carrier angle is not more than 28°. Otherwise when the slat type inter-carrier is replaced by rake type with Donnelley chute, the diameter can be increased up to 80-85%. The UFR is driven by top roller through gear. Surface speed of UFR is 10 % higher than the mill roller speed. The UFR is provided with arrangement to adjust the setting in axial direction by ± 25 mm.

The prepared cane density adopted for setting of UFR for the first mill is about 800 kg/m^3 and it may progressive increased up to 1200 kg/m^3 for the last mill.

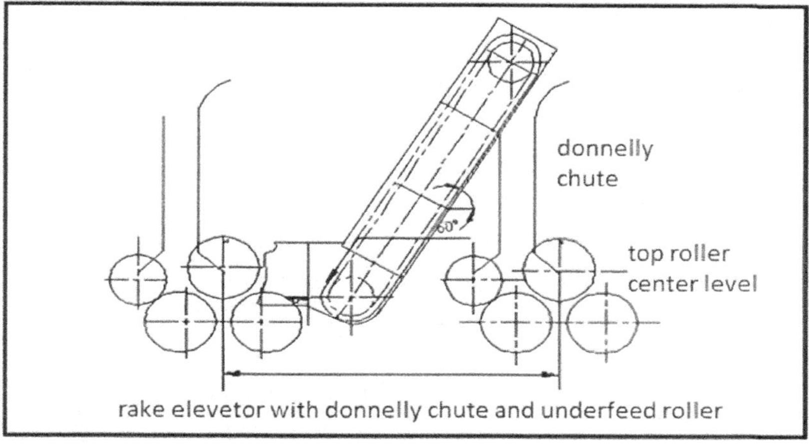

Fig. 23.2 Under Feed Roller with Donnelly chute

Now almost all the old standard three roller mills are provided with UFR. By installation of UFR the crushing capacity can be increased up to 15-25%.

Toothed Roller Feeder (TRF):

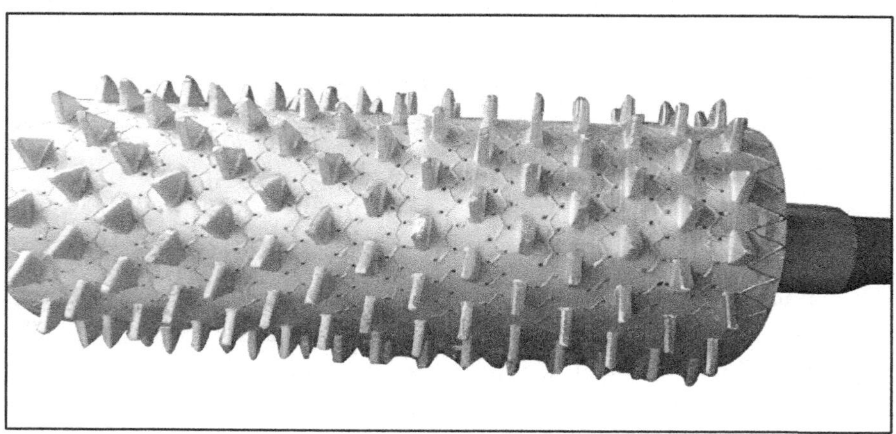

Fig.23.3 toothed roller

Feeder roller with toothed surface gives high grip on prepared cane to improve crushing rate. With this the ratio FWO/DWO which is normally 2.4 can be reduced up to 1.7 depending upon cane preparation, mill condition and position. Therefore extraction in the first pressing of the mill (feed and top roller pressing) is increased. Mill speed can be reduced and juice drainage is achieved.

Toothed Feeder Roller with juice draining arrangement:

Ulka Engineers has patented a design for toothed feeder roller with juice draining arrangement. The TFR is circumferentially perforated like a lotus roller. The extracted juice passes in the roller shell through perforated holes and then flows out from roller sides. This improves juice drainage and juice flooding is almost eliminated.

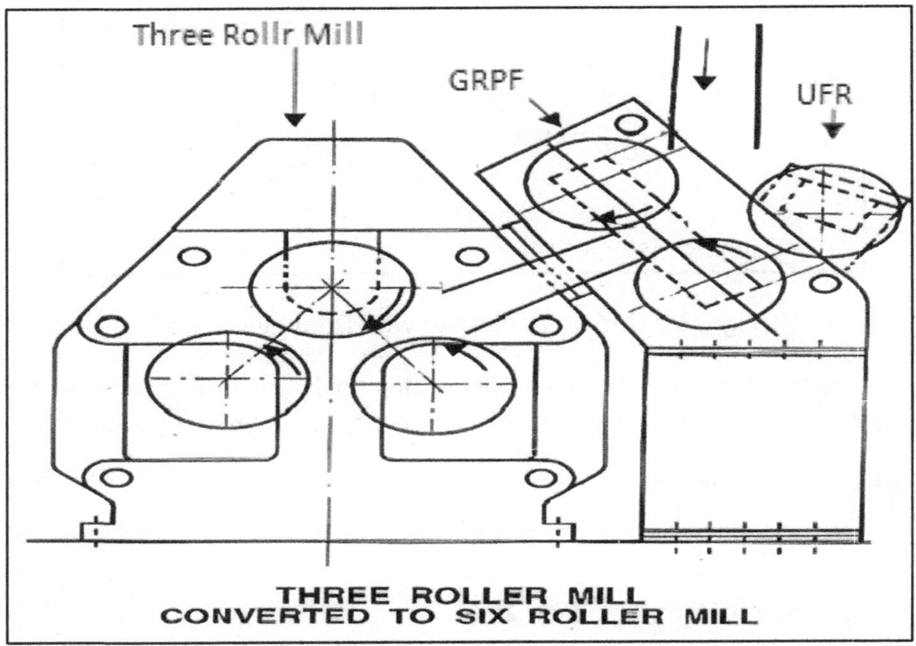

Fig.23.4 A three roller mill with GRPF + UFR

Grooved Roller Pressure Feeder (GRPF):

GRPF is a pair of grooved cast iron rollers having similar in diameter of that of mill rollers. These have linear speed about 25-30% higher than that of the mill rollers. These can develop much higher pressure and are suitable for higher crushing rate. The opening between these rollers should ascribe volume of 1.5 times than that of feed and top roller opening. There is a closed pressure chute between GRPF and mill. Messchaert grooves and Chevron grooves are also provided on roller. A Donnelley chute of 3.0 m height is provided. A separate DC variable drive is preferred.

In GPRF a radial pressure on the bagasse is developed by the rollers that extracts about 25-30% of the juice and develops a feed force on the bagasse to get it into the mill. The setting of GRPF rollers and closed chute is important for getting required feeding force and high density of the bagasse. The mill equipped with GRPF can run at a lower speed and hence better juice extraction is obtained.

GRPF is sensitive to lower cane preparation and high imbibition. It consumes 0.6-0.7 hp/tch power. However for smooth operation of GRPF optimum fibre index is required. Low fibre index develops low feeding force, which may result in jamming of the closed pressure chute and it takes lot of time to clear the chute.

Toothed roller pressure feeder (TRPF):

TRPF consists of a pair of rollers having typical shape teeth with equal spacing all over circumference. The roller teeth intermesh with each other having space to grip the loose prepared cane/bagasse. It compresses the bagasse and pushes it into closed pressure chute provided between TRPF and mill. A Donnelley chute of a moderate height - 1.5m is provided.

For light duty TRPF drive is provided through chain from top roller or feed roller. Light duty TRPF drive can be installed in short space with specially designed rake type intercarrier. For heavy duty TRPF an independent individual drive of electric motor through gears or hydraulic motor drive is provided. A dc variable drive is always preferred. It is preferred as it is supplementary power to mill for higher crushing. The teeth cut the bagasse blanket grab it and push into a closed pressure chute. There is no radial pressure built up in bagasse between two rollers like GRPF and therefore there is no grinding action like GRPF. It is sturdy to cane preparation and imbibition. When TRPF is used the FWO/DWO ratio can be reduced up to 1.5 to 1.6.

These feeders have saw like teeth and develop positive feeding pressure. A combination of compressive and shear stresses are induced in the direction of feed. The feed forces as high as two times that of GRPF are possible with these toothed rollers.

At higher crush rate the feeding pressure requirement is high and GRPF cannot develop such pressure and often problems of choking before GRPF occurs. TRPF is better solution for such conditions.

The high axial load on the bagasse causes buckling of the column in the chute, leading to excessive load on the chute and may thus lead to chute failure. This condition generally occurs when condition of mill rollers is not maintained properly. The roughening of the mill roller should therefore be done regularly. The S.S. Engineers, Pune developed light duty a short space TRPF suitable to old mills which have been installed in many factories.

The success of GRPF or TRPF depends on their proper designing, fabrication and installation. In order to obtain better performance of the plant in terms of crushing rate and extraction, milling plant maintenance is very important. Even best design mill, adopting optimum operating parameters cannot yield good results if maintenance is neglected. The breakdowns only are not the indication of poor maintenance. Even there are no breakdowns, with poor maintenance particularly extraction hampers. Blunt knives and hammers do not result in breakdown but seriously affect cane preparation.

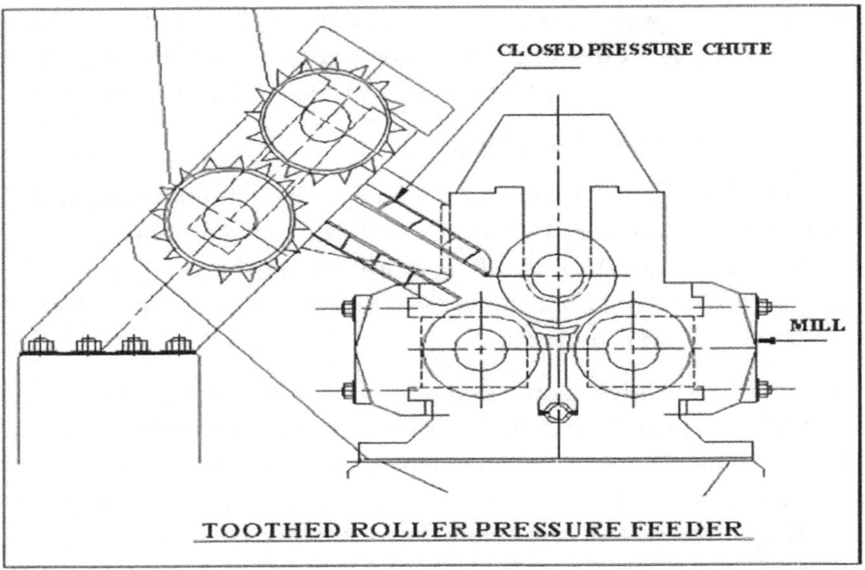

Fig. 23.5 Toothed roller pressure feeder

References:

1 Singh V K, All India Seminar at VSI 1994,
2 Ramkumar, All India Seminar at VSI 1994,
3 Patil A R DSTA convention
4 Engineers S S, brochure,
5 Ulka Engineers Brochure,
6 ThyssenKrupp, Pune India, brochure,

-OOOOO-

24 *IMBIBITION*

24.1 General:

The mill extraction – without dilution only by repeated high pressings can be as high as 78 – 80 percent. Fibre holds the moisture therefore the moisture content of the bagasse cannot be brought down below 46-47 % or in other words the bagasse contains 50 % juice and 50 % fibre. It becomes the limit of extraction. Therefore in order to obtain high extraction, it is necessary to dilute the juice remaining in the bagasse with water before pressing at the next mill. It is done in several stages so that the 50 % juice remaining in the final bagasse is much more dilute and sugar loss in bagasse is reached to a minimum level. The water applied for dilution as well as the procedure is called Imbibition.

Fig. 24.1 Imbibition applied at penultimate mill

Simple imbibition in which water is applied to all secondary mills is now not followed, as it requires more water that gives extra load on the evaporator station. Therefore compound imbibition system, in which comparatively less water is required, is followed.

In compound imbibition water is applied to bagasse going to the last mill. The diluted or thin juice extracted from the last mill is used as imbibition before second last mill. The juice from that mill applied before the previous mill and so on. Thus after each mill the juice in the bagasse is diluted with a juice of lower concentration. Thus the Process becomes nearly a counter current extraction. In this process the quantity of juice extracted by each of the secondary mill is approximately equal to the quantity of imbibition juice applied before it.

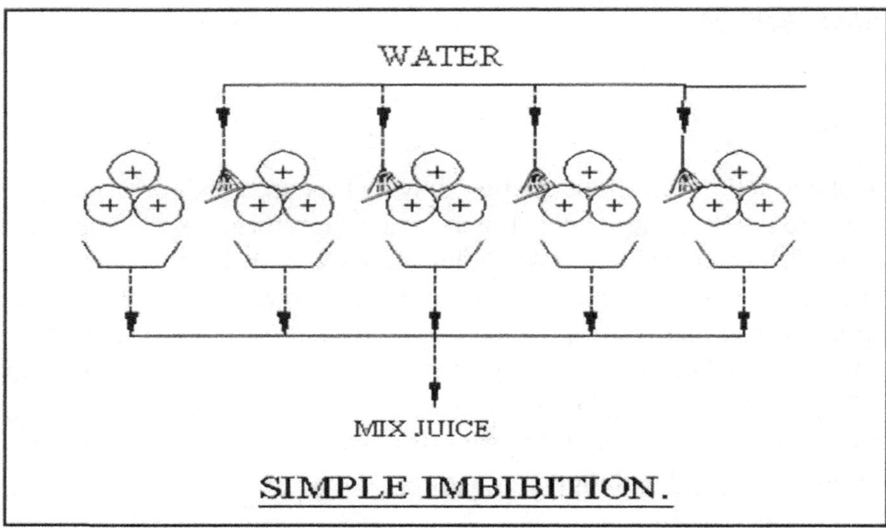

Fig. 24.2 Simple imbibition

The percentage of imbibition water varies from 20 to 45 % on cane or 150 to 330 % on fibre. The distribution of imbibition over the bagasse should be uniform and it should be thoroughly mixed with the bagasse.

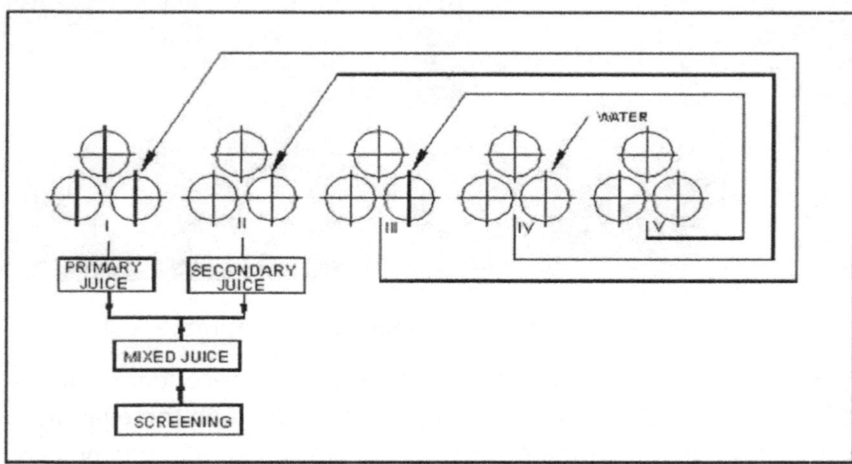

Fig. 24.3 Compound Imbibition system

It is the fibre that takes along with it diluted juice and therefore effectiveness of imbibition depends on imbibition % fibre and not on imbibition % cane. The percentage of imbibition % fibre varies considerably with countries, the capacity of the mill and relative cost of fuel and sugar and accordingly evaporator station capacity is given. When water % fibre used above 300 %, it is generally called true maceration or bath maceration or simply maceration. Temperature of imbibition water is maintained in the range of 60 to 85°C. Tromp has recommended imbibition water temperature of 85°C to destroy the protoplasm of the unbroken cells. Hot imbibition reduces surface tension of the fibre and thus reduce amount of juice retained in the bagasse.

To make the imbibition more effective it is necessary that imbibition water/juice should be applied at higher pressure. A perforated pipe either at the discharge roller or in the next intermediate carrier generally fitted for spraying imbibition water over the bagasse. A pump of 60 m head is used for imbibition. Bagasse can absorb water up to 7 to 10 times the weight of fibre in it. Since quantity of water used for imbibition is about 2 to 3 times the weight of fibre thorough mixing of water whole bagasse is not possible. Therefore

- In many factories imbibition is applied from top as well as from the bottom of the bagasse blanket to ensure effective mixing.
- In some factories for proper mixing of the imbibition and the bagasse, revolving macerators or bagasse spreaders are installed. These are useful for thorough mixing of imbibition with bagasse.
- Hugot describes steam aided imbibition (SAI) followed in some factories. Super-heated steam is used[1] with imbibition. It is claimed that this system does not allow the bagasse to absorb air while emerging from mill (We do not agree with this statement). However this also makes the imbibition more effective.

Vikramsingh[2] while discussing his experiences says *"with increase in imbibition up to 250 on fibre the bagasse pol decreases. Above 250 reduction in bagasse pol is not always observed however recovery is increased indicating higher mill extraction"*.

J P Mukharjee[3] has described bath maceration – *"a simple system of recirculation of maceration fluid followed at Kalamia sugar mill in Australia in 1965. The last – fifth mill was a five roller mill. Distance between 4th and 5th mill was kept 104 ft. So that very long bagasse carrier was available. The underneath the bottom plate of this carrier was the drainage tray, which was divided into five compartments roughly of equal length. The process resembles the working of diffuser. The factory obtained 97 % extraction by using 130 to 140 percent imbibition on fibre"*.

J P Mukherjee has recommended another recirculation system. The feed roller juice and discharge roller juice are collected separately. The feed roller juice of the last mill is mixed with imbibition water and is applied on the bagasse being fed to the same mill. Similarly the feed roller juice collected separately from the penultimate mill is mixed with discharge roller juice from the last mill and applied to the bagasse being fed to the penultimate mill.

According to Douwes-Dekker[4] the time of contact of imbibition with the bagasse before the following mill is not much important. The mixing of the imbibition with the juice in bagasse takes place effectively when the layer of bagasse enters the next mill and pressure is applied over it.

24.2 Optimum Imbibition:

In India now most of the factories are maintaining imbibition % fibre above 250% and some factories above 300%. It is assumed that quantity of imbibition has no effect on the moisture of final bagasse. But this is not very correct, actually by experience it is observed that moisture of final bagasse increases slightly with quantity of imbibition. Therefore considering steam balance, price of bagasse and price of sugar, optimum value for imbibition should be determined.

24.3 Degree of mixing the imbibition and dilution of residual juice:

There are several expressions that give *'Efficiency of Mixing of Imbibition'*. If the imbibition mixing is 100% perfect, the front roller, back roller and residual juice brixes should be equal. However the juice extracted is always of higher purity than the residual juice therefore pol percentage of the residual juice will be slightly lower. This is applicable to all mills. Noel Deer[5], Parr[6] and Hugot[7] have given detail mathematical studies of imbibition process. We do not going to discuss here of that studies.

Hence, the juice extracted by the front roller is of lower concentration than the juice extracted by the back roller. Further, the juice extracted by the back roller is of lower concentration than the juice remained in the bagasse passing out from the mill.

Douwes Dekker[8] gives simple equation to estimate the dilution. He gives the formula

$$\text{Dilution ratio} = \frac{BPJ - BRJ}{BPJ - 0.15\ BPJ} \times 100$$

$$\text{Or} \quad \text{Dilution ratio} = \frac{BPJ - BRJ}{0.85\ BPJ} \times 100$$

Douwes Dekker takes brix of the residual juice in final bagasse as 15 % of the brix of the primary juice as an arbitrary target value during the milling process in the above formula. When brix of the residual juice reaches to 15 % of its original value the dilution ratio becomes 100%.

Douwes Dekker in his paper demonstrates that dilution ratio can be used as an important tool for investigating milling performance. He points out that the dilution ratio depends on

- The primary extraction – the higher the primary extraction, less the juice left in the primary bagasse and imbibition becomes more effective.
- The quantity of imbibition water applied.
- Number of imbibition steps
- The extent to which imbibition water / fluid mixes with the residual juice at each step.

Brix curve:

The most useful and simple method of assessing the effect of imbibition is to plot a graph of brixes of juices of all the mills. This method was put forward by W E Smith[9] it involves only sampling and brix determination of juices. J H Haldane[10] extended the method by giving ratio of actual brix to ideal brix as calculated by Noel Deerr's formula. Gundurao[11] has discussed about juice brix curve in detail. Gundurao emphasise that a factor of prime importance is the ratio of water to fibre.

Generally back roller juices are taken for comparison. The actual brix curve is compared with theoretical curve. The actual brixes are usually higher than the ideal brixes. It is because ideal extraction at mills is seldom achieved. Therefore following mills give juices extracted of higher brixes. However higher deviation indicates fault in operation. There are several variables that can influence the shape of the graph. Careful interpretation of the graph can detect the cause.

24.5 Use of surfactant in imbibition water:

Some surface active substances (surfactants) have remarkable wetting properties and can useful in increasing imbibition efficiency. Ramaiah[12] and his co-workers have found out some specific surfactants that can be used at mills for this purpose. A mixture of non-ionic and anionic surfactants, which gave maximum pol extraction in the laboratory experiments at NSI, was referred to as Sushira.

The effect of Sushira on reduction of pol and moisture was tried in number of factories with complete milling tandem and with milling cum diffuser system. At mills the Sushira was applied along with the imbibition water onto the bagasse coming out of the penultimate mill. The quantity of Sushira found to decrease the pol in bagasse and decrease in moisture of bagasse.

The surfactants to be used have to be chosen carefully. Primarily they should be non-toxic. The surfactants are so sensitive to the presence of calcium salts. Therefore imbibition water to be used should be either condensate water or water of very low hardness.

References:

1 Hugot E ISSCT 1980
2 Vikramsingh (Seminar at VSI 1994)
3 Mukharjee J P DSTA 1982 Bath Maceration p M 39-49
4 Douwes-Dekker ISSCT 1959, p 86
5 Noel Deer, Cane Sugar, p 232-238, ISJ 1928, p 247-249
6 **Parr** P H, ISJ 1921 p562
7 Hugot E, Handbook p 224-237
8 Douwes Dekker SMRI report no.17 1961, p 39
9 Smith W E, ISJ 1929 p 673.
10 Haldane H, ISJ 1934 p 304
11 Gundurao, STAI 1939 p 55
12 Ramaiah ISSCT 1980 p 1113-17

-OOOOO-

25 MILL SANITATION

25.1 General:

Microorganisms grow very fast in cane juice as it contains all the required food for growth. Therefore proper attention is required for cleanliness and sanitation of milling tandem. The microorganisms consume sugars (sucrose, glucose and fructose) and if growth of the microorganisms is not controlled, sugar loss in large proportion may occur at mills. The pol loss at mills is not accounted in chemical control because the pol in cane is assumed as pol in mixed juice plus pol in bagasse. To control the growth of the microorganisms, application of high pressure steam or hot water is used for washing and cleaning of the mills. In addition to cleanliness use of biocides is also essential. Deterioration and sugar takes place in the following pathways -

- **Chemical inversion of sucrose:** The juice is acidic and is of pH 5.0 to 5.4 in nature. Due to acidity of the juice sucrose inversion takes place.
- **Enzymatic inversion of sucrose:** The active cell free enzyme invertase is present in the cane and is responsible for enzymatic inversion of sucrose.
- **Sucrose loss as a result of microbial growth:** The sugar in the sugarcane is exposed to contamination of microorganisms from the time the cane is cut to the time the extracted juice is heated for clarification.

Control of microorganisms is impracticable at fields. From cutting of cane to its crushing, we cannot control the attack of microorganisms. The only solution to check the sucrose loss due to microorganisms in this period is to minimize the time between harvesting and crushing.

Observations made by chemists at different sugar factories reveals that the loss of sugar at mills varies from 1 Kg to 3.0 Kg of sugar per tonne of cane depending upon cleanliness and sanitation maintained at mills. Further the microorganisms not only consume sugar but throw out unwanted exterior products that give trouble in the process.

25.2 Problem caused by microorganisms:

Slime production – The gummy slime i.e. dextran and levan produced by bacteria give problems in the sugar factory by clogging of pipes strainers and pumps

Acid production – The formation of acids such as lactic acid, acetic acid, butyric acid by microorganisms leads to lowering pH of the juice which ultimately increases the rate of inversion of sucrose. It increases consumption of lime in juice clarification process. The increased lime salts remained in dissolved state and passed in the clear juice. The lime salts are precipitated further and cause heavy scale in evaporator.

Viscosity- The gummy slime i.e. dextran and levan produced by bacteria increase viscosity of the juice, syrup and massecuite that affects process as under

- The rate of settling of mud in the clarifier is decreased
- The muddy juice filtration becomes difficult and loss in filter cake is increased
- Massecuite becomes viscous, sticky and rate of crystallisation is reduced
- Due to higher concentration of dextran, 'C' axis elongation of sugar crystals are observed
- The sticky low grade massecuites are difficult to cure in centrifugals. This results in higher sugar loss in final molasses
- Quality of sugar is also affected.
- Dextran is highly dextrorotatory having specific rotation + 195. This results in false sugar content and false higher juice purity. Therefore the apparent loss which is included in undetermined loss is increased.

25.3 Preventive Measures:

- To prevent the ill effect of the growth of microorganisms, matured, clean and fresh supply of cane is very much essential.
- There should be smooth working of the mills without frequent stoppages. During mill stoppages juice remains stagnant in the troughs, gutters, whirlers and strainers and microbes consume sugar in the meantime.
- The milling tandem should be well constructed from the sanitation point of view. Juice should not remain stagnant at any corner. All surfaces over which juice flows should be constructed of non-corroding metals. There should not be rough surface as of brick work and tiles.

Physical Cleanliness:

Physical cleanliness is the basic need of the mill section. The physical cleanliness and use of chemicals restricts growth of microorganisms and are killed to some extent.

The physical methods of destruction of microorganisms are use of live steam and hot water. Regular and thorough cleaning of the milling tandem – troughs, whirlers, gutters and strainers with ample use of hot water is important. Microorganisms hidden in depth may not be much get affected by superficial exposure of steam. Therefore only live steam is not sufficient to arrest the growth of microorganisms. Application of live steam and hot water around mills can be only 50-60 % effective, the remaining should be taken care by use of biocide.

Criteria for using chemical mill sanitation:

- The chemical must confirm to the foods regulations regarding addition in the process, the chemical should not be excessively volatile,
- The chemical should be able to withstand for long enough at mills to be effective,
- There should not remain any traces of the chemical in the sugar
- The FDA in America has approved the use of QAC and these chemicals are used worldwide.

Different types of biocides: The biocides at present commonly used for mill sanitation are

Chlorine: The most commonly used disinfectant is chlorine. Chlorine releasing compounds are applied for the deactivation of most microorganisms and are relatively cheap and convenient. The active ingredient of some chlorine based compound is chlorine dioxide which is known as best sanitation and disinfection chemical. These provide steady bactericidal efficacy within a broad pH range of pH 4 to pH 11. These work by denaturing the protein molecules.

Quaternary ammonium compounds: Quaternary ammonium cations, (also known as QAC) are positively charged polyatomic ions of the structure NR^+_4, R being an alkyl group or an aryl group. Unlike the ammonium ion (NH^+_4) and the primary, secondary, or tertiary ammonium cations, the quaternary ammonium cations are permanently charged, independent of the pH of their solution. Quaternary ammonium salts or quaternary ammonium compounds are salts of quaternary ammonium cations. The QAC are soluble in water and are stable within a wide range of temperature (0°C to 100°C) and pH (3 to 11) and are mainly bacteriostatic at a low concentration.

These are highly effective for broad spectrum micro-biocides that can be effective in controlling the growth of algae, bacteria and fungi. These are effective against both the gram positive and gram negative bacteria. Low dosage levels are required to maintain control. The chemicals neutralize the acidic group of the bacteria and disturb its metabolism. They are non-corrosive in nature. Ammonium bifluoride is commonly used in India in many factories.

Dithiocarbamate: A dithiocarbamate is a functional group in organic chemistry It is the analogue of a carbamate in which both oxygen atoms are replaced by sulphur atoms (when only 1 oxygen is replaced the result is thiocarbamate). A common example is sodium diethyldithiocarbamate. Dithiocarbamates specifically ethylene bidithiocarbamates (EBDCs), in the form of complexes with manganese (maneb), zinc (zineb) or a combination of manganese and zinc (mancozeb), have been used extensively as fungicides in agriculture from the 1940s. The dithiocarbamate based powerful biocides are nonoxidising chemicals. These are used to inhibit or reduce the bacteria and fungi at mills. These are effective at very low dosage. These are stable over wide range of pH and temperature, controls drop in purity and are cost effective, self-degradable and eco-friendly.

In the above based chemicals, formalin or chlorine/ chorine based compounds are now rarely used. Nowadays quaternary ammonium compounds / ammonium bifluoride and dithiocarbamates are commonly used. These are available in market under different trade names. Biocides such as heavy metal salts, quaternary compounds, detergent-based biocides, highly volatile biocidal substances, and so on, would be unacceptable, inefficient, or expensive to use.

The mill sanitation chemicals are generally used continuously in imbibition water in most of the factories. In some factories apart from continuous application booster dose is given for half an hour once or twice in a day.

The dose of the biocide has to be carefully monitored. The concentration that is used at the mills is usually in the range of 2 –20 ppm on cane crushed depending upon the activity of the biocide. At this concentration most of the biocides act as biostatic and only restrict the activity of the microbes but does not really reduce the microbial count in the mixed juice. The minimum dose required to restrict the activity of microbes is called as minimum inhibitory concentration (MIC), which is different for different chemicals. The MIC for any chemical is to be found out by trials and then that dose should be continued. In hot summer days mostly dose is to be increased.

Efficacy of mill sanitation chemicals:

The efficacy of mill sanitation can be done on the following criteria.

- Purity drop from primary juice to mixed juice,
- Purity drop from primary juice to last mill juice,
- Ratio of Reducing Sugars of primary juice to mixed juice
- To measure the pH of PJ, MJ and LMJ,
- Microbial count in primary juice and mixed juice,

These parameters can be guidelines in assessing the efficacy of mill sanitation chemical. With good mill sanitation the juice quality remains better, the materials at pan station are less viscous and loss in final molasses is controlled.

-OOOOO-

26 AUTOMATION AT MILLS

Automation of cane preparatory devices and cane feeding is essential for steady working of mills. Automation helps in steady operations right from the cane equalizer to the first mill feeding and further at mills.

Automation involves,

- Automation of cane feeding control system
- Automatic mill speed control system
- Automatic imbibition water control
- Automatic juice flow control
- Automatic interlocking system
- Alarms and emergency trip / stop

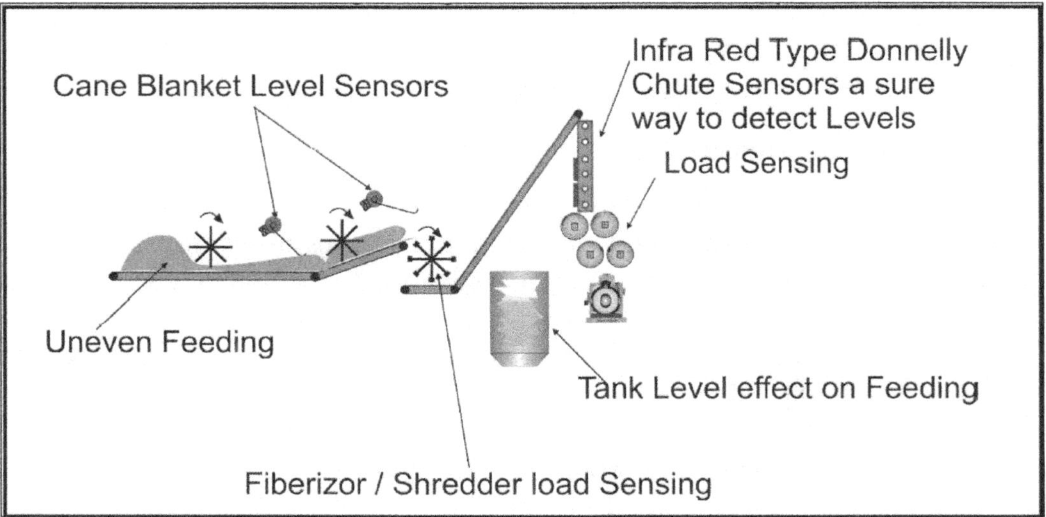

Fig.26.1 Arrangement of sensors at cane carriers and Donnelly chute

Generally cane level on the cane carrier is measured by sensors that work on Hall's effect. Correct level of cane in the carrier is measured. Upon the measurement of cane level the carrier speed is controlled so that the cane level in the cane carrier remains more or less regular. The prepared cane level in Donnelly chute of the first mill is measured by infra-red sensors. It indicates exact level of prepared cane in the Donnelly chute. True level detection through all the shocks, vibrations, moisture and dirt is possible. The speed of the inclined cane carrier or rake type carrier is controlled on the level of prepared cane in the Donnelly chute. Therefore prepared cane level in the Donnelly chute remains almost constant. Hence steady positive feed at first mill is maintained. Consequently steady feeding remains at all mills. This helps in improving primary extraction as well as mill extraction.

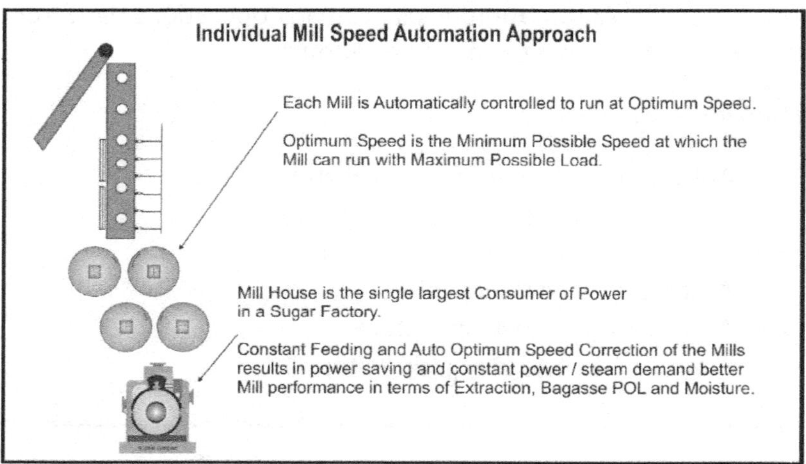

Fig.26.2 individual mill automation

Further due to automation-

- Exact top roller lift is detected and displayed.
- Pressure transmitters are used for sensing chest pressure of turbines.
- The mill speed is controlled with respect to individual mill loading and Donnelly chute level.
- Imbibition water flow is controlled is controlled on the load at penultimate mill and its top roller lift.
- Juice flow feedback is given to adjust the crushing rate.
- Imbibition water gets cut-off automatically when mill runs empty.
- Juice level sensors at mixed juice tank gives signals to avoid juice overflow or empty pump run.
- Bearing temperature of cane carriers preparatory devices and all mills are sensed and alarms are provided if temperature of any bearing rises above prescribed limit.
- Automatic interlocking system between main carrier, rake carrier, mills, inter-carriers, and bagasse carrier and also preparatory devices is now followed in automation systems.

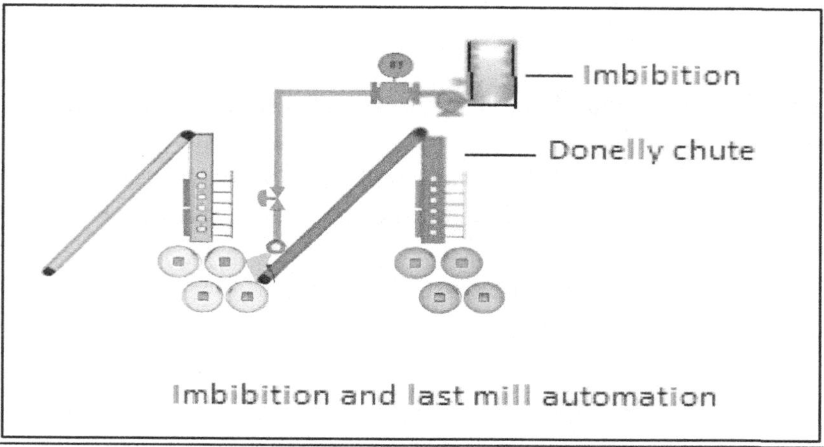

Fig. 26.3 Imbibition water and last mill speed automation

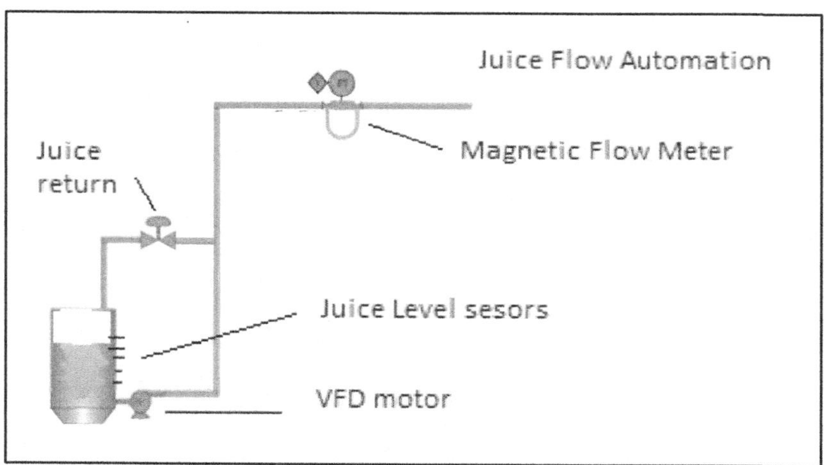

Fig.26.4 Juice flow automation

Advantages of automations

- With automation there are no fluctuations in working. Underfeeding and overloading at preparatory devices as well as at mills is avoided. Jamming and choking problems are almost eliminated. High crushing rate can be maintained.
- The flow of imbibition remains steady.
- The juice flow at the whirlers and raw juice going to boiling house remains almost constant.
- Bagasse pol and moisture remains steady and minimum and loss of pol in bagasse is reduced.

- Comparatively bagasse of low moisture content passes to boilers. This improves boiler operation and boiler working remains steady. Boiler efficiency is also increased.
- The Steam / power demand at mill house remains steady. Fluctuations are almost eliminated. Consumption at preparatory devices and mills is reduced.
- The stoppages are reduced. Generally faults are detected earlier and therefore preventive measures can be taken in time to avoid stoppages. Equipment safety is taken care off.
- Wear and tear of equipment are mostly reduced.
- Possible human errors are eliminated.

-OOOOO-

27 CANE DIFFUSER

27.1 General:

In diffuser the process is percolating water / thin juice through a layer of prepared cane / bagasse that is deposited on a perforated moving bed or conveyor and passes slowly from feed end to the discharge end as the conveyor moves. The cells of the cane are almost completely ruptured therefore the comparatively concentrated juice oozes out and mixes with water or thin juice that is percolating through the prepared cane / bagasse layer. Thus it is somewhat like washing or the lixiviation process. But we loosely call it as cane diffusion process.

FIG.27.1 De Smet Diffuser

In case of sugar beet the beets are cut into cossettes so that cells are opened. The cell wall walls are permeable for sucrose molecules. Therefore, when water or thin juice percolates through a layer of cossettes, the sugar (of higher concentration in solution inside the cells) pass through the cell walls into the water / thin juice. Thus sugar in beet is extracted by true diffusion. But in case of cane the cell walls are and the cell are to be ruptured completely so that the cells are opened and juice oozes out. Thus it is lixiviation. Therefore the words used for operation as "diffusion" and the equipment "diffuser" in cane sugar factory are not appropriate.

The extraction is carried out by lixiviation process. Therefore rupturing of almost all the cells of the cane in cane preparation is very much essential when diffuser is employed for cane juice extraction.

27.2 Diffusion:

The diffusion process for extraction of cane juice is successfully being used for last 100 years in many countries. There are benefits of higher extraction at low cost. Roughly, about 80% of the cane in South Africa is now processed in diffusers.

There are two ways of extracting the juice by diffusion process.

- Cane diffusion or pure diffusion – it is the process of extracting juice from prepared cane. Mill is not provided for primary extraction. This was pioneered in Hawaii.
- Bagasse diffusion or diffusion after primary extraction by mills.

In bagasse diffusion process, mill is employed for primary extraction and the primary bagasse from the mill is send to the diffuser. With bagasse diffuser, only 30 % of the juice goes to diffuser. Therefore length of the diffuser is shortened; further the inversion loss occurring due to higher temperature is also effected on 30 % of the juice. Therefore it is reduced in comparison with whole cane diffusion process.

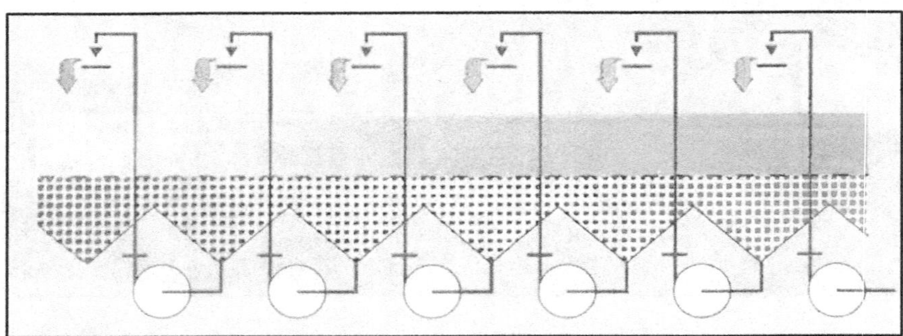

Fig. 27.2 circulation of juice in diffuser

The spent bagasse coming out of diffuser usually called 'Megasse' is saturated with water and contents about 75-85% water. Two mills or a five roller mill – called dewatering mill, is employed to extract the thin juice and reduce the moisture of the magasse to 48%. Then it is called as bagasse.

A cost-effective method of expansion followed in South Africa is to install a diffuser and use some of the existing mills after diffuser (as dewatering mills) or otherwise use the mills as first mill for primary extraction and last mills as dewatering mills.

All the diffusers are moving bed type diffusers. These diffusers are stage wise counter current solid liquid extraction devices. Juice is pumped on to moving bed of prepared cane or bagasse. The diffuser is about 50-65 m long with 10 to 18 stages of circulation. The main diffuser fabricators in the world are - De-Smet, BMA, Silver Ring, Saturn, and F.S. van Hengel.

27.3 De-Smet diffuser:

In India the Andhra Company had collaboration with De-Smet for fabrication of diffuser in India. These diffusers are installed in some factories in India. The diffuser consists of a long tank enclosing a horizontal conveyor on which primary bagasse (or prepared cane in case of cane diffuser) is placed in a layer of uniform thickness. Throughout its travel from feed end of the diffuser to the discharge end, this layer is continuously irrigated with thin imbibition juice. Below the conveyor, the bottom of the tank consists of 11 (or 18) adjacent trays. These trays receive juice percolated from the bagasse. The length of the trays (in the direction of travel of the conveyor) is 2.1 -2.4 m. The apron of the conveyor is a stainless steel grill that allows the juice to drain through. The cane diffuser width is 4 to12 m. and length 50 to 65 m. A pump takes juice collected in each tray and returns it to a distributor placed above the preceding tray. In such a way the juice moves backward from tray to tray from discharge end to feed end.

Silver Ring diffuser is similar, but the screens move in a circle instead of along a straight line. The BMA and Tongaat-Hulett diffusers have fixed screens with a series of chains, which transport the cane layer over the screen. This generally results in a cheaper diffuser for the same screen area.

The last distribution plate above the discharge end tray receives imbibition water. The distribution plate next to the last one receives thin juice from the dewatering mills. The thin juice extracted from dewatering mills is heated, limed and clarified before it is send to diffuser for imbibition. Part of the juice from the tray below the bagasse entry is returns through a juice heater to the bagasse entering the diffuser. The remaining juice is mixed with primary juice and is send to boiling house.

A rake type carrier transports the magasse coming out from the diffuser to the dewatering mills. The bulk density of the layer of bagasse is 570-620 kg /m^3. The bagasse layer thickness in the diffuser is normally 1.8 m. The bagasse while being transported over the apron gets compacted. To prevent this, vertical hanging screws to agitate the bagasse are provided from the top.

The temperatures of juice heater and diffuser is controlled at about 75-80°C. The conveyor operates at variable speed and regulated from o.4 to 0.8 m / min. The retention time of bagasse is normally 30-45 minutes.

The imbibition water applied is about 300 to 370 % on cane. The supply of imbibition water is controlled by the input of bagasse by means of continuous weigher installed on the conveyor of the diffuser.

The grills normally have 32% of open space but now new grill are having 57% open space are bang used.

The time of passage of juice through the bagasse bed is 2.5 – 3.5 min. or speed of percolation 0.7 M/min. This remains very close to the speed of the conveyor.

The rate of percolation of the juice through bagasse layer is in the range of 15-25 m^3/m^2 per hour with an average of 17-18 m^3/m^2 / hr.

The prepared cane is feed uniformly across the width of diffuser with adjustable side gates. The prepared cane is heated to 75°C by hot circulation juice having temperature of 85°C. This circulating juice is called as scalding juice.

The pH of juice in the diffuser is maintained at 6.5 pH by adding milk of lime at two places with the help of lime splitter boxes.

An arrangement is provided to maintain the temperature of diffuser by blowing evaporator body vapours in case the temperature goes down below 70°C.

The whole operation of diffuser is monitored on computer screen and controlled through control panel instruments. The control logic consists of starting and stopping of all prime movers like, conveyors, pumps, screws, valves etc. There are indications for running of electrical drives, temperature, flow, pH of juice, rpm, speed, levels etc. All the prime movers are interlocked. Imbibition is controlled as per the set ratio based on quantity of cane feed. The monitoring and controlling operations of diffuser is much easier as compared to mills. Only one operator is required for controlling the whole operation of diffuser due to automation through D.C.S. control system.

27.4 Treatment of Thin Juice:

The thin juice from the dewatering mills is screened, limed to 7.5 - 8.2 pH, heated to 103°C and then allowed to settle in a clarifier. The clarified juice from overflow is cooled down up to 75 -80°C and then is send to the penultimate distribution plate to use for imbibition. Therefore, the thin juice can percolate easily through the layer of bagasse without any trouble.

Some milk of lime is also added to imbibition water to maintain pH of the thin juice to a pH of 6.0-6.5 to control sucrose loss due to inversion and also to avoid corrosion of mild steel components.

Cane diffuser in South Africa has been consistently capable of achieving higher extraction than milling tandems[1]. *The average extraction is about 97.5+. The retention time of cane in the diffuser varies depending on the desired extraction levels. Typically retention time of about 55, 70, and 90 minutes are required to achieve extraction level of 96, 97, and 98% respectively.*

With increase in pH the colour of the juice increases. If pH is increased above 7.0, the percolation characteristics of the cane layer are adversely affected[2]. Over liming of juice in diffusers is resulted into hydrolysis of acetyl groups from the hemicellulose of the fibre with formation of acetic acid. The acetic acid in diffuser juice may then be volatilised at the evaporator. These vapours cause corrosion of vapour and condensate piping.

In general the acetic acid average level is around 200 ppm. In diffuser under good control this may be around 300 ppm. The value may increase as high as 1000 ppm in poorly controlled diffusers. It is therefore important that good reliable pH control system is required to be used for diffusers.

In Diffuser extraction of silica in juice is generally higher. The silica passes in clear juice and gives scaling problem in evaporator. Silica levels are found to be increased at high pH values in diffuser and when cane with high quantities of trash and tops are processed[1].

27.5 Microbial Control:

In diffuser where temperature is a considerably higher mesophilic microorganisms becomes inactive but hyper thermophiles are active. They are generally lactic acid producing bacteria and are active at temperature up to 70°C. The losses of sugar due to microbial activity are high if temperature is not kept well above 70°C. Therefore generally diffuser temperature is kept about 75-80°C so that in no case the temperature comes below 70°C.

The pH range from 5.0 to 6.5 does not have significant effect on micro-organism activity. The use of biocides at diffusers is very expensive and has seldom proved to be cost effective. On the other hand maintaining temperature is rather simpler, more effective and cheaper way of controlling microorganisms.

27.6 Comparison with Milling:

- The capital investment cost of diffuser is lower than that of the milling tandem.
- The maintenance cost is reduced by 50-60%.
- Steam consumption in boiling house is increased by about 3-4% on cane.
- The power requirement for mills is about 90-100 kw/tfh.

However for diffusers it is only 45-50 kw/tfh. Therefore diffuser requires less steam for prime mover and less exhaust steam is produced. Therefore the deficits in the exhaust steam consumption and production is increased. Therefore more live steam is to be bleeded through PRD (pressure reducing and desuperheating) unit. However, the factories with co-generation plant the power saved in diffuser can be exported. Hence the surplus power for export is increased which gives additional revenue to the factory. On the other hand for the factories with no scope for export of power low-pressure boiler can be used.

- In diffusers the filter cake percentage is considerably reduced which is about 50 % to that of milling plant.
- The diffuser extraction is higher however because of higher nonsugars extraction the boiling house extraction is reduced. The overall recovery slightly increases or remains the same depending upon the boiling house efficiency.

27.7 Effect on Juice:

The suspended solids in diffuser juice are always less than mill juice and therefore for diffuser in most cases screening of juice is not required. Therefore juice screens are not installed.

On an average colour of juice from diffuser is about 25% higher than mill juice due to higher temperature in the diffuser. However, no significant change in colour of raw sugar is observed. Increase in trash and tops by one percentage increases the colour of the juice by 15 %. Sugar colour at the start and at the end of the season is always higher than sugar colour at the middle of the season.

Starch content of the diffuser juice is much lower than mill juice. Starch level in raw sugar from diffuser factories are found to be 25 % lower and gum levels 12 % lower than those in sugar from conventional milling tandem factories.

References:

1 Rein P W ISJ Apr.1999 p 192 – 197
2 Lionnet ISJ 1988, 90 11-16

-OOOOO-

28 JUICE CLARIFICATION PROCESSES

28.1 Screening of Juice:

At milling operation fine bagasse which is generally called as bagacillo or cush cush falls down in the mill troughs along with the juice. The bagasse should not get accumulated at the troughs and should pass easily along with flow of juice. For this purpose the troughs are kept sharply inclined. They are about 55° inclined to the horizontal. Quantity of bagacillo coming along with the juice varies from 5 to 10 g of (dry) fibre per litre of juice. Higher cane preparation increases bagacillo quantity in the juice.

The secondary juices are pumped by chokeless pumps. The chokeless pumps are centrifugal pumps with single sided impellers designed with wide passage so that bagacillo can pass with juice through these pumps. Therefore screening of secondary juices is not required. The primary juice and secondary juice mixed together to get mixed juice. The mixed juice is screened before sending it to boiling house. Now with higher preparation index (85-90 %) the percentage of fine bagacillo in the mixed juice is increased.

Effect of bagacillo:

Higher percentage of bagacillo in the mixed juice is detrimental to the clarification process. Bagacillo contains cellulose, hemicelluloses, pentosan and lignin. These are high molecular weight compounds. These contain a pigment 'saccharetin' and polyphenols which are soluble or remain in colloidal state.

When juice is limed the yellow coloured saccharetin dissolves in the juice in alkaline medium. It increases the colour of the juice as well as colloidal particles in the juice. Turbidity of the juice is also increased.

The heating of juice in alkaline medium causes decomposition of cellulose. The products partly pass along with clear juice and are molassigenic (or melassigenic – molasses forming) in nature and increase loss in final molasses.

Therefore efficient screening of juice is very much essential. However bagacillo is rich in P_2O_5 and it is released on heating which helps in clarification reactions. The increased in bagacillo content increases colour of clear juice. There are two types of screens, 1) stationary type DSM (Dutch State Mines who invented the screens) screens and 2) Rotary screens that are used for mixed juice screening.

DSM Screen:

The DSM screen consists of straight prismatic (wedge shaped) bars that are arranged horizontally in series in a concave form (as shown in the figure). The opening between the bars is usually 0.7 mm. The bars - which are brazed in a 45° arc of a circle, are made up of stainless steel or bronze.

Stainless steel bar has more life whereas in presence bronze growth of microorganisms is controlled. The overflow box and the juice receiver behind the screen are of stainless steel. The length of the screen along the curve is fixed (1.6 m.). The horizontal width is kept from 0.9 to 2.1 m (3-7ft.).

Fig. 28.1 DSM screen used for mixed juice screening

The juice flows from the top overflow box over the screen. The suspended bagacillo particles are retained on the screen and the mixed juice almost free from the bagacillo is passed through the fine openings between horizontal prismatic bars. The retained bagacillo is pushed downward by the juice flow.

In factories of small capacity, the DSM screen is usually fitted over the mills so that the bagacillo retained on the screen falls down directly into the intercarrier. In large factories it is fitted by the side of the mills. In that case a screw conveyor is provided to convey the bagacillo to the intercarrier. Fibre content in the juice after screening is usually 1.5 to 2.0 g/l of juice on dry basis. According to standard specification two screens each having 0.7 mm gap and 1800 mm width are recommended for a 2500 TCD plant.

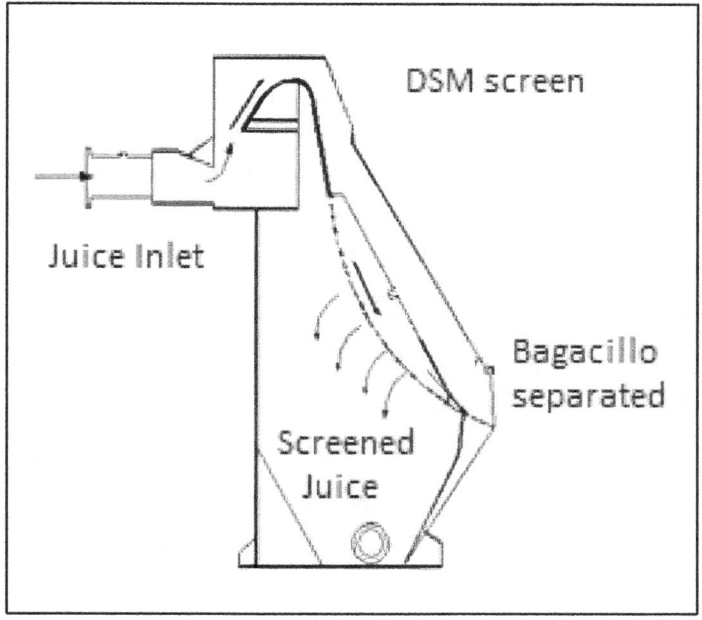

Fig. 28.2 DSM screen

DSM screens has no moving parts, therefore it is almost maintenance free. However the interior part of the screen has to be regularly cleaned to control the growth of microorganisms on that side.

In some factories the cane preparatory index is maintained above 88%. In such case lot of powder is formed and bagacillo percentage is increased which passes through the screen and give trouble in settling.

Rotary Screens:

The second type of screens is of conical or cylindrical form rotating screens. In case of cylindrical form it is fitted with its axis 5-10° inclined to the horizontal, which is not needed for conical rotary screens.

Rotary screen drum is generally made up of two or three screened drums joined together. The screen drums are fabricated of wedge wires with 0.5 – 0.7 mm slot width. Complete screen drum assembly is supported on four support roller assemblies. The screen drum is rotated at approximately 10 rpm by sprocket and chain assembly and is driven by 5.0 – 7.0 kW 1500 rpm / 30 rpm gear motor. Feed box with feed inlet pipe is positioned at the feed end inside the screen drum. Juice is first received in a distribution tray. Then it flows on the screen so that the juice falls on screen tangentially in a direction of rotation of screen. Juice tray below the screen is provided with an outlet pipe. Separated bagacillo travels down the screen drum and is discharged into a screw conveyor via a discharge chute. The unscreened juice contains 14-18 gms/l fibre. When this juice is screened it contains only 1.2-1.8 g/l of fibre.

Water spray system is provided to spray hot water/steam through jet nozzles. This can be used during operation for cleaning and sanitation of the screens. It is provided with automatic on /off. Pressure of water is to be maintained about 4 kg/cm^2. Therefore screen drum can be washed periodically with hot water/steam spray through nozzles without stopping the screening process.

Fig. 28.3 rotary screen

Water spray system is provided to spray hot water/steam through jet nozzles. This can be used during operation for cleaning and sanitation of the screens. It is provided with automatic on /off. Pressure of water is to be maintained about 4 kg/cm^2. Therefore screen drum can be washed periodically with hot water/steam spray through nozzles without stopping the screening process.

All the parts - juice tray, splash guards, inlet feed box, drum shells, at inlet and outlet, cush-cush discharge chute, that come in contact with juice are made up of stainless steel 409 M.

Generally the rotary screen is mounted at right angles to the mills so as to discharge the cush-cush can be discharged directly into the rake elevator.

Double screen of mixed juice with two screens fixed on one rotary drum, one screen of 1mm opening and the other of 0.35 mm opening have been tried in some factories but frequent blinding of the 0.35 mm opening screen is observed. Therefore these have not been continued.

The screens get choked with the particles of size close to the screen opening size. Therefore screens are to be cleaned frequently with hot water/steam. Now vibro-screens are being used in some factories. A new design of vibro-screen which vibrates in three directions – vertical, horizontal and tangential, which can be used after DSM or rotary screen, has been installed in some factories. With such system it is observed that the bagacillo in mixed juice can be reduced up to 0.2 to 0.5 % (on dry basis). The vibro-screens can also be used for filtrate to remove bagacillo from the filtrate before the filtrate is mixed in mixed juice or sulphited juice. This can reduce recirculation of bagacillo.

In clarifier major part of bagacillo floats as scum in flocculating chamber and some part settles in the muddy juice. When the screens are worn or torn and screening is not proper, the higher percentage of bagacillo choke the juice heater tubes and also retard the rate of settling of mud. The fine bagacillo particles passing along with clear juice can act as nucleus for sugar crystals to be developed and give high colour and high ash content in the sugar.

28.2 Juice Weighing:

At clarification station the juice is first weighed by automatic scale. The weighing is necessary because it a basis of sugar balance and technical control in a factory. There are number of different automatic juice weighing scales in use. One commonly used in Indian Sugar Industry - Maxwell - Boulogne automatic weighing scale.

Now recently mass flow meters are being used for accounting of the mixed juice. The advantage of mass flow meter is that variations in the flow rate are controlled to some extent.

Juice Weighing Scales:

In design of Maxwell Boulogne weighing scale a weighing tank is suspended from a beam and hanging up because of counter weight. The bottom valve is closed while the inlet valve is open so that the juice falls into the tank.

When the tank becomes almost full with a certain weight the hanging weighing tank comes down, the inlet valve from the upper tank is closed and the discharge valve is opened. The juice is discharged to the bottom weighed juice receiving tank. When the weighing tank becomes empty it goes back to its original position, the bottom valve is closed and the inlet valve is opened automatically. Thus it delivers a constant weight of juice at each tip.

Accurate weighment is most essential for proper technical control, therefore check weighment arrangement is provided to check accuracy of the weighing scale.

Juice flow meter:

Now in guidelines on standard specifications for Indian sugar industry (2004) mass flow meter is recommended and new factories erected after 2005 mostly installed mass flow meters. Some old factories which have gone for modernization and /or expansion also have installed mass flow meters.

The juice measuring system is now computerized. Meter shows the flow rate in TPH, current hour and last hour flow rate in tonnes. The accuracy of mass flow rate is 0.15%. Effect of ambient temperature is only 0.001% per °C.

An online calibration facility for check weighment is provided for random checking of juice delivered by mass flow meter without stopping the crushing. A separate display is provided for measurement of juice during check weighment.

A stainless steel tank with stirrer is provided at the weighed juice receiving tank for dilution and storage of phosphoric acid. A dosing pump is also provided to regulate the addition of diluted phosphoric acid in mixed juice.

28.3 Juice Flow Stabilization System:

After weighment of juice the mixed juice or raw juice (at clarification station generally called as raw juice) is pumped from juice weighing receiving tank and send to juice heaters for first heating. Steady working conditions always give better results.

With batch weighing, the level in the weighed juice tank varies considerably and because of characteristics of pump, the juice flow rate passing to juice heaters varies considerably. This affects in maintaining steady temperature of the raw juice in the first heating. Also the fluctuations in the flow rate can affect reactions in sulphiter. The efficacy of clarification process depends on stable flow of mixed juice. The juice flow rate should be steady with minimum variations for good control over the clarification process.

Therefore to stabilize the flow of raw juice, 'Juice Flow Stabilization (JFS) System' is installed at raw juice pumps and flow rate of the juice is automatically regulated. In JFS system there is no actual flow measurement. The raw juice flow from receiving tank to juice heaters is controlled by a pneumatically operated control valve on taking feedback from level of raw juice in the juice receiving tank. The juice level in the tank is continuously displayed through a bar graph and the percentage of valve opening is digitally displayed on the panel. Thus Juice from the Mill is measured using mass flow meter and constant flow is maintained to the inlet of Juice heaters. The Juice pumps are operated by VFD and speed is controlled in DCS.

When the crushing rate is steady, the juice level in the receiving tank remains more or less in a marginal range. When crushing rate increases or decreases, the juice level in the tank slowly increases or decreases. The sensors of juice flow stabilization system give signals to the system accordingly and to maintain the juice level in a marginal range, the valve to the delivery line gets automatically controlled. Similar types of systems are now in use in many of the factories in India.

Although many factories use the juice weighing scales, now mass flow meters have been developed and magnetic flow meter are being used for mixed juice flow measurement. VFD motors are now being used for juice pumps that can eliminate valve for manipulating of juice flow.

28.4 Objects of Clarification process

The objects of clarification process are

- To remove all the suspended impurities,
- To remove maximum amount of colloidal matter and impurities from the juice by precipitation and settling,
- The dissolved nonsugars that are precipitable by juice treatment should also be precipitated.
- To adjust the pH of the juice close to neutrality so that inversion of sucrose as well as destruction of reducing sugars would be minimum,

Criteria of good clarification:

The following criteria are considered for good clarification

- The clarified juice should be light in colour and transparent,
- There should be rise in purity from mixed juice to clarified juice,
- The pH of the clarified juice should be 6.8 to 7.0

Generally phosphoric acid is added in mixed juice at the weighing tank to maintain a certain level of P_2O_5 content for good clarification. When juice is treated by milk of lime, phosphoric acid a floc of precipitated solids is formed. Coagulation of fine particles takes place. The floc entraps fine suspended and colloidal matter. In clarifier when juice remains steady the precipitated and coagulated solids settle down by gravity in the mud form and clear transparent juice is obtained.

Formation of optimum floc in the juice is necessary to remove the impurities by precipitation, coagulation and settling. Clarification is an important unit operation of sugar manufacturing process and effect of clarification has direct impact on the severity of scale formation, quality of sugar and loss of sugar in final molasses.

28.5 Clarification Processes:

The three main processes followed for cane juice clarification are:

Defecation:

Lime is the only main clarifying agent used for this process. The lime consumption in this process is about 0.06 -0.08 % on cane. This process is followed for raw sugar manufacture.

Sulphitation:

In this process lime and sulphur dioxide gas are used as main clarifying agents. A precipitate of calcium sulphite is formed. The lime consumption is about 0.12 -0.18 % on cane. The sulphur consumption is about 0.04 -0.08 % on cane. This process is followed in most of the developing countries including India and China for plantation white sugar manufacture.

Carbonatation:

In this process with lime carbon dioxide gas is passed through the juice and calcium carbonate precipitate is formed. The limestone is burnt in lime kiln and the lime formed and CO_2 gas evolved in the kiln are used for clarification. The limestone requirement for this process is about 2.5-3.5 % on cane. Now this process is not being followed in most of the countries. Therefore this is not described in detail in this book.

A new modified process of carbonation is being tried. In factories with distilleries, CO_2 gas evolved in fermentation can be used for carbonation process. The lime is to be used is about 20 to 30 % more as compare to sulphitation process. With this process plantation white sugar can be manufactured. Research work is going on In India.

The white sugar produced by sulphitation / carbonation process for direct consumption is commonly called as plantation white sugar or mill white sugar. Few decades back sulphitation process was being used for clarification of juice in raw sugar manufacture but now it is seldom used. In all the processes of clarification addition of milk of lime and heating the juice above the boiling point are common.

-OOOOO-

29 DEFECATION PROCESS

29.1 General:

The defecation process of juice clarification is a very old process being followed. It is simple and effective process for raw sugar manufacture. In this process lime is the only main clarifying agent used.

Milk of lime is added in the juice to precipitate impurities and neutralize acidity of the juice. The limed juice pH is so adjusted that clarified juice obtained should be of pH 6.8 to 7.1. The milk of lime of 10-16 brix is prepared in lime slacker as described later. Usually quick lime is used in most of the factories however some factories use powdered hydrated lime. Powdered hydrated lime is usually of higher CaO content and therefore is more effective in clarification. The addition of milk of lime is continuous. The optimum pH of limed juice varies considerably according to cane quality and juice retention time in clarifier. However liming the juice above 8.4 pH should be avoided as it generally gives alkaline clear juice that causes –

- Excess lime passes in clear juice as soluble lime salts and gives high scale formation in evaporator.
- High colour formation in clear juice may increase colour of sugar.
- Higher destruction / decomposition of reducing sugars may take place. The destruction of RS decreases exhaustibility of molasses and causes higher sugar loss in final molasses.

When the juice is treated with addition of milk of lime and then heated to 104°C, a precipitate of complex composition is formed. This contains insoluble lime salts, coagulated proteins, fats, waxes and gums. The floc formed entraps most of the finely suspended solids in the juice. Large portion of this precipitate is heavier than juice and a small portion is lighter than juice.

The addition of milk of lime is either controlled manually or automatically. An old method of repeated testing the juice for pH by litmus paper and regulating the addition of quantity of milk of lime manually is still being followed in many factories. Such system depends on skill of operator. Now in modern sugar factories microprocessor based pH indicating, recording and control system is being used.

29.2 Different procedures of Lime Addition:

The juice clarification process comprises of chemical reactions, precipitation of dissolved nonsugars, floc formation and coagulation of impurities. Detail and thorough studies by many technologists in different countries have showed that variations in juice composition and change in colloidal constituents make significant effect on precipitation of non-sugars in the juices. Large differences are found in the behaviour of juices from region to region.

The behaviour of the juices also changes greatly on the time lag between harvesting and transportation. Stale and deteriorated cane juice contains lot of colloidal substances formed as exterior products of microorganisms. These adversely affect the clarification process than any other factor. The purity of the juice also plays an important role in effectiveness of clarification process. High purity juices can be readily clarified and the quantity of clarifying agents required is also considerably reduced.

Modifications in process treatment also plays important role in effectiveness of clarification. The modifications may be in

- Temperature at which chemicals are added,
- Quantity and sequence of chemicals being added,
- Mixing of the chemicals with the juice

The optimum pH of limed juice varies considerably according to climatic conditions of the region, cane variety, its maturity and juice retention time in the clarifier. For e. g. At Kenana (Sudan, Africa) there are two parallel clarification phases and cold liming is followed. For phase I with 444 clarifiers of residence time 2.5 hr. the limed juice pH that has to be maintained to 8.1-8.3. While for phase II with short retention clarifiers of residence time 1 hr. the limed juice pH has to be maintained to 7.5-7.7. Therefore it is not possible to give any standard procedure that will give the best result for all juices all the time. Hence sugar factories follow different procedures for addition of milk of lime. Some commons procedures of these are described here under.

29.3 Cold Liming:

In this process milk of lime is added in cold juice and then the juice is heated to above boiling point. In many factories cold liming is followed to avoid double heating and make process simple.

One more advantage of cold liming is that the juice is not required to cool down for on line pH measurement and control of limed juice pH. However In cold alkaline juice microorganisms particularly leuconostoc develop very rapidly.

Author has observed frequent choking of pH sampling pipe because of slime produced by leuconostoc in Kenana (Sudan) where this procedure was being followed. Therefore where cold liming process is followed special care of mill sanitation has to be taken. Further cleaning of limed juice mixing tank, receiving tank, sampling pipes, pump strainers etc., should be frequently and properly attended. The use of biocide is also very much essential. So that growth of microorganisms at these equipment will be controlled.

29.4 Hot Liming:

In this method the juice is first heated up to 70-75°C, then limed and again heated to above boiling point. In Indian sugar industry this method is followed for raw sugar manufacture as in this process equipment installed for sulphitation process can be used without any modifications.

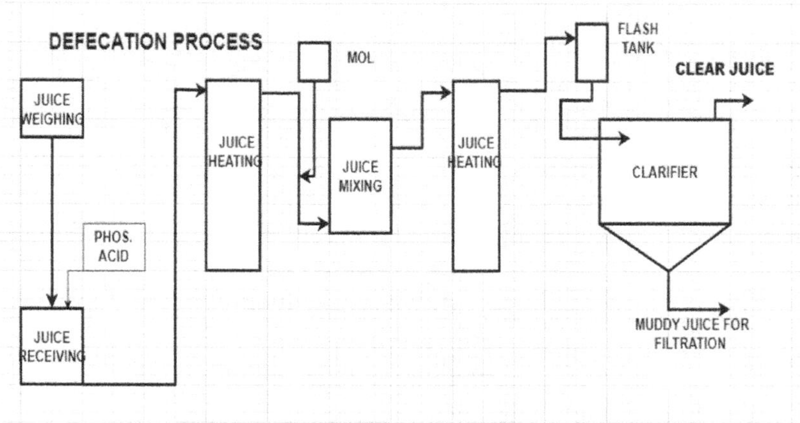

Fig. 29.1 Defecation Process

Heating the juice up to 85-90°C followed by liming and then reheating to 104°C is also followed in some countries. In this case the inversion loss is not much increased as the juice remains acidic at 85-90°C only for few minutes.

In some raw sugar manufacturing factories juice is first heated above the boiling point and then it is limed (see the figure).

In this process certain colloids such as proteins and colloidal silica are precipitated before liming and probably because of this the lime requirement is less than cold liming. With hot liming the difference in pH between limed juice and clarified juice remains almost constant and therefore pH control becomes easier.

29.5 Fractional Liming:

In this process the cold juice is limed up to 6.6-6.8 so that most of the acidity is neutralized. Then it is heated to 90-95°C, again limed to 7.6-8.2 and again heated above the boiling point. Advantages of this process are –

- Lime requirement is reduced by about 20 -25% of that of cold liming,
- This avoids the disadvantages of hot liming – inversion of sucrose before liming when the pH of the juice is 4.9 to 5.4 at higher temperature for few minutes.
- Settling of mud is rapid.

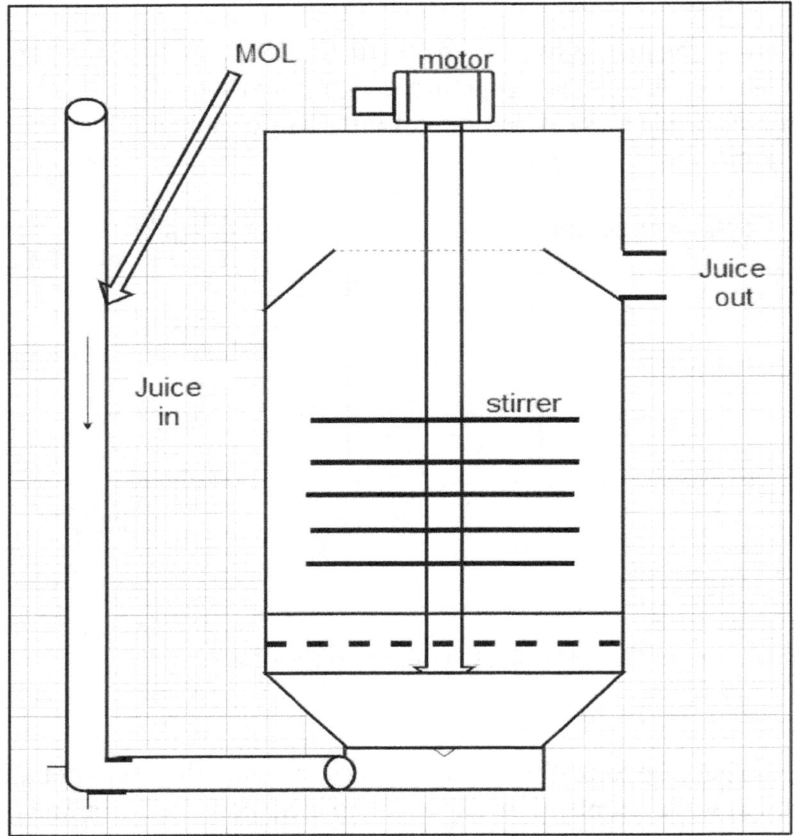

Fig. no. 29.2 Liming Tank

With hot liming sampling for automatic pH control is more complicated. A small heat exchanger is required to cool the sample juice flow below 40°C before it comes in contact with pH electrode. A reaction tank with stirrer and overflow arrangement is usually provided to give a reaction time minimum 5 to maximum 15 minutes for well mixing of the milk of lime and the juice. This improves flock formation.

29.6 Hot or cold liming:

Several lime clarification systems have been developed over the years including cold liming, hot liming, fractional liming and saccharate liming.

Most of the world sugar industry follows hot liming[1, 2]. However in USA cold liming is followed mostly due to simplicity in operation. In Louisiana alone in 1998, 17 out of 19 factories still utilize cold liming[3]. In some factories where cold liming is followed, hot liming is used only when juices are difficult and cold liming does not give satisfactory clarification. In an intensive study made in two factories in Louisiana Eggleston[3] came to following conclusions –

- In hot liming the juice chemistry is dramatically changed. Coloured compounds, dextran and oligosaccharides are generally removed.
- In cold lime tank increase in dextran by about 4% is observed.
- As compare to cold liming the destruction of reducing sugars and consequent colour formation is less in hot liming.
- There is no evidence of high inversion in hot liming.
- In hot liming about 25% less lime is required than cold liming.
- In hot liming process calcium and ash level in clarified juice is less than cold liming process.
- In hot liming turbidity removal is higher as compare to cold liming.
- Mud volume is higher in hot liming.

29.7 Reactions in cane juice clarification:

Phosphate:

When lime is added soluble phosphate in the juice reacts with calcium ions and a precipitate is formed. The precipitate is a mixture of calcium magnesium phosphate compounds. This phosphate precipitate formed entraps or adsorbs colloidal nonsugars of the juice.

Generally it is thought that phosphate content of about 300 ppm in the juice gives for good clarification. In the juice the phosphate is present in organic and inorganic form. In organic form it is present in phosphoproteins phospholipids nucleophosphates etc. The inorganic phosphate is free phosphate ions and mainly this constituent takes part in clarification reactions. About 70-80% of phosphate is precipitated by lime.

The compounds formed are complex compounds of calcium phosphate. Precipitates of Si, Al, and Fe are also formed along with Calcium phosphates.

The advantages of phosphate in clarification are as under:

- It gives higher elimination of colloids,
- It increases rate of settling of mud,
- Filtration quality of mud is improved,
- It gives brighter clear juice,
- It reduces the formation of organic acid salts. The increase in CaO level from mixed juice to clear juice is less,
- Decomposition of reducing sugars is inhibited,
- There is reduction viscosity,
- Sugar quality is improved,
- Exhaustibility of molasses is increased.

In factories of Deccan and south region of India the phosphate content is on lower side. It varies from 150 to 250 ppm. In Maharashtra and north Karnataka where mixed juice purities very high, author has observed that in sulphitation factories the juices can clarify well even when the P_2O_5 level is 225-250 ppm.

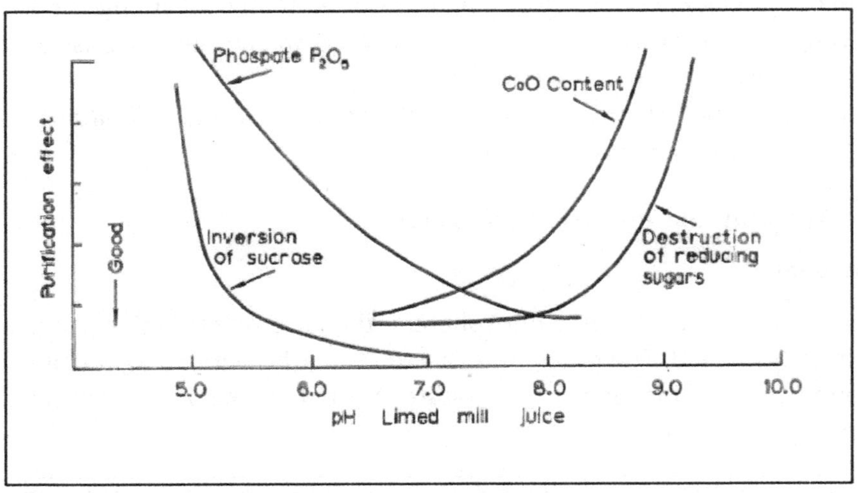

Fig. no. 29.3 Reactions at different pH values

In factories of north region of India the phosphate content in mixed juice is normally 350 to 450 ppm. When phosphate content is high, lime consumption and mud volume is increased.

When cane juice is deficient in phosphate contents, diluted orthophosphoric acid of purity about 80-85% is added in the raw juice. Addition of diluted phosphoric acid in clear juice up to 10 to 15 ppm is followed in some factories. It reduces the scale formation in evaporator bodies by increasing the solubility of calcium sulphite and calcium organic salts and also softens the hard-scales in the evaporator tubes.

Reactions of calcium hydroxide:

Calcium hydroxide added reacts with organic acids viz. aconitic acid, oxalic acid, tartaric acid, citric acid etc., and calcium salts – aconitate, oxalate, tartarate, citrate etc. are formed. These are precipitated to some extent but part of these compounds remain in solution and pass along with clarified juice.

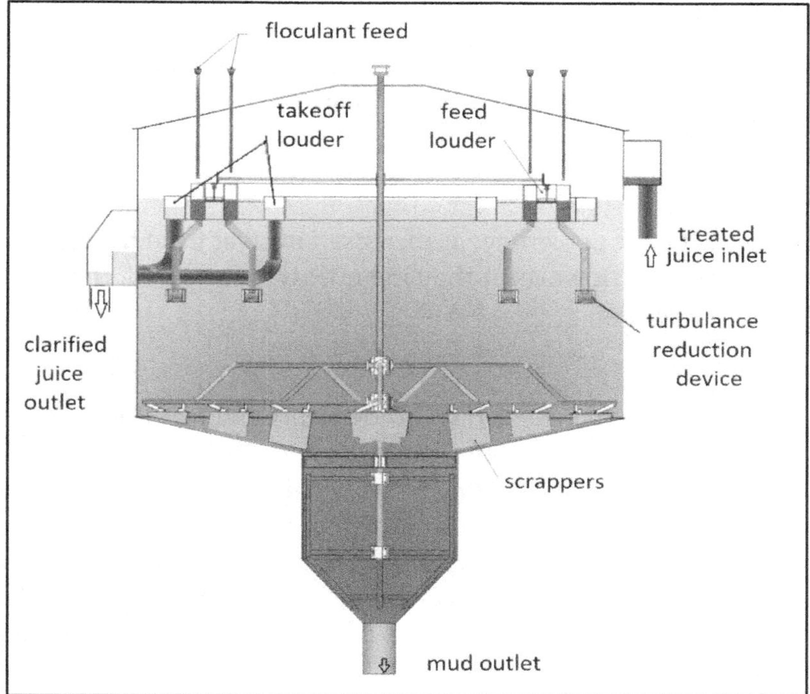

Fig. 29.4 Short Retention Clarifier

CaO content:

CaO content in mixed juice in Deccan region varies from 700 to 1100 mg/lit. In north India it is much less. During clarification CaO content in juice is increased. This increase in CaO content is attributed to the neutralization of acids present in the juice with formation of organic acids salts of calcium. These are not precipitated in clarification and pass in dissolved state in clear juice. Some acids are also formed due to decomposition of reducing sugars during heating and treatment of juice.

Silica:

Silica may be present in dissolved, colloidal and suspended form. Solubility of silica in mixed juice is extremely low and the juice may be saturated with silica. About 40 to 60 % of the silica mostly – colloidal and suspended is removed in clarification. The silica dissolved in the juice passes with clears juice and gets deposited in the form of scale in evaporator.

Magnesium:

Magnesium content in the cane juice may considerably be varied. It precipitates in clarification as $Mg(OH)_2$ but mostly an equivalent amount of calcium remains in solution.

Iron and aluminium:

Iron and aluminium occur in dissolved state at very minute quantities. They also may be present in suspended form in soil particles. These are almost completely removed in clarification.

Ash:

The conductivity ash percentage is increased in clear juice by about 5 to 10 % in sulphitation factories. However if the lime quality is not good, the ash percentage may be further increased.

Inversion of sucrose:

Rate of inversion of sucrose is high at lower pH and high temperature. To keep the sucrose loss due to inversion at a minimum level the pH of the clear juice is kept close to neutrality. The final heating is to be maintained above 101°C for good clarification. Retention time is also one of the parameter for inversion loss. The retention time of juice in the clarifier should be less. Therefore short retention clarifiers have been designed. In conventional clarifiers the capacity of the clarifier given is of three hours retention time whereas in short retention clarifier it is 1 hour.

Reducing sugar:

Reducing sugars remain almost unaffected in acidic juices but decompose rapidly at high alkalinity and higher temperature. They are easily oxidized in alkaline solutions. Therefore destruction of reducing sugars occurs especially when the pH and temperature of the juice are maintained high for longer time. Fructose is more sensitive and readily decomposed in alkaline solution than glucose. The degradation products of reducing sugar are mainly acids and brown colour compounds. The pH of the juice drops because of the acids formation. Highly coloured complex organic compounds are formed that increase colour of sugar.

Higher percentage of reducing sugars gives better exhaustibility of molasses. But the decomposition products of reducing sugars are mostly molassigenic in nature. Therefore the destruction of reducing sugars should be avoided as much as possible. Reducing sugars may also combine with amino acids to form undesirable highly coloured products. This is called as browning or Maillards reaction.

Amino acids:

Amino acids are having both acidic carboxylic groups and basic amino groups. They are amphoteric in nature and hence act as buffers. They are optically active however their concentration is much less and hence the effect on pol reading is almost negligible. They mostly remain in solution and pass with clear juice.

Proteins:

In juice the proteins content is up to 0.5% of total dissolved solids. Albumin is the main protein present. When juice is heated proteins are mostly denatured, precipitated and settled in muddy juice. But any proteins that remain in colloidal state are detrimental to sedimentation because they may act as protective colloids and tend to stabilize the suspended matter.

Gums:

Gums are composed of pentosans. They are present in small amount. These are hydrophilic colloids, which increase the viscosity of the juice. About 50-60 % of the gums are removed in clarification. They may also act as protective colloids and adversely affect the mud-settling rate. If they are not properly removed in clarification they increase viscosity of syrup. Stale cane contains considerable amount of gums. Therefore stale cane usually gives trouble in clarification.

Waxes and fats:

Waxes and fats occur in very small quantities in suspension and colloidal state. These partly float in clarifier in the form of scum (in flocculating chamber of clarifier) and partly are carried down in the precipitated mud.

Colloids:

In clarification colloids may prevent the coalescence of suspended particles. Colloids tend to increase the hydration of particles by making them gelatinous in character so that they settle slowly. This increases mud volume in clarifier. The removal of colloids is important to obtain good transparent light colour clear juice without any turbidity. If they are not removed in clarification they increase the viscosity and colour of the syrup and molasses. They retard the rate of crystallisation.

Colloids are removed by precipitation and also by adsorption. Adsorption is the attachment of molecules or colloidal particles to the surface of solid matter. Precipitates with large surface area may have high adsorption capacity.

The raw juice contains different compounds in colloidal state. Each group of these colloids requires different conditions to obtain an optimum flocculation. Therefore there may be a definite pH value at which maximum quantity of colloids gets adsorbed.

Colloidal particles after being flocculated may return again to the colloidal state. This is called peptisation. Flocculated colloids which peptise easily are called reversible colloids. A change of pH in the medium can cause peptisation.

In the process of clarification colloids may also be formed in the precipitation of inorganic salts and in the decomposition of the organic substances. Nature of precipitate formed depends on cane quality – variety and maturity and freshness of cane, also brix of mixed juice and clarification procedure followed. Overmatured and stale cane contains higher percentage of colloids in that case mud settles slowly. Largely irreversible colloids are removed in clarification by defecation process, but the reversible colloids generally pass with the clear juice without any effect.

References:

- DuPont and Saint Antoine ISJ 1969 Jan. P 40-44,
- Simpson SASTA 1996 p 267-271,
- Eggleston G. ISJ Aug.2000 p 406-416, Sept 2000 p 453-457

-OOOOO-

30 SULPHITAION PROCESS

30.1 General:

In this process excess lime is added than required for defecation process and it is neutralized by sulphur dioxide gas.

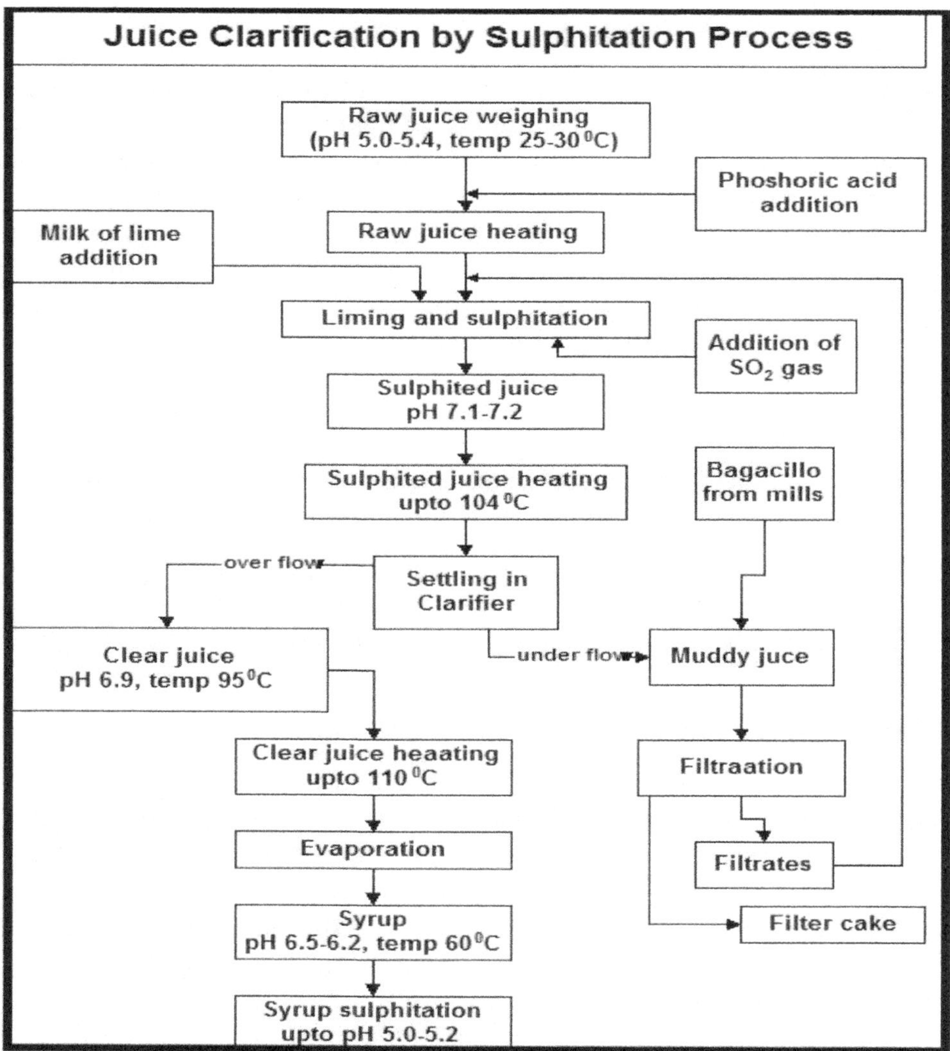

Fig. 30.1 Flow chart of sulphitation process

30.2 Different methods of sulphitation

The sulphitation process comprises of mixing the milk of lime and SO_2 gas with raw juice by different methods in a specific retention time at a particular temperature. Technologists have tried numerous different combinations to achieve best results in clarification. A number of modifications are followed in the system of applying lime and sulphur dioxide gas. In general there are methods i) acidic sulphitation ii) neutral sulphitation and iii) alkaline sulphitation. Further there can be fractional procedures that are also followed. There are also modifications in temperature such cold or hot sulphitation etc.

Acidic sulphitation or presulphitation:

In acid sulphitation or presulphitation the juice is first sulphited up to pH 4.2 - 3.8 and then it is limed in the next vessel to get a final pH 7.1-7.2. The sulphitation is carried out in a vertical cylindrical tower of 4.5-5.0 m height. The diameter varies with capacity of a sugar factory. Grids are fitted in an upper two third portion of the tower. The raw juice is introduced from the top of the tower. It falls though grills hence it is sprayed in fine droplets and comes in contact with rising current of SO_2 gas that is introduced in the tower at the bottom under pressure. The sulphited juice is drawn from the conical bottom of the tower and is then taken to liming tank and neutralized with it lime.

Simultaneous liming and sulphitation:

In this lime is added and sulphur dioxide gas is passed simultaneously in one vessel or sulphiter. The final pH of the juice is maintained slightly alkaline.

With some difference in the above procedure the juice is first neutralized up to 6.6.-6.8 pH in a preliming tank and then it sent to sulphitation tank where simultaneous liming and sulphitation is followed simultaneously to get a final pH. 7.1-7.2.

Alkaline sulphitation or liming followed by sulphitation:

In the first case after heating the juice up to 70-75°C milk of lime is added to the turbulence flow of juice in the pipeline just before the juice enters the sulphitation vessel so that the pH of the juice is increased up to 8.6 to 9.2 in one step. The retention time of limed juice is only for about 10 seconds. Therefore the destruction of reducing sugar will not be much. Then sulphur dioxide gas is passed to neutralize the excess lime. The final pH of the sulphited juice is maintained to 7.1 -7.2. In India in most of the factories this process is being used.

In the second case the juice is prelimed up to 6.6-6.8 pH in a preliming tank of retention time of about one minute. Then it is passed to the main sulphitation tank where first shock liming up to 8.8 -9.2 pH done. Then the juice is sulphited to 7.1 – 7.2 pH. Thus the lime is added in two steps.

30.3 Cold or Hot sulphitation:

The process of cold liming and sulphitation results in heavy scale formation in juice heaters because of delayed precipitation of calcium sulphite at normal temperature. The reactions and precipitation is fast at about 70-75°C.

Cold liming followed by heating and then sulphitation causes more destruction of reducing sugar as the juice remains at higher pH for longer time. Therefore now the juice first heated to about 70-75^{0}C and then liming and sulphitation is followed.

Calcium sulphite is more soluble (4.3 mg/100ml at 18°C) in water at low temperature and the solubility decreases with temperature. Therefore at higher temperature more complete precipitation of calcium sulphite occurs. Here again temperature of raw juice should not be raised to more than 75^{0}C because any higher temperature will increase the rate of inversion of sucrose since the pH of the juice is about 5.0-5.4 before treatment.

The hot simultaneous liming and sulphitation is called Horloff's process. In continuous sulphitation milk of lime and SO_2 gas are continuously added to a constant flowing stream of juice. Baffles are provided to ensure proper circulation and mixing.

30.4 Mixing of milk of lime and sulphur dioxide gas distribution:

Thorough mixing of milk of lime with the juice and distribution of sulphur dioxide gas throughout the cross section of the sulphitation vessel is very much essential. Further good circulation of the juice in the vessel is also important to get good results of clarification. To control destruction of reducing sugars high alkalinity at higher temperature for more time is avoided. Poor mixing may cause localized over liming. Temperature above 75°C is not advised. Some factories in South India keep the temperature below 70°C. The sulphited juice pH is maintained in a range (7.1-7.2) to give a clarified juice pH 6.9-7.0.

Steady crushing rate without frequent minor stoppages and further by controlling the juice flow by stabilization system is very much essential therefore juice flow remains steady. Each cubic centimetre of juice should undergo with the same change in temperature and pH in a scheduled time span. This results in good clarification. The clarifying agents i.e. milk of lime and sulphur dioxide gas should be well mixed with the juice within short time and the mixture should become homogeneous.

The bubbling of the gas into the juice give a force for circulation of the juice as a result of which mixing is attained. It is estimated by technologist that one m^3 of gas can bring one complete circulation of three m^3 of juice and one m^3.of juice normally requires about 4 to 5 m^3 sulphur dioxide gas of approximate 10 % concentration. Therefore in a well-designed sulphitation vessel with retention time of juice about 7 to 9 minutes, the natural circulation caused by bubbling of gas is sufficient for good mixing. Therefore no stirrer is required. However in many designs of sulphitation vessels a stirring arrangement is provided.

There are different arrangements for distribution of gas into the juice. The sulphur dioxide gas should be well distributed throughout the cross section of the vessel into the juice for which more number of holes is required. However if there are more number of small holes, there is possibility that these holes may get choked with scale. Therefore limited numbers of medium size holes are provided for the gas outlet.

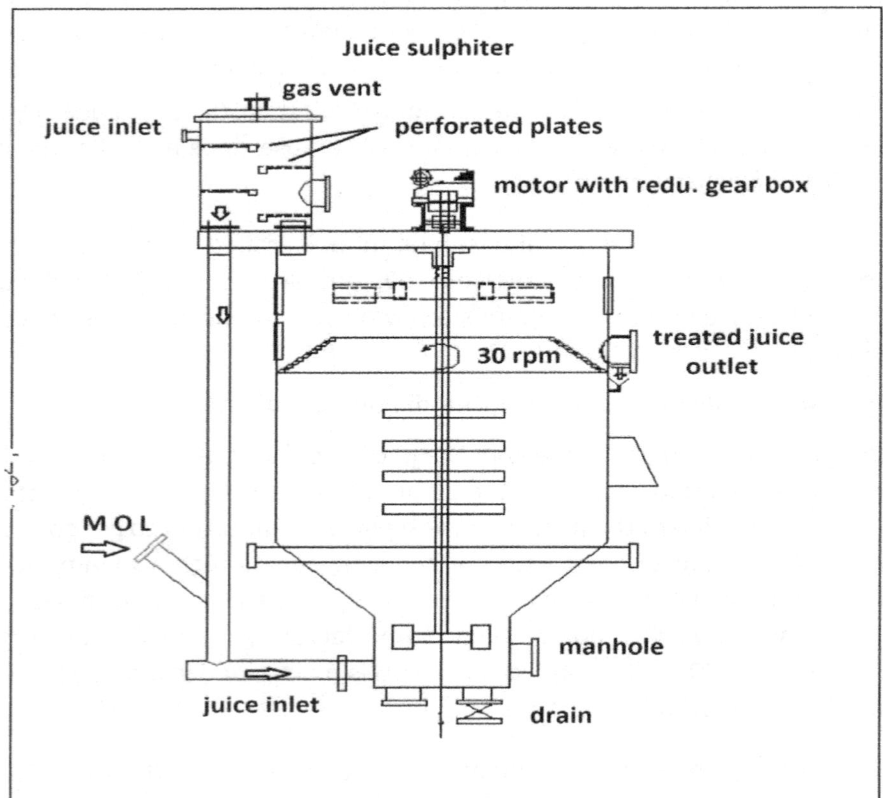

Fig. 30.2 juice sulphiter

30.5 Action of Sulphur Dioxide:

When sulphur dioxide gas is bubbled through the limed juice immediately sulphurous acid is formed. In the limed juice the concentration of OH^- ions is more i.e. the pH of the juice is high because of dissociated Ca $(OH)_2$

$$Ca\,(OH)_2 \;=\; Ca^{++} + \;2\,OH^{--}$$

This causes dissociation of sulphurous acid into H+ and SO^{--} ions.

$$H_2O + SO_2 = H_2SO_3$$

$$H_2SO_3 = 2H^+ + SO_3^{-}$$

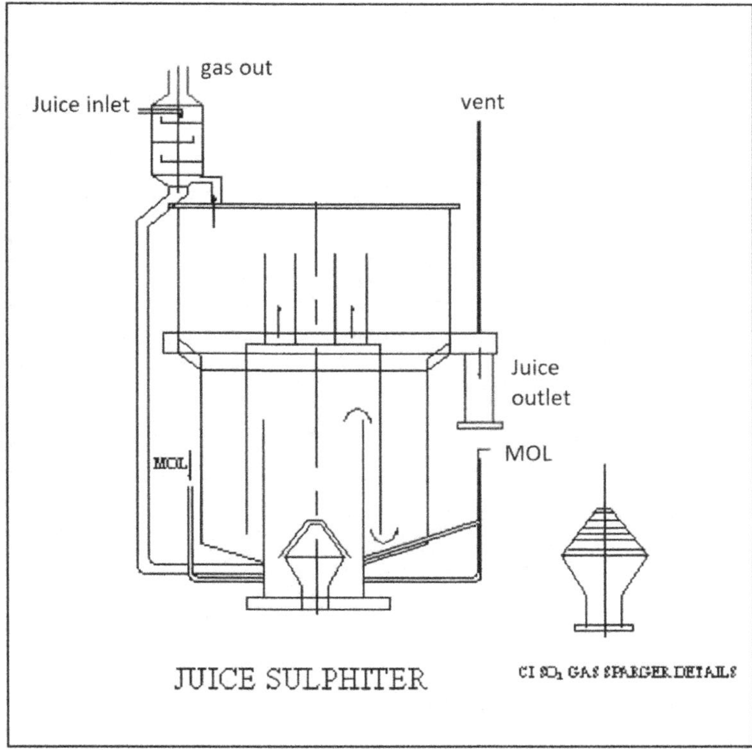

Fig. 30.3 juice sulphiter

The concentration of Ca $^{++}$ ions and SO_3^{--} ions is increased and precipitation of $CaSO_3$ starts. Higher concentration of Ca $^{++}$ and SO_3^{--} is favourable for a more complete precipitation but at the same time the high concentration give rise to a large number of nuclei precipitation of $CaSO_3$. In factory working limited number of coarser $CaSO_3$ precipitate is required in order to get good filterability of the precipitate and therefore the concentration of Ca^{++} and SO_3^{--} ions should be optimum during precipitation.

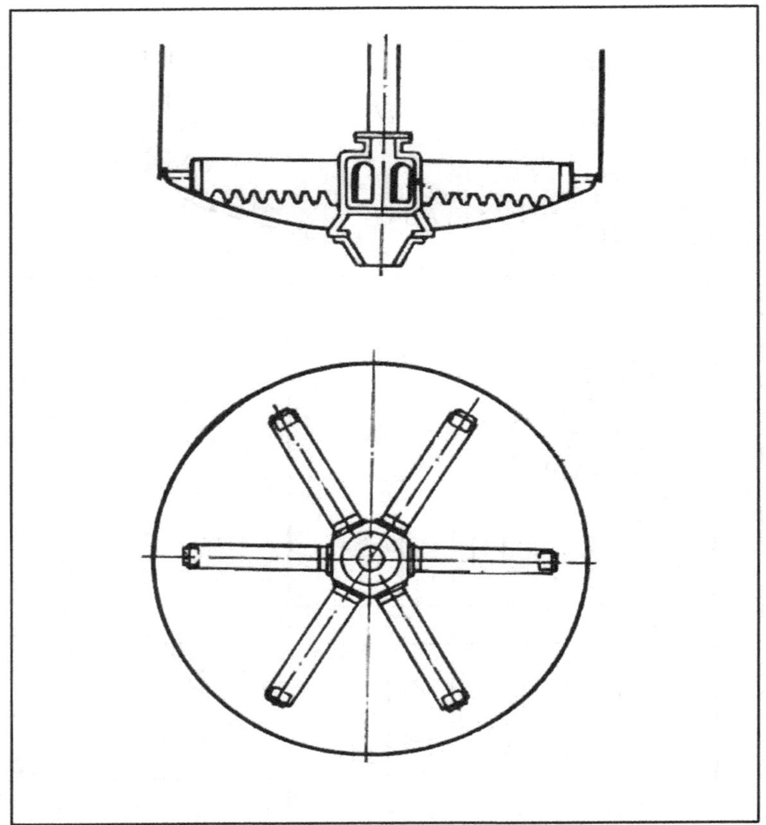

Fig 30.5 star sulphiter

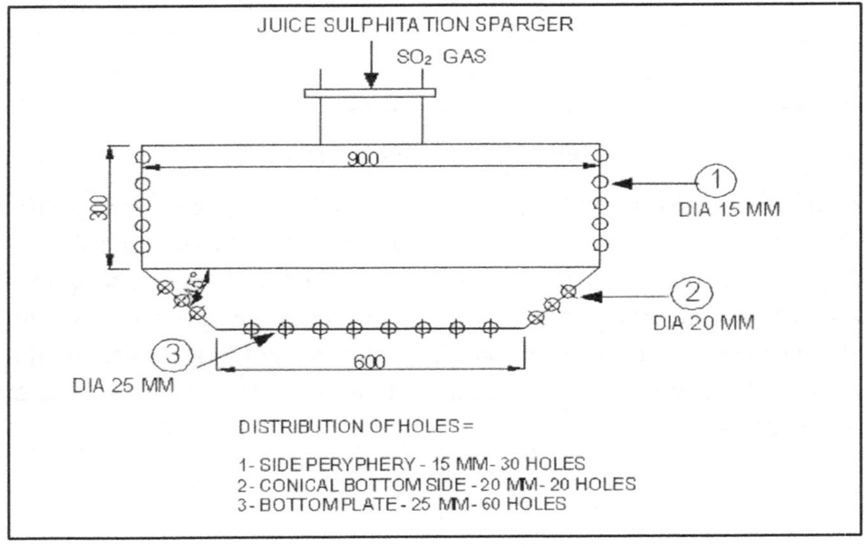

Fig. 30.6 sulphur gas sparger

$$Ca^{++} + SO_3^{--} = CaSO_3$$

Solubility of $CaSO_3$ is 30-40 ppm in experimental sugar solution (15% sucrose + 1.5% glucose solution). The H^+ ions combines with OH^- ions to form undissociated water molecules.

$$H^{++} + OH^{--} = H_2O$$

As the sulphitation continues, more and more CaSO3 precipitates that results in gradual decrease in Ca^{++} ion concentration and at last the precipitation is completed.

$$Ca(OH)_2 + H_2SO_3 = CaSO_3 + 2H_2O$$

After the reaction the concentration of the ions in the solution is very small and complete neutralization is reached.

Sulphur dioxide when passes in the juice immediately dissolves to form a weak sulphurous acid that neutralizes the excess lime and prevents colour formation in further process of evaporation and crystallisation.

According to Zerban probably sulphur dioxide combines with the reducing sugars and blocks the carbonyl functional group that is essential for caramels and melanoidins formation.

Viscosity of the juice is also reduced by action of sulphur dioxide gas.

The sulphited juice when exposed to air darkens to the contact surface due to oxidation.

Thus sulphur dioxide plays an important role in juice clarification.

- Neutralization of unreacted lime,
- precipitation of calcium sulphite and
- settling down the colloidal matters from the juice

These are the main functions of sulphur dioxide. The reactions are associated with removal of colouring matters.

One more role of SO_2 in the process is to inhibit non-enzyimic browning reactions like Maillards reaction and degradation of invert sugars.

Sulphur dioxide also inhibits colour formation due to enzyimic reactions by inhibiting enzyme polyphenol-oxidase, thereby preventing the formation of intermediates which are colouring compounds or colour precursors.

Every precaution is to be taken that the gassing should not be excessive and pH of the juice should not come below 7.0 pH. If passing of gas is further continued, juice becomes acidic. H_2SO_3 in the second stage dissociates as follows:

$$H_2SO_3 \quad = \quad H^+ \quad + \quad HSO_3^-$$

The precipitated $CaSO_3$ rediscovers and reacts with sulphurous acid to give calcium bisulphate,

$$CaSO_3 \quad + \quad H_2SO_3 \quad = \quad Ca\,(HSO_3)_2\ (dissolved)$$

The calcium bisulphite thus formed remains in dissolved state and passes along with clear juice. In evaporator when the juice gets concentrated, bisulphite is decomposed into sulphite and gets precipitated and deposits on the heating surface forming scale. This gives heavy scale in evaporator bodies. Therefore care should be taken that the pH of the sulphited juice should not go below 7.1-7.2.

30.6 Precipitation and Coagulation:

The precipitate formed depends on juice quality and procedure followed for clarification. The desirable results are

1) Precipitation should be as complete as possible
2) There should be heavy floc formation
3) The floc formation is followed by coagulation
4) Rapid settling of the coagulated solids
5) Dense mud formation
6) And clear transparent juice

The flock once formed should not be again damaged. Therefore the juice after treatment should pass to clarifier without much turbulence in the flow. It means after treatment and heating the juice should pass from the treatment vessel to the clarifier in the laminar flow i. e. by gravity. In practice it is almost impossible. However in some factories for e.g. Kenana (Sudan) where defecation process is followed, the juice heaters are erected at the highest floor and juice after heating above boiling point passes to flash tanks by gravity.

30.7 Continuous sulphiter:

There are different types of Continuous juice sulphitation vessels in factories. Heated juice continuously flows in the sulphiter while milk of lime is added continuously to the entering juice. A continuous flow of SO_2 gas at higher pressure enters the sulphiter at the bottom of the tank. The gas while passing through the juice circulates the juice so that mixing of the gas is attained. Baffles are provided help in distribution and efficient mixing.

Nowadays automatic control system is used for maintaining the juice flow steady and to control pH at juice sulphitation vessel. Automatic control systems based on microprocessors and further computerization are playing very important and useful role in improvement the performance of juice sulphitation process.

- As per standard specifications a continuous sulphiter of capacity 15 m^3 is provided for a 2500 tcd plant.
- The working height of juice column above the gas distribution is minimum 2.0 metres.
- Stack gas recovery tower is provided to the sulphiter.
- As per specifications a reaction tank after sulphiter is provided in the sulphitation unit, residence time of which may be 3 minutes. Then the juice overflows to the receiving tank.
- The speed of the stirrer of the sulphiter is kept 16 rpm.
- In most of the sulphiter designs an arrangement for shock liming and simultaneous liming is provided.

Quarez type sulphiter:

The Quarez sulphiter consists of a holding tank, a circulation pump, and a ventury tube and sulphur burner. The level in the tank is kept constant by means of an overflow. Juice from the holding tank is circulated by the pump and is forced through an injector creating a vacuum, so that the SO_2 gases from the burner is sucked in and mixed with the juice.

30.8 Automation

Now automatic control systems are used for juice flow stabilisation and pH control at juice sulphitation in clarification process. CEERI (Pilani) has first developed a microprocessor based pH control system for juice clarification in 1992. It is a three loop system. The figure of which is given herewith. Now in most of the factories two loop pH control system is installed. The pH electrodes used are ruggedized industrial grade with automatic temperature compensation. Suitable valves for controlling the flow of lime and SO_2 gas with proportional control are provided. The microprocessor based is with LCD display arrangement for pH and temperature values. An arrangement for connecting PC and printer is provided. A high and low value alarm is also provided.

30.9 Addition of Milk of Lime:

The addition of milk of lime (MOL) is done either in one step (shock liming) or two steps (pre liming and shock liming). The MOL flow rate is controlled by taking feedback from pH. In many factories in north India the MOL addition is based on pre-set V/V ratio of MOL to juice. This system also gives good results when pre-set ratio is adjusted according to trials taken in laboratory and MOL brix is maintained constant.

Automatically washing arrangement to clean the electrodes after fixed interval is also provided. The valve for MOL flow is

- Pneumatically controlled by taking feedback from pH and juice flow rate
- In some automation stepper motors are used to control the flow rate of MOL.

In factories where film type sulphur burners are used the final pH can easily be controlled by regulating the burning rate of sulphur.

30.10 Precautions to be taken during operation:

To get good result of clarification some precautions are to be taken at clarifier.

- Brix of the mixed juice should not be more than 16.00.
- Dose of addition of phosphoric acid should be decided every 2/3 day after observing the P_2O_5 content in mixed juice before addition and it should be controlled accordingly,
- Juice flow stabilisation system is very much essential for good result of clarification process.
- The mixed juice is to be heated to 72 – 75°C. At this temperature solubility of calcium sulphite is less hence better precipitation occurs.
- For normal juices, heating the sulphited juice up to 101°C can clarify juices well. Higher heating increases unnecessary steam consumption,
- Lime should be checked for quality as it arrives, the active CaO content should not be less than 70 %,
- The milk of lime should be grit free. Screening of milk of lime and diluting it to a fixed brix say 14.00 is very important.
- Sulphur burner temperature should be observed regularly.
- The automation of sulphitation unit should be observed for its smooth working.
- The pH electrodes used for measurement of pH at juice clarification should be replaced and cleaned regularly.
- The flocculent dose if addition is necessary should be calculated and exact dose well dissolved /mixed in hot water should continuously be added.
- The clarifier should be operated by overflow,
- There should be continuous withdrawal of muddy juice from all compartments with steady flow,
- The flocculating chamber should be overflowed every 2/3 days to remove the scum collected in the flocculating chamber,
- The withdrawal of clear juice from both the sides should be equal,
- Air venting of each compartment should be on both sides and that should not get choked. When it gets choked by mud should be cleaned by passing hot water,

- Generally the mud level should be just up to 2nd cock in each compartment, so that the suspended solids from the incoming juice are easily trapped by the muddy juice in short time.
- The coil provided for collection of clarified juice should be close to the top of the compartment,
- In some factories when lumps of mud are formed and they choke up the mud overflow pipe, muddy juice has to be withdrawn by liquidation line. This has to be followed in case of trouble only. It should not be made as regular practice.
- There should be practice of measurement of colour and transmittance of clear juice every day.

30.11 Results of Clarification:

The clear juice is transparent, light in colour without any turbidity. In sugar factories the clarified juice is generally analysed every two hours only to observe clear juice purity and to find out rise in purity from mixed juice to clear juice. Generally there is rise in purity from mixed juice to clear juice. A statistical data for rise in purities in 300 Indian sugar factories have been collected. It is shown in the pie graph below.

Table 30.1 statistics of purity rise from mixed juice to clear juice

Sr. no.	Rise /fall in purity range	Percentage of factories
1	Fall in purity	8.5
2	Purity rise 0.00 to 0.25	41
3	Purity rise 0.25 to 0.50	28.5
4	Purity rise 0.50 to 0.75	14.0
5	Purity rise 0.75 to 1.00	5.5
6	Purity rise above 1.00	2.5

From the above table it can be seen that 69.5 % of the factories have purity rise in the range of 0.0 to 0.50 %.

The CaO content is observed twice or thrice in a week to check the rise in CaO content from mixed juice to clear juice.

Formation of sulphuric acid in traces due to SO_3 formation in the sulphur furnace also forms insoluble calcium salt. This all leads to increase calcium in clear juice. There is an increase in CaO content from mixed juice to clear juice by about 350 to 500 ppm. The rise in CaO content from mixed juice to clear juice should be minimum. Around 350 ppm rise in CaO content from mixed juice to clear juice is considered as excellent figure and the cane quality and control over the process is considered as very good.

The P_2O_5 is also observed along with CaO content. The P_2O_5 level gives indication whether externally phosphoric acid is to be added in mixed juice or not, and if to be added to how much quantity. Further the P_2O_5 content in clear juice can indicate the control over the process. The P_2O_5 content less than 40 ppm indicates more attention should be taken over the process.

30.12 Results of clarification of burnt harvested cane:

Burnt harvested cane gives slightly inferior clarified juice. It gives more trouble in clarification if crushing is delayed. The setting rate is decreased. Turbidity and colour of the clear juice is increased. The massecuites and sugar colour is also increased.

The dextran content in burned harvested cane is generally higher by about 20 % and it increases rapidly with time. In burnt cane juice clarification chemical consumption is considerably increased.

30.13 Filtration of clear juice:

Even after taking all cares in clarification process, there are some chances of passing fine bagacillo in clear juice. To remove this bagacillo use of pressure filters have been tried in some factories. At Sanjivani sugar factory Kopargaon, Maharashtra, pressure filtration system for filtration of clear juice has been installed. The results of the clear juice filtration at Sanjivani has are given in the table below.

Table 30.2 Results of clear juice filtration

Sr. no.	particulars	Clear Juice	
		Before filtration	**After filtration**
1	% transparency at 560 nm	52	53.8
2	Turbidity 900 nm	4.3	3.7
3	% turbidity removed		13.95
4	Colour (IU) 420 nm	9907	9503
5	% colour removed		4.07
6	Bagacillo mg/lit	88	56
7	% of bagacillo removed		41.6

Ref. - STAI, DSTA joint convention 2011, p no. PP 59-68.

30.14 Syrup sulphitation:

The syrup is sulphited up to 4.8 to 5.2 pH value. So that the colouring matter is decolorized and viscosity of the syrup is also reduced. Retention of the syrup in sulphiter is kept 15 minutes.

-OOOOO-

31 *CARBONATION PROCESS*

31.1 General:

Carbonatation or carbonation process is an old process. This process was originally developed in beet sugar industry and it was followed in cane sugar industry for production of plantation white sugar from 1876. However in this process large quantity of limestone and coal is required. A separate kiln was necessary to burn the limestone. The lime and CO_2 gas formed in the kiln is taken for carbonation process. The number of equipment required is more and the process is also complicated. The production cost is higher than that for sulphitation process In India this process was being followed till 1990 in some factories. At that time about 10 to 15 % of India's sugar production was from carbonation process. However due to high investment, complicated process and large quantities limestone and coal this process have been stopped in India.

31.2 Demand for sulphurless sugar:

About 65-70 % of the sugar produced in India is used by pharmaceutical, food and beverage industries. These industries demand is sulphur free sugar. Sugar produced by double sulphitation contains 40 to 60 ppm sulphur and does not give very clear floc free solution which is basic need of these industries. This floc formation is mainly due calcium sulphate and polysaccharides. Therefore this sugar is not sold in bulk quantities to these industries in domestic as well as international market. In international market also demand for sulphurless sugar is increasing. Therefore plantation white sugar produced by double sulphitation process is not well accepted in International market.

Further there are some drawbacks with the sulphitation process.

- The control of SO_2 gas production is difficult. The gas does not get completely absorbed in the juice and passes in the air and pollutes the air,
- It is toxic and poisonous; health of workers working close to the burner gets affected.
- It is corrosive, the equipment get corroded rapidly,
- Heavy scale is formed at evaporator, therefore lot of chemicals and time is required cleaning the vessels.
- Calcium sulphate passes in final molasses and creates problems in fermentation,
- The spent wash disposed of from distillery contains sulphur. The spent wash digestion to biogas generates H_2S which has to be removed when the gas is to be used as fuel in boilers.

Efforts are being made to stop the use of SO_2 gas in the manufacture of plantation white sugar. A commercially viable environment friendly process has to be developed for production of plantation white sugar.

31.3 Experiments for developing new carbonation process:

Use of CO_2 from the distilleries the carbonation process with some modifications can be followed for manufacture of plantation white sugar. The fermenter gas contains about 95% CO_2. Experiments are being carried out to use CO_2 from fermenters of distilleries for clarification of juice. In the new simplified process milk of lime is to be added to the juice almost comparable to sulphitation process. The CO_2 gas collected from the fermenter is to be compressed and passed through the juice to neutralize the excess lime.

Liuthixia[1] et al modified double carbonation process which consumes 3-4 time less lime than old conventional double carbonation process.

Silver zossi[2] has compared quality of clear juice by defecation, sulphitation and single carbonation, and concluded that quality of juice obtained by single carbonation process was almost equal to clear juice by sulphitation process. However lime consumption was 20 % higher than sulphitation process.

NSI Kanpur[3] has made lab scale trials of simplified single carbonation process. After treatment juices were analysed for pH, purity, RS, CaO content, turbidity and colour. The NSI team concluded that the process can be followed for production of sulphurless plantation white sugar.

It is claimed that in carbonation process the raw juice is treated with milk of lime and carbon dioxide at pH 8.8 to 9.5 and temperature 80°C at about 20-40 minutes. The results are significant. There is always better rise in purity, removal of starch, phosphate and turbidity and substantial colour removal. However, the trials taken are not consistent and satisfactory. Further trails should be continued. In our opinion the results of the lab trials conducted by NSI are not much encouraging however further trials of clarification of juice by carbonation using CO_2 gas from fermenter should be continued with different process combinations. Further conclusion may be drawn only after pilot scale plant trials.

References:

- Liuthixia et al Proc. SIT 2009, New Orleans p 978,
- Silver Zossi Proc. SIT 2009, New Orleans, p 987,
- Narendra Mohan STAI, DSTA joint convention 2015, p FP 490.

-OOOOO-

32 JUICE HEATERS

32.1 General:

Tube and shell type Juice heaters is consist of cylindrical shell with tube plates from both the sides – 10 to 14 inches inside from the shell ends with tubes fitted in two tube plates inside it. The end side shell portions outside the tube plates are divided into compartments by baffles. These baffles make passes of the tubes from top to bottom and from bottom to the top. Both ends of the shell are closed with steel doors. The juice inlet and outlet compartments are of single pass. The rest of the compartments are provided two passes one for upward juice flow and the other for downward juice flow. If there are 10 tubes per pass, then there will be 20 tubes for each compartment (except first and the last) 10 for downward flow and 10 for upward. Juice enters the juice heater through inlet and passes through all the compartments.

Fig. 32.1 Juice heater

High head pumps provided for juice so that the juice can pass through the passes of the juice heater. The tubes are generally of 4.1 m in length. Juice heaters are provided with brass tubes of 42 mm ID/ 45 mm OD or annealed stainless steel tubes as per IS 13316, of OD 45 mm OD and 1.2 mm thick. The juice heater ligament of the tubes is 12 mm or more. In case of vapour line juice heaters and dynamic juice heaters it is 15 mm.

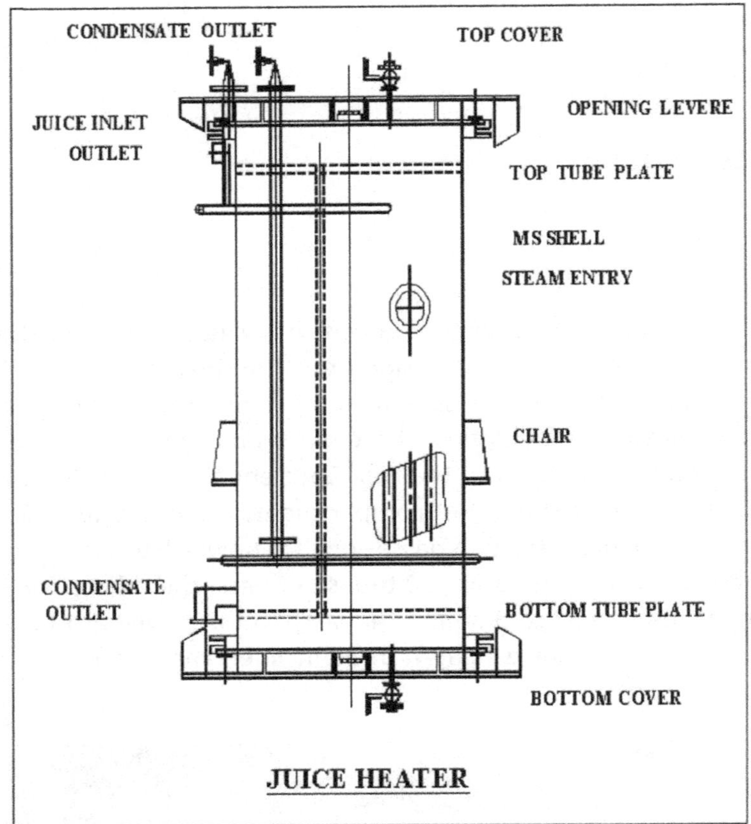

Fig. 32.2 Juice Heater

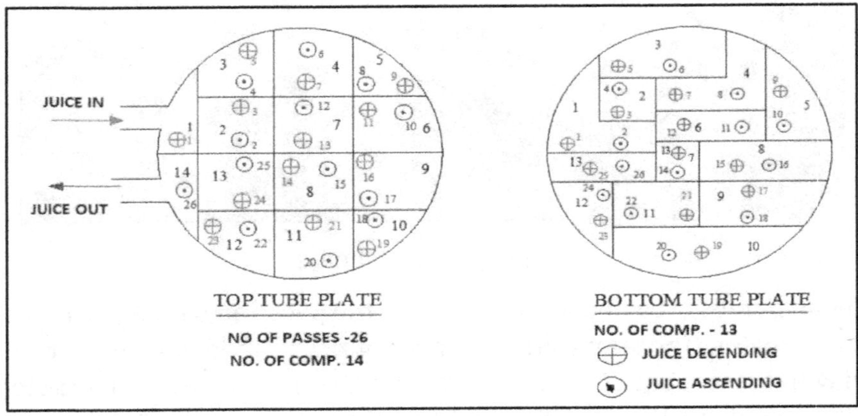

Fig. 32.3 Arrangement of Juice Heater Passes

The exhaust steam or vapours contains some incondensable gases. These incondensable gases are bad conductor of heat and if not removed from the steam side, affect the heat transfer and juice may not be heated as expected. Therefore the incondensable gases entering along with steam / vapour, in the juice heater are to be removed efficiently. Incondensable gas removal piping is provided for this purpose. Some incondensable gases are heavier and some are lighter than steam. Therefore the incondensable are withdrawn from the top as well as from the bottom of the juice heater shell. The incondensable gas withdrawal taps at the bottom are about 100 mm above from the bottom tube plate so that the condensate collected at the bottom does not come up to these taps.

The incondensable gases piping are provided with two outlets one open to atmosphere and the other connected to evaporator last body vapour pipe going to condenser. When exhaust is used the incondensable gas piping is kept open to atmosphere and when bled vapours are used and the vapour pressure is less than atmospheric pressure, the piping outlet is connected to vacuum.

While calculating the steam / vapour pipe diameter velocity of vapour is considered as 30-35 m/sec. The vapour entry is placed about one third of the length down from the top of the heater. This arrangement avoids excessive vibration of tubes. Condensate outlets from the heater should be sufficient so that the velocity of flow of the water does not exceed 1 m/s.

Heaters are tested for vapour side at 5 kg/cm^2 and for juice side at 8.5 kg/cm^2.

In most of the factories the juice heaters for raw and sulphited juice are interchangeable. To these juice heaters double beat valves having common housing are provided. To the calandria of juice heaters exhaust and first / second effect vapour connections of evaporator set are provided with valves. A safety valve is provided to the calandria.

Now in many factories dynamic juice heaters have been installed. They are provided in vapour pipe going from one effect to the next effect of evaporator set. To all the juice heaters arrangement is provided to drain the juice heaters. The juice drained goes to mixed juice receiving tank of weighment tank. Below the juice heaters a platform is provided for draining, opening the covers and repairing etc. Each juice heater is provided with dial type thermometers at inlet, outlet and for calandria. Each juice heater is provided with separate condensate receiver and pump.

32.2 Juice heater heating surface calculations:

While calculating the heating surface of juice heaters, first we will calculate the heat required for juice heating per degree centigrade. Thus we will calculate specific heat of juice first.

Specific Heat of juice (Cj):

Let us consider the raw juice has *B* brix. It means 100 Gms. of juice contains

(100 – B) gms of water and *B* gms of dissolved solids. Or one gm. of juice contains 1 – (0.01 X B) gms of water and 0.01 X B gms of dissolved solids. Now the dissolved solids contain about 78 – 86 % sucrose, specific heat of which is 0.38. The reducing sugars are about 5-12 %, the specific heat of which is also very close to specific heat of sucrose. Therefore, assuming average specific heat of dissolved solids as 0.4, the specific heat of the juice will be

$$Cj = 1 - 0.01B + 0.4 \times 0.01B$$

$$Cj = 1 - 0.01B + 0.004B$$

$$Cj = 1 - B\,(0.01 - 0.004)$$

$$Cj = 1 - 0.006B$$

Considering Brix of the juice 15.0 the specific heat of juice will be

$$Cj = 1 - 0.006 \times 15$$

$$Cj = 0.90$$

Coefficient of Heat Transfer:

The heat transmission from steam or vapour side to juice side or in technical language the coefficient heat transfer of juice heater depends on following factors

- film of condensate outside the tube
- scale formation on outside the tube
- presence of incondensable gases
- vapour / steam temperature
- logarithmic mean temperature difference between steam and juice
- metal of the tube – brass or stainless steel
- diameter and thickness of the tube
- scale from inside the tube
- Juice velocity inside the tube.

Table 32.1 Coefficient of heat transfer for Juice heating At different velocities:

Velocity of juice	Coefficient of heat transfer
m/sec	Kcal/m^2/°C/hr.
1.29	490
2.35	975
2.523	1060

From the above table it can be seen that the coefficient of heat transfer at juice heaters varies widely upon velocity of the juice. With lower velocity the coefficient of heat transfer falls rapidly. Juice heaters are designed to keep the velocity of the juice in heaters about 2.0 to 2.2 m/sec or higher. While calculating heating surface for juice heaters, coefficient of heat transfer is generally taken in the range of 500 to 750 Kcal /m^2/°C/hr. The heat transfer coefficient increases with increase in steam or vapour temperature and increase in juice velocity. In India juice velocity in juice heaters is usually taken as 2 m/sec. French manufacturers take it as 2.8-m/sec.

Mean temperature difference:

The temperature difference between the heating steam and juice being heated is not an arithmetic mean but this logarithmic mean and is calculated by the equation

$$\Delta Tm = \frac{\Delta Ti - \Delta To}{ln \frac{\Delta Ti}{\Delta To}}$$

Where,

ΔT_m = logarithmic mean temperature difference
ΔT_i = temperature difference between incoming juice and steam
ΔT_o = temperature difference between outgoing juice and steam

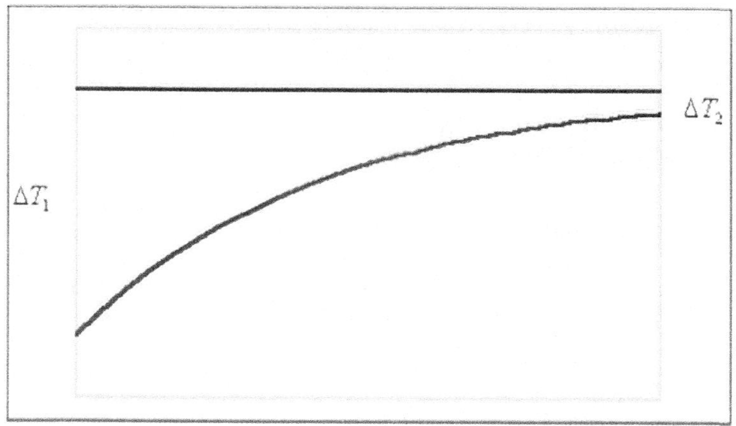

Fig. 32.4 Graph showing vapour and juice temperature in juice heater

Example:

For a 2500 TCD plant, at a crushing rate 115 t/hr. with juice % cane 103 %, and filtrate return to mixed juice tank 15 % on cane, the raw juice 2nd heating is done from 42°C to 72°C by the vapours are from 3rd effect of evaporator of temperature 98°C, then

$\Delta T_i = 98 - 42 = 56$

$\Delta T_o = 98 - 72 = 26$

$$\Delta T_m = \frac{56 - 26}{Ln\ 56 / 26} = \frac{30}{Ln\ 2.158}$$

$$\therefore \Delta T_m = 38.96$$

Heating surface required:

Heating surface required for juice heater

$$= \frac{\text{Quantity of heat required per hour to heat the juice}}{\text{Heat transfer coefficient X logarithmic mean temp. diff.}}$$

Quantity of heat required per hour

= Quantity of juice per hr. X Sp. Heat of juice X Rise in temp.

$= Qj\ Cj\ \Delta T$

Where, Qi = raw juice quantity including filtrate (15 % on cane)) per hr.
Raw juice quantity – 115 X1.03 = 118.45 MT/hr.
Filtrate return 15% on cane – 115 X 0.15 = 17.25 MT/hr.
Total juice passing through juice heater = 135.7 MT/hr.
Qj = 135.7 MT/hr. or 135700 kg/hr.
K = heat transfer coefficient 600 Kcal/m²/°C/hr.

Putting all the value in the equation we get

$$A = \frac{Qj\ Cj\ \Delta T}{K\ \Delta T_m}$$

$$A = \frac{135700 \times 0.90 \times 30}{600 \times 38.96}$$

$$A = 156.7\ m^2$$

The recommended heating surface for juice heaters for 2500 tcd plant is 170 m^2.

Total Number of Tubes:

Most of the manufacturers keep standard tube length of juice heaters 4 m with

42-mm ID and 45-mm OD. Considering the average diameter in calculating the heating surface

H.S. of one tube $= \pi D L$

$= 3.142 \times 0.0435 \times 4.00$

$= 0.5467$

Thus for 170m^2 heating surface

$$\text{Number of tubes} = \frac{170}{0.5467} = 310.96 \text{ say } 311$$

Number of tubes in each pass:

Now to maintain the velocity of the juice, the number of tubes in each pass (n) is calculated as follows

Volume of juice per sec.

= Cross sectional area of one tube X No. of tubes in each pass X juice Velocity per sec.

$$V = \frac{\pi D^2}{4} \times n \times v$$

Or $n = 1.273\, V / v D^2$

Where n = *Number of tubes per pass*

V = *Volume (m^3) of juice passing per second*

D = *I. D. (m) of the tube*

v = *Velocity (m/sec.) of juice*

$$V = \frac{\text{Wt. of juice (kg) passing per sec}}{\text{Density of juice}}$$

$$V = \frac{135.700 / 3600}{1.06}$$

$V = 0.03556$ m^3/sec

Hence

$$n = 1.273 \times 0.03556 / 2.0 \times 0.042 \times 0.042$$

Thus number of tubes per pass n = 12.83 or say 13,

Now number of passes in juice heater

$$= \frac{\textit{Total no of tube in juice heater}}{\textit{No. of tubes per pass}}$$

$$= \frac{311}{13} = 23.92 \text{ or say } 24,$$

Hence total number of passes in the juice heater will be 24 and number of tubes per pass being 13 the total number of tube in the juice heater will be 312.

32.3 Steam Consumption at Juice Heaters:

The steam consumption at juice heaters varies mainly with mixed juice percentage cane and initial temperature of mixed juice. It also varies with filtrates recycled % cane but that variation is less.

1. The mixed juice percent cane generally varies from 95 to 110. In South Africa in some factories with diffusers, the mixed juice percentage cane is about 120 % on cane but in that case, the diffuser juice temperature is higher.
2. The initial raw juice temperature varies as per ambient temperature and temperature of imbibition water used. While calculating the steam consumption at juice heaters the initial temperature of raw juice is generally taken as 25 – 30°C.
3. The filtrate returned varies from 12 to 18 % on cane.

The steam requirement at juice heaters for sulphitation factories is calculated hereunder with the following assumptions.

Table 32.2 assumptions

Cane Crushing rate	100 tch
Raw juice % cane	105 %
Raw juice heating	From 30° to 75°C
Sulphited juice % cane including filtrate returns mixed with it.	120 %
Sulphited juice heating	From 73° to 103°C
Clear juice % cane (assumed as equal to mixed juice % cane)	105 %
Clear juice heating	From 97° to 112°C
Specific heat of juice	0.91 Kcal / kg
Latent heat of steam (at 0.75 kg/cm^2 g pressure)	530 KCal / kg

Steam required for raw juice heating:

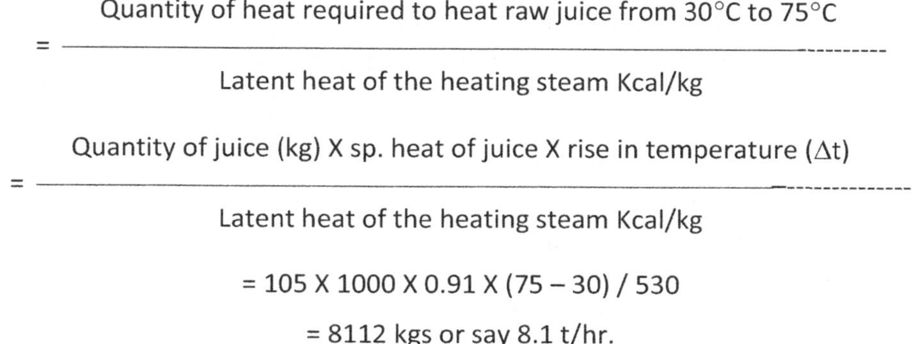

$$= \frac{\text{Quantity of heat required to heat raw juice from 30°C to 75°C}}{\text{Latent heat of the heating steam Kcal/kg}}$$

$$= \frac{\text{Quantity of juice (kg) X sp. heat of juice X rise in temperature } (\Delta t)}{\text{Latent heat of the heating steam Kcal/kg}}$$

= 105 X 1000 X 0.91 X (75 – 30) / 530

= 8112 kgs or say 8.1 t/hr.

Steam required for sulphited juice heating:

Consideri3g 15% filtrate recycling, the sulphited juice quantity will be 120 % on cane and the steam requirement for sulphur juice heating will be

= 120 X 1000 X 0.91 X (103 – 73) / 530

= 6181 kg or say 6.2 t/hr.

Steam required for clear juice heating:

Considering clear juice % cane equal to mixed juice % cane the steam required for clear juice heating will be

=105 X 1000 X 0.91 X (112 – 97) / 530

= 2704 kg

Or 2.7 t

Table 32.3 Total steam required for juice heating

1	Steam required for raw juice heating	8.1 t/hr.
2	Steam required for sulphited juice heating	6.2 t/hr.
3	Steam required for clear juice heating	2.7 t/hr.
4	Total steam requirement for juice heating	17.0 t/hr.

Thus, the steam requirement for juice heating is about 17.0 t/hr. for 100 t crushing rate or 17% on cane. Here we can say, the steam requirement for raw juice and clear juice heating is 0.18 t/°C/100 tch and for sulphited juice heating is 0.2 t/°C/100 tch.

32.4 Condensate tubular juice heater:

Now in some factories for ultimate steam economy raw juice is being heated by condensate. Therefore the calculations of heating surface of condensate tubular juice heater are given here under. Heating surface for horizontal model of cross – counter current, shell and tube type condensate juice heater for 100 tch plant –

The features of the heat exchanger considered are as follows

1) horizontal model,
2) Condensate to shell side and juice in tubes,
3) Baffles 75 % shell side,
4) Distance Between Baffles 500 mm,
5) Tubes – ID 42 mm, OD 45 mm,
6) Length 4/5 metres.

We know the common formula of heat transfer Q = UA ΔTm Or A = Q / U ΔTm

Where

- A = Heating surface Area
- Q = quantity of heat to be transferred in unit time
- U = Overall heat transfer coefficient (OHTC) kcal/m2/°C/hr. which is 650 for aqueous solutions (at this temperature range),
- ΔTm = To The log mean temperature difference will be given by

$$\Delta Tm = \frac{\Delta Ta - \Delta Tb}{ln\frac{\Delta Ta}{\Delta Tb}}$$

Where

ΔTa = Temperature difference in hot condensate inlet temperature and cold juice outlet temperature.

ΔTb = temperature difference in hot condensate outlet temperature and cold juice inlet temperature.

However it is not simple counter current flow. The hot condensate passes in zigzag way and it is partly in cross direction with the heating surface and partly tangential or parallel. In such complicated case apart from estimating temperature difference by LMTD method it is needed to follow heat exchanger design calculation given under TEMA (Tubular exchanger manufacturers association, USA) for correction factor to the LMTD. However not going in detail calculation, we shall consider here a factor as 0.65. Therefore the figure will be 0.65 x ΔTm. Thus the correction factor reduces LMTD and increases heating surface considerably.

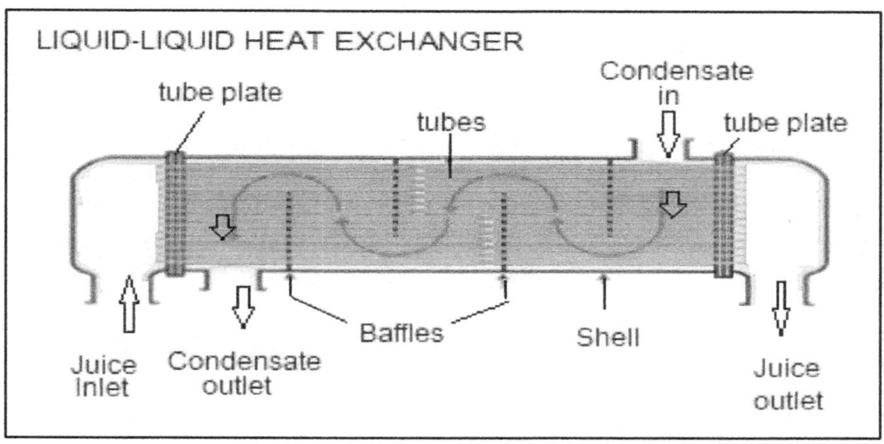

Fig no. 32.5 Horizontal condensate juice heater

Example: Cold raw juice is heated from 46°C to 62°C by hot condensate of 100°C. The hot condensate water is getting cooled from 100°C up to 57°C.

Table 32.4 raw juice heating by condensate

Particulars	**Temperature in °C**		
condensate	Inlet	outlet	Decrease in temperature
	100 (T1)	57 (T2)	43
juice	outlet	Inlet	increase in temperature
	62 (t2)	46 (t1)	16
Temperature Difference	38 (Ta)	11(Tb)	simple mean ΔT - 27

Putting the values for raw juice 2nd heating-

$$\Delta Tm = \frac{(38-11)}{ln\frac{38}{11}}$$

Logarithmic mean temperature difference $\Delta Tm = 21.77$

$$A = Q / U X 0.65 \Delta Tm$$

$$= (102 \times 1000 \times 0.91 \times 16) / 650 X 0.65 x21.78$$

$$= 161.39 m^2$$

However in factory working the flow rate and temperature of condensate and juice may vary considerably. Considering these variation we safely can take heating surface as 200 m^2.

33.5 Reducing the steam consumption by using condensate:

Above we saw that, the steam consumption at juice heaters is 17.00 % on cane. Now instead of vapours if raw juice in the 2nd stage is heated by hot condensate from all the sources, the steam can be saved. The raw juice heating is done as follows-

Table 32.5 steam consumption using of condensate

Sr. no.	Particulars	Steam consumption
1	Raw juice first heating from 30 to 45°C in vapour line juice heater	2.7
2	Raw juice second heating from 45 to 60°C by condensate of the 3rd effect and pans (part)	----
3	Raw juice third heating from 60 to 75°C by condensate of the 2nd effect and juice heaters	-----
4	Sulphited juice 1st heating from 73 to 87°C by vapours from the 3rd effect.	2.8
5	Sulphited juice second heating from 87 to 101°C by vapours from the first effect	2.8
6	Clear juice first heating from 97 to 107°C by vapours from the first effect	1.7
7	Clear juice second heating from 107 to 117°C by exhaust steam	1.7
	Total steam consumption for juice heating	11.7

Thus by heating the raw juice by condensate in two stages after 1st heating in VLJH, the steam consumption at juice heaters which was 17.0 % in previous case, can be brought down to 11.7 % on cane. Hence there can be net saving in juice heating by 5.3 % on cane.

32.6 Direct Contact Heater:

Direct contact heaters (DCH) or condenser heaters are now being used in sugar factories. The advantage of DCH is its simplicity in design and low cost investment. The direct contact heater (DCH) is constructed of stainless steel body. DCH is a heat exchanger in which the juice and vapours are in direct contact in counter current way. The vapours - when come in contact with lower temperature juice, get condensed and latent heat as well as sensible heat of vapours are given to the juice. As there is direct contact the heat transfer is most efficient. The advantage of DCH is approach temperature - temperature between heating vapour and heated juice is minimum. Therefore low temperature vapours from evaporator can also be used. The juice gets diluted as vapours are condensed in the juice. Juice entry is provided with specially designed tangential box.

In conventional or dynamic Juice heaters we need a temperature difference between heating media and heated media of about 6-12^0 C depending upon vapour temperature. In the case of DCH the ΔT could be maintain as low as 1-2^{0}C. In DCH there is no heating surface and therefore no brushing and cleaning is required. Hence there is no need to provide stand by heater. In direct contact heater head loss become zero so it reduces the power requirement of the pump. With the use of direct contact heaters for all juice heating, the quantity of juice increases after each heating. By total heating the juice quantity is increased by about 17 to 20% on cane. This increases steam consumption, also gives extra load on evaporator set.

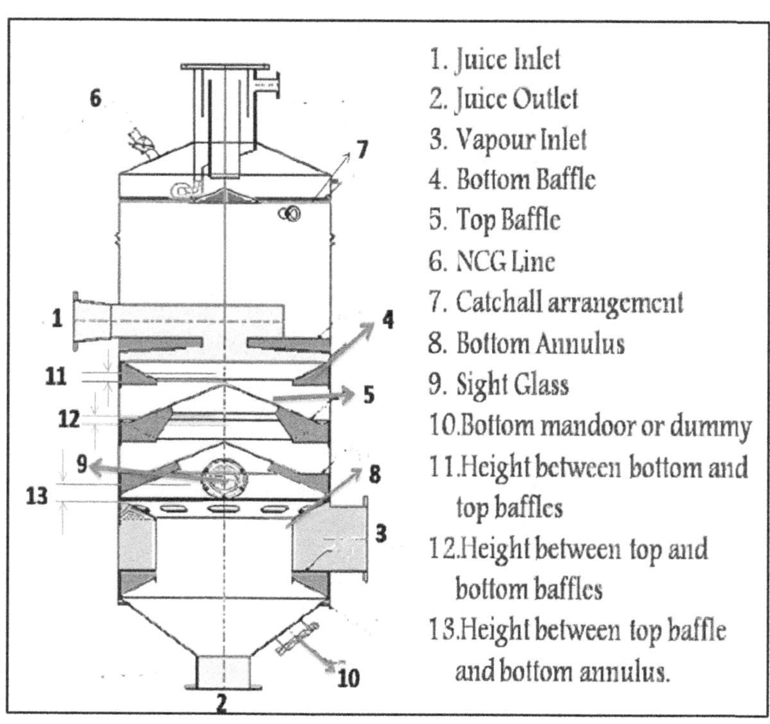

Fig. 32.6 Direct contact heater

Advantages:

- SS construction leading to longer life.
- Requirement of lesser space
- Piping system is simple and installation is easy.
- Consumes low pressure vapour effectively
- The system does not require any maintenance like mechanical cleaning.
- Heating of juice in DCH removes dissolved gases in the juice very efficiently.
- DCH does not require standby unit.

Disadvantages:

- The condensate of the heating media (vapour) is added along with heated media (juice) so it was extra load on evaporation system.
- While in operation care has to be taken to avoid entertainment. Particularly when low pressure vapours (3rd or 4th body vapours) are used.

-OOOOO-

33 CLARIFIER

After treatment and final heating the juice is send continuous settling tank or continuous settler which is generally called as clarifier. In the clarifier the light weight precipitated solids which are much less in quantities, float on the juice as scum and the heavier precipitated solid settles down.

33.1 Theory of settling of precipitates:

The suspended and precipitated solids from the treated juice are separated from clear juice by sedimentation or settling. Sedimentation or settling is the process in which suspended matter from the liquid is separated by gravity. In clarifier velocity of the juice is brought down to extremely low value so that the coagulated precipitates settle down by gravity.

The settling rate of particles depends on

- difference in densities of the particle and liquid
- shape and size of the particle
- viscosity of the liquid

The velocity of the particle is given by Stokes law

$$V = \frac{D^{\wedge}2(d1 - d2)g}{18U}$$

Where V – velocity of fall of the particle (cm/sec)
D – Diameter of the particle (cms)
d1 – density of particles g/cm^3
d2 – density of liquid g/cm^3
U - Viscosity of the liquid in poise

The above equation is for spherical particles. However all the particles are not of spherical shape and the constant (1/18) changes.

The precipitated particles include organic as well as inorganic constituents. The organic matters such as fats and waxes are of low densities (around 1.0 or even less) whereas inorganic matters such as calcium phosphates are of higher densities (around 2.2 –3.0). The average density of particles is about 1.5-1.8. The density of the juice varies from 1.04-1.07. However the temperature at which the settling takes place in clarifier is about 98°C and at this temperature the juice density is still lower. For rapid settling the important factor is size and shape of the particles. Large particles with spherical shape give faster settling. Formation of large size spherical particles depends on proper flocculation and coagulation of the precipitate and colloids. Viscosity of the juice is the internal friction that reduces the rate of settling of particles.

33.2 Flash tank:

The treated juice heated above 103°C, if send directly to the clarifier, as it becomes open to atmosphere while entering the clarifier flashing or self-evaporation starts and vapour bubbles are evolved. These vapour bubbles pass upward and disturb the precipitate. Therefore the settling mud by gravity gets disturbed.

Therefore before sending the juice to clarifiers it is passed through a flash tank. In the flash tank the flashing effects occurs rapidly due to whirling velocity of the juice in the flash tank and the temperature of the juice comes down below the boiling point of the juice. Then the juice is send to the clarifier at a slow rate. Now in some factories heating of the juice is done up to 97-99°C only. In such case flashing of the juice does not occur.

Fig. 33.1 flash tank

Flash tank size:

We will calculate the flash tank inlet outlet diameter and tank size for a 5000 TCD (230 tch) plant here under

Table 33.1 assumptions

Crushing rate TCH	230
mixed juice % cane	110
Mixed juice quantity	253
Filtrate return (17%)	39
Total Juice quantity MT/hr.	292
Juice density at 99°C	1.02
Juice volume m^3/hr.	286.3
Juice volume m^3/sec	0.080

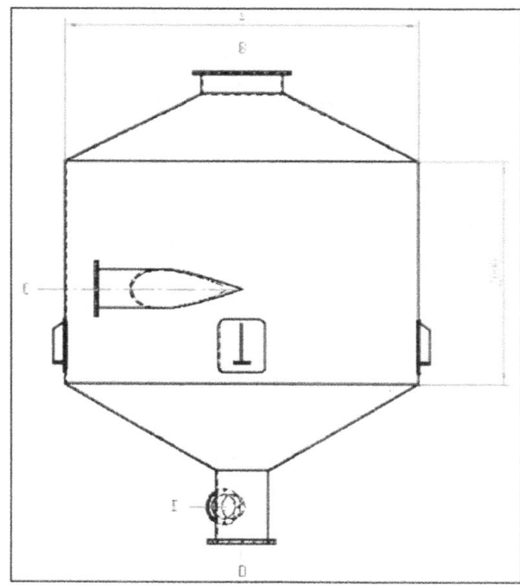

Fig. 33.2 flash tank

Flash tank juice inlet pipe diameter:

In calculating the inlet pipe diameter the velocity of the juice is assumed as 1.5 m/sec. The diameter can be calculate by the following formula

$$V = \frac{\pi D^2}{4} \times v$$

Where

V = Volume of juice flow (m^3/sec)
D = Diameter of the inlet pipe
v = Velocity of juice (m/sec.)

Putting the value of V and v

$$0.080 = \frac{3.142\ D^2}{4} \times 1.50$$

Solving the above equation we get D = 0.261

The standard pipelines available are of sizes 10" or `12" in diameter. Considering higher crushing rate 12" diameter pipeline can be used.

Flash tank dimensions:

The flash tank volume should be about 6 m^3/100 tch for venting out vapours. For as 5000 tcd (230 tch) plant it should be 13.8 m^3 say 14 m^3. Now considering height to diameter ratio of the tank as 1.35 we will calculate height and diameter of the tank.

$$V = \frac{\pi D^2}{4} \times h$$

Where

V = Volume of flash tank
D = Diameter of the flash tank
h = Height of flash tank = 1.35D

Putting the value of V and v

$$14.0 = \frac{3.142\ D^2}{4} \times 1.35\ D$$

$$\text{Or} \quad D^3 = \frac{14.0 \times 4}{3.142 \times 1.35}$$

Solving the above equation we get D = 2.37 or say 2.40

Thus the diameter of the flash tank will be 2.40 m and height will be 3.24 m.

Outlet pipe diameter:

The velocity of the juice pass from flash tank to clarifier through a U siphon should be around 0.5 m/sec. So that the floc formed will not be disturbed. Therefore considering the juice velocity 0.5m/sec. we will calculate the diameter of the outlet and siphon.

$$V = \frac{\pi D^2}{4} \times v$$

Where

V = Volume of juice flow (m^3/sec)
D = Diameter of the inlet pipe
v = Velocity of juice (m/sec.)

Putting the value of V and v

$$0.080 = \frac{3.142\ D^2}{4} \times 0.50$$

Solving the above equation we get D = 0.450 m.

Thus the diameter of the flash tank outlet and the siphon will be 0.45 m.

33.3 Construction and working of clarifier:

A continuous clarifier is a large cylindrical vessel. Treated juice is fed in it continuously. As the vessel is large juice velocity becomes very low and precipitated solids can be easily settled down. Conventional clarifier is divided into four compartments to increase the area of settling. It is provided with a central hollow shaft (generally called as central tube) rotating very slowly (about 3 rph). The centre tube carries arms (in each compartment). Metal sheet scrappers are fitted to the arms. The scrappers - as they rotate, brush the bottom tray and slowly push mud settled on the trays towards the mud boot.

There are many different designs of clarifiers. The most widely used is Dorr 444. Dorr 444 is the design of Dorr Oliver Company. Graver clarifiers are also used in some factories.

A flocculating compartment of half the diameter of main clarifier and of one meter height is provided with skimming blades and feed opening to the centre tube. The skimmers or skimming blades are hanging at a level of juice. They push the scum formed into a gutter from where the scum flows to mud box.

The clarifier has separate clear juice and muddy juice outlet from each compartment. The clarifier has clear juice and mud withdrawal gravity boxes with sleeves, telescope pipes. Two clear juice pumps of suitable head are provided to pump the juice through clear juice heaters to 1st effect of evaporator.

Four manholes, one for each compartment is provided with platforms. A cat's ladder is provided from ground to top of clarifier to give access to the manholes.

At the clarifier an arrangement for preparation and dosing of flocculent is provided. The juice to be clarified enters tangentially at the top into a feed compartment or flocculating chamber. Here some scum rises to the surface of the juice. In the flocculating chamber hanging arms are fitted to the centre tube. Scrappers or skimming blades are fitted to these arms. These skimming blades rotate as the centre tube rotates and scrap off the floating scum outer side into a foam canal

The juice enters from the foam compartment into the centre tube. The central tube has openings in each compartment. Each compartment works as an independent clarifier. The juice flows down through the centre tube and enters in each compartment and in the compartment the juice flows horizontally.

Then the clear juice flows upward and finally horizontally to the juice outlet. The precipitated solids having higher densities descend downwards and further settle down on the bottom plate or tray. The muds deposited on the bottom plate are pushed slowly by scrappers to the central into the mud boot.

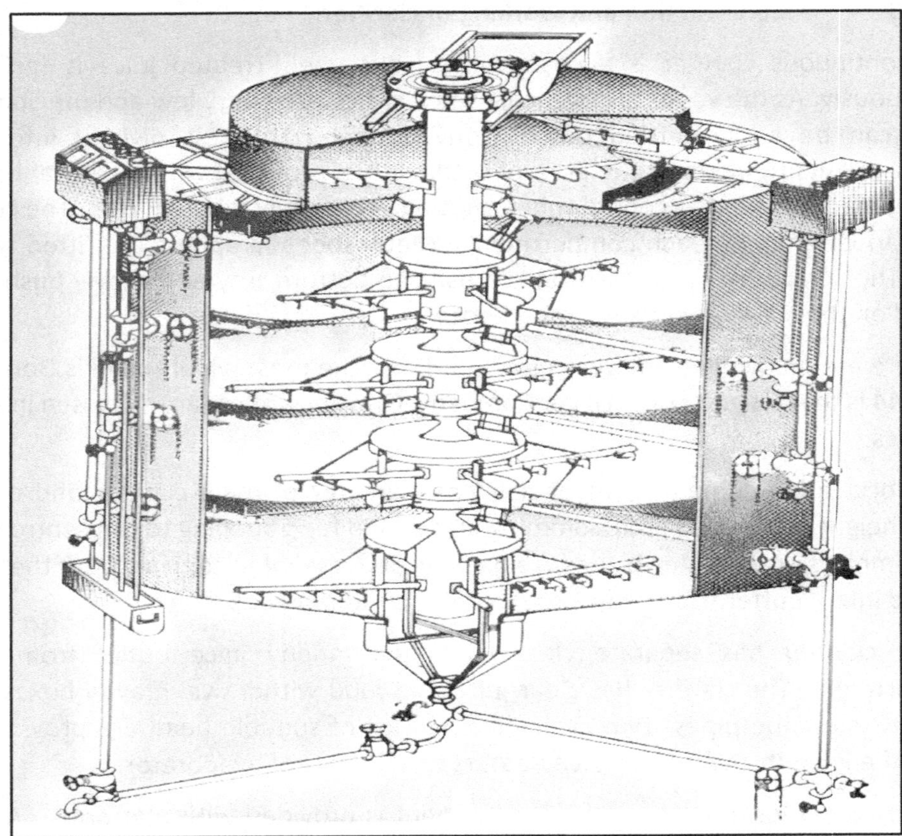

Fig. 33.2 Dorr clarifier

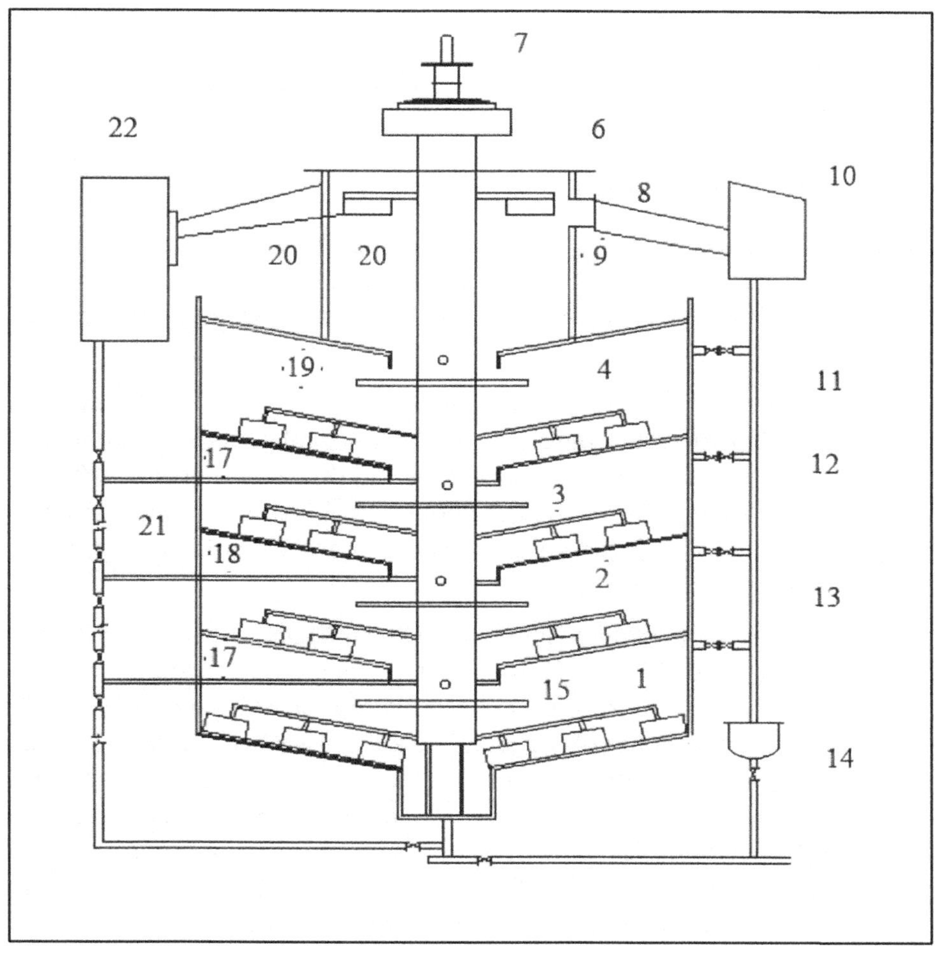

Fig. 33.3 Clarifier 444 design

Sr. no	Particulars	Sr. no.	particulars
1	Compartment A	12	C J withdrawal pipe
2	Compartment B	13	Mud scrapper plates
3	Compartment C	14	Sump tank
4	Compartment D	15	baffles
5	Flocculating chamber	16	Feed holes
6	Central shaft	17	Tray plates
7	Bypass gate valve	18	Mud boot
8	Skimmer plates	19	Cover plate
9	Feed	20	Foam channel
10	Clear juice box	21	Mud piping
11	Stirrer arms	22	Mud box

The clear supernatant juice is withdrawn from each compartment by circumferential internal pipes with several openings fitted close to the roof of the compartment. The juice then passes to overflow box, by vertical pipes fitted with sliding sleeves, which permit regulation of rate and overflow level.

The mud collected at the mud boots are extracted by mud piping which are terminated at the mud outlet box.

Vent pipes are provided at the roof of each compartment. These vent pipes allow to escape gases - if any formed in the clarifier, to atmosphere. The clarifier is enclosed, except for a door giving access to the flocculating chamber. It is completely lagged.

When removal of scum is not carried out regularly, the scum thickness increases and further the scum passes along the juice into the main compartments and through it pass out with the clear juice. To avoid this, the level of overflow height can be adjusted by sliding sleeves. These should be adjusted in such a way that the skimming blades readily push the scum formed in the flocculation chamber into the foam canal without taking liquid juice with it.

For 2500 tcd plant a Dorr 444 or similar clarifier of 4 compartments each of height 5ft (1,524 metre) and 30 ft. (9.144 metre) diameter (volume 400 m^3) is recommended. At a crushing rate of 115 tch, mixed juice % cane 105, and filtrate return 15 %, the retention time of the juice is clarifiers is 3 hrs. It may come down as crushing rate and/or imbibition increased. In the areas like south Maharashtra and north Karnataka, where juice qualities are very good and process is also technically fairly controlled, juice retention time in Dorr 444 clarifier can be brought down up to 2 hours 15 minutes. In such case use of settling agent is required. However in region like costal Andhra Pradesh, even with the use of settling agents the retention time cannot be brought down below 3 hrs.

33.4 Short Retention time Clarifier:

The Sugar Research Institute, Australia is pioneer in designing a short retention time clarifier. This design is called as SRI clarifier. This or similar type of designs are now widely being used the cane sugar factories in the world. It is called as short retention clarifiers or trayless clarifiers or single tray clarifier.

The short retention ranges in size from 5 to 14 m. diameter. The retention time of the juice in this clarifier is normally one hour. In past multi-tray clarifiers were being installed in all sugar factories, but now in most of the sugar factories the SRI clarifiers are being installed.

There is no difference in basic principle such as upward flow velocity of juice should be less than the settling rate of mud particles. However use of flocculent is necessary in case of SRI clarifier to maintain the high settling rate of mud.

Fig. 33.4 short retention time clarifier

In designing short retention clarifier upward flow velocity is calculated as half the initial settling rate of the mud in the juice. The initial velocity of settling is usually taken as 30 mm/min considering the upward velocity and the rate of incoming juice volume the cross sectional area of the clarifier and hence the diameter of the clarifier is calculated. The residence time of juice is taken as one hour. In Australia with high quality juices the retention time in SRI clarifier could be brought down to as low as 28 minutes in some factories.

The advantages of the single tray or SRT clarifiers are:

- Retention time is less hence less loss of sucrose and less colour formation,
- Lower capital cost,
- Lower maintenance cost,
- Easy to liquidate and hence regular cleaning is possible.

Online brix measurement of clear juice:

Now in modern factories an online brix meter for clear juice is installed for measurement of clear juice brix. It is 'Coriolis Density Sensor' with dual U type design. The accuracy of which is +/- 0.0005 gm. /cc. the ambient temperature effect is 0.001% per °C. The brix meter has facility to give alarm in case brix value exceeds desired settings. It can be connected to PC and printer.

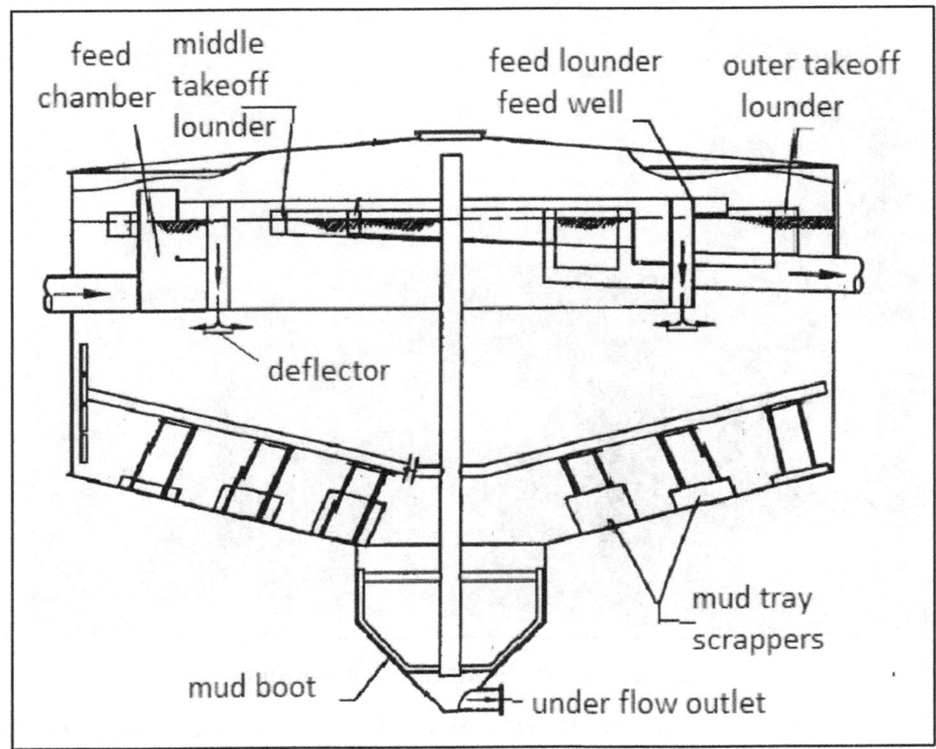

Fig. 33.5 SRT clarifier

33.5 Inversion loss in clarifier during cleaning:

During periodical cleaning the clarifier may not always be liquidated. Many times the juice is stored for about 24 to 36 h. and sucrose loss occurs. The loss is due to inversion or due to activity of microorganisms. Inversion loss is more marked at temperature above 90°C whereas loss due to activity of microorganisms is higher below 70°C. Thus 70° to 90°C is optimum range of temperature when juice is to be store in clarifier in general cleaning.

Therefore in case of conventional clarifier, it is always better to regulate temperature of treated juice to about 90-95°C during the last 2-3 hrs. crushing, before the shutdown. This lower temperature will not affect the quality of outcoming clear juice when flocculent is used. The loss is also closely related to the pH and increase when the pH of the juice falls below 6.7. The following table gives observation made during shut down in one factory. The temperature of the juice in clarifier was not observed. However it was 98°C at the start of the shutdown.

Table 33.2 Purity drop and increase in reducing sugar Observed in clarifier juice during periodical cleaning in a factory

Sr. No.	Hours	Clarified juice from clarifier				
		pH	Brix	Pol	Purity	R S /100°brix
1	00.0	7.00	17.22	14.63	84.96	3.34
2	4.00	6.9	17.26	14.66	84.92	3.50
3	8.00	6.9	17.29	14.60	84.44	3.53
4	12.00	6.8	17.46	14.67	84.02	3.63
5	16.00	6.8	17.68	14.73	83.31	3.68
6	20.00	6.7	17.74	14.72	82.97	3.87
7	24.00	6.7	17.78	14.67	82.51	3.98
8	28.00	6.5	16.76	13.76	82.10	3.99
9	32.00	6.5	16.84	13.78	81.83	4.18
10	36.00	6.4	17.18	14.05	81.78	4.48
11	40.00	6.4	17.04	13,91	81.63	4.78
12	44.00	6.3	17.54	14.29	81.47	4.92
13	48.00	6.3	17.64	14.34	81.29	5.18

CLARIFICATION STATION

-OOOOO-

34 FILTRATION OF MUD

34.1 Construction and working of rotary vacuum filter:

Rotary vacuum filter is made up of a hollow horizontal drum. It is partly submerged in the muddy juice to be filtered and rotates about its horizontal axis.

Fig. 34.1 Rotary Vacuum Filter

The periphery (or deck) of the drum that serves as a filtering surface is normally divided into 20 or 24 equal independent sections. Each section is connected individually to a vacuum distribution system (disc) by a small metal pipe. The disc carries three different zones.

- The first zone connecting to a chamber where low vacuum (150-250 mm of Hg) is maintained.
- The second zone connecting to a chamber where high vacuum (350- 500 mm of Hg) is maintained.

The third zone is not connected to vacuum but is connected to atmosphere. The deck of the drum- all 20 or 24 sections, is covered with stainless steel screens of very fine perforation (100 holes per sq. cm.) supported by back frames that are made up of plastic. The lower portion of the filter drum (about 20 % of the filtering surface) is submerged in the muddy juice to be filtered in a hollow half-cylindrical trough.

As the filter rotates the section entering the muddy juice is connected to the low vacuum zone. The vacuum sucks the muddy juice. The juice passes through the fine perforations whereas the bagacillo and coarse mud solids are retained on the screen.

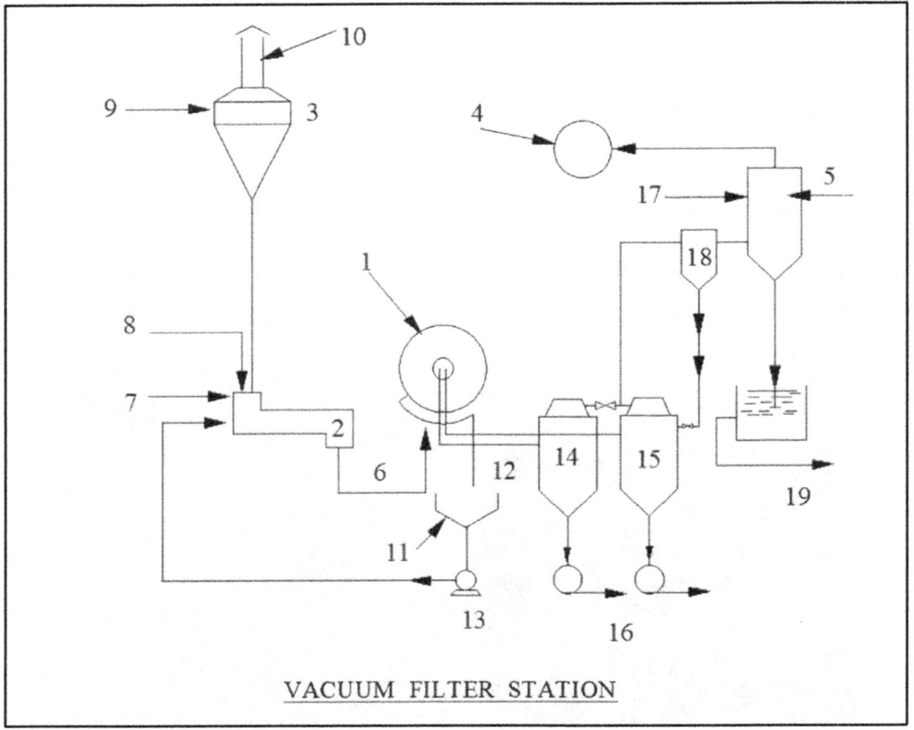

Sr. No.	Particular	Sr. No.	Particular
1	Filter	11	Overflow tank
2	Mixer	12	Overflow
3	Cyclone	13	Recirculation pump
4	Vacuum pump	14	Low vacuum receiver
5	Condenser	15	High vacuum receiver
6	Filter feed	16	Filtrate receiver
7	Muddy juice	17	Condenser water
8	Settling	18	Trap
9	Bagacillo	19	Barometric seal
10	Air vent	20	

Fig 34.2 Vacuum Filter Station

Thus a layer of mud / solids is formed. It continues to build until the section emerges out of the muddy juice. The first juice that passes through the screen when the layer of mud solids is being built is much poorly filtered and contains fine suspended matter. This is called cloudy or heavy filtrate and is collected in heavy filtrate receiver.

The section as emerges out of muddy juice, the vacuum pipe of the section at the disc gets connected to high vacuum zone. Juice from the cake is sucked inside by the high vacuum. The filtrate received in this zone is comparatively clear and is called as light filtrate.

Hot water spray pipes are fitted to wash the cake. When the section passes through the high vacuum zone wash water is sprayed over the cake. The water dilutes the juice in the cake and the vacuum sucks the diluted juice through the cake. Thus the layer of cake is washed off. This is called as *desugarisation of the cake*. After the sprays the washing is continued by dripping pipes with sheet metal distributors that pass water to drip on the cake. After the last drip pipe drying of the cake is done.

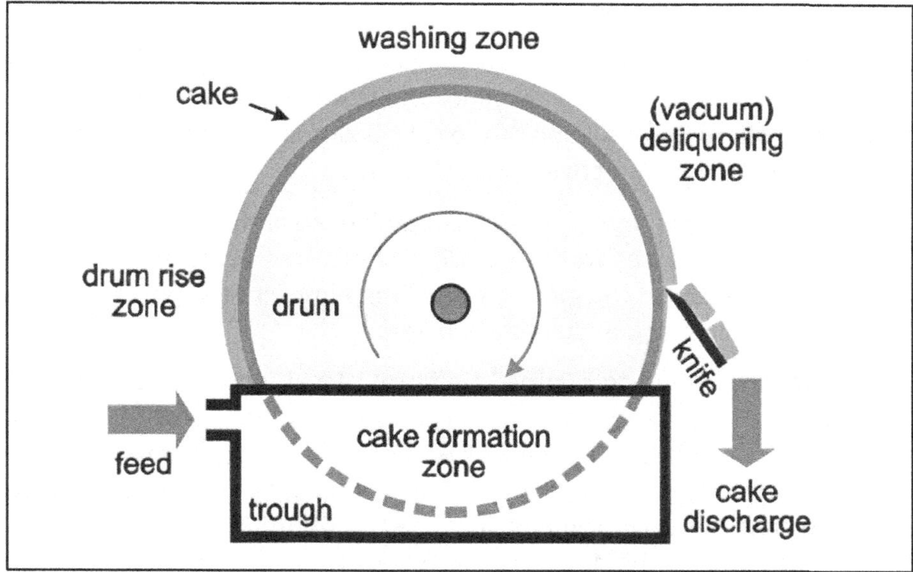

Fig. 34.3 Vacuum Filter Station

At the end of the cycle the section come to a place where a rubber scrapper is fitted. When the filter section is about to reach the scrapper the distributor disc breaks the vacuum of the section. Then the scrapper as reaches to the scrapper, the scrapper detaches the cake. The section freed from the cake again enters the muddy juice. The detached is drops on a belt conveyor that carry it outside building.

The recommended proportion of solids on dry basis in the muddy juice going to the filtration is about 7 %. In working conditions it should not be less than 5.0%.

In normal working the temperature of the muddy juice after addition of bagacillo is be about 85°C. At this temperature waxes in the juice do not solidify. But when the section emerges out of muddy juice the cake comes in contact with muddy juice and temperature of cake may come down and the waxes may get solidify at the screen and blind the screen. To avoid cooling of the muddy juice, it's over flow from the vacuum filter tray to the bottom mud receiving tank and recycling the same to tray should be avoided. Further the vacuum filter should be stopped and the screens should be cleaned after every 3-4 days.

An agitator is fitted in the trough on the axis of the filter. It is driven by a separate small motor is provided. It oscillates to and fro and keeps the muddy juice in constant agitation. This avoids settling of mud solids of the muddy juice in the trough. To avoid rupture of flocculent mud particles the muddy juice should flow by gravity to the mud mixture and further to the filter.

The speed of rotation of the drum may be regulated between 12-rph to 18-rph. All an average 15-rph is usually used.

The amount of wash water percentage used for desugarization of filter cake is 100 –150 % on filter cake or 4-6 % on cane. The thickness of the cake varies from 5 mm to 13 mm. Optimum exhaustion of the cake is obtained with a thickness of 8-10 mm since washing is most effective at this thickness.

Two separate receivers one for heavy cloudy filtrate and one for light filtrate are provided. The filtrate passes from the disc valves to the receivers and then flows down to the pumps fitted at the ground floor. The pumps lift the filtrates and deliver to mixed juice receiving tank or sulphitation vessel.

In case of failure of filtrate pumps, the filtrate level increases and the receiver gets full of filtrate. In such case there are chances of entrainment of the filtrate to the condenser. To avoid this vacuum brake valves with high tension springs are provided to each receiver. A floating pot which is hanging to the spring keeps the spring tight so that the valve remains in closed position. In case juice level in the receiver increases the pot floats on the juice and spring tension gets released. The valve gets opened and vacuum is released. The weights of the pots are given by the manufacturer in the equipment manual. However if weight of a pot is not known, then keeping the pot floating on water it is filled with sand till 80 % of its height gets dipped in the water when it floats.

A filtration rate of 200-300 $l/m^2/h$ may be expected and 70-85 kg of filter cake quantity is discharged per sq. metre/hr. Therefore normal capacity of the rotary vacuum filter should be about 60 to 65 m^2 screening surface area per 100 tch.

The filtration by rotary vacuum filter is not pure filtration. Both the filtrates contain lot of fine suspended solids and are highly turbid. Therefore both heavy and light filtrates cannot be send to evaporation along with clear juice. Hence are recycled to the process. Generally the filtrates are added in the raw juice receiving tank or sometimes before sulphitation after heating of the raw juice.

The retention of the filter i.e. the percentage of suspended solids that are passed through the filter and are recycled in the process varies between 25-35 %.

The filter cake contains 1.0-3.0 % of sugar and 72 to 84% moisture. The filter cake percentage on cane varies between 3.00-4.50 % and the sugar loss in filter cake varies in the range of 0.05 –0.14% on cane.

34.2 Additional equipment required for mud filter:

Bagacillo Screens:

To filter the muddy juice at rotary vacuum filter it is necessary to mix bagacillo in it. Rotary vacuum filters do not work properly unless optimum quantity of bagacillo is well mixed with the muddy juice. Screening of bagasse is required to separate bagacillo from it. This is generally done by fitting screens at the bottom of bagasse elevator. The bagacillo from the screen is conveyed with high velocity current of air by a fan through a galvanized iron duct to a cyclone. The cyclone is fitted high exactly above the mud mixer. The air conveying bagacillo passes through the vent of cyclone and the bagacillo falls down into the mud mixer and mixes with mud.

The quantity required is about 6-8 kg of per tonne of cane. Generally 4/5-mm size hole are provided to the bagacillo screens. The area of bagacillo screen required is about 7-8 m^2 per 100 tch. 6-7 m^3 of air can convey one kg of bagacillo, the velocity of air being 20-25 m/sec.

Condenser with vacuum pump:

The quantity of air entering through the filtering area is a considerably large amount and therefore jet condenser cannot be used for making vacuum at the filters. Therefore a barometric condenser along with a vacuum pump of suitable capacity is used.

34.3 Precautions for better working of RVF:

The bagacillo requirement is 6-8 kg/t of cane. The ratio of bagacillo/ mud solids (on dry basis) should be 0.8 to 1.0. Powdered bagacillo may cause frequent choking of screens of RVF and sucking of juice/ water gets reduced. This results in higher sugar loss in filter cake.

The optimum, lower vacuum for building of cake (200-250 mm. 8-10") and higher vacuum for desugarization of cake (350-450 mm, 14-18") vacuum should be maintained

The thickness of filter cake depends on speed of rotation of drum and vacuum maintained. The lower speed develops thick cake and higher speed gives thin cake. 7 -10 mm is the optimum thickness. Fluctuation in vacuum and lower vacuum may result in thin cake formation. With thin cake, recirculation of mud solids may be increased. With higher thickness of cake the pol % filter cake may increase.

At the top portion of the filter drum there are generally three perforated pipes to apply wash water over the filter cake. Uniform water distribution is very much essential for maximum desugarization of the cake. The wash water pressure should be 2.0 to 2.7 kg/cm^2 (30-40 psig). At lower pressure the complete area of filter cake may not be washed properly and pol% filter cake may increase. The wash water should be of 75-80°C.

At lower vacuum zone cloudy filtrate is obtained. Once cake is developed as the section passes to high vacuum zone, the filtrate is comparatively clear.

34.4 Screening of filtrate:

The bagacillo and mud particles level in the filtrates should be as low as possible so that colour formation in the juice is minimized. However some quantity of very fine bagacillo cannot be retained on rotary vacuum screens and pass along with filtrate. This fine bagacillo increases turbidity and colour of juice. Chhote Lal[1] has described screening of filtrate through 100-mesh screen to remove this fine bagacillo from the filtrate. However the trials have not been continued as a general process.

34.5 Use of Filter Cloth:

In some factories in India instead of stainless steel screens thick filter cloth have been tried to filtrate the muddy juice (See the figure below). The filtered juice obtained was clear but not very shining. The filtered juice was send directly to evaporator and recirculation of filtrates is avoided. However sending filtrate for further process may adversely affect sugar quality in plantation white sugar manufacturing. Therefore this has not been popular. But in our opinion this can be used for raw sugar.

34.6 Filtrate Clarification:

In conventional process of juice clarification the filtrate from rotary vacuum filter, which contains still lot of suspended solids, are recycled to the raw juice-receiving tank or to the sulphiter. This increases load on sulphiter and clarifier. Of course the capacity of the sulphiter juice heaters and clarifier is given taking into consideration of the recycled filtrate. However due to recycled filtrate mud volume in clarifier is increased. The rise in calcium content from mixed juice to clear juice is higher. The colour of clarified juice is also higher. The ash content in clear juice is also slightly increased and the clear juice is not much brilliant. Therefore to maintain better quality of clear juice now in some factories the separate treatment of filtrate has been tried. This improves clear juice quality. The quality of clear juice obtained from the filtrate is also good and therefore it is send to evaporator.

The filtrates are mixed together in a buffer tank and then treated with milk of lime and phosphoric acid. Then the filtrate is send to sulphiter for sulphitation, then juice heater to heat it to boiling point. Then it is passed through a flash tank to clarifier. All the process is kept separate with separate equipment. The clear juice obtained is send to evaporator.

In this process separate buffer tank, sulphiter, juice heater, clarifier are to be provided and process becomes somewhat complicated, elaborative and clumsy and difficult to handle. Very few factories have installed this system. The factories which have installed this system are not much satisfied and most of these factories have not continued its use.

Decanter Centrifuge for separation of mud

34.7 Decanter Centrifuge General:

Application of decanter centrifuges is to separate large amounts of high density solids from liquid continuously. The decanter centrifuge is also known as solid bowl centrifuge. A high rotational speed is used for separation of the solids. A characteristic feature of the bowl is its cylindrical /conical shape. The fast rotation generates centrifugal force up to 4000 x g. Under this force, the solid particles of higher density are collected and compacted on the wall of the bowl.

Fig. 34.4 Decanter Centrifuge Station

A scroll (screw conveyor) rotates inside the bowl at a slightly different speed. This speed difference is called the differential speed. The scroll /screw conveyor transport the settled solids on inside wall of the bowl up to the end conical part of the bowl. Hard surfacing and abrasion protection materials are required for the scroll to reduce wear.

The dewatered solids leave the bowl through discharge opening. The clarified liquid flows to the cylindrical end of the bowl in the decanter centrifuge, from where it runs out through openings in the bowl cover. These openings contain precisely adjustable weir discs/weir plates by means of which the depth in the bowl can be set. Decanter centrifuge are being used in the chemical, oil and food processing industries as well as for waste waters.

34.8 Use of decanter centrifuge in sugar industry:

Since 2007 horizontal design decanter centrifuges are being used in some sugar factories for separation of solids in muddy juice. The rotating assembly is mounted horizontally with bearings on each end to a rigid frame. The feed enters through one end of the bearings, while the gearbox is attached to the other end. The horizontal machine is arranged in a way such that the muddy juice can be introduced at the centre of a rotating horizontal cylindrical bowl from the smaller end. Five decanters of standard capacity are required for a 5000 tcd plant. Two for the first stage two for the second stage and one as standby.

34.9 Operation of Decanter:

Muddy juice from the clarifier is taken to a receiving tank where milk of lime is added to adjust the pH to 7.0 to 7.2. Then the muddy juice is transferred to a mud conditioning tank for proper mixing. Then the conditioned muddy juice is transferred to the first stage decanter through a progressive cavity pump. For safety purpose pressure relief valve is provided to the delivery line for each transfer pump. Muddy juice flow is measured by magnetic flow meter. The feed is equally devided for each decanter. At muddy juice entry of decanter, polymer solution dosing connection is provided. Polymer is very much essential in operation. Anionic grade polymer with 0.05 % concentration solution is used.

The standanrd decanter rotates at a speed of 3250 rpm. The filtered juice (called first stage centrate) separated in the first stage decanter is collected in a separate tank. This juice is not very clear but comparatively contains less solids than the filtates from RVF. It be circulated to sulphiter.

The first stage decanter mud solids are mixed with hot water and the mud slurry formed is transferred to the second stage decanters. The flow rate is precisellly controlled for which magnetic flow meter and autocontrol valve is provided on delivery line. At the second stage decanter inlet pipe a separate connection is provided for dosing of polymer to each machine.

The second stage centrate is taken to imbibition. Therfore the load on evaporator is reduced.

Solid cake with 60 to 70 % moisture and 30 to 40 % dry solids is dischagred from the second stage decanter. It is transported by conveyor to tractor troelly for disposal.

The decanters can cleaned by flushing hot water.

The polyelectrolyte consumption is around 1.25 kg/tonne of dry solids for the first stage and 0.5 kg/tonne of dry solid for the second stage. This is about 6gm/tonne of cane in the first stage and 2.4 g/tonne of cane in the second stage.

34.10 Results of use of decanter for muddy juice fitration:

The cake obtained is only 1.5 to 2.0 % on cane. Thus the cake pecentage is reduced by about 50% in comparison with RVF. The analysis of centrate observed as follows:

Table 34.1 Centrate analysis

Particulars	Brix	Pol	purity
Centrate I	10-12	8-10	80-82
Centrate II	3-4	1.5 -2.5	50-55
Suspended solids in filtrate		0.5 to 1.0 % w/w	

Table 34.2 Cake analysis

Particulars	Value
Pol % cake	1.0 to 1.5
Moisture % cake	65 to 75
Cake % cane	1.5 to 2.0
Pol in cake % cane	0.02 to o.03

34.11 Advantages of decanter centrifuge:

- The total system of decantation separation is simple, the systm is totally enclosed,
- Vacuum pump, condenser like vacuum system is not required, therefore chances of entranment are nil.
- Bagacillo is not required as in case of RVF, therefore equipment like bagacillo mesh, fan,conveyor,cyclon etc. is not required.
- The operation is simple,
- About 80 to 90 % separation is achieved, In RVF only only 66-70 % separation is achieved,
- The retention time is much less,
- The loss in filter cake which is about 0.07 to 0.10 % on cane in case of RVF is reduced to 0.03 to o.05 % on cane. thus the loss in filter cake is reduced by about 50%, or less.
- The power requirement is less.

Some technologist claim that with decanters the colloids in the juice are not removed much effectively as in case of RVF, but no detail obsrvations are given.

34.12 Recycling of muddy juice at mills:

In many factories sometimes RVF cannot be in operation for various reasons and filter station becomes idle, but the factory cannot take mill stoppage. In such difficult and complicated situation the muddy juice from the clarifier is taken to penultimate mills and crushing is continued. The muddy juice is spread on bagasse going to the penultimate mills.

However at Chadha sugars unit Kiri (Punjab)* where defeco-melt process is being followed for production of refined sugar, recycling of muddy juice to mills was followed in 2014-15 throughout the season.

The muddy juice was discharged on the bagasse going to the 3rd and 4th mill through the imbibition juice chute. A dose of little quantity of milk of lime was added in muddy juice buffer tank to keep the pH of the muddy juice close to neutrality. The imbibition % fibre was maintained same as that of previous year. The results obtained are as follows,

The percentage of muddy juice remained normal. The pol% bagasse and moisture% bagasse was increased. However the pol loss in bagasse was equal to the pol loss in bagasse plus pol loss in F C on % cane in the previous year. That means the pol loss was not increased.

The sugar quality did not get affected. The final molasses purity was decreased by 1.17 units. There was steam saving by about 1.25 % on cane. Electrical power consumption was reduced. The power sale and bagasse sale was increased. The factory claimed that by closing RVF station, there was net saving of Rs. 659.47 lakh in terms of power, steam, contractor cost, chemical cost, etc.

There was no clinker formation at boilers. Operation of boilers remained satisfactory. Some factories which have tried to follow this process found trouble at boilers of clinker formation.

In our opinion before recommending it as a regular process it should be tried at least in three more sugar factories and the results should be very keenly observed by NSI and VSI. Further the economics also has to be studied carefully. Recycling the mud should be tried in the factories only where the milling tandem is of minimum 5 mills, so that the loss in the bagasse can be controlled.

References:

1 Chhote Lal - STAI 2001 p mfg. 56 – 62
2 Paul A K, proc. STAI DSTA Joint convention 2015, p FP 613-626.

-OOOOO-

35 LIME SLAKER

35.1 Construction and working of a lime slacker:

A horizontal rotary cylinder is used for slacking of lime. It rotates slowly. It's called as lime slacker. Lumps of quick lime are fed from one end of the lime slacker. Hot condensate water is poured into the slacker to prepare milk of lime from the quick lime. To get high slaking temperature only enough water is added to the lime fed in the slacker. Then some water is added for easy flow of the slaked lime to the discharge side.

Fig. 35.1 Lime Slacker

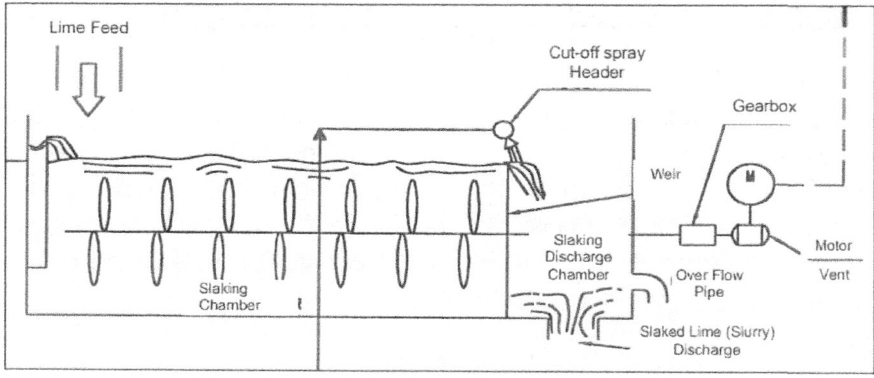

Fig. 35.2 Lime Slacker drawing

Preparation of milk of lime comprises of following steps,

- Slaking of the lime in lime slacker,
- Removing of coarse impurities like sand, stones etc. from the slacker,
- Removing of finer insoluble impurities like grit by classifier,
- Bringing the milk of lime to a desired density.

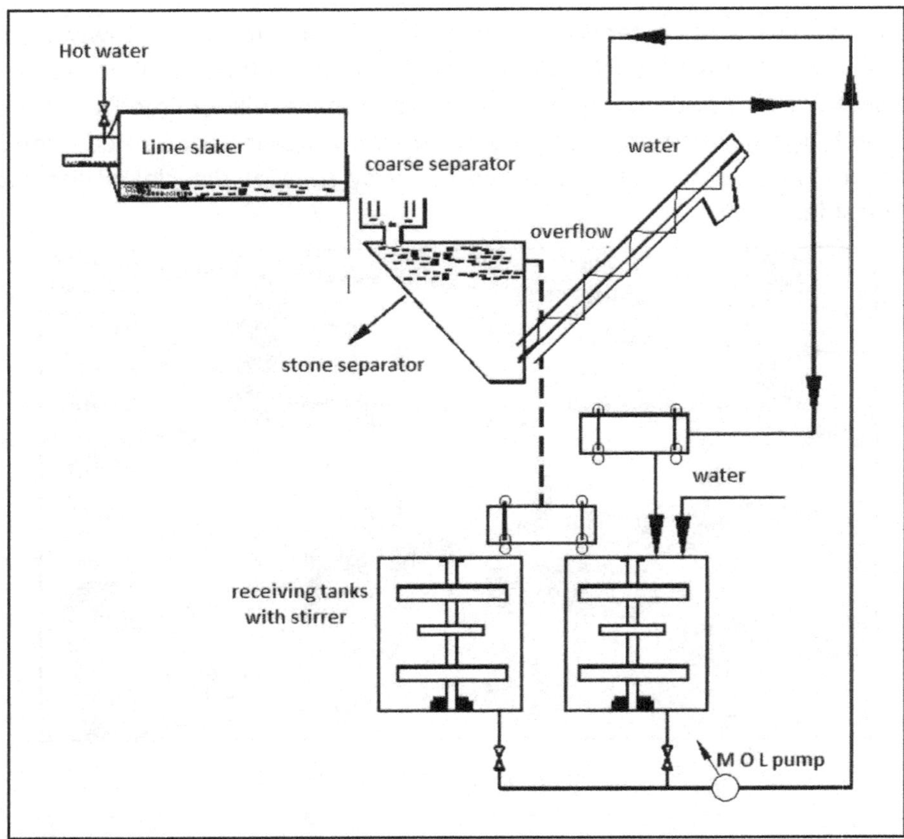

Fig. 35.3 Milk of lime preparation Station

The milk of lime is first passed through a coarse mesh to remove large undissolved pieces or granules. But it still contains small grits and hydrated pieces of lime and coke. Then De-grit of milk of lime is done by equipment called classifier. The milk of lime is passed through a lime classifier. There are different designs of classifier. A scroll conveyor type design is more in use. It is fixed in an inclined position and rotates slowly so that there is no stirring effect. Milk of lime is delivered by an overflow to the storage tank while grit that settles at the bottom is removed by the scroll conveyor. The lime storage tank is a cylindrical tank provided with stirrer that prevents settling of lime slurry at the bottom. The final dilution of the milk of lime is carried in the storage tank.

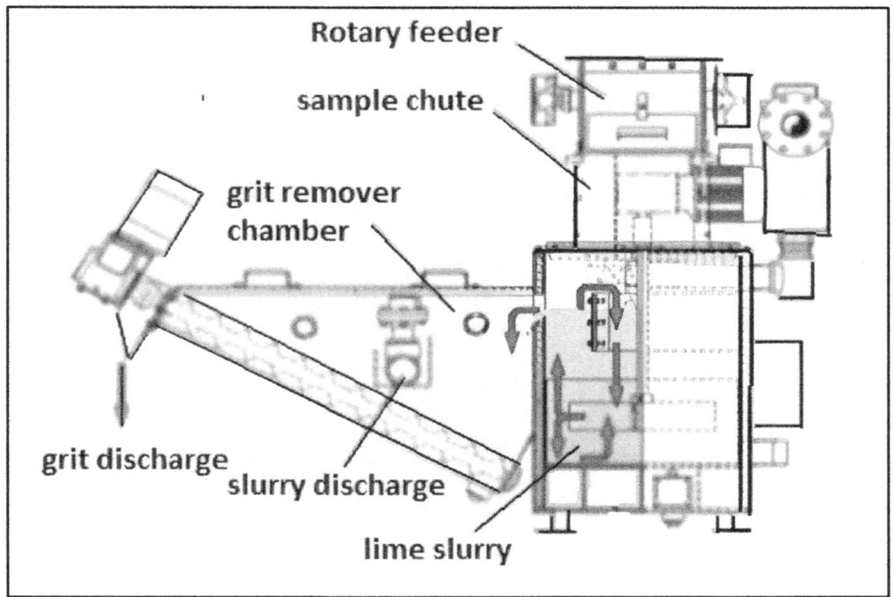

Fig. 35.4 grit separation from milk of lime

Supplementary equipment called hydro-cyclone for effective recycling of coarser particles is also used in some factories and has been observed beneficial.

For 2500 tcd plant a lime slacker capable of slaking about 1200 kg/hr. is provided. The slacker is provided with motor and reduction rear to give 6 to 10 rpm speed.

Generally a Koran flash tank type lime classifier with grit remover or rake type lime classifier is provided.

Two milk of lime storage tanks of about 200 HL capacity each provided with stirrers. A common open gutter of suitable size is provided over the MOL storage tanks. The milk of lime from slacker and return from sulphiter comes in gutter and is taken in the tanks.

Two pumps with grit catchers / strainers (one as standby) each capable of delivering 6 m^3 of milk of lime at 20 m head are provided.

A delivery pipeline is provided from MOL tank to sulphiter, mud over flow tank and to condenser outlet channel.

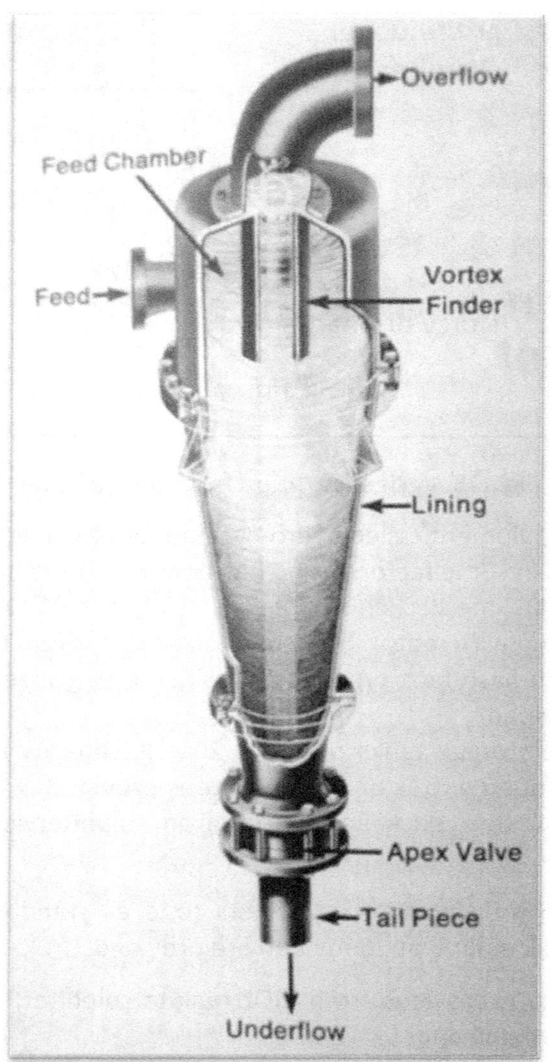

Fig. 35.5 Hydro cyclone - Fine grid separator

-OOOOO-

36 SULPHUR BURNER

36.1 General:

In sulphitation sugar factories sulphur dioxide is generated by burning sulphur. Octahedral sulphur in powdered form is used. It melts at 119°C and boils at 444°C. It sublimes easily. Molten sulphur is dark red in colour above 200°C. The density of sulphur is 2g/cm3. It burns with blue flame. Sulphur dioxide has strong suffocating and irritating odour. Sulphur is soluble in carbon disulphide.

Fig. 36.1 sulphur burner

There are two types of sulphur furnaces closed type and open type. The open type sulphur furnaces are used with Quarez type sulphiter. For closed type burner compressed air of 0.4-0.5 kg/cm^2 g pressure is provided. In the design and operation of the sulphur burner some important points are to be considered:

1. Formation of SO_3 should be minimum
2. Sublimation of sulphur should not occur.
3. The burner operation should be continuous.
4. The SO_2 formation should be regulated.

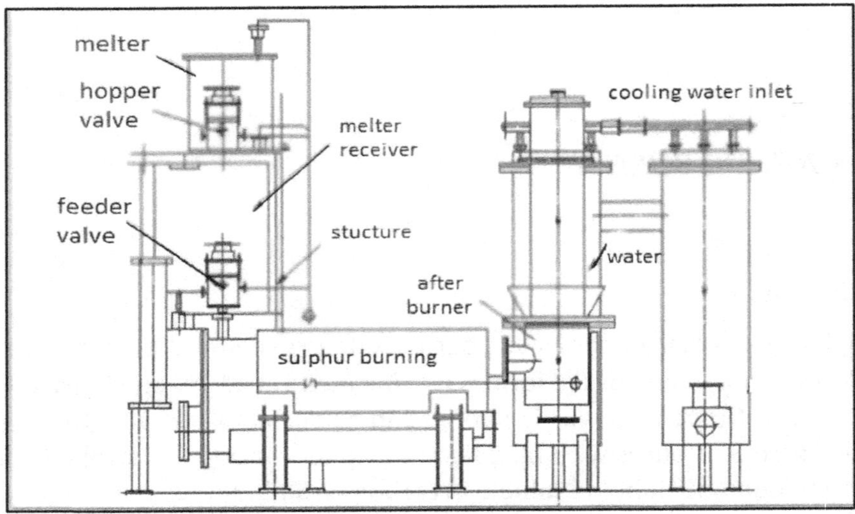

Fig. no. 36.2 continuous sulphur burner

36.2 Combustion of sulphur:

Sulphur dioxide is a gas resulting from the combustion of sulphur. The oxidation of sulphur is a reaction in gaseous state. Therefore the solid sulphur must be melted and evaporated before its molecule reacts with oxygen to form sulphur dioxide gas.

$$S + O_2 = SO_2 + 71.7 \text{ Calories}$$
$$32 + 32 = 64$$

Thus one kg of sulphur requires one kg of oxygen. The reaction release 2217 k calories of heat per kg of sulphur. Combustion takes place at 360-380°C. The air contains 23.15 % of oxygen by weight. Therefore the theoretical air requirement is

100 / 23.15 = 4.3 times the weight of sulphur.

It means to burn one kg of sulphur theoretically 4.3 kg of air is required. When only theoretically required air is provided, the SO_2 content of sulphur burner gas will be 21 %. But it is not be possible to obtain complete combustion of sulphur with the exact amount of air and excess air is necessary to ensure complete combustion. Generally 100 to 110% excess air is provided and the concentration of SO_2 gas in the gas mixture after burning the sulphur usually comes down to 10-12 %. Thus the weight of air is to be used will be about nine times the weight of sulphur.

Formation of small amount of sulphur trioxide cannot be avoided in the sulphur burner. Conversion of SO_2 into SO_3 of about 1-2 % of the burned sulphur can be considered as a low level of SO_3

$$2\ SO_2 + O_2 = 2\ SO_3$$

The disadvantages of SO_3 formation are

- Sucrose in the juice is destroyed by sulphuric acid.
- Because of SO_3 and H_2SO_4 the gas pipeline and the scrubber gets corroded rapidly.
- Any SO_3 or H_2SO_4 coming into the juice is transformed into $CaSO_4$ is fairly soluble and increases lime salt content of clarified juice and that forms scale in evaporator.

High temperature and catalyst like iron oxide promote Sulphur trioxide formation. In order to reduce the SO_3 formation the following steps are to be taken -

The moisture in the air causes the formation of sulphuric acid, which corrodes the piping rapidly. Therefore air supplied to the sulphur burner is dried before going to the sulphur burner. The usual way of drying the air is to pass the air through a layer of unslaked lime. Quick lime is sprayed on perforated trays placed in the cabinet that is provided at the suction side of the air compressor. The air passes through the trays to the compressor. The quick lime absorbs moisture from the air when air passes through it. The quick lime gets saturated of moisture after few hours. Therefore it should be replaced by fresh lime in every shift. This has to be strictly followed where the humidity in the atmosphere is high.

Formation of SO_3 much reduces at temperature. Therefore rapid cooling of the gas mixture below 200°Cis necessary. This cooling is done in vertical outlet pipes, which are surrounded by water jackets.

Sometimes vapours of sulphur without combustion pass from the furnace along with the gas mixture. While cooling the gas these vapours condense to fine particles and deposit on through inside piping and scrubber. This is called sublimation. Sublimation becomes apparent when the air supply is less and percentage of SO_2 is raised to 16 %. Therefore the air supply should always be sufficient in proportion to the sulphur vapour production. In practice sublimation cannot be avoided completely and therefore the gas produced is passed through a scrubber or sublimator.

The rate of vaporization should also be fairly constant. The rate of vaporization of sulphur is proportional to the furnace temperature.

Therefore by controlling temperature of sulphur furnace production of sulphur vapours can be controlled. For this the excess of heat being generated by burning the sulphur is taken out by cooling the sulphur furnace from outside by water. To circulate cold water over the outside surface of the furnace, tray is provided to the surface of the furnace.

A scrubber is a cylindrical tank partly filled with broken bricks. The sublimated sulphur gets deposited over the bricks. The gas piping from furnace to scrubber and the scrubber are water-jacketed to cool down the gas rapidly. The gas temperature near the sulphiter is about 80-90°C.

Air compressor:

For a 2500 tcd plant two compressors of capacity 600m^3 /hr. capacity each are recommended. The delivery pressure of the compressors should be 1 kg/cm^2 (g). An air receiver with adequate capacity of filter is provided to the compressor. A suitable a relief valve is also provided.

36.3 Construction of a Continuous Sulphur Burner:

A closed type sulphur dioxide gas generation station consists of

- Air drying cabinet with trays
- Air compressor
- Sulphur furnace
- Scrubber and cooler

Two separate containers with steam jackets one above the other are provided to the sulphur burner. The upper is called as 'Hopper' and the lower is called as 'Feeder'. The bottom valve of the hopper is closed and sulphur is fed into the hopper. Steam is applied to the hopper and feeder. Sulphur is melted in the hopper by the heat from steam jacket. Bottom valve of the feeder is closed air tight and then the bottom valve of the hopper is slowly opened. The molten sulphur passes down in to the feeder. Then the hopper valve is closed tightly and feeder valve is opened slowly. The molten sulphur from the feeder slowly passes to the vapour production and burning chamber through a drain pipe that is dipped in the molten sulphur. The level of molten sulphur in the burner remains almost constant.

Air from air drier is sucked by compressor and is passed into the furnace from one end. It is spread by baffles over the surface of the molten sulphur. Part of the air - secondary air is provided at the gas outlet of the burner. Working of the burner is continuous and with uniform flow of the gas.

The burner operation and control is not much easy. In the conventional sulphur burner the rate of vaporization of sulphur depends on surface area of molten sulphur coming in contact with air current and temperature of the molten sulphur. The surface area remains constant whereas the temperature of the molten sulphur cannot be increased or decreased rapidly. One way of controlling the vaporization is to control primary air and release extra air to atmosphere.

Table 36.1 Furnace temperature
And % of SO_2 in the gas observed in a factory

Furnace temperature °C	Percent SO_2 in gas
440	5.28
480	8.50
520	11.01
530	11.96
560	13.73

In standard specification three (one as standby) continuous sulphur furnaces of 0.6 m^2 or higher sulphur burning area capable of 70 kg sulphur per hour is recommended for a 2500 tcd plant.

- A lining of refractory bricks / tiles is given inside the burner.
- The furnace is water jacketed.
- Sulphur gas pipelines and crosses are of stainless steel 304. Suitable glass lined rubber diaphragm valves are provided.
- For cooling of gas a counter current arrangement for gas pipes and scrubber is provided.
- The SO_2 gas pipeline provided is of SS 304 grade.

36.4 Film type sulphur burner:

In this sulphur burner molten sulphur in injected and sprayed by pump into a sulphur burning cylindrical tank. Sulphur is melted in a coffin type closed melting vessel. Steam coil with condensate drain is provided to melt the sulphur. Two pumps (one as standby) are provided pump the molten sulphur and inject into the sulphur burning tank.

A steam jacketed delivery pipe carries molten sulphur. The cylindrical vessel is filled with high temperature refractory bricks. The molten sulphur falls from the top on the bricks forming thin film. The sulphur immediately vaporizes and comes in contact with air and burns immediately. The checkered brick work ensures proper mixing of air and sulphur vapours so that complete combustion of the takes place. The gas mixture is withdrawn from the bottom outlet.

A molten sulphur recirculation arrangement is provided to calibrate the sulphur pump outlet and for mill stoppages. The burning rate is regulated according to the final pH of the sulphited juice. To control the burning rate the flow of the quantity of molten sulphur as well as air is controlled.

Fig.no. 36.3 Film type sulphur burner

The molten sulphur, burner temperature and after cooler temperature are continuously digitally displayed on the panel. The burning rate can be accurately maintained without any change in the concentration of SO_2. The automation is fully compatible. Sublimation of sulphur does not occur and no wastage of sulphur. It is not required to start the burner before starting of the crushing. In conventional coffin type sulphur burner the rate of vaporization depends mainly on temperature of the molten sulphur and that cannot be controlled easily within short time.

-OOOOO-

37 MULTIPLE EFFECT EVAPORATOR - THEORY

37.1 General:

After extraction and clarification of juice the next operation in the manufacture of sugar is evaporation. In juice clarification all the suspended matter and mostly all colloidal particles are removed. As far as possible the dissolved nonsugars are precipitated and removed. Thus the clarified juice obtained is clear transparent light in colour and at pH close to neutrality. The clarified juice contains about 84 to 87 percent water and 13 to 16 percent dissolved solids. Now the next operation is to concentrate the clarified juice to convert the sucrose into solid crystal form with impurities remaining in mother liquor or molasses. It is done in two stages

- Evaporation and
- Boiling and Crystallization.

Evaporation: The clarified juice is boiled to evaporate the water content of it and is concentrated to thick juice or syrup. In sugar technology, this process is called as evaporation. A 'Multiple Effect Evaporator' is employed for evaporation process. This is a continuous operation. The multiple effect evaporator set consists 4 to 5 number of evaporator bodies interlinked together to work in series. The sugar percentage in syrup is well below the saturation point therefore crystals do not form in the syrup. The syrup is then send to pan station for crystallization process.

Sugar boiling and crystallization: At pan station, when the syrup is further boiled in vacuum pan, small quantity of water remained, cannot hold all the sucrose in the solution. It means concentration of sucrose in the highly concentrated syrup passes in supersaturation phase. At this time fine microscopic sugar crystals (seed) that act as nuclei to develop crystals are fed to this boiling supersaturated syrup in the pan. The boiling is continued to grow the crystals up to required size. The process is called 'Sugar Boiling and Crystallization' or simply 'Sugar Boiling' and it is carried out in a vessel called 'vacuum pan'. The vacuum pan is also a specially designed evaporator body that works for boiling the thick mass (massecuite) of sugar crystals and mother liquor or molasses.

Vacuum boiling essentiality: Sucrose and reducing sugars in the clarified juice are sensitive to high temperature. Sucrose undergoes inversion or decomposition at high temperatures. The inversion and decomposition of sucrose is a direct loss. Destruction of reducing sugars increase colour formation and also adversely affects the exhaustibility of molasses. Therefore in evaporation operation the juice should remain at high temperatures for minimum time. The evaporation should be carried out at as low temperature as possible. This is done by boiling the juice under reduced pressure / under vacuum because boiling point of any liquid falls down under vacuum. Therefore vacuum boiling is necessary in cane sugar industry.

37.2 Definitions and terminology:

The definitions and description of simple terms that are used in evaporation are common in science and technology. It will be advantageous to go through brief explanation of these terms.

Evaporation: Scientifically Evaporation is the phenomenon in which molecules of liquid on the surface pass into the gaseous state. This takes place continuously at the surface of all exposed liquids. The molecules at the surface have little attraction for the other and tend to fly off into the air or space. The energy required to pass the liquid into gaseous state is being taken by the neighboring molecules. The evaporation increases with temperature; with warm water it is higher. In closed vessel the evaporation continues till vapour phase is saturated with evaporative liquid.

Boiling: Boiling is a rapid evaporation of a liquid which occurs when the liquid is heated to its boiling point – the temperature at which vapour pressure of the liquid is equal to the pressure exerted on the liquid by the surrounding environmental pressure. Boiling occurs throughout the mass of the liquid unlike evaporation that occurs at the surface only. Further boiling can also take place when the vapour phase is saturated.

The boiling point of liquid varies depending upon pressure over it. Boiling point at atmospheric pressure is called as normal boiling point, or atmospheric pressure boiling point or atmospheric boiling point. The liquid under vacuum has lower boiling point than atmospheric boiling point.

Latent heat: When water or any liquid passes in to the form of vapour a certain quantity of heat energy is required to make this change. This quantity of heat is called as latent heat of evaporation. To convert one kg of water at 100°C into steam at 100°C requires 539 kilocalories (kcal) of heat. In this process the temperature remains same but phase or state of material is changed, i.e. from water to steam. As there is no change in temperature, it means the heat taken is hidden. This quantity of heat to pass the molecule from liquid state to gaseous state is called as latent heat of evaporation. Conversely when steam is condensed to water it gives out the heat which was required to convert the water to steam.

Sensible heat: Sensible heat is heat exchanged by a body or thermodynamic system that has as its sole effect a change of temperature. It is the energy that is indicated by thermometer. The heat energy required to raise the temperature of one gram of water through one degree Celsius is called one calorie and to raise the temperature of one kg of water through one degree is called one kilocalorie. Sensible heat and latent heat are not special form of energy. Rather they describe the exchange of heat under conditions specified in terms of their effect on a material or a thermodynamic system.

Total heat: A quantity often used in calculations is the total heat of steam. This means the quantity of heat required to convert one kg of water from freezing point (0°C) to steam at a given temperature.

Pressure: pressure is force per unit area applied in a direction perpendicular to the surface.

Atmospheric pressure: The natural pressure of the atmosphere is 1.034 kg/sq. cm. and barometric height of 0.76 m of mercury (Hg). A manometer is an instrument that is to measure the atmospheric pressure. Nowadays different pressure measurement instruments are available.

Absolute pressure: The absolute pressure is zero referenced to perfect vacuum so it is equal to the gauge pressure plus atmospheric pressure.

Gauge pressure: Gauge pressure is zero referenced against ambient air pressure. So it is equal to absolute pressure minus atmospheric pressure.

Negative pressure or vacuum: When the pressure is less than atmospheric pressure is appended with the word vacuum and the gauge vacuum gauge.

Differential Pressure: Differential pressure is the difference in pressure between two points.

Wet steam: when the steam contains some droplets of water it is called as wet steam. In the first stage the steam generated contains some water droplets.

Dryness factor of steam: The dryness factor indicates the water content of the steam as for instance dryness fraction of 0.98 indicates 2% water content of steam.

Dry saturated steam: when the wet steam is further heated in boilers the water content of the steam is completely vaporized and the steam becomes dry. Ideally the dry saturated steam is at the boiling temperature of a given pressure.

Superheated steam: when the dry saturated steam is further heated its temperature increases above the saturation temperature to the given pressure, thus the steam becomes supersaturated or super-heated. The superheated steam as it is free from any traces of moisture is required to use for turbines.

Coefficient of heat transfer: Coefficient of Heat transfer is the heat transmitted per unit heating surface per degree difference in temperature per hour (kcal/m^2/°C/h).it is also called as 'coefficient of transmission of heat' or 'heat transfer coefficient'.

The specific evaporation coefficient: The specific evaporation coefficient is the weight of vapours produced by unit area of heating surface per degree temperature difference in heating medium and the juice per hour. It is expressed as kg/m^2/°C/hr.

The evaporation coefficient: It is the weight vapours produced per unit area of heating surface per hour (kg/m^2/hr.). It is also called as specific evaporation, however in factory working it is known as evaporation rate.

37.3 Multiple Effect Evaporation:

Multiple Effect Evaporator' is a set of specially designed 4 to 5 vessels interlinked from first vessel (body or effect) to the last vessel. When juice is boiled in the first effect, the vapours produced are passed by vapour pipe to second effect to boil the juice in the second effect and so on. This is possible because the following vessel is under lower absolute pressure.

Table 37.1 Relation between Pressure upon the Water and its Boiling Point

Gauge pressure kg/cm^2	Absolute pressure Kg/cm^2	Boiling temperature of water °C	Vacuum in cm of Hg	Absolute pressure Kg/cm^2	Boiling temperature of water °C
1.0	2.034	120.5	5	0.965	98.1
0.9	1.934	118.6	10	0.897	96.1
0.8	1.834	116.7	15	0.830	94.0
0.7	1.734	115.1	20	0.761	91.7
0.6	1.634	113.4	25	0.693	89.2
0.5	1.534	111.4	30	0.625	86.5
0.4	1.434	109.3	35	0.557	83.6
0.3	1.334	107.4	40	0.489	80.3
0.2	1.234	105	45	0.421	76.7
0.1	1.134	102.8	50	0.353	72.5
0.0	1.034	100	55	0.285	67.5
			60	0.218	61.5
			65	0.150	53.5
			66	0.136	51.6
			67	0.122	49.4

Table 37.2 steam pressures and super-heated steam temperatures Commonly used in sugar industry

Boiler Pressure	Saturation temperature of steam °C	Temperature of superheated steam °C
21 kg/cm^2	215	300
32 kg/cm^2	240	380
45 kg/cm^2	260	445
67 kg/cm^2	285	480
86 kg/cm^2	300	510

Boiling point of water: The boiling point of water / juice depends on the pressure over it. Water/ juice boiling point goes on decreasing as the pressure over it decreases. For example at atmospheric pressure water boils at 100°C but when it is boiled under vacuum of 650 mm of Hg it boils at 53.5°C only. This property is used in operating the 'Multiple effect evaporator'.

Vacuum at evaporator set: To operate the set, vacuum of about 630 to 660 mm of Hg (or 0.13-0.18 kg/cm^2 absolute pressure) is created in the last vessel by using condenser. The absolute pressure difference from the first vessel to the last vessel is equally distributed in decreasing order from the first effect to the last effect.

Fig. 37.1 Multiple Effect Evaporator

When the following vessel works under lower pressure than the preceding vessel, then the juice in following vessel boils at temperature lower than the temperature of the vapours from the preceding vessel. Therefore the temperature difference required between the vapours from each vessel and the boiling juice in the following vessel is obtained. As a result the vapours produced from the juice are used to concentrate the same juice in the following vessel. The syrup boiling temperature in the last vessel is only 57 60°C.

The clear juice entering the first vessel is boiled in it by the exhaust steam applied. The juice is concentrated and passes to the second vessel. Vapours produced in the first vessel pass by vapour pipe to the vapour chest drum or calandria of the second vessel. The second vessel works at lower pressure than the first vessel. Therefore here the juice boils at lower temperature. The vapours coming from the first vessel boils the juice in the second vessel and the juice is further concentrated. In this way the juice gets progressively concentrated from the first to the last vessel. It should be noted that from the second effect to the last effect juice enters in each body at a temperature above the boiling point corresponding to the vapour pressure in the vapour space of that body.

The concentrated juice – thick juice or syrup is withdrawn from the last vessel. In this operation the exhaust steam coming from the prime movers is used in the first vessel only and vapours from the juice are used in the following vessels. In this way large quantity of steam can be saved. Thus by boiling the juice under vacuum following advantages are achieved.

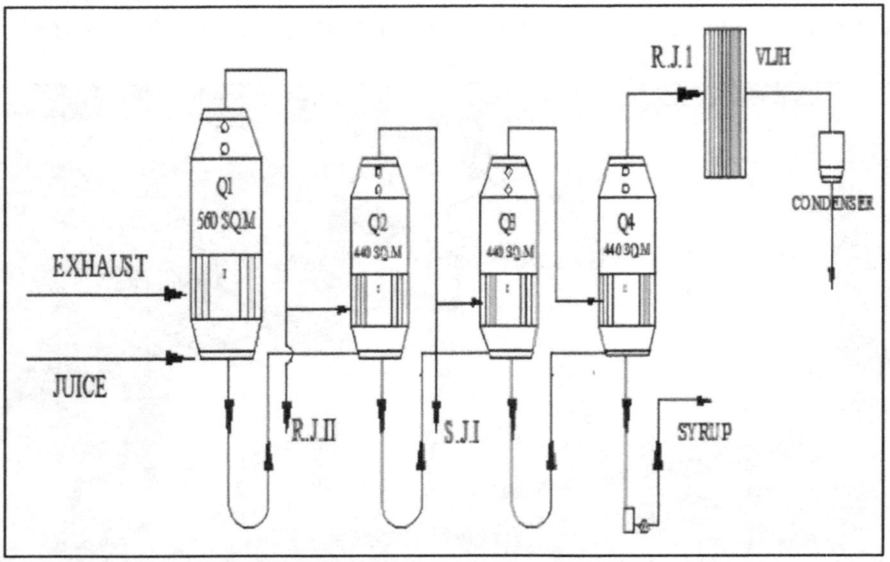

Fig. 37.2 Schematic diagram of evaporator set

The required temperature difference between the heating vapours and the boiling juice in the successive vessel is obtained. Therefore steam economy can be achieved.

The boiling temperature of juice goes on decreasing from the first vessel to the last vessel. Therefore the rates of inversion of sucrose, destruction of reducing sugars and colour formation in the juice goes on decreasing from first vessel to last vessel. It is almost negligible in the last two vessels.

37.4 Norbert Rillieux Principle:

Norbert Rillieux has invented the multiple effect evaporator. Rillieux has not written any papers, however he developed following three general principles applicable to multiple effect evaporator.

1) In multiple effect evaporator one tonne of steam will evaporate as many tonnes of water as there are effects. For example by use of one t of exhaust steam in quadruple 4 t of water will be evaporated and in quintuple 5 t.

2) If vapours are withdrawn from a given effect of multiple effect evaporator and used outside the evaporator system in place of steam then,

The steam saving will be

$$\frac{N \times W}{M}$$

Where

N = Number of vapours withdrawal effect
W = Weight of vapours withdraw
M = Total number of effects

For example if 2.5 tonnes of vapours withdrawn per hour from third effect of a quintuple set and used in place of exhaust for heating raw juice, then the saving in exhaust steam will be

Steam saving = 3 X 2.5 / 5 = 1.5 t / h.

3) The steam / vapours get condensed. But the incondensable gas or air that is with steam / vapour does gets condensed. The condensate as well as incondensable gases/air must continuously be removed so that they will not get accumulated in the steam / vapour chest.

37.5 Highest limit for syrup brix:

The syrup from the last effect of evaporator is of purity 78 to 86 and temperature 57-60°C. In this range of purity and temperature the syrup becomes saturated of sucrose at about 77° to 82° brix and spontaneous crystallization of sugar starts at about 82° to 86° brix. Hence from the point of view of steam economy the juice may be concentrated up to 75 to 78 Brix. But in practice in factory working it is not possible to concentrate the syrup to such a high brixes for the following reasons.

At higher brixes if the evaporation operation could not be controlled the grain formation may start in the syrup itself.

Scale deposition at the evaporator is so rapid that the heating surfaces foul in only few days. In such case standby evaporator set will be needed.

In sulphitation process syrup is sulphited. The absorption of SO_2 gas at brixes higher than 60 is poor.

With lower brixes of syrup, the circulation of massecuite in vacuum pan with natural circulation remains normal. Therefore size and shape of the crystals remain more regular.

The false grain if any formed in the sugar boiling gets dissolved immediately upon feeding the syrup of brix 60-65 to the boiling massecuite. With very high brix of syrup it is not possible.

The higher brix syrup may produce crystals of irregular shape and size. With higher brix of syrup the possibility formation of false grain increases.

Therefore the syrup brix in raw sugar factories is kept to about 63-65° brix. It general observation that at higher brix, syrup does not get sulphited satisfactorily up to a desired pH. Therefore in factories where double sulphitation process is followed, the syrup brix is kept about 56-60° brix.

37.6 Quantity of water to be evaporated in evaporator:

Let us assume

E = Weight of water to be evaporated % cane
J = Clear juice % Cane
S = Syrup % Cane
Bj = Brix % Clear Juice
Bs = Brix % Syrup

In multiple effect evaporator - or generally called as simply evaporator, water is evaporated and the solids in clear juice pass to syrup. Some of the non-sugars deposit in the form of scale on the evaporator tubes. Some of the volatile substances get evaporated. However these quantities are much less and hence are neglected in calculating quantity of water to be evaporated. It means solids in clear juice and solids in syrup remain same or in factory working we can say the tonnes of brix in the juice and syrup remain the same.

Therefore

$$Bj \times J / 100 = Bs \times S / 100$$
$$\therefore S = Bj \times J / Bs$$
$$\text{Water to be evaporated } E = J - S$$

Putting value of 'S' we get

$$E = J - (Bj \times J / Bs)$$
$$E = J (Bs - Bj) / Bs$$

Example: If J = 105 % on cane, Bj = 14.00 and Bs = 62 .00

Then $E = 105 (62 - 14) / 62$

$E = 81.29$ % on cane

In factories clear juice is not weighed, but it is almost equal to mixed juice % cane.

Therefore for calculation purpose clear juice % cane is assumed as equal to mixed juice % cane. Here in the above calculation the clear juice % cane is taken as 105% on cane. It may vary from 95 to 110% depending upon the imbibition water % cane used at mill and accordingly water to be evaporated will vary.

37.7 Upper and lower limit of juice boiling temperature:

The reducing sugars and sucrose are sensitive to temperature and therefore the juice should not be boiled at higher temperature for longer time. However to achieve maximum possible steam economy by profuse vapour bleeding, the juice boiling temperature in the first vessel may be as high as 115-118°C. This juice boiling temperature corresponds to about 0.8 kg/cm^2 (12 psig) vapour pressure in the first effect. Hence the maximum exhaust pressure should be about 1.2 to 1.3 kg/cm^2 (18 - 19 psig) and superheated temperature 140°C ± 10°C.

In the operation of multiple effect evaporator, vacuum in last effect is generally maintained up to 630-660 mm of Hg. At this vacuum water boils at about 53-56°C. With increase in temperature due to dissolved solids in the juice or brix of the juice, by 2.0-2.5°C (at average syrup brix 50-56°) the syrup boiling temperature in the last effect at the top of the tubes will be about 55- 59° C. At the lower side of the tubes it will be higher due to hydrostatic head.

37.8 Heat transfer in evaporator:

Evaporator vessel consists of seam / vapour chest drum or calandria having tubes that serves as heat exchanger. The heating steam is outside the tubes and the boiling juice passes through tubes. The general expression for heat transfer can be given as, When two fluids of temperature difference ΔT are located on opposite sides of surface or tubes, the quantity of heat transmitted from one to other is given by the equation –

$$Q = UA\,\Delta T$$

Where

Q = Quantity of heat transferred

U = Coefficient of heat transfer from steam / vapour to boiling juice

A = Heating Surface Area of the tubes

ΔT = Temperature difference in heating medium (Steam/ vapour) and the boiling juice

The coefficient of heat transfer will depend upon the followings

- Temperature of the steam or vapour

- Condensate film around the tube
- Scale outside the tube
- Metal of tube and thickness of the tube
- Scale inside the tube
- Juice film inside the tube

Steam / vapour temperature: The coefficient of heat transfer depends on the steam / vapour temperature. Classen has showed that at higher temperature the coefficient of transfer of heat is much higher if we take two cases- In one case temperature on one side is 115°C and the other side 114°C. In second case on one side 60°C and other side 59°C. Then the coefficient of transfer of heat in the first case will be much higher. According to Classen by increasing high vacuum above 60 cm (23.7 in), the decrease in boiling temperature also decreases the coefficient of transmission of heat and hence the benefit by increasing the temperature difference is neutralized.

Condensate film around the tubes: Heat transfer coefficient of condensate is only 0.0019 kcal/m^2/°C/hr., which is almost equal to scale. However the film of condensate is very thin. Equipment with horizontal tubes gives better heat transfer as the condensate drops off from the tubes directly. Some 80 years back horizontal juice heaters were being commonly used in sugar industry.

Evaporator sets with horizontal tubes were also being used in the beginning of the sugar industry. But practically using equipment with horizontal tube is complicated. Cleaning of the horizontal tube is also difficult and hence using equipment with horizontal tubes is almost stopped in the industry.

Thickness of metal: The coefficient of heat transfer decreases with the thickness of the tubes. But the thickness does not vary much. Therefore in practice the change is negligible.

Metals of the tubes: Silver and copper are best conductors of heat. In practice copper tubes were being used some 80 years back however now brass and stainless steel are used. If conductivity of copper is considered as unit, then conductivity of various materials is as under-

Scale deposited: The scale deposited on the heating surface is most adversely affecting material layer for the transfer of heat. In Australia tests were made for thermal conductivity of scale from evaporator. The average value of heat transfer coefficient was found 0.45 kcal/m^2/°C/m.

Film of juice: The thin film of juice inside of the tube also has an effect on the heat transfer from the vapour to the juice.

Distance of particle from heating surface: The average distance between the juice particle and the heating surface also have an effect on heat transfer. Therefore with smaller diameter tubes heat transfer is better. In case of falling film evaporator bodies in which juice falls in the form film have better heat transfer coefficient.

Table 37.3 Comparative thermal conductivity of materials (When thermal conductivity of copper is considered as unit)

Comparative thermal conductivity	
Cooper	1.000
Aluminium	0.546
Brass	0.275
Steel	0.118
Water	0.0019
Air	0.00008
Scale	0.002

Velocity of juice / Circulation of juice: The table given Prokong has studied the effect of velocity of juice on the heat transfer coefficient. The data is as follows

Table 37.4 Coefficient of heat transfer for Juice Heating

Velocity of juice	Coefficient of heat transfer
m/sec	Kcal/m^2/°C/hr.
1.29	490
1.35	975
1.523	1060

From the above table it can be seen that the coefficient of heat transfer at juice heaters varies widely upon velocity of the juice. With lower velocity the coefficient of heat transfer falls rapidly. Juice heaters are designed to keep the velocity of the juice in heaters about 2.0 to 2.2 m/sec or higher. In factory working the coefficient of heat transfer also varies upon the scale deposited in the tubes.

Similarly at evaporator with circulation pump, velocity of juice close the heating surface can be increased by circulation of the juice in the bodies. So that heat transfer at evaporator can be improved. The system was being followed in some old factories However at present such system is not commonly used.

Viscosity of the juice: The coefficient of heat transfer is also adversely affected by viscosity of the juice. The viscosity of the juice increases with increases in brix. In the last body the high brix juice is of higher viscosity. It retards transfer of heat. Therefore heat transfer coefficient in last bodies is low.

When the heating surface are clean the heat transfer coefficient is high however as evaporator operation is continued and days pass, slowly scale gets deposited over the inside surface of the tubes. The scale as its heat transfer coefficient is very low adversely affects the overall heat transfer coefficient. Therefore there is wide variation in the coefficient of heat transfer reported by expert.

Jelinek and Classen give separately the heat transfer coefficient for quadruple under normal working conditions as follow:

Table 37.5 heat transfer coefficient for quadruple

Effect	**Coefficient of heat transfer**
	Kcal/m²/°C/hr.
Semikestner/FFE working as pre-	3,000
First	1680 / 2700
Second	1560 / 1800
Third	1200
Fourth	360/720
Sugar boiling and crystallization	
Graining and High grade massecuite	600/1080
Low grade massecuite boiling	300/420
Final tightening	180/420

Webre gives coefficients of heat transfer for Evaporator Set ranging from 6800 Kcal/m^2/°C/hr. for the first effect heated by steam at 115°C to 2200 Kcal/m^2/°C/hr. for the last effect heated by vapours of 65°C.

The temperature difference in steam/vapours and boiling juice: while designing an evaporator set, differential pressure in between the calandria and the vapour space of each body is considered and temperature difference in between the steam/ vapours and boiling juice for each effect is calculated.

37.9 Exhaust steam conditions:

Low-pressure steam produced by the prime movers called exhausted steam, exhaust steam or simply exhaust is used to heat, boil and concentrate the juice and massecuite in the boiling house. In most of the old factories the steam turbines installed are designed to withstand at the exhaust steam backpressure of 1 kg /cm^2 and the safety valves set at backpressure 0.9 kg/cm^2. These factories work at the exhaust pressure 0.5 -0.8 kg/cm^2 (7.0 to 12 psig). In modern sugar factories (erected in India after 1985) turbines installed are designed to work at 1.5 kg/cm^2 backpressure and these factories works at 0.9 to 1.2 kg/cm^2 back pressure.

In factory working the ideal situation is that the live or high pressure steam consumption of the prime movers and the exhaust steam demand in the boiling house are equal. This situation is now automatically controlled in sugar factories with cogeneration plant where the exhaust pressure is maintained constant by controlling the exhaust production from the extraction turbine. The remaining live steam in the turbine passes to surface condenser. In remaining factories where only back pressure turbines are in operation both these quantities vary in factory operation.

When exhaust production is more than its demand in the boiling house, the excess exhaust steam has to be blown off to atmosphere. In such situation lot of heat is lost. Further the exhaust condensate quantity is reduced. Hence more vapour condensate is to be diverted to boilers. To avoid such situation the exhaust production should be about 80 to 85 % of the exhaust demand and the balanced 15 -20 % demand is to be fulfilled by bleeding high-pressure live steam in the exhaust steam.

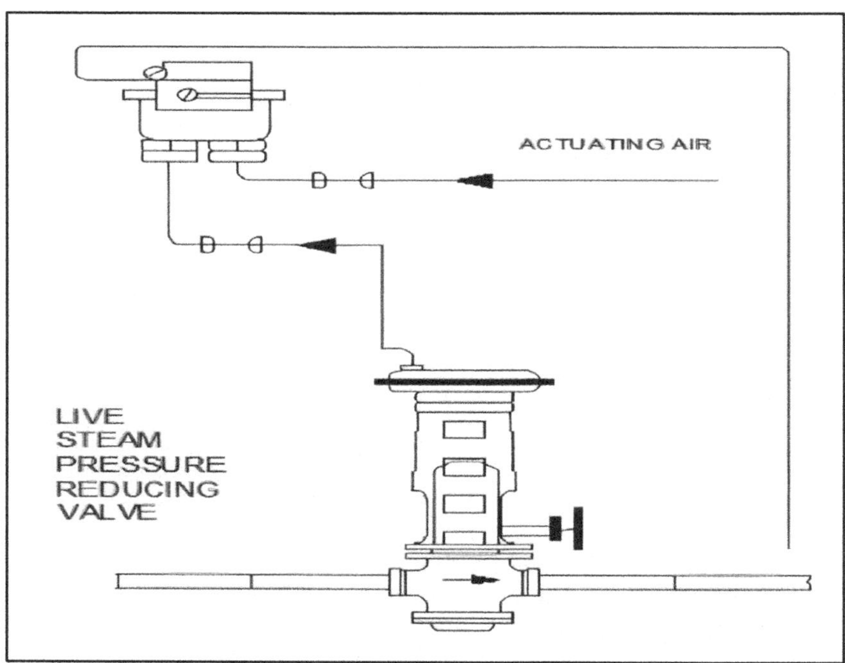

Fig.37.3 Automatic Pressure Reducing Valve.

The exhaust produced by the prime movers is generally of temperature 150-180°C and is supersaturated. Further on bleeding the live steam into exhaust, the exhaust becomes highly supersaturated. The supersaturated steam is much worse conductor of heat than saturated steam. (Therefore loss of heat in the pipe with supersaturated steam is much less than saturated steam.) But for this property the highly supersaturated steam cannot be satisfactorily used at evaporator where high heat transfer rate and rapid condensation is required.

Heat transfer coefficient of saturated steam is high. Supersaturated steam up to 10 -30°C also gives good performance at evaporator. Therefore the exhaust mixed with the bleeded live steam should be desuperheated up to 10-30°C above the saturation temperature. In practice a set point of automation for the exhaust temperature should be about 15°C higher than the saturation temperature. For e g the exhaust steam at pressure 1 kg /cm^2 has saturation temperature 120°C. It should be desuperheated up to 135° ± 5°C.

To bleed the live steam as make up steam demand in boiling house and to desuperheat the exhaust, a pressure reducing and desuperheating system (PRDS) is installed. In PRDS the live steam is bled and condensate water sprayed is mixed with the exhaust steam to maintain the exhaust steam pressure and temperature at a desired level.

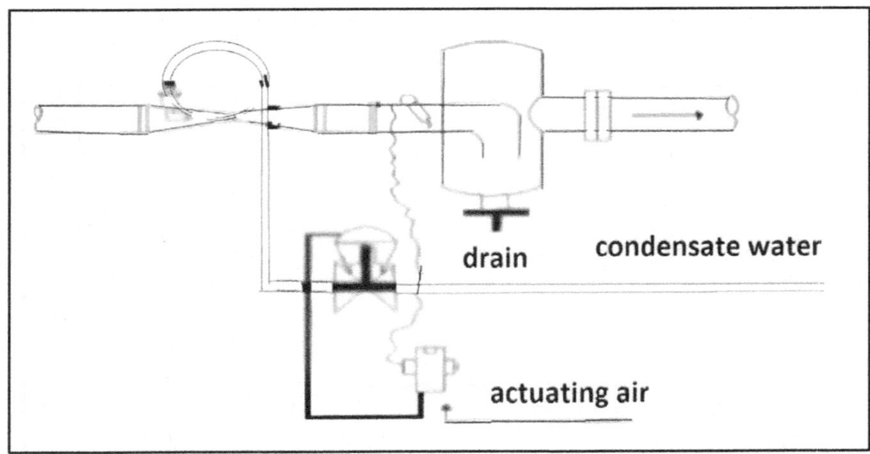

Fig 37.4 Pressure reducing and desuperheating

Automation of PRDS helps in maintaining the parameters within tolerable limits. One more advantage of bringing down the exhaust temperature is that the destruction of reducing sugar due to high temperature of the exhaust is reduced hence colour formation and increasing viscosity is avoided to some extent.

37.10 Boiling Point of Juice:

Boiling point of juice depends on

1) Vapour pressure above the juice in the evaporator body;
2) Brix of the juice;

3) Hydrostatic head pressure of juice column upon the juice.

Vapour pressure above the juice:

Boiling point of water / liquid depends upon atmospheric pressure or the vapour pressure (in case of closed vessel) over it. When vacuum is maintain in a closed vessel, the boiling point of water / liquid is reduced. At atmospheric pressure pure water boils at 100°C, however when water boiled in a closed vessel in which the vacuum of about 300 mm of Hg is maintained, the water boils at 85.5°C. This property of liquids is used in the operation of multiple effect evaporator. Vapours formed from the juice used to boil the same juice under vacuum at lower temperature in the next vessel. The steam table gives relation between pressure/vacuum and water boiling temperature/dry water vapour temperature.

Elevation of Boiling Point due to dissolved solid:

Boiling point of water is increased due to dissolved solids in it. Boiling point of juice under a given pressure is higher than water. For example under atmospheric pressure water boils at 100°C but clear juice boils at 100.25°C. This increase in boiling point is known as elevation of boiling point (e_1) or boiling point elevation (BPE) or rise in boiling point (BPR). The BPR increases with increase in brix of the juice / syrup. The BPR may be different for different compounds of the same concentration. Therefore BPR for juice / syrup depends on concentration of sucrose as well as concentration of impurities. The BPR is higher for low purities juices / syrup. As the boiling temperature changes with atmospheric pressure over it, the BPR also changes but it is negligible.

The following table gives elevation of boiling point (°C) of sugar solution, cane juice and molasses of different purities at atmospheric pressure (by Classen and Thieme). When juice of purity 80 and an average brix of 20, boiling at atmospheric pressure (as the case may be in the second effect of evaporator set) as per above table the e_1 (°C) will be 0.3°C. However by formula given above it is 0.5°C. It indicates that the formula gives slight variation from e_1 (°C) given in the table. However the difference is little and can be overlooked.

At different pressures as the boiling temperature of pure water changes, the BPR for juices also changes but the change is very small and can be neglected.

Table 37.6 Elevation of boiling point (e_1) in °C Due to dissolved solids at atmospheric temperature

Purity › Brix ⩡	100	90	80	70	60	50
15	0.2	0.2	0.2	0.2	0.3	0.3
20	0.3	0.3	0.3	0,4	0.4	0.5
25	0.4	0.5	0.5	0.6	0.7	0.8
30	0.6	0.7	0.7	0.8	1.0	1.1
35	0.8	0.9	1.0	1.1	1.3	1.4
40	1.0	1.1	1.3	1,5	1.7	1.9
45	1.4	1.5	1.8	2.0	2.2	2.4
50	1.8	1.9	2.2	2.5	2.8	3.1
55	2.3	2.5	2.8	3.1	3.5	3.9
60	3.0	3.2	3.6	4.0	4.4	4.9
65	3.8	4.1	4.5	4.9	5.5	6.0
70	5.1	5.5	6.0	6.5	7.1	7.7
75	7.0	7.5	8.0	8.6	9.4	10.1
80	9.4	10.0	10.5	11.3	12.3	13.1
85	13.0	13.7	14.4	15.3	16.4	17.4

Hugot[1] gives the following formula to calculate e_1 (°C) for juice / syrup of high purities at different brixes at atmospheric pressure.

$$e_1\ (°C) = 2\ B/\ (100 - B)$$

37.11 Effect of Hydrostatic Head:

Boiling point of liquid increases with increase in pressure above it. Now suppose juice particles are boiling at a certain depth below the juice level, then the total pressure exerted over the boiling juice particles will be equal to the vapour pressure above the juice plus the pressure of the juice column (hydrostatic head) over the juice particles. Thus the total pressure exerted over the boiling particles is increased due to the hydrostatic head and hence the boiling point of these particles is also increased. Hence the juice particles at different depths boil at different temperatures.

This increase in boiling point of juice due to hydrostatic head decreases the temperature difference (ΔT) in boiling juice and heating vapours and adversely affects the rate of evaporation. Therefore while calculating the heating surface, the increase in boiling point of juice has to be taken into consideration. In the operation of evaporator the incoming juice is at a higher temperature and therefore flashes even before it reaches the bottom of the calandria. The juice is agitating continuously. Therefore the effect of hydrostatic head is fluctuating. Here in the following table a case of quintuple is considered. The first effect is semikestner and the next Robert type evaporator bodies.

Table 37.7 Calculation of elevation boiling point (e_2) due to hydrostatic head

	Evaporator Body	Q 1	Q 2	Q 3	Q 4	Q 5
		Semi-Kestner	Robert type	Robert type	Robert type	Robert type
1	Ab. Pressure in vapour space Kg/cm^2	1.66	1.29	0.92	0.55	0.18
2	Juice inlet brix	14.30	22.30	38.53	49.54	54.30
3	Water evaporated kg/hr.	37660	28460	8560	2660	2660
4	Juice outlet brix	22.30	38.63	49.54	54,30	60.08
5	Average juice brix	18.30	30.47	44.09	51.92	57.19
6	Height of calandria tubes m.	4.0	2.0	2.0	2.0	2.0
7	Height of juice column in the tube m	0.8	0.6	0.6	0.6	0.6
8	Density of juice* for average juice brix	1.008	1.076	1.149	1.201	1.249
9	Weight of juice column per sq. cm. (gms) Height in cms X density	80.64	64.56	68.94	72.06	74.94
10	Hydrostatic head of juice column kg/cm^2	0.0806	0.0646	0.0689	0.0721	0.0749
11	Total ab. pressure at the bottom of the tube kg/cm^2	1.7406	1.3546	0,9889	0.6221	0.2549
12	Vapour temperature °C in the vapour space	113.84	106.33	96.78	83.24	57.41
13	Water boiling temperature at the bottom of the tube	115.46	107.76	98.77	86.38	65.00
14	Increase in boiling point due to hydrostatic head (e_2)	1.62	1.43	1.99	3.14	7.59

*The densities of juices taken in the above table are at corresponding boiling temperatures for more accuracy. However the densities of juices given in the following two tables are at normal juice temperatures.

Now in modern sugar factories of higher capacities falling film evaporator in which hydrostatic head is totally absent as juice falls from top to the bottom are being preferably installed.

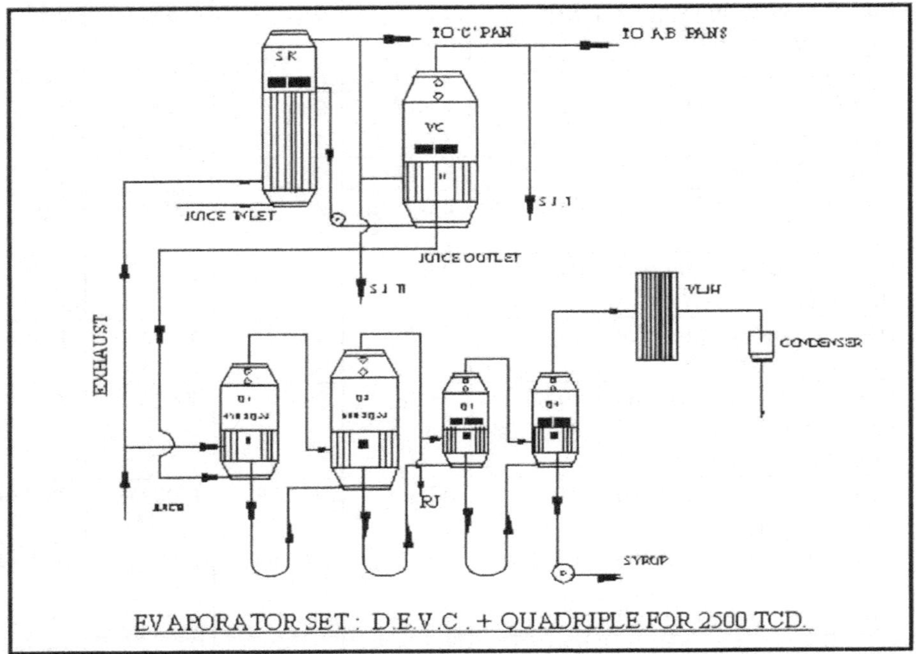

EVAPORATOR SET : D.E.V.C. + QUADRIPLE FOR 2500 TCD.

DEVC + quadruple recommended in India 1982 specifications

-OOOOO-

38 EVAPORATOR - CONSTRUCTION

38.1 General:

There are three types of evaporator vessels or bodies used in sugar industry

- Conventional - Robert type evaporator body,
- Long tube vertical rising film evaporator (LTVRF); commonly called as Kestner or Semikestner;
- Long tube vertical Falling film evaporator (LTVFFE or simply FFE).

According to number of effects connected in series the multiple effect evaporator set is called as triple, quadruple or quintuple effect etc.

38.2 Construction of Robert type evaporator body:

Robert type evaporator is a closed vertical cylindrical type body having saucer bottom and a dome at the top. The body is fabricated of mild steel.

1) Calandria: Steam drum or Calandria is lower portion of the body below which there is a bottom saucer. The calandria has two horizontal tube plates one at the top and the other at the bottom. These tube plates are m s plates (of 25 -50 mm thick depending upon size of the vessel) with large number of holes drilled to fit tubes in it.

Brass or stainless steel tubes of 2 to 2.25 m height and of 35 to 42 mm inner diameters are fitted vertically in these tube plates at both the ends by expanding the ends and are made leak proof. The length of the tube is kept 8 mm greater than the outer distance between the tube plates so that the tubes can be fitted properly in the tube plates. The thickness of the tubes varies from 1.5 mm to 2.0 mm.

Generally brass tubes are used for Robert type evaporator bodies. The tubes are annealed at the ends to prevent cracking after expansion. These tubes work as heating surface of the body. Generally length and diameter of the tubes are kept same for all evaporator bodies. When smaller diameter tubes are used juice velocity in the tubes remains higher. The mean distance between the juice particles and the heating surface is less. Therefore the coefficient of heat transfer is better. With smaller diameter tubes, the vessel diameter also can be reduced. But cleaning of smaller diameter tubes is very difficult. Therefore tubes below 35 mm diameter are not recommended.

According to Indian Standard Specifications for sugar machinery, the recommended height of the calandria for Robert type evaporator body is 2 m and the I D and O D of the brass tubes are 42 mm and 45 mm respectively. The brass tubes are of 70% copper and 30% zinc, or 70% copper 29% zinc and 1% tin.

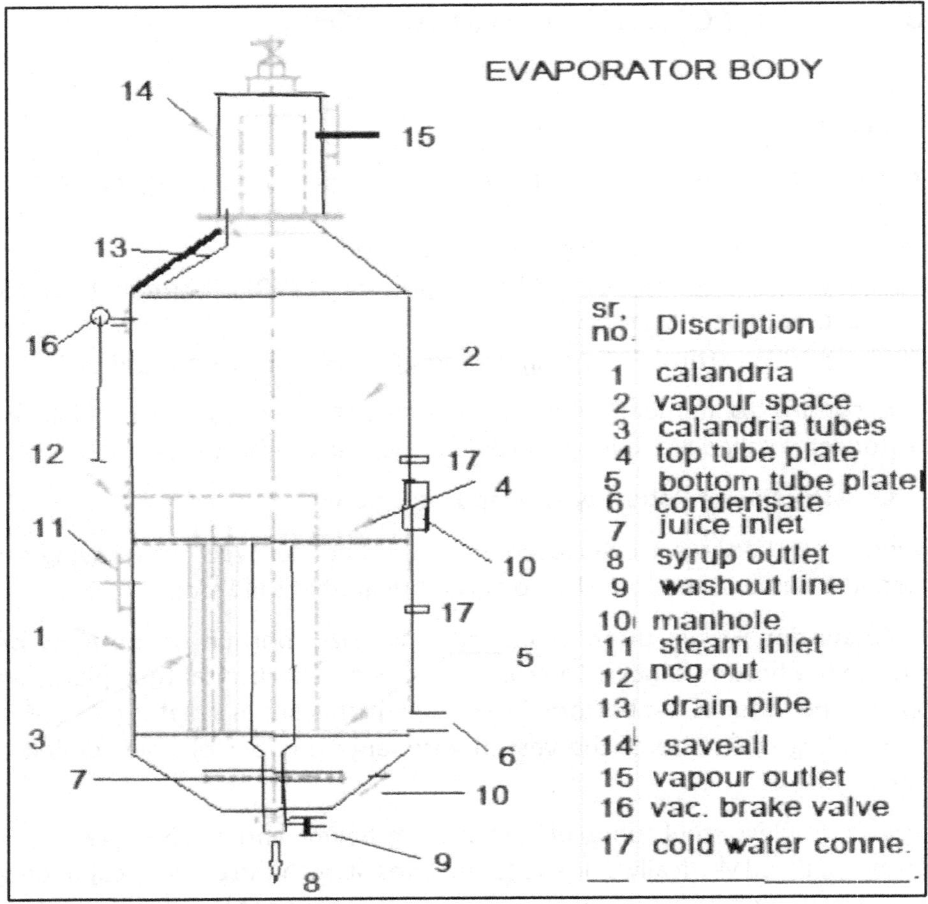

Fig.38.1 Robert type Evaporator Body

Annealed Stainless steel tubes confirming IS 13316 of 45 mm O D and 1.2 mm tube thickness are also recommended for Robert type evaporator bodies. However stainless steel tubes are generally used for semikestners and falling film evaporators.

The minimum distance between to two neighbouring holes of the tube plate is called as ligament. It is kept 10 mm. The distance between the centres of two neighboring tubes is called as pitch. The tubes are fitted in the tube plate in staggered arrangement. With this arrangement maximum number of tubes can be fitted per unit area for a given tube diameter and ligament between the tubes.

In between calandria and the bottom saucer, a collar of about 100 mm is kept. It is required to expand the tubes to fit the same to the bottom tube plate. The bottom saucer is fitted by bolts or welded to the calandria. The bottom saucer has juice inlet and in most of the designs outlet connections.

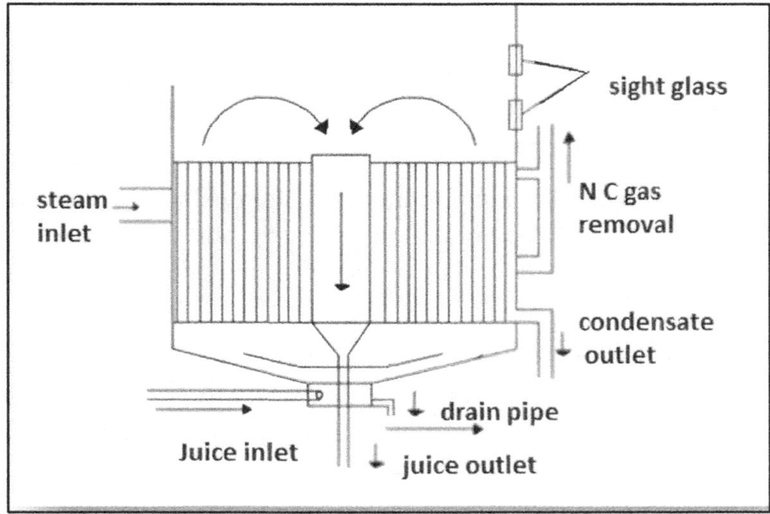

Fig. 38.2 Calandria and downtake

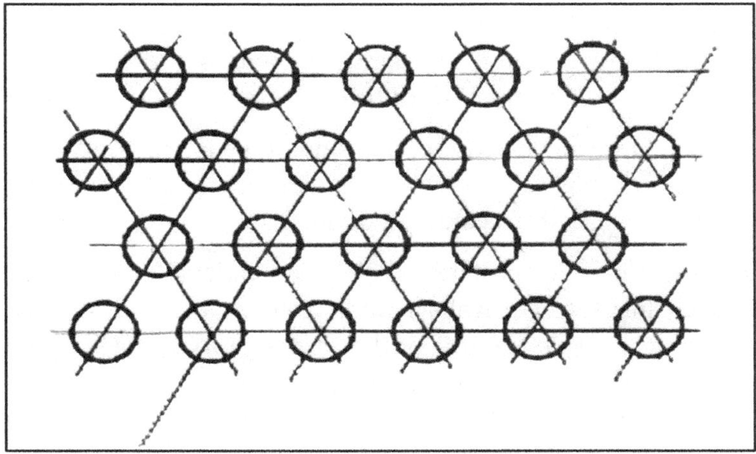

Fig. 38.3 Arrangement of tubes

Juice is continuously fed at the bottom of the vessel and steam is admitted to the calandria. The juice passes through the tubes from the bottom to the top above the calandria. The steam / vapours are heating the juice from outside the tubes. The heat from the steam / vapours is conducted to the juice boiling at lower temperature than that of steam / vapours. The steam / vapours are condensed and latent heat of the vapours is given to the juice.

2) Downtake: A well called 'central downtake' is provided at the centre of the calandria (Fig. 3.3). The juice flowing into the downtake goes out of body through the outlet pipe and passes to the next body. The diameter of the central well is about 20 % of the vessel diameter. In some designs, annular downtake is provided. Large vessel may be provided with number of small downtakes instead of one central downtake.

In some old designs semi-sealed downtake were provided. But the downtake should be completely sealed at the top tube plate. Semi sealed downtake provides easy way for short circuiting of juice flow.

With sealed downtake and U shape inverted siphons in the juice cut lines, the vapours never pass through downtake to the next body and short circuiting of vapours is avoided Therefore working of evaporator becomes more smooth and steady.

The juice piping from one effect to the next effect is generally called as cut–over-line or cut-line. An inverted siphon of approximately 4 to 5 m height is provided so that the pipe line remains filled with the juice and short circuiting of vapours from previous effect to the next effect takes place.

Most of the evaporator bodies installed are of central downtake however few evaporator bodies with lateral downtake or peripheral downtake are installed.

Lateral downtake some manufacturers provide downtake at the periphery of the calandria. For such lateral downtake designs juice enters at the opposite side of the downtake has to pass through number of tubes before coming to the downtake.

Annular downtake: in some designs annular downtake is provided. In these designs vapour shell diameter is higher than calandria diameter as the annular gutter width form both sides is added to the calandria diameter.

Experimental and 'Computational Fluid Dynamics' (CFD) studies carried out by Sugar Research Institute Australia reveals that a vessel with fully distributed peripheral feed and a single central outlet is preferable. It gives optimum flow pattern (plug flow) without internal recirculation or stagnant regions. This helps in increasing the throughput capacity of an effect ranges from 4 % (for the no. 1 effect) to above 30 % (for the last effect).

3) Steam or vapour entry: For efficient operation of the evaporator the steam / vapours should reach the far end portion of the calandria from the steam / vapour entry without any obstruction. The steam / vapours must reach at each tube every moment. For this purpose in some designs vertical baffle plates are provided at the vapour entry. These baffle plates give definite direction to the vapours so that the vapours are evenly distributed throughout the calandria and pass to the opposite portion from the entry. However after some years of use, these baffles get corroded and the path of steam is changed. This adversely affects working of the body and evaporator set. Therefore the corroded baffles should be replaced by new ones but it is a very difficult work.

In most of the designs vapour lanes / passages are provided between the tubes from vapour entry up to some length for even distribution of vapours throughout the calandria. It is generally observed that more tubes are damaged near the vapour entry. This gives sugar traces in condensate.

4) Incondensable gas withdrawal: The steam or vapours contain some incondensable gases or noxious gases. The incondensable gases are bad conductors of heat and if accumulated in the Calandria adversely affect the rate of heat transfer. Therefore, continuous removal of the incondensable gases from the calandria is very much essential.

The locations of incondensable gas withdrawal points/taps are very important. These taps are located at the opposite side at the farthest end from the steam /vapour entry. This also helps in drawing vapours from vapour entry to the far end. With incorrect locations of incondensable gases removal of taps, the steam / vapour may pass through the taps but the incondensable gases may remain behind inside the calandria at the farthest end resulting in poor evaporator performance.

The taps are provided to the calandria shell few centimetres above the bottom tube plate and below the top tube plate. These taps are connected to a common pipe outside or inside the vessel. Though no ammonia escapes from cane juice (as in case of beet juice) these connections are commonly called as ammonia connections or ammonia pipes. A common pipe of these ammonia piping is connected to the vapour pipe of the last body that passes to condenser.

In some designs the ammonia pipe is vented to the vapour space of the same body. This helps in reducing the vapour loss but the concentration of the incondensable gases in the next effects goes on increasing and in the last effect it would increase to undesirable level that may affect heat transfer. The incondensable gases are removed continuously by these ammonia connections.

In some designs, some perforated tubes in place of heating tubes are fitted in the holes of the calandria tube plates to remove the incondensable gases

The concentration of the incondensable gases should not exceed than 2% expressed as relative pressure of the vapour pressure in the calandria. It means while withdrawing the incondensable gases 98% vapours pass along with 2% incondensable gases.

When the ammonia connections from the second effect onward are connected to the vapour pipe of the last effect going to condenser, the high temperature vapours from the 2nd , 3rd and 4th (in case of quintuple) pass to the condenser. However these connections are provided before the vapour line juice heater (VLJH) installed in the vapour line. The vapours are mostly used at VLJH to heat the raw juice.

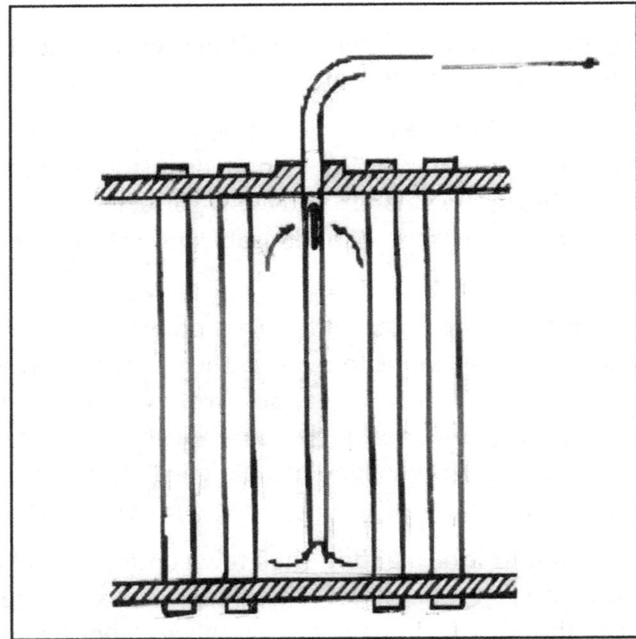

Fig. 38.4 Incondensable gas withdrawal

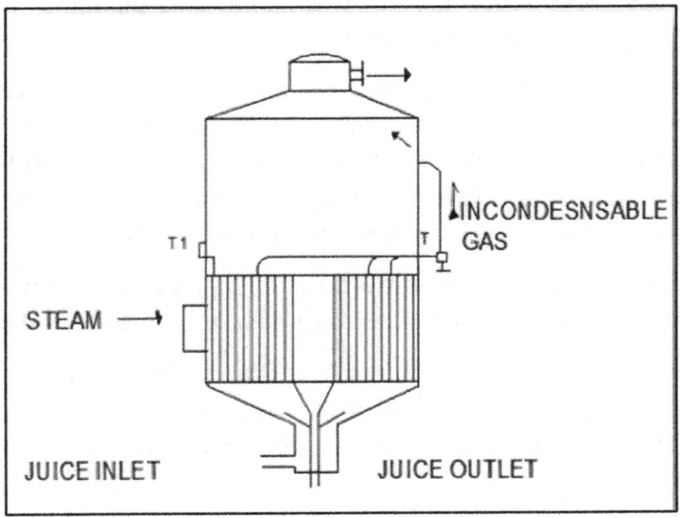

Fig 38.5 Incondensable gases removal

5) Film of condensate: A film of condensate on the outer side reduces the rate of heat transfer. The greater the thickness of condensate film, the less is the rate of heat transfer. The thickness of the condensate film increases from the top to the bottom. Therefore the overall heat transfer coefficient is not uniform and is maximum at the top and minimum at the bottom.

In some designs the steam entry is at the centre of the tubes. The steam flow in the upper half of the tubes is in upward direction (counter current with condensate) while in the lower half it is in downward direction (Co-current with condensate). The condensate flow is restricted by the upward flow of steam in the upper half of the tubes. Chhote Lal has described sweeping of exhaust and vapours. Modifications made to make the steam flow and condensate in the same (downward) direction helps to remove the condensate film promptly. It is mentioned as sweeping effect.

It is assumed that the juice film inside of the tubes also affects the rate of heat transfer. The film is not of uniform thickness along with tube length. It is maximum near the bottom and minimum near the top.

6) Evaporator body plate thickness: The evaporator body is fabricated of mild steel plates. The thicknesses of the plates recommended in the 'Guidelines to specifications of cane sugar plants" given by Indian Sugar Industry are as follows

Table 38.1 evaporator body plate thickness

Heating surface Sq. m.	Shell Thickness mm	Saucer thickness mm	Tube plate Thickness mm
Up to 300	12	16	25
300 to 1350	16	22	32
1350 to 2100	18	25	36
Above 2100 and LTE	22	28	50

7) Height of body: Height of body above the calandria i. e. Vapour space height for vapour cell, first and last effects is generally 2.5 times the height of the calandria. The vapour space height for second or third effect is 2.2 times the height of the calandria. However, in some designs the height is kept more. The higher vapour space height reduces the chances of entrainment.

8) Entrainment: During boiling very light juice droplets are entrained by high velocity vapours and are carried over to the vapour side of calandria of the following vessel or to the condenser. This is called entrainment.

To avoid entrainment save-all or entrainment separator is provided inside the dome at the top of the body. The entrained juice droplets are entrapped in the save-all and are returned in the boiling juice. There are different types of save alls.

When vapour space is sufficient and design of save all is efficient at normal operating conditions generally does not occur, and if any occurs it is so small that is not detected. If entrainment separators of proper design are not installed and /or care is not taken in maintaining the juice level, entrainment may occur frequently.

If entrainment occurs from any evaporator body except last body, the juice droplets pass into the calandria of the next body and mix with the condensate of that body. The condensate is thus gets contaminated. If entrainment occurs from the first effect the condensate of the second effect – which is generally send as make up boiler feed water gets contaminated. If the contaminated condensate is fed to boilers it adversely affects the boilers and turbines working.

Sugar at high temperature decomposes into acids which attack the boiler tubes. Therefore passing of contaminated condensate to the boiler should be immediately stopped and the condensate should be diverted to overhead hot water storage tank. The contaminated condensate from storage tank is used at mills and boiling house and does not give any trouble in the process.

However the overflow of the hot water tank should not be taken to the injection water outgoing channel. Otherwise in case, contaminated condensate passes to the overhead tank and the tanks gets overflowed to the channel water, then spray pond water gets polluted. There should be a practice in routine analysis that the condensates of all vessels should be tested for sugar (alpha napthol test).

9) Vapour pipe: At the top of the evaporator body there is dome. Vapour outlet is provided through it. The vapour outlet is connected to the calandria of the next body by a vapour pipeline. The vapour pipe from the last body is connected to a condenser. The vapours that pass through the vapour pipe to the condenser are condensed by cold water and vacuum of about 630-660 mm of Hg is created in the vapour space of the last body.

The exhaust valve to the calandria of the first vessel is always kept fully open when in operation. Due to this the pressure in vapour space of the first effect is maintained and the vapours are more effective at pans and juice heaters. When vapours are withdrawn from the first effect for use at pans or juice heaters, a valve in vapour pipe is generally provided to control the vapours going to the second effect. Valve is also provided in vapour pipe going from 2nd to the 3rd effect.

10) Juice flow: The juice flows from the downtake of one body to the bottom of the next body because of lower absolute pressure in the next body than the first body. The connection pipeline in between the two bodies is generally called as cut over line or cut line. The cut line is provided with inverted siphon and a cut valve.

The absolute pressure difference in successive evaporator bodies is in the range of 0.30 to 0.45 kg/cm^2 which corresponds to 3.0 to 4.5 m of water column. Considering margin for fluctuations in the pressure difference a siphon of vertical height of 4 to 5.5 m is required. The cut line diameter is kept to maintain the velocity of the juice 0.6 m/sec.

The cut valves can be operated from the evaporator floor. As juice passes through the cut line to the next body, the difference in boiling temperatures of the juice in the two bodies causes flashing or self-evaporation of the juice. The flashing slowly starts in the siphon itself. But when the juice enters in the next body the flashing becomes vigorous and it gives upward movement to the juice. If the incoming juice is not distributed properly at the bottom, then frequently 'swelling up' of juice upward occurs at some portion. This disturbs the boiling and also may cause entrainment.

Therefore to prevent entrainment possibilities the entering juice must be distributed over a large area. Manufacturers provide different arrangement for juice inlet. A perforated coil close to periphery is most suitable. The holes should be provided downwards about 30° to the horizontal towards the periphery. The diameter of the hole should not be less than 25 mm. With smaller diameter choking of the holes by scale pieces may occur.

There is a continuous flow of juice from the first vessel to the last vessel. Syrup from the last vessel is continuously withdrawn by a pump. Two syrup extraction pumps (one as standby) of 30 m head are provided to transfer the syrup to storage tanks at the pan station. A strainer is provided before the syrup pump to catch solid material if any passing through the syrup. A continuous syrup sampler is provided to the delivery line of the syrup.

11) Sight glasses and other fittings:

To observe the working of evaporator body it has provided three circular windows with toughened sight glasses. The lowest sight glass is at a height of 1.5 m above the working platform and from inside it is 250 mm above the top tube plate.

There is also 'light glass' with electric bulb arrangement in the back to allow light to pass into the vessel.

A vacuum brake valve is provided to each body.

A cold water connection of pipe size 75 – 100 mm dia. is provided to both calandria and the body to check the leakages. Generally, leakages are checked at 3 kg/cm^2 pressure.

A suitable pressure relief valve (spring loaded / dead weight lever type) is provided to the first body calandria and vapour space.

Suitable connection for soda washing and draining of juice is provided.

A live steam connection at reduced pressure and vapour outlet to atmosphere at the dome is provided for soda /acid boiling.

Two manholes – one 300 to 350 mm above the calandria and the other to the bottom saucer, are provided to go inside the body for mechanical cleaning and physical inspection. Manhole is also provided at the horizontal portion of the vapour line.

Main platform of evaporator station is at a height of 6 to 8 m above the ground level. Under the evaporator set a working platform is provided for cleaning, repair and maintenance.

A soda preparation tank of capacity 20 m^3 arrangement is provided at the ground level near the evaporator set. A transfer pump of 15 m head and 20 l/sec capacity is provided to this tank.

Syrup /juice sampling arrangement are provided to each body.

The body is well lagged with glass wool and plaster of paris. When the vessels are completely well lagged heat loss is reduced.

38.3 Entrainment separators or saveall:

By entrainment the juice / syrup may pass to the calandria of the next body and further to boilers or condensate storage tank, otherwise if entrainment occurs from last body to the condenser and spray pond.

Generally, condensate from the second effect or sometimes from the pans is partly sent to boilers as make up water. If condensate contaminated with entrained juice droplets is passed to boilers, it gives trouble at boiler. When the condensate contaminated with entrainment enters the boiler, due to high alkalinity and high temperature the sugar decomposes forming organic acids. The boiler water pH goes down which is detrimental to the boiler tubes and shells. High foaming is also observed. Sugar contamination leads to dark coloration to the boiler water and gives characteristic odour. In this condition quick action has to be taken. The load on the boiler is to be reduced. Maximum possible blow downs of boiler water are to be given replacing the contaminated boiler water with quality clean water.

If entrainment is occurred from the last body then the syrup is mixed with injection water. The spray pond water becomes dirty with dark colour and heavy large foam is formed at the outlet channel of the injection water. The pH of the water also is lowered down and to maintain the pH generally lime or pH booster is added in the spray pond water.

To avoid entrainment and hence loss of sugar and other problems, entrainment separator, generally called as saveall or catchall are provided. The savealls are operating on the principle of change in velocity and direction of vapours so that the droplets entrained along with vapours strike on surface of the savealls and return to the evaporator body. There are various types of savealls. In designing a saveall the following two things are considered.

- The pressure drop due to change in direction of the path should be minimum
- The separated juice should be returned down in the boiling juice.

Helmet type saveall: In helmet type save-all (Fig.3.9) rising vapours hit the vane of saveall, the direction of the vapours is reversed and the juice droplets fall down due to the abrasion against the surface.

- **Centrifugal type saveall:** In these savealls vapours entering from the bottom pass tangentially through the vanes. The whirling action of vapours causes abrasion of the entrained droplets with the curved portion. This results in separation of the liquid drops.

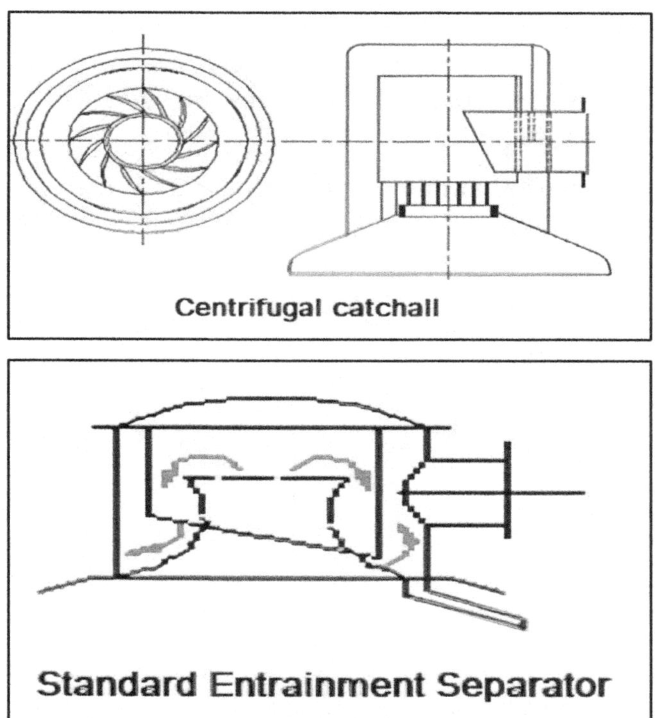

38.6 Different types of savealls

Umbrella type saveall: In umbrella type saveall the rising vapours hit the curved portion of the umbrella and are deflected downwards.

The juice entrapped in the saveall should be drained back into the vessel immediately. If it gets accumulated in the saveall for longer time, then that may further pass into the vapour pipe along with the vapours. The drain pipe is usually of 50 mm diameter so that it does not choke up and also can be cleaned easily. The drain is usually open to the side of the vessel.

Wire-mesh eliminator: Another device is consisting of layers of fine stainless steel wires mesh loosely packed in steel frames in small segments. It is also called as demister. The thickness of the meshing is about 150-200 mm. It is installed below the dome of the vessel and covers almost the whole cross section of the body. The segments can be removed and taken out of body in the off season for physical cleaning.

When vapours pass through the layer of wire mesh, entrained droplets get scrubbed with the wire layers and are arrested in it and are return into the boiling juice. About 80 to 90 % droplets are arrested. These are more effective in the first and second vessels.

The mesh eliminators foul up after few weeks of evaporator operation and therefore should be cleaned at the time of chemical cleaning. For this purpose spray nozzles are provided above the mesh. A soda / acid / water connection is provided to these nozzles. In the off season, these are to be removed out of evaporator bodies and required to be cleaned physically.

Corrugated poly baffle entrainment catcher: The corrugated poly baffle entrainment catcher is quite different from the conventional entrainment catcher design.

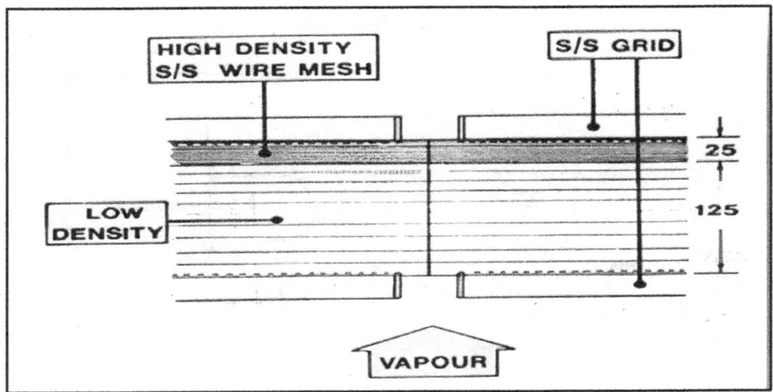

Fig 38.7 Stainless steel eliminator or demister

It comprises of stainless steel multi-baffle plates designed in the form of corrugated sheet having a vortex at the top and steep angle between two layers. When vapour pass through it strike on the surface of the baffles and changes its direction which results in falling of the droplets of juice / syrup. These droplets coalesce and become heavy cannot be again carried by the incoming vapours. The bigger drops formed, easily flow and drop down the corrugated shape plates.

An advantage of this type of arrester is that there is little restriction of the vapour flow path and pressure drop across poly baffle assembly is very small[6] (3 to 5 mm of water).

However boiling and flashing of juice should not reach up to the baffle. The baffles cannot control high volume of liquid. In case of evaporator the vapour space height is generally 4.0 – 4.5 m. in case of pan the vapour belt height should be minimum 2.75 m.

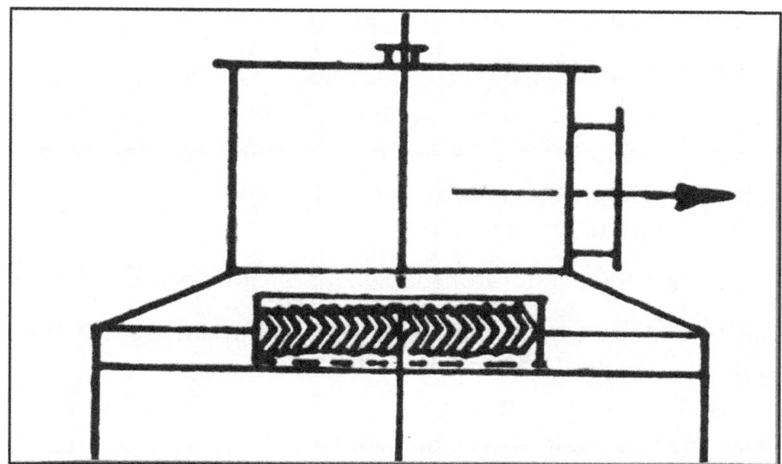

38.8 Corrugated poly fable entrainment catcher

External catcher: In addition to save-all an external catcher is always provided to the last body and sometimes to the Semikestner. External catcher is usually a vertical cylinder with diameter bigger than vapour pipe fitted in between it. In the catchall grills formed of 25 - 40 mm size angles are fitted vertically. The cross sectional area of the external catcher is 3 to 4 times that of vapour line. The cylinder has conical bottom with drain pipe having siphon arrangement to collect the arrested juice / syrup droplets and return the same to the body. Horizontal recovery bottle can also be installed.

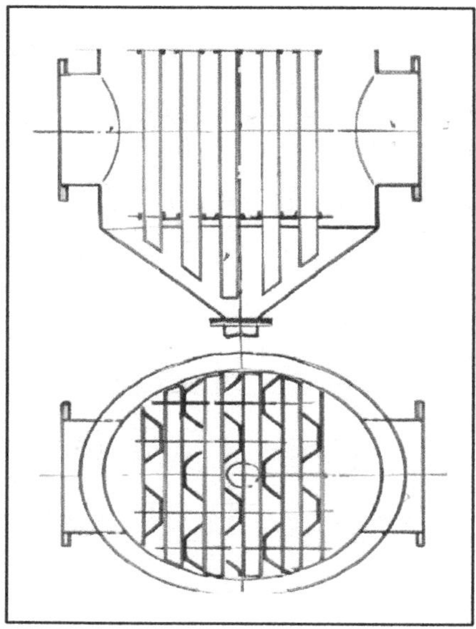

38.9 External catchall

Precautions against entrainment:

1) While designing an evaporator body the following points are to be considered:
   - The juice entry of the evaporator body should be at the bottom;
   - The height of the vapour space should be sufficient;
   - The save-all should be efficient.
2) While operating the evaporator following precautions are to be taken
   - Juice level in each body should be at an optimum level. It should not rise unnecessary high;
   - Forcing the multiple effect above its normal capacity increases the juice level in the bodies. It should be avoided;
   - Vacuum in each body should be maintained at an optimum level. Unnecessary high vacuum in the last body should be avoided;
   - The exhaust pressure should be steady, any fluctuation should be avoided.

38.4 Calculation of vapour pipe diameter:

The vapour pipe diameter should be minimum, for which following points are to be considered.

- The vapour velocity should be optimum,
- The risk of entrainment should be minimum;
- The pressure drop in between vapour space of the first effect and the vapour side of the calandria of the next effect should be minimum.

Taking above points into consideration, the approximate vapour velocities for each effect of the evaporator set are given below.

Table 38.2 vapour velocities at each effect

Source of steam / vapour	**Quadruple m/sec**	**Quintuple m/sec**
Exhaust Steam	25 - 30	25 - 30
1st effect	30 - 35	30 - 35
2nd effect	35 - 40	30 - 35
3rd effect	40 - 45	35 - 40
4th effect	50 - 60	40 - 45
5th effect	----	50 - 60

While calculating vapour pipe diameter highest evaporation rate is considered. Further 10% excess vapours may be considered as a margin for safety.

Volume of vapours passing (m^3/sec)	=	**Cross sectional area of vapour pipe (m^2)**	X	**Velocity of Vapours (m/sec)**

$$V = \pi D^2 \times v/4$$

Where

D = Diameter of the vapour pipe
V = Volume of vapours per sec.
v = Vapour velocity

Hence

$$D = \sqrt{1.273\, \mathbf{V}/v}$$

Volume of vapours produced (**V**) is given by

V (m^3/sec) = $E_{sp.} \times S \times V_{s.} / 3600$

Where

$E_{sp.}$ = Evaporation rate in the body (kg/m^2/hr.)

S = Heating surface of the body (m^2)

$V_{s.}$ = sp. volume of vapours in the body (vapours m^3/kg)

Example:

We will take an example of the first effect of the quintuple

Heating surface (S)	1250 m^2
Specific evaporation ($E_{sp.}$)	30 kg/m^2/hr.
Absolute pressure in vapour space	1.66 kg/cm^2
Specific volume of vapours	1.065 m^3/kg

Then $V = 30 \times 1250 \times 1.065 / 3600$

$\therefore V = 11.09\ m^3/sec.$

In practice 10 % extra vapour volume is taken as margin. Therefore, **V** will be equal to 12.20. Putting this value of **V**, the diameter of the vapour pipe will be

$$D = \sqrt{1.273 \times 12.20 / 30}$$

Hence D = 0.719 m or say 0.720 m.

This diameter of vapour pipe will be at the outlet of the body. Accordingly, diameter of branches of vapour lines to pans, juice heaters and the next effect can be calculated. The vapour pipes are of 8 mm thick mild steel. When vapours are withdrawn from the first body of multiple effect evaporator for use at pans or heaters, a valve in vapour pipe is generally provided to control the vapours going to the second effect. The exhaust valve to the first effect is kept wide open. Due to this vapour pressure is maintained required for pan boiling. Similarly valve is also provided in vapour line going from second effect to the third effect.

References:

- Steindler R J ASTA 2003
- Chhote Lal SISSTA 1999 p 205 – 2135
- Bhojraj S K STAI 1991 p E 1 – 14
- Honig P vol. III p 160
- Chaturvedi P P IS June 1976 p 123 – 130
- Sandera – L A Tromp – Machinery and Equipment of cane sugar factory (1936) p 408

-OOOOO-

39 CONDENSATE AND NONCONDENSABLE GASES

39.1 Condensate from evaporator:

The steam / vapour applied in the calandria gives its latent heat to the juice boiling inside the tube and get condensed. The condensate thus formed descends down over the tube surface and get collected on the bottom tube plate of the calandria. If the condensate is not removed efficiently, it gets accumulated in the calandria and covers the lower part, which adversely affects the working performance of the body. Therefore, it is necessary to remove it as quickly as possible.

To remove the condensate one, two or some time if the body is large, three condensate drain pipes are provided to the bottom tube plate. The condensate pipe is usually joined to the calandria through a cone. It gives more cross sectional area at the outlet for easy flow of condensate. The diameter of the condensate pipe is kept such that the velocity of condensate will be about 0.5-0.6 m/sec. Generally condensate vertical receivers are provided at the ground level before the condensate extraction pumps. A sight glass is provided to the receiver / pipe to observe the flow of condensate. At the delivery line of the condensate a sampling cock for testing of condensate is provided.

To detect contamination of condensate an online conductivity measurement device can be used. In "Guidelines to specification of cane sugar plants" for Indian Sugar Industry an on line conductivity measurement of condensate is recommended for condensate water going from $1^{st}/2^{nd}$ effect to boiler with output 4-20mA DC for remote indication and recording and alarm for high and low values.

Flashing of Condensate:

The absolute pressure in the calandria of the multiple effect evaporator goes on decreasing from first to last effect. Therefore temperature of condensate from the first vessel to the last vessel goes on decreasing. If condensate from first effect is opened to the calandria of the second effect, spontaneous boiling of the condensate takes place. Small amount of vapours are produced and the condensate temperature comes down just below the boiling point corresponding the absolute pressure in the calandria of second effect. It is also called as self-evaporation or flashing of the condensate.

The vapours thus produced by flashing of the condensate mix with the vapours in the calandria of the second effect and get used to boil the juice in the second effect.

Therefore the vapours used in the calandria of the second will be increased by a small amount. Consequently evaporation in the second effect also will be increased. This can be successively followed up to last effect. Hence, there is gain in evaporation. In this way if condensate is circulated by flashing from the first effect to the last effect the gain in evaporation can be about 4.5%.

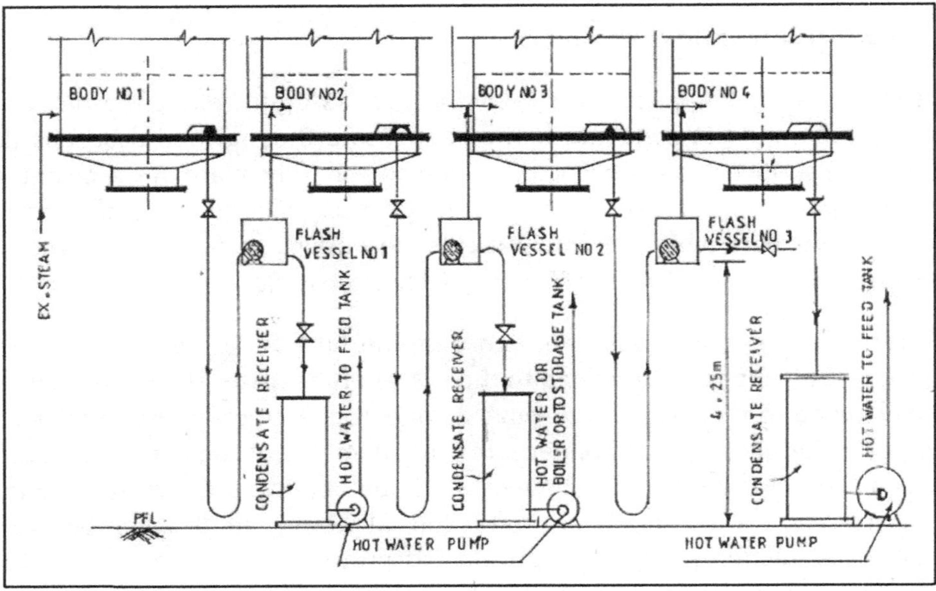

Fig. 39.1 flashing of condensate

Flashing pot: To open the condensate from first effect to the calandria of the next effect an inverted siphon before a flashing pot (Fig. 40.2) can be used. The pressure difference between the two calandria can be get balanced by the difference in the level of condensate in the pipe of the siphon to next side calandria.

Flash tank: A common flash tank (with partitions for each body) under vacuum in which condensate from all evaporator bodies are taken through siphons can also be installed for this purpose. It is comprising of different flash chambers recovering heat from condensate of each effect. All the chambers work at different vacuum/ absolute pressures.

Generally sparged entry is provided which facilitates efficient separation of condensate and vapours. Condensate is transferred to the next chamber either by control valve or U siphon. This provides centralized condensate removal system using single condensate pump. This reduces steam consumption due to maximum flash generation. However when whole the condensate is circulated by flashing from first effect to last effect, in case of entrainment the entire condensate may get contaminated.

Fig 39.2 Condensate receiving flash tank

It is always better from operation point of view of boilers and turbines that all the exhaust condensate (expect that used at juice heaters) should be send to boilers.

It is advisable that the condensate from the first effect (exhaust condensate) after flashing to the second effect calandria should not be mixed with the condensate of the second effect. Because in case entrainment occurs from the first effect the mixed condensate (of first and second effects) gets contaminated that gives more trouble at boilers. However if the exhaust condensate is kept separate then there is surety that boilers will get maximum quantity of pure condensate without any contamination. In case the condensate from the second effects gets contaminated, instead of sending it to boilers, it can be diverted to hot water storage tank. But the supply of exhaust condensate remains intact. When after flashing the hot water condensate of lower temperature is send to boiler some additional bagasse will be consumed at boilers.

Composition of the condensate:

Generally people think that the condensate as it is obtained from vapours is pure water but it is not so. Volatile nonsugars constituents of the juice of which boiling point is below that of water get evaporated and pass with vapours. Further these volatile compounds may get condensed along with vapours and pass with the condensate or are dissolved in the condensate. Therefore traces of aldehydes, organic acids, methyl and other alcohols, O_2, CO_2, SO_2 and NH_3, etc. are found in condensate.

pH: Depending upon the quantity of the different types of impurities in the condensate, the pH of the condensate varies from 5.5 to 8.0

Acetaldehyde: Acetaldehyde is formed by alkaline decomposition of reducing sugars. The condensate formed from the vapours from the first effect has higher aldehydes content than those of the next effects. It is found in mixed juice as well as in clear juice. In alkaline solution it is transformed into paraldehyde[1] ($C_6H_{12}O_3$), this at higher temperature polymerizes to form complex organic compounds called aldehydes resin. This is not volatile; it remains in boiler water, and is responsible for coloration. The aldehydes concentration is higher in immature cane as well as cane harvested a few days after heavy rains.

Acids: The nature of the acids is not known. They may be lower aliphatic acids such as formic or acetic.

O_2: The amount of CO_2 in the condensate varies from 7 to 25 mg/lit.

Table 39.1 Analysis of condensate

Sr. No.	Particulars	Effect				VLJH
		1st	2nd	3rd	4th	
1	pH	6.8	7.1	7.0	7.2	7.5
2	Conductivity μS	75	40	30	30	25
3	TDS (ppm)	60	40	35	25	20
4	Total hardness (ppm)	12	02	N D	N D	N D
5	Total alkalinity (ppm)	65	55	68	70	75
6	Chlorides (ppm)	15	2.5	--	--	--
7	Sulphates (ppm)	N D	--	--	--	--
8	Phosphates(ppm)	5	3	2	N D N	N D
9	Silica (SiO_2) (ppm)	0.018	0.016	0.015	0.012	N D
10	Iron (ppm)	0.56	0.26	0.20	0.12	0.14
11	Dissolved oxygen	--	--	--	--	--
12	Turbidity (NTU)	3.5	2.4	2.2	1.8	1.5

Manoj Goel[2] (STAI 2008, p M82-90) gives the above analysis of condensate of evaporator set and of vapour line juice heater at Simbhaoli, in which defecation process is followed for clarification of juice. The analysis done was from the point of view to use the condensate for boiler after polishing with resins.

Utilization of condensate:

All exhaust steam condensate (except of juice heaters) is send to boilers as boiler feed water. The temperature of exhaust condensate is high. The exhaust condensate is not contaminated because of entrainment. In case of first effect even if in case the tubes are leak, because of higher pressure to the steam side, heavy leakage of juice does not take place. If sugar test is observed that can be only due to traces of sugar and remedial action can be taken without any high risk of trouble at boilers. In addition to the exhaust condensate additional vapour condensate is always needed to be send to boilers as feed water make-up for the following reasons –

- Exhaust condensate of juice heaters is not send to boilers.

- Boiler water is lost due to boiler blow downs hence steam generation is always less than water feed to boilers.
- Losses occurs in the circuit due to leakages.
- High-pressure steam is used for centrifugals, sulphur burners; cleaning etc. is lost to atmosphere.

Condensate from second effect is partly used as make up water for boilers. There are two advantages in using the second effect condensate first - there is minimum possibility of entrainment from the first body (as there is no vacuum in the vapour space of the first body) and second its temperature is high.

The remaining condensate from the second effect and the following effects is used for process such as –

- Imbibition
- Wash water for rotary vacuum filter
- Preparation of milk of lime
- Seed washing in the pans
- Molasses conditioning / dilution at pan station,
- Movement water for pans,
- Washing while massecuite curing at centrifugals
- Melting of sugar

Contamination of condensate due to tube leakages:

In evaporator operation when tube leaks arise the contamination of condensate is generally not much serious. When the leakages are close to top tube plate of the calandria, it is not even being detected until it is cleaned and tested by filling the calandria with water. It is because when the evaporator set works the juice side of each vessel is always at lower absolute pressure than its vapour side.

In case of leakages, near the top tube plate vapours from the calandria pass to the juice side and mix with the vapours from the boiling juice and pass through the vapour space to the calandria of the next body. Along with vapours some incondensable gases also pass. Juice does not pass to the vapour side and if passed only in traces that may not even be detected. In case of stoppages the juice boiling ceases and juice level comes down to 1/3rd of the tube height and passing of juice into the condensate does not occur.

If leakages occur at the bottom tube plate and vapours pass inside, that are condensed in contact with juice/syrup. If condensate being collected above the bottom tube plate passes to the juice side it, dilutes juice /syrup so that syrup brix lowers down. As a result of which steam consumption at the pan station increases. However for any reason if evaporator operation is stopped and vacuum of that particular effect lowers down and vapour side pressure becomes equal to or even lower than the juice side, then juice passes into the condensate and sugar can be detected by alpha napthol test. Therefore the condensate of all evaporator bodies should essentially tested for sugar when the evaporator operation is stopped for any reason for some time. Consequently testing of calandria of evaporator bodies at the time of periodical cleaning is necessary.

Contamination of Pan Condensate: Similarly at pan station with tube leakages the contamination of condensate by sugar is generally not observed or detected during pan operation. But in doubts the condensate can be tested for sugar when pan is dropped.

Contamination of Juice Heaters condensate: On the other hand in case of juice heaters the steam/vapour side pressure is always lower than the juice side pressure and therefore if any tube leakage occurs juice at that moment passes to the vapour side and is mixed with condensate. The contamination is also heavy and can be gauged only by the colour of the condensate. Therefore the condensate from juice heaters is not send to boilers. Otherwise in case of leakage there will be more serious problem at boilers. When condensate contamination by juice is high and that passes into the boilers a characteristic odour is smelled to the steam passing through the drain pipes.

Condensate pump:

While installing condensate pumps following points are to be taken into consideration

The suction pipe from the calandria to pump should be short with minimum bends;

A vacuum equalizing connection is to be provided from the suction side of condensate pump to the top of the calandria. Otherwise leakage if any occurred, the air entering hinders the water flow to the pump.

The suction side of the condensate pump may be under vacuum. Therefore, pressure difference in the atmosphere and calandria is to be taken into consideration while deciding head of the pump.

39.2 noncondensable gases:

The steam / vapours are contaminated with air and gases. The quantity is very small but if they are not removed fast and get accumulated in calandria, heat transfer coefficient is drastically reduced and the boiling may get ceased. It is because air and gases evolved are bad conductors of heat. The origin of air is from -

Generally, when raw water which contains some dissolved air, is taken as make up water for boilers, the air dissolves in the raw water passes along with steam.

However with high-pressure boilers the feed water is send through de-aerator to the boilers and the possibility of passing air along with water is least.

Some air may get entrapped in the clear juice while the juice gets overflowed from the clarifier. This air gets released while boiling the juice in evaporator.

Air may pass through joints, pump glands and sight glasses inside the evaporator bodies that are under vacuum.

As the juice is boiled at high temperature because of chemical reactions some gases and volatile compounds may be formed and pass along with vapours.

Table 39.2: the effect of incondensable gases in steam on the heat transfer coefficient as noted by Claassen-Langen and Kerr.

Percentage of air in Steam	Heat Transfer coefficient Cal/m^2/°C/h)	
	Classens – Langen	Kerr
0	3150	2600
0.1	2750	2125
0.2	2550	1800
0.3	2400	1530
0.4	2250	1370
0.5	2100	1220
1.0	1775	-
2.0	1300	-
3.0	1050	-

Incondensable gas withdrawals:

To remove incondensable gases small taps - opposite to steam / vapour entry, are provided to the calandria. The taps are connected together to a single pipe. These pipes / connections are also called as ammonia pipes / connections.

- For the calandria of first and in some cases second effect which work on higher vapour pressure than atmospheric pressure, the ammonia connections are vented to atmosphere.
- For the calandria which work more or less at atmospheric pressure or under vacuum, the ammonia pipes are connected either to vapour space of the same body or to the vapour pipe going from last body to the condenser.

When the incondensable gases are removed from the calandria to the vapour space of the same body, they pass along with vapours to the calandria of the next body. Thus, the concentration of the incondensable gases in the vapours goes on increasing to the following bodies.

The incondensable gases passing through the ammonia connection contain only 2-3 % incondensable gases and the remaining are vapours. These vapours of high temperature are lost to condenser without use. Moreover, the high temperature vapours give extra load to condenser. However now VLJH are used and these high temperature vapours get utilized.

Air and carbon dioxide, which are heavier than vapours pass at the bottom while some other gases being lighter passes at the top. Therefore, the ammonia taps are provided above the bottom tube plate and below the top tube plate. The taps are about 10 cms above the bottom tube plate. This is because if for any reason condensate gets accumulated in the calandria it should not be sucked by the taps.

The temperature of the section of evaporator body where the incondensable gases are accumulated falls below the temperature of the juice[3] by 1.5 to 2.5°C. However in factory working to regulate the ammonia vent by experience is easy. The ammonia vent of the first effect is open to atmosphere and can be seen. For the next bodies, ammonia valves openings should be in increasing order from second to last,

Usually more tube leakages are observed at a portion from where the incondensable gases are not removed efficiently. This problem can be controlled by providing additional withdrawal points for incondensable gases from that particular portion. Vapours from juice contain more incondensable gases. More over the vapour pressure at the next effects goes on decreasing. Therefore the incondensable gas removal at the farther effects as well as at juice heaters and pans are made bigger in size.

Honig says that if the temperature of incondensable gases (passing through the ammonia pipe) is lower by 1.5°C than the temperature of vapours in the calandria, then the venting of the incondensable gases is good. Moreover, wastage of heat is negligible.

References:

1. Honig P vol. III p 160
2. Manoj Goel[2] (STAI 2008, p M82-90)
3. Chaturvedi P P IS June 1976 p 123 – 130

-OOOOO-

40 KESTNER AND FALLING FILM EVAPORATOR

40.1 Kestner (LTVRF):

French engineer Paul Kestner designed long tube vertical rising film (LTVRF) evaporator body in 1906. Therefore, it is generally called as Kestner type evaporator or simply Kestner. It works on 'climbing juice film' principle. In this vessel, juice enters at the bottom at a boiling temperature. The juice is boiling when it enters the tubes. The vapour bubbles formed increase in diameter as they rise through the tubes and become almost equal to the diameter of the tube. The juice rises in the form of climbing film over the surface of the tubes. Therefore distance in the juice particles being heated and heating surface is least. Hence heat transfer of the Kestner is higher than conventional Robert type body.

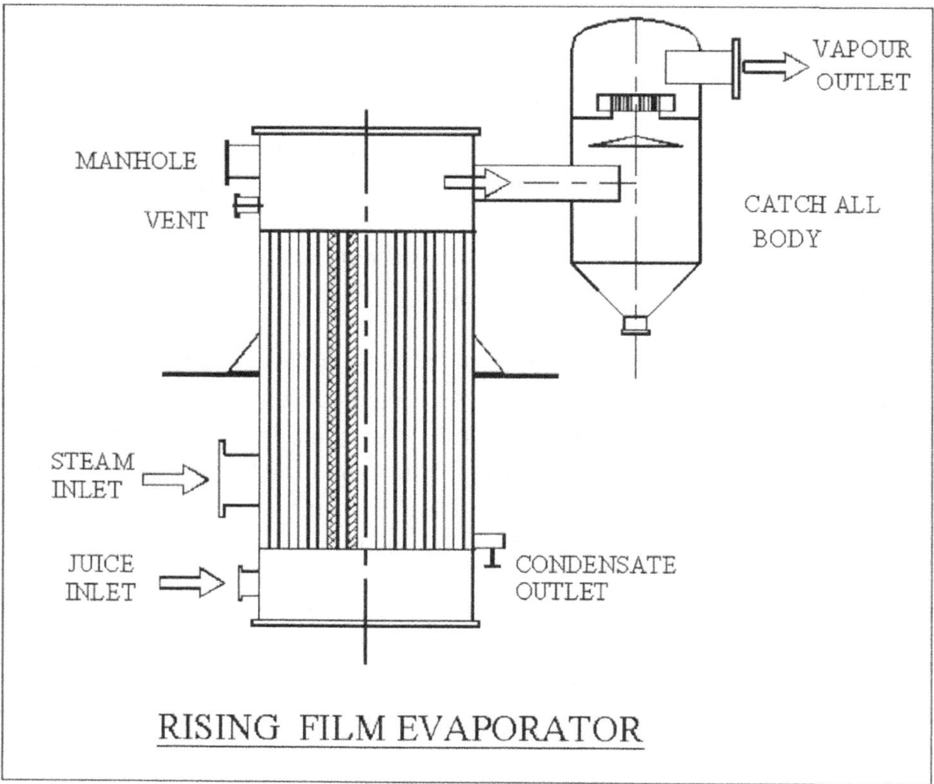

Fig.40.1 Semikestner with external vapour separator

Length of the tube: In this equipment length of the tube is 4 to 9 m and diameter 28 to 35 mm. When the tube length is kept above 8 m, the vessel is called Kestner and when the length is kept 4 to 5 m., it is called Semikestner. Classen[1] has shown that the maximum heat transfer can be achieved at 4.9 m tube length.

Diameter of the tube: The less the diameter the better is the performance. Therefore it better to keep the tube diameter less than 37 mm. The diameter of the tube is greater for higher tube lengths. It is necessary to maintain the climbing film effect. According to guidelines to specifications of cane sugar plants for Indian sugar industry for semikestners stainless steel tubes (IS13316) of 4 m length, 45 mm dia O D, and 1.2 mm thick are recommended.

Vapour space and vapour separator: An external vapour separator cum save-all is provided to Kestner. However, semikestners are now generally provided with vapour space height equal to the length of the tube and save-all above it. This design has advantage that the mechanical cleaning and changing the leaky tubes can be attended easily. The boiling juice emerging with high velocity from the tube tops tends to atomize to mist like particles on hitting head-on any deflector[4]. This then escapes through the usual savealls and pass with vapours to calandria of the next body. Thus frequent minute entrainment may occur. Therefore specially designed saveall are to be installed.

Juice level in the tube: While Semikestner is in operation, most of the volume of the tubes is occupied by vapours with film of juice over the heating surface. Therefore, effect of the hydrostatic is considerably less as compare to Robert type evaporator body. The optimum height of liquid column is about 20% of the tube height. This helps in increasing the heat transfer in this equipment.

Retention time: The Kestner or Semikestner type bodies are of less diameter with shallow bottom. Therefore, it has much less holding volume as compare to Robert type body of same heating surface. Moreover, the juice level is also only about 20 percent as against 30-35 percent in conventional Robert type body. Therefore retention time in Semikestner is much less and hence in this equipment inversion and colour formation is less as compare to Robert type evaporator body.

Heat transfer: The LTVRF evaporator calandria are found to have approximately 1.2 / 1.5 times the overall heat transfer coefficient over standard calandrias[2]. Perk[3] has reported that in case of a Semikestner the overall heat transfer was 50 % higher as compare to the Robert type body. Hence, 33 % less heating surface was required. The retention time was one fourth of the retention time for Robert type with same capacity.

Juice inlet temperature: Kestner / Semikestner are fast or sensitive equipment and require careful monitoring of all the operating factors. The juice entering in the vessel should be at about 3°C above the boiling point of the juice in the vessel. Then only it can work on climbing film principle. If the juice enters at a lower temperature certain lower portion of heating surface works for heating the juice. Therefore heating surface used for boiling the juice is reduced and the equipment gives poor performance.

Superior for first effect: The juice forms a climbing film only when it is thin or of lower brix. Thick and viscous juice does not climb easily. Therefore, Kestner gives good results for the first effect of the evaporator set but does not give good results for the further effects. Hence, it is generally employed for the first effect, however in few factories it is being used as second effect of the evaporator set.

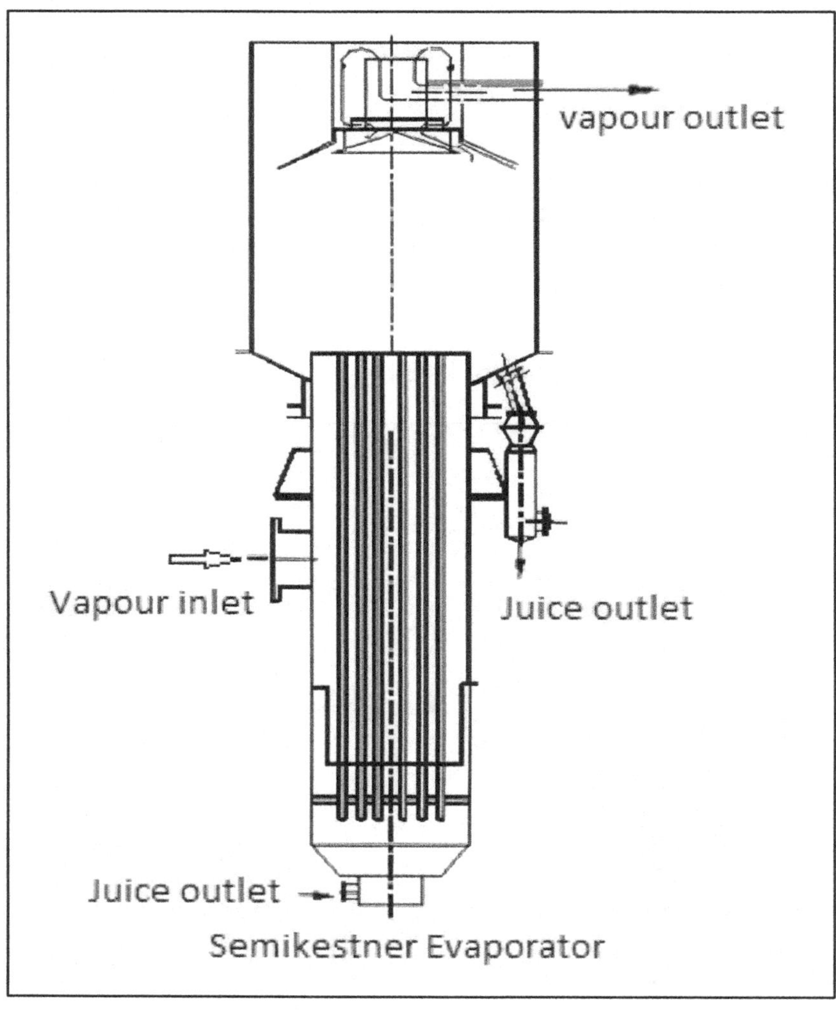

Fig.40.2 Semikestner

Less scale formation: The formation of scale in Semikestner is comparatively less than Robert type evaporator body as juice passes rapidly through the tubes.

At the top of LTE a suitable platform is provided for operation and maintenance. Semikestner takes less floor space when vapour space and vapour separator are above the calandria as in the Robert type, however if external vapour separator is provided, space is not saved.

At Pandurang sugar factory the semikestner was giving poor performance. Following modifications were made to improve the performance[5].

- The heavy and light incondensable gas pipes were isolated. The bottom taps for incondensable gas were raised by 200 mm from the bottom tube plate.
- The exhaust entry valve seat was increased from 425 mm to 635 mm. The steam chest for distribution was modified for easy entry of exhaust in the calandria.
- Two additional drains were provided hence the maximum travel path of the condensate was reduced from 2.8 m to 1.6 m. Thus for getting good performance of evaporator body due considerations are essential towards proper steam distribution, complete removal of incondensable gases and shortest possible travel of condensate inside the calandria.

41.2 Falling film evaporator (FFE):

Fig.40.3 FFE evaporator set supplied by ISGEC to sugar factory in Peru.

Now many factories have installed FFE in various configurations in evaporator set. In India most of the FFEs are employed as first or second effect of evaporator sets. However now in some newly erected factories FFE are installed for all effects. In this vessel juice falls from top and a thin film of juice passes over the heating surface therefore it is called as FFE. The tube length of FFE is 7 to10 m, most commonly 9m. The tubes are arranged in staggered spacing and generally are of 42 mm diameter and 1.6 mm thickness. The juice with vapours produced is collected in the lower chamber.

The vapours pass from the lower chamber through the vapour separator to the next vessel. The heating steam / vapours enter at the upper 1/3 of the calandria. At this position, the tubes are generally surrounded by sleeves, which protect the tubes from the impact of steam. Incondensable gases are drawn at the top and bottom of the calandria.

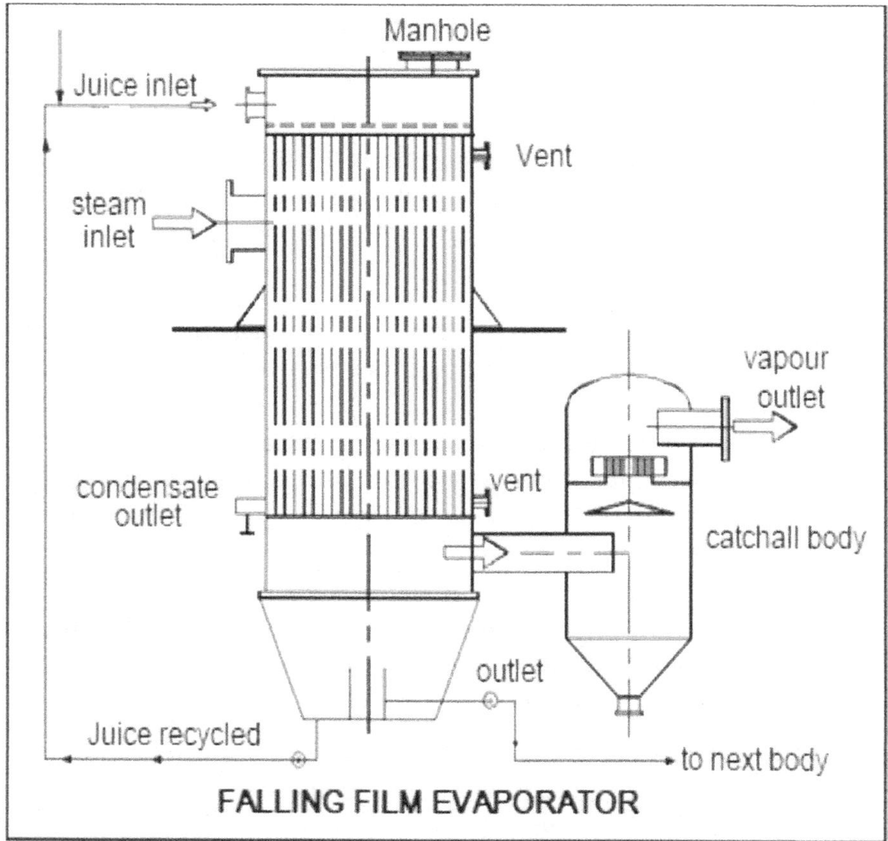

Fig. 40.4 falling film Evaporator

The advantages of this equipment are that the hydrostatic head upon the boiling juice is totally absent and retention of juice is considerably less as compare to Robert type evaporator or Semikestner of the same capacity.

In the working of FFE continuous juice flow at specified flow rate is very much essential. Proper juice distribution to all the tubes is also very much important. Each and every tube should receive the required volume of juice per minute, so that no section of any tube remains dry at any moment.

If juice is distributed unevenly, some tubes get so little juice that the total water from the juice is evaporated, the solids are charred and the tubes are blocked. In such case, FFE gives poor performance. To avoid the drying of the tubes juice of about 4 to 6 time the incoming juice is re circulated. To circulate and transfer the juice two recirculation pumps and two transfer pumps of suitable capacity and suitable head are provided. A hot water connection to the juice distribution system is provided with automatic control to feed in water in case of emergency.

Juice distribution systems:

Juice distribution system is an important design feature in an FFE and has to be arranged carefully to ensure uniform distribution of juice to all the tubes so that at any moment all the heating surface remains wet. There are different designs of juice distribution systems

- In one design, the juice is distributed through a five-tier distributor. Number of perforated plates one above the other are fitted above the top tube plate.
- In the second design, the distribution of juice is made through specially designed distributing nozzles inserted at the top end of each tube. The juice is distributed to the walls with a spiral motion, forming a film. A juice distribution tray is fitted above the nozzles.
- In one design the top end of the tubes are covered with conical caps that work as umbrella and therefore juice cannot pass directly into the tube. Juice enters the tube at the periphery of the tubes. Hence a film of juice is formed on the tube surface.
- In some designs, the total heating surface is separated in two or three sections. Incoming juice is passed through a section and the recirculated juice is passed through another section. Therefore, the juice being circulated does not mix with incoming juice of lower brix.

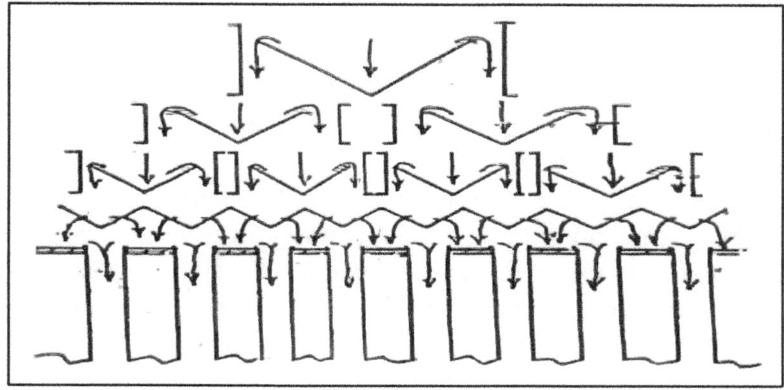

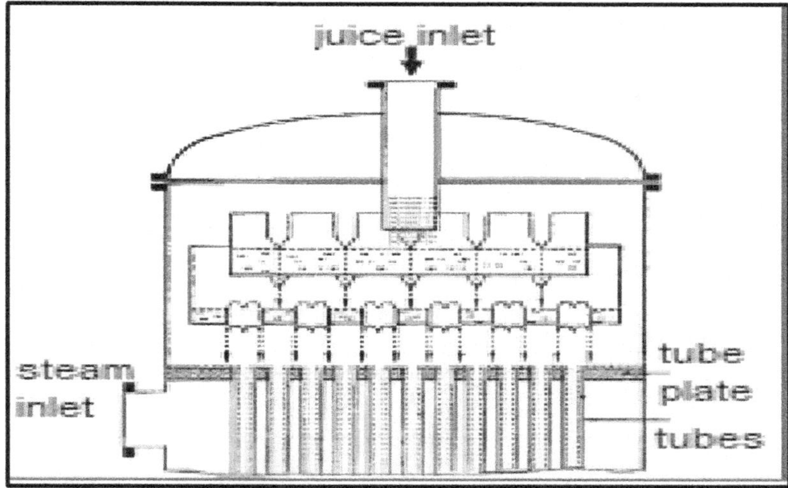

Fig 40.5 Juice distribution system of FFE

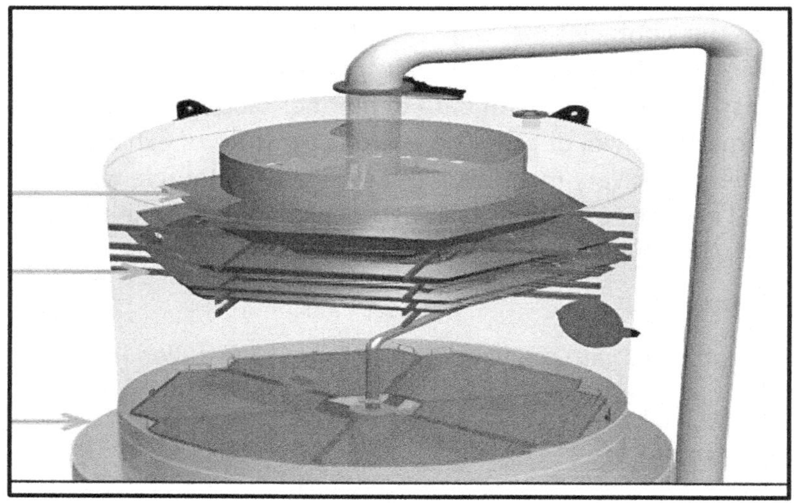

Fig. 40.6 Cascade distribution

Heat transfer:

The coefficient of heat transfer of FFE is higher than semikestner or Robert type evaporator body. Saleem Sajid[6] gives in-depth theoretical study of FFE considering the parameters such as juice film thickness, Reynolds's number. His study reveals that specific evaporation coefficient in the range of 29 kg/m^2/hr. to 43 kg/m^2/hr. is attainable during normal operation depending upon tube length. Therefore, FFE can work well at lower pressure and lower temperature difference. To ensure proper distribution of steam, some factories have made modifications for a sweeping effect of steam in the calandria and have reported superior performance[7].

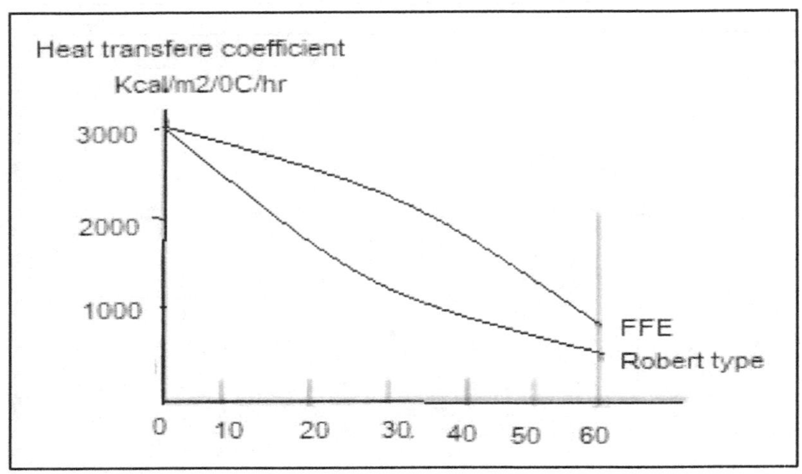

Fig 40.7 Heat transfer coefficient of FFE and Robert type

Contact time:

In FFE, the contact time of juice with the heating surface is very short. Therefore if juice flow is steady and juice distribution is proper, inversion and colour formation is less or negligible. With data of number factories, Bhargawa[8] concludes that with automatic control of pressure and temperature of exhaust and pH control of juice, the inversion is very much less. The system of quintuple with FFE is superior to conventional evaporator set for decreasing inversion loss when first two or three or all effects of a quintuple are FFE.

Working results:

Bhagat[9] gives rate of evaporation of for FFE installed in some factories as under –

Table 40.1 rate of evaporation observed in some factories

Factory	Effect in the evaporator set	Evaporation rate $kg/m^2/hr.$
Malegaon	1	24.65
	2	20.80
Manjara	1	29.60
Mula	1	20.56
Ponni	1	24.12
Rosagaon	1	24.61
	2	21.60
	3	10.54
Anna	1	21.52

The ultimate evaporation rate of 29 $kg/m^2/hr.$ to 43 $kg/m^2/hr.$ as given by Saleem Sajid[6] is not attained. The table does not give any information about working parameters such as exhaust pressure; vapour pressure, temperature difference in heating steam and boiling juice. The reason for low evaporation rates may be due to improper juice distribution that results in non-uniform juice film thickness and inefficient steam distribution in the calandria. But the table gives the idea about the rate of evaporation generally attained for FFE in factory workings. Now in many factories FFE have been installed. In this equipment care has to be taken that the full heating surface remains wet every moment for which sufficient juice should be passed through each tube continuously. For which juice circulation with proper distribution of juice is very much essential. Failure of this, results into choking of many tubes. The efficiency of the equipment and consequently of evaporator set is reduced. Many factories have faced this problem. During factory operation for some reasons like mill stoppage, tripping of juice pumps, the incoming juice flow is stopped but steam is continued. In such conditions because of starvation, caramelization or further charring of the remaining juice and drying and choking of the tubes may occur. To avoid this, immediately hot water inlet to the juice distribution system should be opened. It should be preferably automatic; otherwise the operator at the station should most attentive to start the water immediately.

Gurumurthy[10] while discussing his experience with FFE as first effect of quintuple says, *"We realized that adequate importance was not given for recirculation of the juice and the vessel was not provided with any instrumentation for control levels. The manual operation and the percentage of circulation level must have been improper. Therefore we found some of the tubes fully choked when we opened at the end of the season. Modifications were made in recirculation arrangement so that to restrict the flow of juice into the next vessel and to increase recirculation of juice. An automatic level-controlling device was installed to ensure constant availability of juice for recirculation. It helped smooth operation".*

Cleaning of FFE

The scale deposition in FFE is less as compare to conventional Robert type evaporator body. Therefore, there is less drop in overall heat transfer coefficient throughout the run. It is difficult to remove the cumbersome juice distribution arrangement for mechanical cleaning of the tube. However as the scale is less and soft and can be removed by chemical cleaning only. All the factories therefore resorted to chemical cleaning of FFE with the juice distribution system in situ during the crushing season. Mechanical cleaning is done only in the off-season.

The main problem with FFE is the perfect working of juice distribution system. However, FFE are not much popular in India as the uneven juice distribution gives trouble.

40.3 Plate Evaporators:

Plate heat exchanger (PHE) is invented by Dr Richard Seligman in 1923 and revolutionized methods of indirect heating and cooling of fluids. The plate evaporators (PE) are of compact designs and operate in rising film mode similar to Plate type heat exchanger (PTHE).

In plate heat exchanger thin corrugated metal plates used to transfer heat. These plates are gasketed, welded or brazed together depending on the application of the heat exchanger. Generally in larger plate heat exchangers gaskets are used between the plates, for smaller version brazing is done.

The forms of corrugation are important to get high efficiency of heat transfer. There are two types: inter-mating and chevron type corrugations. In general, greater heat transfer enhancement is produced from chevrons type corrugation. Stainless steel plates are commonly used as these are corrosion resistant. They have enough strength and can withstand at high temperature. The plates are interlinked with the other plates that form the channel with gaps of 1.3–1.5 mm between the plates. In some redesigns it is wide. The plates are fitted into a rigid frame / manifold system with two manifold headers.

A turbulent flow is formed in the liquid as it passes through the thin gaps between the corrugated plates at high velocity. The majority of the volume of the liquid comes in contact with the plate. Therefore heat transfer in the plate heat exchanger is higher. As compared to shell and tube heat exchangers, the temperature approach in plate heat exchangers may be as low as 1-2°C whereas shell and tube heat exchangers require an approach of 5-6 °C or more.

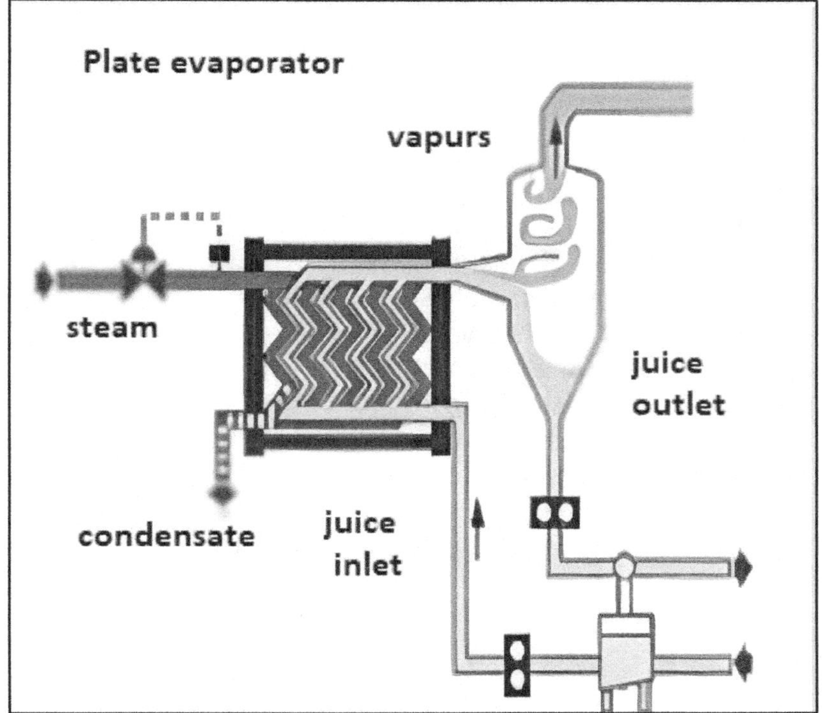

Fig. 40.8 Plate Evaporator

Increase and reduction of the heating surface area is simple in a plate heat-exchanger, through the addition or removal of plates from the stack. Size of the plate heat exchanger is smaller as compare to tubes and shell type heat exchangers. The post-commissioning service and spare parts availability is important in installation of plate heat exchanger.

Plate evaporators are successfully being used in the European beet sugar industry. In Australia PE has been installed at Harwood Sugar Factory in 1992 to boost the capacity of the second effect of the sextuple[11]. A plate evaporator is also installed at Raceland LA in 2001 as a second effect booster. The juice residence time is about 30 sec or about one tenth that of typical Robert type vessel. The heat transfer coefficient is about 40 % greater than that of Robert type.

It has not yet been accepted in cane sugar industry as regular equipment. Author is also doubtful about its successful use in sulphitation factories in which mechanical cleaning of heating surface in the running season is unavoidable and cleaning the plates by mechanical means is very difficult thing.

References:

1 Webre A L ISJ 1939 p 218
2 Mukharji J P SISSTA Seminar Nov. 87 p 16 –23
3 Perk C G M SASTA 1963, p 761 – 771
4 Ghosh S K Co-op. Sugar July 84, p 659 – 665
5 Pulate S M DSTA 1999 p M 19 -27
6 Saleem Sajid STAI 2000 p M 100-113
7 Bhagat J J IS May 1997 p 101 – 110
8 Bhargawa J C STAI1989 p 41-62
9 Bhagat J J IS May 1997 p 101 – 110
10 Gurumurthi,
11 Wright P G ISSCT / STAI Workshop 1994 p 189 – 200

-OOOOO-

41 VAPOUR BLEEDING AND STEAM ECONOMY

41.1 General:

The cane sugar industry is self-sufficient for energy as bagasse is considered as a free fuel available from the cane. The bagasse is burned in boilers and high pressure steam is generated. The steam is used to run power and mill turbines. Low pressure steam, coming out from turbines called exhausted steam or exhaust steam or simply exhaust is used to heat, boil and concentrate the juice and massecuite in the boiling house.

To conserve the steam the total exhaust production from the prime movers should be less than the actual exhaust steam demand in the boiling house. Otherwise if exhaust production is more than requirement it will be blown to atmosphere and it will be a huge loss of energy. Therefore in all sugar factories (except factories with cogeneration by means of condensing turbines with extraction system) the exhaust demand and production is so balanced that the production is always 15 to 25 % less than demand. The additional low pressure steam required is made available by bleeding high pressure steam into the exhaust through the 'pressure reducing and desuperheating system' (PRDS). The steam used in boiling house is for the following purpose

1. Juice heating
2. Evaporation
3. Pan boiling
4. Miscellaneous use

41.2 Steam consumption at juice heaters:

The steam consumption at juice heaters varies mainly with mixed juice percentage cane and initial temperature of mixed juice. It also varies with filtrates recycled % cane but by narrow range.

The mixed juice percent cane generally varies from 95 to 110. In South Africa in some factories with diffusers, the mixed juice percentage cane is above 115 % on cane but in that case, the diffuser juice temperature is higher.

The initial raw juice temperature varies as per ambient temperature and temperature of imbibition water used. While calculating the steam consumption at juice heaters the initial temperature of raw juice is generally taken as 25 - 30°C

The filtrate returned varies from 12 to 18 % on cane.

Specific Heat of juice (Cj):

Let us consider the raw juice has B brix. It means 100 gms of juice contains 100 – B grams of water and B grams of dissolved solids or one gram of juice contains (1 – 0.01B) gram of water and 0.01B gms of dissolved solids. Now the dissolved solids contain about 78 – 85 % sucrose, specific heat of which is 0.38 and 5-12 % reducing sugars the specific heat of which is also very close to specific heat of sucrose. Therefore assuming average specific heat of dissolved solids as 0.4, the specific heat of the juice (Cj) will be

$$Cj = (1 - 0.01B)\ 1 + 0.4 \times 0.01B$$
$$Cj = 1 - 0.01B + 0.004B$$
$$Cj = 1 - B\ (0.01 - 0.004)$$
$$Cj = 1 - 0.006B$$

Considering Brix of the raw juice 14.30, the specific heat of juice will be

$$Cj = 1 - 0.006 \times 14.30$$
$$Cj = 0.9142 \text{ Cal /gm. of juice Or } Cj = 0.9142 \text{ KCal /kg of juice}$$

The steam requirement at juice heaters metric tonnes per hour for a 100 tch sulphitation factory is calculated hereunder with the following assumptions. In other words steam consumption for juice heating % on cane is calculated hereunder with following assumptions.

Table 41.1 assumptions

Crush rate	100 tch
Raw juice % cane	108 %
Raw juice heating	From 30° to 75°C
Sulphited juice % cane	120 %
Sulphited juice heating	From 73° to 103°C
Clear juice % cane	108 %
Clear juice heating	From 97° to 115°C
Specific heat of juice	0.91 KCal / kg
Latent heat of steam (at 0.75 kg/cm^2 g pressure)	530 KCal / kg

Steam required for raw juice heating:

The raw juice is heated from ambient juice temperature 30°C to 75°C at which temperature the juice is sulphited.

$$= \frac{\text{Quantity of heat required per hour to heat raw juice from 30°C to 75°C}}{\text{-Latent heat of the heating steam Kcal/kg}}$$

$$= \frac{\text{Quantity of juice (kg) X sp. heat of juice X rise in temperature } (\Delta t)}{\text{Latent heat of the heating steam Kcal/kg}}$$

= 108 X 1000 X 0.91 X (75 – 30) / 530

= 8344 kgs, or say 8.3 t/hr.

Steam required for sulphited juice heating:

Considering 15 % filtrate recycling, the sulphited juice quantity will be 123 % on cane and the steam requirement for sulphur juice heating will be

= 123 X 1000 X 0.91 X (103 – 73) / 530

= 6335 kg or say 6.3 t/hr.

Steam required for clear juice heating:

Considering clear juice % cane equal to mixed juice % cane the steam required for clear juice heating will be

=108 X 1000 X 0.91 X (115 - 97) / 530

= 3371 kg or 3.4 t

Table 41.2 Total steam required for juice heating

1	Steam required for raw juice heating	8.3 t/hr.
2	Steam required for sulphited juice heating	6.3 t/hr.
3	Steam required for clear juice heating	3.4 t/hr.
4	Total steam requirement for juice heating	18.0 t/hr.

Thus, the total steam consumption for juice heating percent is about 18.0 t/hr. for 100 t crushing rate or 17% on cane. Here we can say, the steam requirement for raw juice and clear juice heating is 0.185 t/°C/100 tch and for sulphited juice heating is 0.21 t/°C/100 tch. We will make use of these findings while calculating vapour bleeding at evaporator.

Direct contact heater:

The direct contact heaters or condenser heaters are now being used in sugar factories. With the use of direct contact heaters, the quantity of juice increases after each heating. The juice quantity is increased by about 15 to 20% on cane depending upon in how many stages the juice is heated in direct contact heater. This increases steam consumption at juice heaters, also gives extra load on evaporator and steam consumption at evaporator also increases.

Heating of raw juice by condensate:

Now in many factories raw juice is heated by hot condensate.

Raw juice heating (RJ2) from 45 to 57°C by 3rd 4th and 5th effects and pans condensate:

In this system after VLJH the further heating of raw juice from 45°C to 69°C is done in two stages. Raw juice second heating (RJ2) is done by condensate from the 3rd, 4th and 5th effects and pans in liquid-liquid heat exchanger.

Table 41.3 Raw juice 2nd heating

Sr. No.	Source of condensate used	Quantity of condensate MT	Temperature °C
1	3rd effect	13.0	104
2	4th effect	3.0	94
3	5th effect	3.0	82
4	pans	15.0	100
	Total /average	34.0	99

Heat taken by the juice = heat given by the condensate

$108 \times 1000 \times 0.91 \times (57 - 45) = 34.0 \times 1000 \times (99 - T_{c3+})$

$T_{c3+} = 64.3$°C say 64°C

Thus after heating the raw juice up to 57°C, the condensate temperature will be 64°C.

Raw juice heating (RJ3) from 57 to 69°C by 2nd effect, juice heaters and pans condensate:

The raw juice then will be heated from 57°C to 69°C by the condensate of the 2nd effect plus all juice heaters, fourth body and pans (partly).

Heat taken by the juice = heat given by the condensate

$108 \times 1000 \times 0.91 \times (69 - 57) = 42.0 \times 1000 \, (105 - T_{c2+})$

Temperature of condensate $T_{c2+} = 77.0$°C

This condensate will be mainly used for imbibition. About 36 MT of the condensate will be send for imbibition. From the remaining condensate will be send to boiler feed water tank as make up water and /or filter cake wash.

41.3 Steam consumption at pan station:

With conventional 'three massecuite boiling system' the steam consumption is calculated hereunder assuming

Table 41.4 massecuite % cane

Sugar recovery % cane	11.50
A massecuite % cane	26.90
B massecuite % cane	11.40
C massecuite % cane	7.90
Total massecuite % cane	46.20

Assuming syrup brix 60, melt brix 65, and A light molasses brix 70, the average brix of the feed material for A massecuite comes out to be 62.00. Thus it is considered that the feed material for A massecuite is of 62 brix and the feed material (AH -A heavy molasses, BH – B heavy molasses CL – C light molasses) for B and C massecuite after conditioning is of 70°brix. Then the massecuite produced and the water to be evaporated from one tonne of feed material for each massecuite is calculated in the following table.

Table 41.5 Water Evaporated Tonnes

Masse.	Average brix of feed	Tonnes of solids in one tonne feed material	Masse. Brix	Masse. Produced tonnes	Water Evaporated Tonnes
A	62.00	0.62	92.00	0.674	0.326
B	70.00	0.70	96.00	0.729	0.271
C	70.00	0.70	100.00	0.700	0.300

Therefore theoretically water to be evaporated per tonne of each massecuite will be

Table 41.6 theoretically water to be evaporated per tonne of massecuite

Massecuite	water to be evaporated per tonne of massecuite
A massecuite	0.484
B massecuite	0.372
C massecuite	0.428

Considering one tonne of steam evaporates one tonne of water the theoretical steam consumption is equal to the water to be evaporated.

However in practice the actual water to be evaporated is always more than the theoretical quantity. Some water needs to be feed to the pan – for seed washing, for grain hardening, for circulation of highly viscous low grade massecuite (movement water). Therefore, quantity of water to be evaporated is increased. Further because of heat loss some more vapours are required. Webre[1] gives the ratio of actual to theoretical steam consumption for different massecuites. However now with improved equipment, pan boiling, other process techniques and automation of the process we can consider the ratio as follows

Table 41.7 assumptions -actual to theoretical vapour consumption

Massecuite	**Ratio actual / theoretical vapour consumption given by Webre**	**Ratio actual / theoretical vapour consumption in modern techniques**
A massecuite	1.10	1.05
B massecuite	1.15	1.05
C massecuite	1.20	1.10

Table 41.8 actual steam consumption per tonne of massecuite

Massecuite	Theoretically water to be evaporated per tonne of massecuite	Ratio actual / theoretical steam consumption	Actual Steam consumption per tonne of massecuite
A massecuite	0.484	1.05	0.508
B massecuite	0.372	1.05	0.391
C massecuite	0.428	1.10	0.478

Table 41.9 steam consumption at the pan station

Massecuite	**Masse. % cane**	**Steam consumption**	
		Theoretical	Actual (Say)
A massecuite	26.90	13.01	13.67 (13.7)
B massecuite	11.40	4.24	4.45 (4.4)
C massecuite	7.90	3.38	3.71 (3.70)
Total	46.20	20.63	21.83 (21.80)

Thus, the total steam consumption for massecuite boiling can be around 21.80 % on cane. Now we will take round figures of steam consumption given in the last column of the above table while calculating the total steam consumption in the boiling house.

The use of dry seed or B seed saves steam for A massecuite, however it is negligible and therefore is not considered while calculating the steam consumption for A massecuite. In the above example, massecuite % cane is 45.5 %, however massecuite % cane considerably varies from 32 to 53% depending on –

- Quality of sugar to be produced,
- Recovery % cane (raw or white),
- Massecuite boiling scheme followed,
- Exhaustion of massecuite at each boiling,
- Care taken at the time of massecuite boiling in using unaccounted water,
- Operating parameters maintained in massecuite boiling,
- Exhaustion of mother liquor of massecuite in the crystallizers upon cooling,
- Care taken in centrifuging in using wash water.

In some Caribbean countries and South American countries two massecuite boiling scheme is followed for raw sugar manufacture. In such factories, the total massecuite percent cane and therefore the steam consumption at pan station is considerably low.

Ramdasan[2] has given theoretical massecuite % cane for very high recovery and very low recoveries with three massecuite boiling scheme and concluded that for higher recovery the steam consumption at pan station can be 6% higher than that for very low recovery. In high recovery area such south region of Maharashtra state and north region of Karnataka state of India, with high purity juices, the A massecuite purity is above 90 –91. It becomes impossible to bring down C massecuite purity up to 57-58 with three massecuite boiling scheme. Therefore, in this area, in many factories three and half massecuite-boiling scheme is followed. In some factories in India three and half or four massecuite boiling scheme is followed mainly to produce high quality bolder grain sugar and / or sugar of low ICUMSA colour. In comparison to raw sugar production, plantation white sugar manufacture requires higher steam. Thus steam consumption at the pan station can increase or decrease depending upon the massecuite boiling scheme followed. The steam consumption at pan station may vary from as low as 16 % to as high as 26 % on cane.

41.4 Miscellaneous steam consumption:

Now most of the factories are using vapours from the first effect (of about 0.4 – 0.6 kg/cm^2 g pressure) for the following miscellaneous use.

Table 41.10 miscellaneous steam consumption

Molasses conditioning	1.0
Melting of Dust , Rory, C seed, etc.,	0.7
Washings of pan	0.3
Total	2.0

41.5 Steam consumption at evaporator station:

In past until 70 years back all the vessels from multiple effect evaporators were made of equal size, each supplying vapours to the following vessel without withdrawal of any vapours. No much steam economy was considered.

In India some 35 years back the bagasse saving was not economical for most of the factories, because use of bagasse was limited and bagasse was fetching very low price. The cogeneration and supply of electricity to state grid was not come up. Even bagasse handling and transportation cost was not being recovered by sell of bagasse. Therefore, emphasis was not given on efficient use of steam and bagasse saving. Most of the factories were designed with 21 kg/cm^2 boiler pressure and steam driven turbo alternators that provide all the steam and electricity to run the factory. Little surplus bagasse was saved. The bagasse was generally saved only to run the boilers during mill stoppages, general cleaning, finishing the end process and starting the next season.

In last thirty five years the picture has changed. With increase in use of bagasse for paper, hard board, bricks manufacturing and revised state government policies of purchasing electricity from sugar factories, the Bagasse prices gone up and factories can earn money on sale of saved bagasse. Therefore, efficient utilization of steam has become essential to save the bagasse for factories without cogeneration or cogeneration with back pressure route.

In case of factories with cogeneration by 'Condensing Extraction Steam Turbine (CEST) cogeneration system', the reduction in use of exhaust in boiling house increases electricity generation for export.

We have seen in Rillieux's second principle that, there is steam saving when vapours are drawn, robbed or bleeded from evaporator set. Vapours from the last effect going to condenser are lost. When these vapours are used for heating the raw juice by installing a vapour line juice heater (VLJH) in the vapour line going from the last body to the condenser, it recovers direct heat that is being lost in the injection water. There is complete saving of exhaust. It gives highest economy. Further the injection water that has acquired heat in the condenser has to be cooled down in spray pond for which electrical energy is required. This power is saved.

The main unit where the steam consumption can be reduced is the evaporator station. Under vacuum evaporation, some part of the vapours has to be condensed in the condenser. *The steam consumption at evaporator station is the 'Vapours going to condenser from the last effect of the evaporator'.* Therefore, main principle in governing the steam economy measures is to reduce the quantity of vapours going to condenser from the last body of the evaporator. The evaporator steam consumption can be brought down by increasing the vapour bleeding from the evaporator set. This results in lower evaporation load to the last effect.

In deciding vapour bleeding from any evaporator body the following things are to taken into consideration

There should be sufficient vapours can generated to rob the same and use outside the evaporator either at pans or juice heater or for miscellaneous use. Therefore vapour bleeding system and vapour generation load upon each body is first decided. And then the heating surfaces of each body are calculated.

The vapours robed to boil massecuites at pans should be of temperature 103°C or higher. However now low head batch pans or continuous pans can work on vapours of about 100°C for A and B purity massecuites.

While withdrawing vapours from any effect to heat the juice, a temperature difference in between vapours and heated juice temperature is to be taken into consideration. Because the juice heater heating surfaces are calculated accordingly. The table here under gives the difference required.

Table 41.11 temperature difference required in heating media and juice

1	Exhaust steam	6-7°C
2	First effect vapours	9-10°C
3	Second effect vapours	12-13°C
4	Third effect vapours	15-16°C

Now in modern sugar factories the vapours from the first effect are of higher temperature of about 110-115° (depending upon working conditions) can be used for clear juice first heating (from 95° to 102/105°C) and sulphited juice second heating.

In beet sugar industry the exhaust pressure of about 2.7 to 3.0 kg/cm^2 is maintained. However, the cane juice is highly sensitive to higher temperatures. When cane juice is boiled at higher temperature the inversion of sucrose, destruction of reducing sugars and colour formation is very high. Therefore, the system of pressure evaporation is not followed in cane sugar industry. However now some new factories have gone for FFE evaporator set maintaining lower vacuum in last body so that vapour temperature of last body is about 88°C. These vapours are used for continuous pans and vapours going to condenser are totally eliminated. Thus steam consumption at evaporator becomes zero.

Water to be evaporated at evaporator station:

We have assumed here that the clear juice brix is 14.3 and now we will assume that the juice is concentrated in the evaporator up to 60.00 brix. Hence water to be evaporated will be –

Clear juice 5 cane = 108 % on cane,

Clear juice brix = 14.30

Syrup brix = 60.00

Water to be evaporated E = 108 (60 – 14.30) / 60

Or E = 82.26 % on cane

Vapour bleeding system case I

Table 41.12 vapour bleeding system is followed quintuple I

Sr. no.	Particulars	Increase in temp.	vapours drawn from	vapour quantity
1	Raw juice first heating (RJ1)	30°to 45°C	Last effect	2.7
2	Raw juke second heating (RJ2)	45 to 75°C	Third effect	5.6
3	Sulphited juice first heating (SJ1)	73 to 90°C	Second effect	3.6
4	Sulphited juice second heating (SJ2)	90 to 103	First effect	2.7
5	Clear juice first heating (CJ1)	97 to 106	First effect	1.7
6	A massecuite boiling	------	Second effect	13.7
7	B and C massecuite boiling	------	First effect	8.1
8	Miscellaneous	-------	First effect	2.0

The clear juice second heating (SJ2) is to be carried out by exhaust steam.

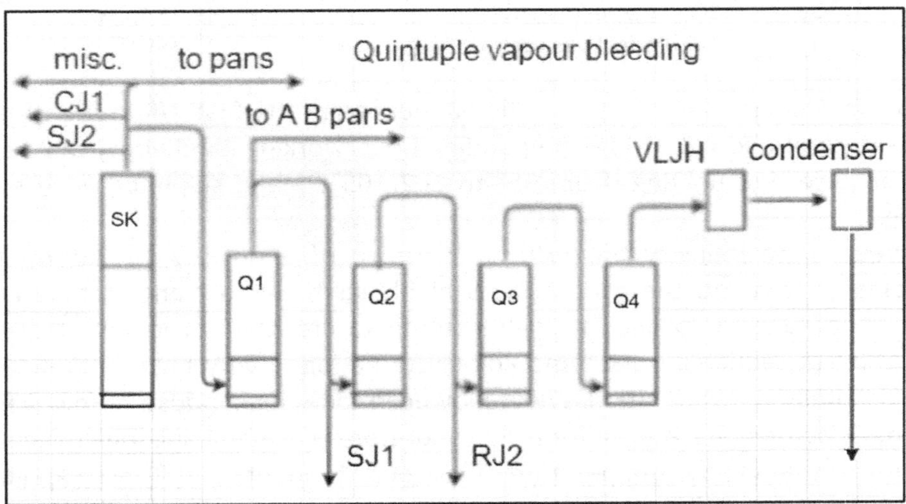

Fig. 41.1. Vapour bleeding system – Quintuple

Table 41.13 Water to be evaporated in each body of quintuple:

No. of Effect	Vapour bleeding	water to be evaporated t/h
	RJ2 SJ1 A masse. SJ2 CJ1 B + C masse. Misc.	
5th	X	3.27
4th	X	3.27
3rd	X + 5.6	8.87
2nd	X + 5.6 + 3.6 + 13.7	26.17
1st	X + 5.6 + 3.6 + 13.7 + 2.7 + 1.7 + 8.1 + 2.0	40.67
Total	5X + 16.8 + 7.2 + 27.4 + 2.7 + 1.7 + 8.1 + 2.0	82.26
Water to be evaporated in the last body X		3.27
Vapours consumed at VLJH		2.70
Vapours going to condenser		0.57

Table 41.14 Total steam consumption percent on cane

1	Steam used for juice heating	18.00
2	Steam used at pan station	21.80
3	Steam required for evaporation	0.57
4	Miscellaneous use	2.00
	Total	**42.37**

Thus with this vapour bleeding arrangement the steam consumption in boiling house remains about 42 to 43 %. However as discussed all the parameters such as mixed juice % cane, massecuite boiling scheme, Massecuite % cane vary considerably from factory to factory and hence the steam consumption also varies. This scheme is now being used in many factories.

The condensate of first effect is superheated of 116-117°C and is directly used for batch centrifugals. When flashing of the remaining condensate from first body and that of all following bodies to their preceding bodies is done. It gives further a steam economy of about 0.4 % on cane. If the total condensate from the first is flashed off at each next effect up to the last effect the steam economy can be up to 0.7 % on cane. Thus the steam consumption in the above case can be brought down to 41 to 42 % on cane.

Vapour bleeding system case II – Quintuple II

In this second case the raw juice is after first heating from 30° to 45°C in VLJH is heated from 45 to 69°C by condensate in second and third stage. Then it is heated by NCG from 69 to 75°C. Therefore there is net saving of about 5.6 % saving. Thus the steam consumption for juice heating which is 18.0 % in the above first case comes down to 12.40 % in this case. Consequently the overall steam consumption also is reduced by about 5.6%. Thus the steam consumption can be brought down to

35 to 37 %.

Vapour bleeding system case II – Quintuple II

Table 41.15 vapour bleeding system is followed

Sr. no.	Particulars	Increase in temp.	vapours drawn from	vapour quantity
1	Raw juice first heating (RJ1)	30°to 45°C	Last effect	2.7
2	Raw juke second heating (RJ2)	45 to 57°c	heated by condensate	------
3	Raw juke second heating (RJ3)	57 to 69°c	heated by condensate	-----
4	Raw juice fourth heating (RJ4)	69 to 75°C	heated by NCG	------
5	Sulphited juice first heating (SJ1)	73 to 90°C	Third effect	3.6
6	Sulphited juice second heating (SJ2)	90 to 103°C	First effect	2.7
7	Clear juice first heating (CJ1)	97 to 106°C	First effect	1.7
8	A massecuite boiling	------	Second effect	13.7
9	B and C massecuite boiling	------	First effect	8.1
	Miscellaneous	-------	First effect	2.0

The clear juice second heating (SJ2) is to be carried out by exhaust steam, the steam requirement being 1.7 % on cane. The total steam requirement for juice heating comes out to be 12.4 t.

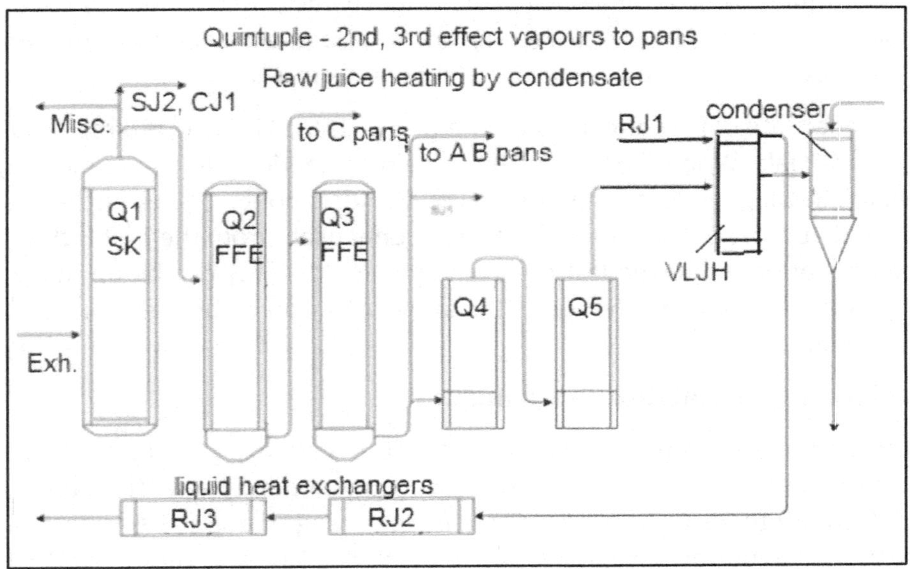

Fig. 41.2 quintuple with 2nd and 3rd effect vapours to pans

Table 41.16 Water to be evaporated in each body of quintuple:

No. of Effect	Vapour bleeding	water to be evaporated t/h
	SJ1 A masse. A/B/ C masse, SJ2 CJ1 Misc.	
5th	X	3.0
4th	X	3.0
3rd	X + 3.6 + 6.5	13.1
2nd	X + 3.6 + 6.5 15.3	28.4
1st	X + 3.6 + 6.5 + 15.3 + 2.7 + 1.7 + 2.0	34.8
Total	5X + 10.8 + 19.5 + 30.6 + 2.7 + 1.7 + 2.0	82.26
Water to be evaporated in the last body X		3.00
Vapours consumed at VLJH		2.70
Vapours going to condenser		0.30

Table 41.17 Total steam consumption percent on cane		
1	Steam used for juice heating	12.40
2	Steam used at pan station	21.80
3	Steam required for evaporation	0.30
4	Miscellaneous use	2.00
	Total steam consumption	**36.50**

This vapour bleeding system is being followed in some newly installed sugar factories. In this case vapour bleeding from the second and third effect can be adjusted according to the pan floor working. The low pressure vapours from the third effect can be used for continuous pan or footing seed washing of A massecuite and graining for B massecuite when the pan level is low.

Further ample evaporator capacity always helps in maintaining higher imbibition at mills which helps in higher mill extraction. With higher imbibition mixed juice brix lowers down at the same time higher syrup brix can be easily maintained. Further the low mixed juice brix helps to improve clarification as settling of mud becomes easy. The steam consumption at pan station is also reduced due higher brix of syrup. Hence total steam consumption is reduced. The run between the two cleanings can also be increased by few days, so that, the general cleaning can be taken at a convenient date. On the other hand with inadequate heating surface the imbibition water % at mill has to be reduced. Many times the lower brix syrup passes at pan station and evaporation has to be completed in single effect in the pan. This results in higher steam consumption at pan station.

References:

1 Webre A L, Meade Cane Sugar Hand book 1960 p 150
2 Ramadasan SISSTA 2001p201-206

Fig. 41.3 modern falling film type evaporator set

-OOOOO-

42 HEATING SURFACE CALCULATIONS

42.1 General:

In the last chapter we have studied two different vapour bleeding systems and worked out the expected steam consumption for each vapour bleeding system. A suitable vapour bleeding system can be decided considering

- Quantity of exhaust production and its pressure,
- Operating conditions of the evaporator set are to be maintained and
- Expected steam consumption.

Once vapour bleeding system is decided, we can work out the quantity of water to be evaporated per hour in each effect of the evaporated set. From the water quantity, the quantity of heat transferred from vapours to juice per hour in each body is calculated.

Now we will go again to the theory part of the heat transfer. In the first chapter we have revised the formula -

$$Q = U\,A\,\Delta T$$

$$\text{Or } A = Q / U\,\Delta T$$

We have seen above how quantity of heat transferred 'Q' is calculated. We can work out the temperature difference 'ΔT' in the heating medium steam / vapours and the juice boiling. The vapour temperature corresponding to absolute pressure in the calandria can be seen from the steam table. The juice boiling temperature can be calculated as in the second chapter.

However 'U' the overall heat transfer coefficient (kcal/m^2/°C/hr.) can be taken from the figures reported by the experts. The table of the overall heat transfer coefficients for multiple effect evaporator given by Jelinek and Classen is given in chapter one.

Webre gives coefficients of heat transfer for evaporator set ranging from 6800 Kcal/m^2/°C/hr. for the first effect heated by steam at 115°C to 2200 Kcal/m^2/°C/hr. for the last effect heated by vapours of 65°C. There is wide variation in the coefficient of heat transfer reported by many experts from different countries for evaporator sets.

We give here under figures with practical consideration.

Table 42.1 Average range of heat transfer coefficients - kcal/m²/°C/h			
		Quadruple	Quintuple
First effect	SK/FFE	3200-3500	3000-3400
	Robert type	2600-3000	2400-2800
Second effect	Robert type	2000-2400	2000-2400
Third effect	Robert type	1300-1600	1100-1400
Fourth effect	Robert type	500-600	700-900
Fifth effect	Robert type	--	400-500

42.2 Evaporator Heating Surface Calculations:

From the above table one can take the figure of heat transfer coefficient for each body and can calculate the heating surfaces of any evaporator set given working conditions. Considering the heat transfer coefficient in the above table we will calculate heating surface for evaporator set.

Let us take example of vapour bleeding system Case no IV quintuple for calculating the heating surface of evaporator set for a 100 tch plant. The operating conditions assumed are as follows

Table 42.2 assumptions for heating surface calculations	
Crushing rate	100 tch
Evaporator set	quintuple
Exhaust pressure	1 kg/cm²
Exhaust temperature	120°C
Clear juice % cane	108%
Clear juice inlet brix	14.30
Syrup brix	60.00
Water to be evaporated	82.26 t/hr.
Absolute pressure in last body	0.18 kg/cm²
Pressure drop (equal for all bodies)	0.37 kg/cm²

Now in all factories clear juice heaters are provided and the clear juice is heated up to or higher than the temperature at which juice boils in the first effect. Therefore no heating surface is utilized to heat the clear juice up to boiling point. However, in case the clear juice is not heated in juice heater, part of the heating surface of the first effect is utilized to heat the clear juice to boiling point. Therefore the evaporation rate of the first effect is drastically reduced which adversely affects the working performance of the next bodies. In calculating the heating surface, equal pressure drop for all bodies is considered. Temperature of vapours from each body corresponding to absolute vapour pressure in that body is taken from the steam table. Here we will take the first case of vapour bleeding calculated for water evaporated in each body in the last chapter.

Table 42.3 Water evaporated in each body (case I)

No. of Effect	water to be evaporated t/h
1	39.67
2	25.17
3	8.87
4	3.27
5	3.27
total	82.26

Table 42.4 Calculations of heating surface

	Evaporator Body	Q 1	Q 2	Q 3	Q 4	Q 5
		S. K.	Robert type	Robert type	Robert type	Robert type
1	Ab. Pressure in vapour space of the body Kg/cm^2	1.66	1.29	0.92	0.55	0.18
2	Juice inlet brix	14.30	22.30	38.53	49.54	54.30
3	Water evaporated kg/hr.	40670	26170	8870	3270	3270
4	Juice outlet brix	22.94	37.52	47.82	53.22	60.00
5	Average juice brix	18.62	30.23	42.67	50.52	56.61
6	Height of calandria tubes m.	4.0	2.0	2.0	2.0	2.0
7	Height of juice column in the tube m	0.8	0.6	0.6	0.6	0.6
8	Density of juice* for average juice brix	1.008	1.076	1.149	1.201	1.249
9	Weight of juice column (gms/ cm^2) Height in cms X density	80.64	64.56	68.94	72.06	74.94
10	Hydrostatic head of juice column kg/cm^2	0.0806	0.0646	0.0689	0.0721	0.0749
11	Total ab. pressure at the bottom of the tube kg/cm^2	1.7406	1.3546	0,9889	0.6221	0.2549
12	Vapour temperature in the vapour space	113.84	106.33	96.78	83.24	57.41
13	Elevation of boiling point due to dissolved solids (e_1)	0.45	0.87	1.58	2.16	2.68
13	Water boiling temperature at the bottom of the tube	115.46	107.76	98.77	86.38	65.00
14	Increase in boiling point due to hydrostatic head (e_2)	1.62	1.43	1.99	3.14	7.59
15	Juice boiling temperature at the top	114.29	107.20	98.36	85.40	60.09
16	Juice boiling temperature at the bottom	115.91	108.63	100.35	88.54	67.68
17	Average juice boiling temperature	115.10	107.91	99.35	86.97	63.86
18	Temperature diff. In boiling juice and heating vapours	4.91	5.93	6.98	9.81	19.36
	Evaporator Body	Q 1	Q 2	Q 3	Q 4	Q 5
19	Coefficient of heat transfer (kcal/m^2/°C/hr.)	3200	2200	1200	700	400
20	Heat transfer per sq. metre per hr. (kcal/m^2/hr.)	15712	13046	8376	6867	7744
21	Latent heat at corresponding ab. pressure	529.0	534.8	540.0	550.0	564.0
22	Evaporation rate	29.70	24.39	15.51	12.48	13.73
23	Heating surface required (m^2)	1369	1073	571	262	238

*Density of juice at corresponding boiling temperature.

In the above calculations the exhaust steam pressure is taken as 1 kg/cm^2 gauge pressure. The saturation temperature of which is 120°C. This temperature is taken in calculating the heating surface of the first effect. However the exhaust even after desuperheating many times remains to some extent superheated.

Further it always varies. It means that the ΔT temperature difference in the exhaust steam in the calandria and the boiling juice in the first effect remains higher than the figure taken for calculation. Therefore in the first effect generally higher rate of evaporation can be achieved. So higher vapours withdrawal from the first effect are possible.

42.3 Dessin's Formula:

French Engineer Dessin has given an empirical formula to calculate the specific evaporation coefficient for any vessel of the evaporator set. It is as follows

C = 0.001 (100 - B) (T - 54)

Where

C = Sp. evaporation coefficient- kg/m^2/°C/h
B = Brix of the juice leaving the vessel
T = Temperature of heating steam in the calandria.

In defecation factories if evaporator sets are cleaned weekly, the factor 0.001 works well. However in sulphitation factories where scale formation is higher, further the evaporator set cleaning is done after three to five weeks, the values of the factor for each evaporator body are less are given in the following table.

Table 42.5 Value of the factor in Dassin's formula for each evaporator body

	Quadruple	Quintuple
First effect	0.001	0.001
Second effect	0.001	0.001
Third effect	0.0008	0.0009
Fourth effect	0.0007	0.0008
Fifth effect		0.0007

Table 42.6 Heating Surface Calculations as per Dessin's formula:

Evaporator Body		Q 1	Q 2	Q 3	Q 4	Q 5
		Semi Kestner	Robert type	Robert type	Robert type	Robert type
1	Water evaporated kg/hr.	40670	26170	8870	3270	3270
2	Temp diff. In boiling juice and heating vapours °C	4.91	5.93	6.98	9.81	19.36
3	Sp. Evapo. Coeffici. as per Dessin's formula (kg/m^2/°C/hr.)	6.74	4.161	2.633	1.645	0.876
4	Evaporation rate (kg/m^2/hr.)	33.09	24.67	18.37	16.12	16.96
5	Heating surface required (m^2)	1229	1060	483	203	193

In most of the factories, there are batch type pans and the pan boiling is a batch process. The vapour demand varies during the boiling of the individual pan. It is highest at the beginning and lowest at the end of the strike. By proper sequencing and scheduling of working of various pans the wide fluctuations in the vapour demand at the pan station can be minimized but cannot be eliminated.

The variation of vapour demand at the pan station adversely affects the working of evaporator. When the vapour demand at the pan station is decreased to maintain the desired syrup brix, more water is to be evaporated in the following effects. Therefore, it is always preferred to give some higher heating surface to the last two effects in case of quadruple and to the last three effects in case of quintuple.

Table 42.7 Comparison of heating surface calculated and recommended

Evaporator Body	Q 1	Q 2	Q 3	Q 4	Q 5
	Heating Surface m^2				
H S for 100 tch as calculated by coefficient of heat transfer assumed	1268	1167	552	213	194
H S for 100 tch as calculated by Dassin's formula	1229	1060	483	203	193
H S 100 tch with practical consideration (margin of safety)	1400	1250	550	260	260
H S for 2500 tcd (115 tch) recommended in Guidelines given by Indian sugar industry	1600	1400	600	300	300

42.4 Practical consideration in deciding heating surface: In deciding the heating surface the following factor are to be considered –

- **The quantity and hardness scale deposition** - The scale may be getting deposited faster and hard in some regions depending upon the soil and weather conditions. Accordingly some more heating surface should be considered in such regions.
- **The process being followed-** the process followed for juice clarification is important. Defecation process gives comparatively less and soft scale that can be removed easily. However in sulphitation process the scale deposition is fast, more and hard. Therefore in sulphitation factories it is always preferred to give some higher heating surface.
- **Percentage of imbibition water used-** The percentage of imbibition water varies considerably, accordingly clear juice percent varies. Higher clear juice percent cane is to be considered.
- **The normal crushing days of run-** The run i. e. duration of days in between two consecutive evaporator set cleanings also varies considerably from one week to five weeks. Accordingly the heating surface is required to be decided.
- **Higher capacity utilization-** Higher capacity utilization in peak recovery period is always considered for higher profits. It is also is to be considered in deciding heating surfaces.
- **Fluctuation in operating conditions-** Fluctuations in the operational parameters such clear juice brix, exhaust pressure, vacuum etc. in factory operation are also to be taken in to consideration.
- **Standby evaporator bodies-** The factories having plan of standby evaporator bodies can take cleaning of the evaporator bodies frequently. In such case keeping less heating surface can reduce the residence time of juice in the bodies. So that the colour formation and inversion loss can be reduced.
- **Method of calculating heating surface-** While calculating heating surface of tubes French manufacturers take the surface in contact with juice i.e. interior surface of the tubes. American/British manufacturer take the surface in contact with steam/vapour i.e. exterior surface of the tubes. However Indian manufactures take mean of the both interior and exterior surface. This point also has to be taken in deciding the heating surface.

-OOOOO-

43 MODERN EVAPORATOR SET

A FFE evaporator set for raw and refined sugar plant with ultimate steam economy

In this chapter we have taken a quadruple set of falling film evaporators and vapours from the last body pass to continuous pans. Thus the steam consumption at evaporator becomes zero. Profuse vapour bleeding is considered that can give ultimate steam economy for a raw sugar plant attached with refinery.

Flashing of first effect condensate to the second effect is followed. For the next effects, all the condensate is circulated for flashing up to 4th effect. For simplicity the calculations are made for 100 TCH plant as in previous chapter. Heating surfaces required are calculated.

Table 43.1 Basic assumptions for evaporator heating surface calculations

Sr. no.	Assumptions:	Quantity
1	Crushing rate	100 tch
2	Maceration % on cane	40.5
3	Mixed juice % cane	113.28 MT
4	Bagasse % cane	27.22
5	Pol % bagasse	1.8
6	Moisture % bagasse	49.00
7	Brix % bag	2.3
8	Fibre % bag	48.7
9	Fibre % cane	13.50
10	Brix % mixed juice	13.30
11	Pol % mixed juice	11.04
12	Mixed juice purity	83.00
13	Pol in mixed juice % cane	12.51
14	Pol in bagasse % cane	0.49
15	Pol % cane	13.00

Table 43.2 Juice quantity and heating

Sr. no.	Assumption /particulars		Quantity
1	Pol % filter cake		1.0
2	Filter cake % cane		4.3
3	Pol loss in F C % cane		0.043

4	Pol in clear juice % cane		12.47
5	Brix in clear juice % cane		14.93
6	Clear juice purity		83.52

Table 43.3 Juice heating

Sr. no.	Assumption /particulars	Quantity
1	Raw juice heating by NCG, increase in juice by 1.12 %	114.40
2	Filtrate recycled	17.00
3	Milk of lime added to mixed juice % on cane	2.0
4	Defection juice % cane	133.40 MT
5	Raw juice 1st heating by condensate in heat exchanger	30 to 36°C
6	Raw juice 2nd heating in VLJH at continuous pan	36 to 46°C
7	Raw juice 3rd heating by condensate in heat exchanger	46 to 66°C
8	Raw juice 4th heating in by NCG	66 to 73°C
9	Defecation juice 1st heating by vapours from 4th effect in DCH	73 to 88°C
10	Defeca. juice 2nd heating by vapours from 3rd effect in DCH	88 to 98°C
11	Defecation juice 3rd heating by vapours from 2nd effect in DCH	98 to 101°C
12	Clear juice heating by vapours from 1st effect in DCH	97 to 112°C

Table 43.4 Defecation juice and clarified juice percentage after heating

Sr. no.	Assumption /particulars	Quantity
1	DJ 1st heating in DCH 73 to 88°C by vapours from last effect	136.794 MT
2	DJ 2nd heating in DCH 88 to 98°C by vapours from 3rd effect	139.139 MT
3	DJ 3rd heating in DCH 98 to 101°C by vapours from 2nd effect	139.855 MT
4	Clarifier under flow % on cane	17.0
5	Clarifier overflow % On cane	122.855 MT
6	Clarified juice % after heating in DCH to 101°C.	126.196 MT
7	Brix % Clear Juice (after heating up to 112°C in DCH)	11.83

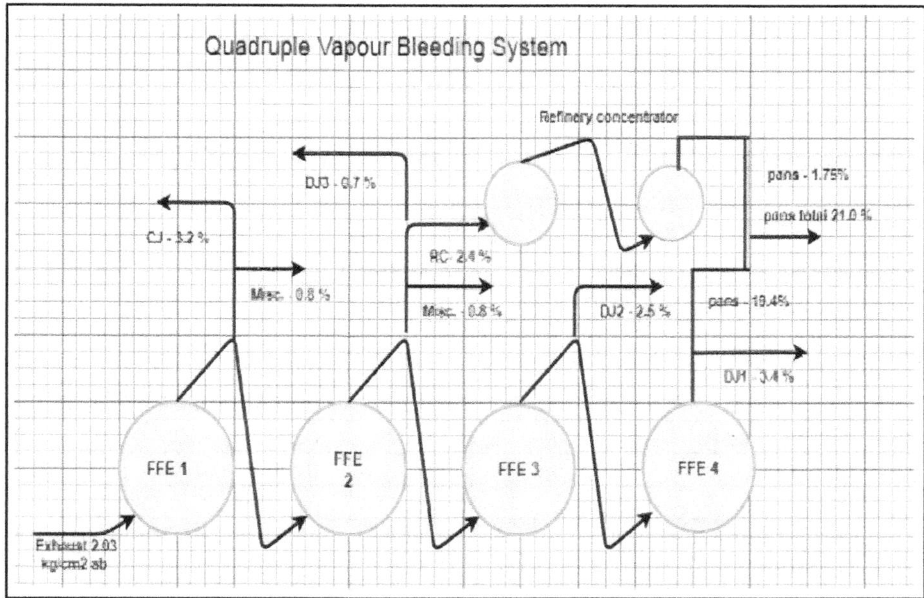

Fig.44.1 Evaporator set - vapour bleeding arrangement

Table 43.5 Evaporator working conditions:

	Evaporator Body	FFE 1	FFE 2	FFE 3	FFE 4
	Calandria vapours working conditions				
1	steam / vapour pressure kg/cm2 abs	2.03	1.70	1.37	1.04
2	Steam/vapour temperature °C	120.5	115.15	108.65	100.96
3	steam / vapour total heat Kcal/kg	647.61	645.69	643.30	640.43
4	vapour latent heat Kcal/kg	526.32	530.14	534.21	539.47
	Vapour space working conditions				
5	pressure kg/cm abs	1.70	1.37	1.04	0.71
6	temperature °C	115.15	108.65	100.96	91.04
7	steam / vapour total heat Kcal/kg	645.69	643.30	640.43	636.60
8	vapour latent heat Kcal/kg	530.14	534.21	539.47	545.45
	Condensate -				
9	water temperature °C	120	115	108	100
10	Quantity kg/hr.	32,500	26,760	23,486	21927

Table 43.6

	Evaporator Body	FFE 1	FFE 2	FFE 3	FFE 4
	Juice -				
11	Inlet juice quantity kg/hr.	126196	95,421	68,247	44,449
12	Outlet juice quantity kg/hr.	95,421	68,247	44,449	22259
	BPR (by Hugot formula)				
13	Incoming juice °C	0.27	0.32	0.44	1.05
14	Outgoing juice °C	0.37	0.55	1.05	3.26
15	Average °C	0.32	0.44	0.75	2.16
16	Juice inlet temperature °C	112	115.52	109.20	101.99
17	Juice outlet temperature °C	115.52	109.20	101.99	93.20
18	Inlet juice brix	11.83	15.65	21.88	33.59
19	Outlet juice brix	15.65	21.88	33.59	67.07
20	Specific heat of inlet juice	0.93	0.91	0.87	0.80
21	Specific heat of outlet juice	0.91	0.87	0.80	0.59

Table 43.7 exhaust / vapours

	Evaporator Body	FFE 1	FFE 2	FFE 3	FFE 4
22	Steam / Vapours supplied to the calandria kg/hr.	32500	26,760	23,486	21927
23	Vapours generated	30,760	27,174	23,798	22,190
24	Vapours bleeded kg/hr.	CJ - 3215 Misc.- 800	DJ 3 - 716 Misc.- 800 R C. 2400	DJ2 - 2521	DJ1 - 3394
25	Total vapours bleeded	4011	3916	2345	3394
26	Vapours to next effect kg/hr.	26764			
27	+ Vapours from flashing of condensate kg/hr.	----	228	474	761
28	Total vapours to next effect (to pans from 4th effect) kg/hr.	26764	23,486	21927	19567

A) Heat balance of evaporator:

Table 43.8 First effect

Heat entering Kcal/hr.

Sr. no.	Particulars	Calculations	Heat in KCal
1	Steam	32500 kg X 647.6 total heat	21047325
2	Juice	126196 kg X 112°C X 0.93 sp. heat	13144575
3	Total		34191900

Heat leaving Kcal/hr.

Sr. no.	Particulars	Calculations	Heat in KCal
4	Loss	12 KCal/kg	390000
5	Condensate	32500 kg X 120°C	3900000
6	(loss + condensate)		42,90,000
	Heat in vapours + juice X_1 – vapour quantity	645.69 X_1 + (126196 – X_1)0.93*115.42 =	2,99,01900
6	Vapours	30775 kg X 645.69 KCal	1,98,71,110
7	Juice	95421 kg X 0.91 (sp. Heat) X 115.42°C	1,00,30,961
8	Total		3,41,92011

Vapours balance:

Sr. no.	Particulars	Calculations	kg/hr.
9	Vapours generated		30,775
10	Vapours bleeded	1) clear juice heating from 97 to 112°C, vapours in - 3215 kg/hr.	
		2) vapours for miscellaneous use - 800 kg/hr.	
		Total vapours bleeded 4015 kg/hr.	
11	Vapours to 2nd effect		26760

Table 43.9 Second effect

Heat entering Kcal/hr.

Sr. no.	Particulars	Calculations	Heat in KCal
1	vapours KCal	26,760 kg X 645.69 total heat	1,72,86,664
2	Juice	95,421 kg X 0.91 (sp. Heat) X 115.42°C	1,00,30961
3	Total		2,73,09625

Heat leaving Kcal/hr.

Sr. no.	Particulars	Calculations	Heat in KCal
4	Loss	10 KCal/kg	2,67,600
5	Condensate	26,760 kg X 115°C	30,77,400
6	(loss + condensate)		33,45,000
	Heat in vapours + juice	643.30 X_2 + (126215 – X_1)0.87*109.20 = Where X_2 – vapour quantity, 0.87 - Sp. heat of juice, 109.20 - Temp of juice	2,39,64,625
6	Vapours	27174 kg X 643.30 KCal	1,74,81,034
7	Juice	68247 kg X 0.87 (sp. Heat) X 109.20°C	64,83,738
8	Total		2,73,09772

Vapour balance

Sr. no.	Particulars	Calculations	kg/hr.
9	Vapours generated		27,174 kg
10	Vapours bleeded	1) Refinery concentrator - 2400 kg/hr.,	
		2) DJ3 heating 98 to 101°C - 716 kg/hr.	
		3) vapours for miscellaneous use - 800	
		Total vapours bleeded - 3916	
11	Vapours to 3rd effect		23,258
12	Vapours by flashing of condensate		228
13	Total vapours to 3rd effect		23,486 kg

Table 43.10 Third effect:

Heat entering K Cal/hr.

Sr. no.	Particulars	Calculations	Heat in KCal
1	vapours KCal	23,486 kg X 643.30 total heat	1,51,08,544
2	Juice	68241 kg X 0.87 (sp. Heat) X 109.20°C	64,83,782
3	Total		2,15,92,326

Heat leaving Kcal/hr.

Sr. no.	Particulars	Calculations	Heat in KCal
4	Loss	8 KCal/kg	1,87,888
5	Condensate	23486 kg X108° C	25,36,488
6	(loss + condensate)		27,24,376
	Heat in vapours + juice	X_3 + (68241 – X_3)0.8*101.99 = Where X_3 – vapour quantity, 0.8 - Sp. heat of juice, 101.99 - Temp of juice	1,88,67950
6	Vapours	23,798 kg X 640.43 KCal	1,52,40,953
7	Juice	44449 kg X 0.8 (sp. Heat) X101.99 °C	36,26683
8	Total		2,15,92,011

Vapour balance:

Sr. no.	Particulars	Calculations	kg/hr.
9	Vapours generated		23,807
10	Vapours bleeded	DJ2 heating 88 to 98°C vapours bleeded - 2345 kg/hr.	
11	Vapours to 3rd effect		21,453
12	Vapours by flashing of conden.		474
13	Total vapours to 4th effect		21,927 kg

Table 43.11 Fourth effect:

Heat entering Kcal/hr.

Sr. no.	Particulars	Calculations	Heat in KCal
1	vapours KCal	21,927 kg X 640.43 total heat	1,40,42,709
2	Juice	44449 kg X 0.80 (sp. Heat) X 101.99°C	36,26,683
3	Total		1,76,69,392

Heat leaving Kcal/hr.

Sr. no.	Particulars	Calculations	Heat in KCal
4	Loss	6 KCal/kg	1,31,562
5	Condensate	21,927 kg X 100 C	21,92,700
6	(loss + condensate)		23,24,262
	Heat in vapours + juice	$636.36\ X_4 + (44449 - X_4)0.59*93.20 =$ Where X_4 – vapour quantity, 0.59 - Sp. heat of juice, 93.20 - Temp of juice	1,53,45,130
6	Vapours	22,191 kg X 636.36 KCal	1,41,21465
7	Juice	22,258 kg X 0.59 (sp. Heat) X 93.20 °C	1,22,3923
8	Total		1,76,69,650

Vapour balance:

Sr. no.	Particulars	Calculations	Heat in KCal
9	Vapours generated		22,191 kg
10	Vapours bleeded	DJ3 heating	3,394
11	Vapours to pans		18,797
12	Vapours by flashing of conden.		761
13	Total vapours to pans		19,568 kg

Table 43.12 Evaporator heating surface calculations:

Sr. no.	Particulars	FFE 1	FFE 2	FFE 3	FFE 4
1	Calandria - Vapour temperature	120.5	115.15	108.65	100.96
2	Vapour space - vapour temperature	115.15	108.65	100.96	91.04
3	BPR	0.32	0.44	0.75	2.16
4	Juice boiling temperature	115.47	109.19	101.71	93.20
5	Temperature diff. In heating vapours and boiling juice °C	5.03	5.96	6.94	7.76
6	Sp. Evapo. Coefficient as per Dessin's formula (kg/m²/°C/hr.)	5.61	4.78	2.90	1.10
7	Evaporation rate (kg/m²/hr.)	28.22	28.49	20.13	8.54
8	Water to be evaporated kg/hr.	30,790	27,184	23,807	22,027
9	Theoretical Heating surface required (m²) for 100 tch (according to Dessin's formula)	1091	954	1183	2579
10	Heating surface recommended for 100 tch sq. metres.	1450	1450	1450	1800
11	Heating surface for 230 tch plant sq. metres.	3400	3400	3400	4000
12	with equal heating surface of all FFE for 230 TCH	3600	3600	3600	3600

*with one additional FFE as standby

B) Refinery double effect concentrator

Table 43.13 First effect - Working conditions:

sr. no.	Particulars	
1	Vapours supplied kg	2400 kg/hr.
2	Clear liquor feed kg	20,250 kg/hr.
3	Clear liquor inlet temperature	70°C
4	clear liquor inlet brix	60.00
5	BPR – inlet liquor	3°C
6	BPR outlet liquor	3.7°C
7	Average BPR	3.35°C
8	Vapour temperature	100.96
9	Liquor boiling temperature	104.31

Table 43.14 First effect - Heat entering Kcal/hr.

Sr. no.	Particulars	Calculations	Heat in KCal
1	vapours KCal	2400 kg X 640.43 total heat	15,43,920
2	Clear liquor	20,250 kg X 0.64 (sp. Heat) X 70°C	9,07,200
3	Total		24,51,120

Heat leaving Kcal/hr.

Sr. no.	Particulars	Calculations	Heat in KCal
4	Loss	8 KCal/kg	19,200
5	Condensate	2400 kg X 108 C	2,59,200
6	(loss + condensate)		2,78,400
	Heat in vapours + clear liquor	640.43 X_5 + (20,250 – X_5)0.61*104.66 = Where X_5– vapour quantity, 0.59 - Sp. heat of liquor, 93.20 - Temp of liquor	21,72,720
7	Vapours	1526 kg X 640.43 KCal	9,77,296
8	Clear liquor	18,724 kg X 0.61 (sp. Heat) X 104.66 °C	11,93,388
9	Total		24,49,084

Vapours

Sr. no.	Particulars	kg/hr.
11	Vapours generated	1526
12	Vapours from flashing	22
13	Total vapours pass to 2nd effect	1548

Table 43.15 Second effect - Working conditions:

sr. no.	Particulars	
1	Vapours supplied kg	1548 kg/hr.
2	Clear liquor feed kg	18724 kg/hr.
3	Clear liquor inlet temperature	104.66°C
4	clear liquor inlet brix	64.89
5	BPR – inlet liquor	3.7°C
6	BPR outlet liquor	5.02°C
7	Average BPR	4.36°C
8	Vapour temperature	95.40
9	Liquor boiling temperature	104.31

Table 43.16 Second effect - Heat entering Kcal/hr.

Sr. no.	Particulars	Calculations	Heat in KCal
1	vapours KCal	1548 kg X 640.43	9,91,386
2	Clear liquor	18,724 kg X 0.61 (sp. Heat) X 104.66 °C	11,93,388
3	Total		21,84,774

Heat leaving Kcal/hr.

Sr. no.	Particulars	Calculations	Heat in KCal
4	Loss	6 KCal/kg	9,288
5	Condensate	1548 kg X 100° C	1,54,800
6	(loss + condensate)		1,64,088
	Heat in vapours + juice	636.36 X_6 + (18724 – X_5)0.57*96.06 = Where X_6 – vapour quantity, 0.57 - Sp. heat liquor, 96.06 - Temp of liquor leaving the effect	20,20,686
7	Vapours	1713 kg X 636.36 KCal	1090085
8	liquor	17011 kg X 0.57 (sp. Heat) X 96.06 °C	931424
9	Total		21,85,597

Vapours:

	vapours	kg/hr.
10	Vapours generated	1711
11	Vapours from flashing	34
12	Total vapours pass to pans	1745
13	Clear liquor outlet brix	71.11

Table 43.17 Refinery concentrator heating surface calculations:

Sr. no.	Particulars	FFE 1	FFE 2
1	Calandria - Vapour temperature	108.65	100.96
2	Vapour space - vapour temperature	100.96	91.04
3	BPR	3.35	4.36
4	Juice boiling temperature	104.31	95.40
5	Temperature diff. In heating vapours and boiling juice °C	4.34	5.56
6	Sp. Evapo. Coefficient as per Dessin's formula (kg/m^2/°C/hr.)	1.55	0.94
7	Evaporation rate (kg/m^2/hr.)	6.73	5.23

8	Water to be evaporated kg/hr.	1526	1711
9	Theoretical Heating surface required (m^2) for 100 tch (Dessin's formula)	227	327
10	Heating surface recommended for 100 tch sq. metres.	290	290
11	Heating surface for 230 tch plant sq. metres.	670	670

Table 43.18 Miscellaneous steam consumption details:

Sr. no.	Miscellaneous vapour consumption from first effect	Percentage on cane
1	Raw sugar melter	0.50
2	Pan washings	0.25
3	Transient heater	0.05
4	Total	0.80
	Miscellaneous vapour consumption from second effect	
1	Molasses conditioner	0.40
2	Raw house melter	0.20
3	Sugar drying	0.20
4	Total	0.80
	Total	**1.60**

-OOOOO-

44 TIPS IN REDUCING STEAM CONSUMPTION

44.1 General:

Miscellaneous use of steam /vapour consumption are such as melting of sulphur at the sulphur burner, melting of seed dust and sweeping sugar, pan washing after each strike. In addition to this, steam consumption is increased by about 0.5 % on cane for general cleaning in the factory.

At sulphur burner Steam of pressure 5-7 kg/cm^2 and temperature 170-180°C is used for melting the sulphur. The consumption of steam at sulphur burner is negligible. This steam also was used at centrifugal section but now with the use of modern centrifugals steam wash is not used and the use of steam has considerably been reduced.

In factories that follow double sulphitation process for plantation white sugar manufacture steam consumption is around 35-44 % on cane. In some modern factories with FFE quadruple set and continuous pans are installed. The pans are boiled on vapours from the last effect of quadruple. All the vapours from the last effect pass to pans and the steam consumption at evaporator becomes zero. Steam consumption of these factories have been brought down to 27-28 % on cane.

In raw sugar factories steam consumption is generally less than 40% on cane and the factories which follow two massecuite boiling system it may come down up to 35 % on cane.

Important points in reducing steam consumption:

1) Capacity utilization: Higher capacity utilization with minimum mill stoppages gives maximum steam economy.

Time lost is also important. If there are frequent mill stoppages, it causes fluctuations in exhaust steam pressure and juice flow rate to the evaporator. The fluctuation disturbs working of evaporator set.

For a 2500 TCD plant every one-hour stoppage during working results in extra steam consumption of approximately 15 tonnes.

2) Imbibition % cane: The higher mixed juice % cane increases the steam consumption at evaporator. Decrease in imbibition decreases the mixed juice percent cane and increases the mixed juice brix. In Taiwan, it is estimated that a 1 % increase in brix of mixed juice leads to a 1% decrease in steam consumption[3].

However, reducing the imbibition percent at mill is not recommended as it directly affects the mill extraction. To avoid unwanted increase in quantity of mixed juice, the imbibition efficiency should be high with optimum quantity of imbibition. In Indian sugar factories the imbibition water % on fibre normally is from 200 to 300 %.

3) Steady exhaust pressure: The exhaust pressure and temperature should be steady now in most of the factories automation of pressure reducing and desuperheating station (PRDS) is installed. And therefore the exhaust pressure and temperature remains more or less constant.

4) Clear juice heating: clear juice heating above the boiling point of the juice in the first effect of the evaporator set is very much essential to achieve good performance of evaporator.

5) Faulty operation of evaporator set: Many times steam demand is increased due to faulty operation of evaporator set. The poor operation of evaporator set always results in thin syrup, which increases steam demand at the pan station. Haleem[2] has discussed numerous ill effects of thin syrup besides increase in steam consumption.

6) Fluctuation in steam demand at pan station: in most of the factories the pan boiling is a batch process. The steam demand varies during the boiling of the individual pan. It is highest at the beginning and lowest at the end of the strike. These fluctuations affect evaporator station working. Syrup of lower brix passes to the pan station. It increases vapour demand at pan station. To reduce these fluctuations the pans should be at different levels in boiling for e g while one is being discharged the other should be for graining or footing and the remaining at levels of second or third sight glass. Thus by proper scheduling of pan boiling it is possible to minimize the wide fluctuations in steam demand at pan station.

In many factories there are continuous pan for B massecuites and in some factories for A massecuite. But only few numbers of factories in which all three massecuite are boiled in continuous pans. Continuous pans considerably reduce the fluctuations in vapour demand at pan station that helps in stabilizing the working of evaporator set. Continuous pans and low head calandria pans can work efficiently at low pressure steam; hence second body vapours from the evaporator set can be effectively utilized for pans that give more steam economy.

7) Use of water at pans and centrifugals: The steam consumption at pan station can be reduced by proper control at the pan station on

- Dilution of molasses,
- Use of movement water
- Timing for grain hardening.

Use of molasses conditioners with automatic control gives fast boiling and improved working of pans resulting into steam economy. Use of higher brix B heavy and C light molasses (71 – 73° Bx.) can reduce the steam consumption at pan station.

Now with high capacity fully automatic batch centrifugals and continuous centrifugals, the steam consumption at centrifugal station is considerably reduced.

Preparation of magma in light molasses or syrup and melting of seed in syrup reduces input of water and hence the load of water to be evaporated at the pan station can be reduced. Use of clear juice for melting purpose in place of water reduces steam consumption at pan station. Further, the load on evaporator is also reduced as feed clear juice quantity is reduced.

The condensate from clear juice heater is of high temperature (112 - 117°C) can be used at centrifugals so that the steam requirement for superheating the water used at centrifugals can be reduced or even can be avoided.

The water used at pans and centrifugals if metered helps to control unwanted extra through put of water in the process and steam consumption can be controlled. Leakage of water into the process apparatus causes increase in steam consumption.

8) Exhaustion of massecuite: Optimum exhaustion of high grade and intermediate massecuite reduces the massecuite percentage, which consequently reduces steam consumption at pan station.

Use of ammonia or high pressure vapours as jigger steam at pans gives better circulation of boiling massecuite in the pans. This improves boiling at lower pressure. Pans boil well on low pressure vapours from the second effect. The exhaustion of the massecuite is also maintained. Low head pans with improved designs with mechanical stirrers and continuous pans can work well on vapours of slightly lower pressure than atmospheric pressure.

9) Instrumentation: Instrumentation at different units is of great help in the control of process parameters.

With installation of automatic brix controller for molasses conditioners and melter, the extra unnecessary dilution is avoided. This reduces the evaporation load at pan station. Similarly, installation of automatic feed control valves to vacuum pans improves working of pans resulting in reducing steam consumption. The measurement of condensate production (with flow meters) at pans and evaporator gives idea of pan and evaporator working in respect of steam / vapour consumption.

10) Use of first body vapours: The high pressure vapours from the first effect of evaporator set can be used for clear juice heating (1st stage), soda boiling in spare bodies of the evaporator set for steam economy[3].

11) Use of antiscalent: Use of antiscalent at evaporator to some extend inhibits deposition of scale on heating surface that helps to maintain higher overall heat transfer coefficient. Therefore, the syrup brix remains higher for longer duration in the run that gives steam economy.

12) Bolder grain and smaller grain: With bold grain sugar production the recirculation of sugar as well as nonsugars is increased. Massecuite % cane is increased. The time required for pan boiling is also increased. Hence, the capacity of the pans is reduced and the vapour consumption is increased.

13) Leakages of steam: the loss of steam /heat due to leaks is seldom realized. Leakages of steam from very small hole or flange joints gives loss[4] resulting in higher steam consumption. If all steam leakages are added and are equivalent to 5 mm dia hole to exhaust pipe, then approximately 50 kg of steam will be lost per hour[5]. It means 1.2 MT of steam per day or say 0.5 MT of bagasse per day. Therefore systematic inspection of pipeline should be done. Most of the leaks are occurred at the flange joints. Sometimes the flanges are not smooth and even after replacing the joints the leakage are continued. In such cases the flange has to be changed.

14) Proper lagging of pipes: At evaporator, some heat is lost by convection, conduction and radiation. W E Kerr estimates the loss of heat in well-lagged quadruple 2.7 % of steam applied to first effect[6]. The loss of heat occurring in the first effect really causes (n-1) times steam loss where 'n' is the number of effects in the evaporator set. Damaged or inadequate insulation leads to heat loss. Proper lagging of all pipes and bodies is essential in order to check the heat loss due to radiation.

References:

1 Webre A L, Meade Cane Sugar Hand book 1960 p 150
2 Ramadasan SISSTA 2001p201-206
3 Noel Deerr cane sugar 1921 p 455.
4 Haleem Sajeed STAI 1995 p M 323 – 333.
5 Dwivedi R K STAI 2002, p M 119 – 128
6 V Singh STAI Seminar Hyderabad June 1983 p 65
7 Mangal Singh
8 Webre A L p 235.
9 Valagatte S A DSTA 2002 p E 37 – E 42
10 Jadhav J T DSTA 2004, p G31-35

-OOOOO-

45 EVAPORATOR OPERATION

45.1 Testing of evaporator set:

Before starting of season all evaporator bodies are to be tested for leakages by filling the bodies completely with water. To start the testing a body is checked from inside and if any unusual material like cotton waste lying inside is to be removed and the bottom saucer is to be cleaned. Then manholes are to be closed and then all the valves of the body (juice inlet outlet valves, wash out valve, condensate valve of the calandria of the next body, vapour bleeding valve to the vapour pipeline etc.) are to be closed.

The top air vent of the body, the air vent to the vapour pipe, and ammonia connection of the next body are to be kept open. It is required to vent out the air. The bottom manhole of the next body should be kept open for checking the cut-valve for leakage.

Then water inlet valve of the body is opened and water is filled inside the body. When the body is filled with water, the water passes through the vapour pipe to the calandria of the next body. When the calandria of the next body is filled completely, the ammonia valves of the calandria are closed. Then the vapour pipe and the dome of the body are filled. When the body and vapour pipe are filled completely the water flows through the vent at the top of the body. Then the air vents are closed.

Further pressure in body is developed corresponding to overhead tank. Sometimes the pressure is developed by service water pump. The pressure is observed. It should remain steady. Then the water inlet valve is closed. The body is checked carefully for any leakages through the joints, flanges, valves, pump glands, sight glasses etc. If any leakages are observed, those are attended after draining the water and the body is again tested.

To test the last body a dummy is fitted at bottom end of the tail pipe of the condenser. The valves of the condenser are closed and the body is tested. The dummy in the tail pipe is removed after satisfactory testing of the last body.

45.2 Starting of evaporator set:

When evaporator is to be first started, all valves are checked and closed. A dummy is fitted on the water pipe going to evaporator set for testing. The clear juice heater is checked for valves and all the valves are closed. While taking the evaporator into operation, the following steps are taken.

- The exhaust valve and the ammonia vent valve of the first effect are opened. The exhaust steam is entered in the calandria and the air is passed through the air vent. The body is heated. It is necessary to heat the body before taking the juice into it, because if the body is cold the hot incoming juice becomes cold and the boiling of juice does not start in short time.

- At the same time the injection pump is started and the condenser valves are opened to create vacuum in the last body.
- Then the ammonia connection of the calandria of the last body is wide opened so that air from the calandria of the last body is removed. Consequently, the air from the vapour space of the preceding body is passed to the calandria of the last body and from there to the condenser. Therefore, vacuum is created in the preceding body. In this way, vacuum is created in preceding bodies.
- As a next step the exhaust valve, the juice inlet valve and juice outlet valve of the clear juice heater are opened.
- After that, the vacuums in the bodies are checked.
- Next, the exhaust valves of the clear juice heater and the first effect are almost completely opened.
- The juice inlet valve of the first body is opened.
- Now the clear juice can be taken to evaporator. The clear juice pump is started. The clear juice pump suction and delivery valves are opened.
- Generally evaporator is started before the clarifier is overflowed. The clear juice from the clarifier is taken by draining it to sump tank and from sump tank to clear juice pump. So that the juice flow can be controlled till the evaporator set operation gets stabilized. When the clarifier is overflowed, by that time evaporation operation is stabilized.
- When clear juice pumping is started and the clear juice pass to the clear juice heater, the air vent valves - provided on the top cover to remove the air in the juice heater, are opened and closed one by one.
- The juice is passed through the heater to the first body.
- The vacuum equalizing connections to condensate pipe of the juice heater and the first effect are opened. The suction valves of the condensate pumps are opened. The condensate pumps are started and the delivery valves are opened. As juice is passed to next bodies, it is followed for the same.
- As juice boiling is started in the second effect, the vapour bleeding of the first effect is to be followed. Similarly when juice boiling in the third effect is observed the vapour bleeding of the second effect is done. While bleeding the vapours, care has to be taken that if the vapours are just at atmospheric pressure, or under vacuum, the ammonia connection of the juice heaters / pans are to be connected to vacuum.

At the starting of the season when mills are started but no juice is available to send it to evaporator, to consume the exhaust, water boiling is done at the evaporator. When the evaporator set starts after shut down the vents are kept wide open for short time to remove air from the bodies and then after some time the openings are gradually reduced. When operation gets stabilized the vent valves are one fourth opened.

As juice passes from the first body to the last body it should observed that the juice boiling level in each body is up to half of the first sight glass. Alternatively it can also be maintained up to 1/3rd of the tube height by observing the juice level gauge (glass tube).

When the syrup level in the last body is attained, the suction valve of the syrup pump is opened. The pump is started and the delivery valve is opened as per requirement. The syrup brix can be judged by the descending syrup drops over the sight glass of the last body.

45.3 Working of evaporator:

The evaporator station finds very simple in operation as long as it operates without problems. But when problems are encountered even the best designed evaporator sets fail to give satisfactory results under irregular working conditions and defective operation. Therefore good understanding of the principles of multiple effect evaporator and knowledge of practical difficulties is very much essential.

The steam consumption at pan station and total steam consumption very much depends of brix of syrup. If syrup brix decreases from 60 to 50, the steam consumption is increased by about 3.5 to 4 %. Therefore, efficient operation of evaporator is very much important.

The performance of evaporator depends on transfer of heat in each calandria from steam/ vapour to boiling juice inside the tube. This transfer of heat depends on temperature difference, juice level, scale inside the tubes and viscosity of the juice.

Unnecessarily forcing the evaporator set to work at higher capacities by maintaining higher crushing rate with high imbibition effects low brix of syrup.

Generally sampling arrangement for taking juice samples for brix measurement is provided for each body of the evaporator. A 50-mm diameter pipe coupling with two valves is usually fitted to the cut-line of each body. With this fitting, juice sample can be obtained without affecting the vacuum. The brix of outlet juice for each body can be checked by taking sample through it.

When there is minimum brix difference in inlet and outlet juice brix than expected, it indicates short-circuiting of the juice circulation in the body. This adversely affects working.

In operating the evaporator set following particulars, are to be taken into consideration.

1) Exhaust steam pressure / temperature:

For good performance of evaporator the exhaust pressure and temperature should be maintained within tolerance limit. The higher exhaust pressure gives better performance at evaporator.

However when the bodies are clean the condensation rate of exhaust is high and less pressure is observed. When evaporator is in operation as days pass scale deposition continues and the coefficient of heat transfer goes on decreasing. Consequently, the exhaust pressure slowly goes on increasing and the temperature difference in the exhaust steam and juice boiling of the first effect goes on increasing. This maintains the rate of evaporation in the body.

To bleed the live steam into the exhaust and reduce the temperature of the bleeded live steam and exhaust, a well-designed auto controlled pressure reducing and desuperheating system (PRDS) is required.

Condensate connection to provide sufficient water for desuperheating is also equally important. At Ponni the pump provided for PRDS was getting condensate from the first effect condensate delivery. The PRDS was not getting sufficient quantity of water. When a separate connection from boiler feed water tank to the pump was given, working of the PRD unit was improved[1].

2) Clear juice heating:

The clear juice is heated up to or above the boiling point of the juice in the first effect. The heated clear juice temperature 2-3°C higher than the juice boiling temperature in the first effect gives good results in the first effect. It gives flashing effect as juice enters the first effect. Therefore, the juice velocity and turbulence within the tubes is increased. It improves the heat transfer coefficient in the first effect.

3) Optimum vacuum in each body:

The vacuum in the last body should be about 650 (±10) mm of Hg. The vacuum in the preceding effects gets automatically adjusted as vapours from each body get condensed in the calandria of the next body. However, the incondensable gas removal should be properly taken care off. At excessive high vacuum the boiling point of syrup decreases but at the same time viscosity of syrup increases and heat transfer and hence rate of evaporation in the last effect of the evaporator set decreases.

Maintaining steady vacuum is very important for efficient working of evaporator set. Fluctuations in vacuum result in unsteady heat transfer. It may also result in entrainment. Vacuum fluctuations may occur due to air leakages and it should be checked carefully.

Low vacuum results in high temperature in the bodies. Hence, temperature difference in vapours and the boiling juice is decreased that reduces heat transfer.

- Insufficient vapour bleeding gives extra load on the following bodies and to the condenser that may reduce vacuum.
- Air leakages should be avoided 0.5% air in the steam reduces the heat transfer coefficient by about 10-14 %.
- The low vacuum may be due to high temperature of injection water. The higher temperature of injection water is generally because of inefficient working of spray pond section like short-circuiting of water channels, inefficient working of spray pump, less number of nozzles or the nozzle fallen down.
- Low vacuum may be due to inefficient working of injection water pumps. Sometimes strainers of injection waters pumps get jammed by some foreign material and pump may not get sufficient water to lift. Therefore, the strainers should be periodically cleaned.
- After years of use, because of acidic injection water the impellers of the pumps get worn out and the efficiency of the pump gets reduced. It cannot lift the water as per capacity.
- Inefficient working of condenser may also be the cause of low vacuum. The condenser should be checked periodically for air leakages. For efficient working of condenser, it should get optimum quantity of water of lower temperature (maximum 35°C) and at a desired pressure (as per design of the condenser).

Now in new designs of multijet condensers a strainer at the entrance of the water connection is provided. It arrests foreign material to pass to the nozzles of the condenser, so that the nozzles do not get choked. Further the foreign / odd material arrested in the strainer can be removed easily by opening it in short time.

The tail pipe of the condenser should be sealed by the water level in the injection water channel.

In working, leakages may be developed at the joints of evaporator bodies, the condensate, vacuum equalizing and ammonia piping. The vapours and the condensate are acidic and therefore the piping may get corroded and leakages are developed. This is often observed particularly at the bends of the piping. Air passes through the leakages and vacuum is lowered down. Therefore the joints and piping are to be checked carefully in the off season and at every general cleaning for leakages and the worn out joints, and corroded piping's should be replaced.

When the sight glasses break, crack or they are not well adjusted, air leakage affects vacuum. There is also, danger of chemical solution leaking out while the evaporator is being cleaned. The sight glasses gaskets should be of standard material and in good condition.

4) Vapour bleeding:

Vapour bleeding followed according to the designed configuration of the evaporator set always gives better performance of the evaporator.

Use of vapour line juice heater gives net saving of steam of about 2.2 to 2.8 % on cane and load on condenser is also reduced.

Operation of vacuum pans should be coordinated in such a way that there should be a sequential order in discharging and initiating the strikes. If two strikes are discharged at a time, the vapour demand at the pan station is considerably reduced that affects the working of the evaporator set. When the discharged pans are started, the vapour demand is drastically increased. Then more vapours are drawn to the pans and the following vessels of evaporator set get less vapours. Hence, again the working of evaporator set is adversely affected. These fluctuations in steam demand at pan station are observed more in small plants with four or five number of pans.

5) Removal of condensate:

The condensate formed in the calandria of each body has to be removed as quickly as possible. If it is not removed efficiently, it gets accumulated in the calandria and heat transfer coefficient of the heating surface of the body is reduced. If the condensate gets accumulated in the calandria and its level is increased. It gives hammering.

6) Removal of incondensable gases:

The incondensable gases if not removed effectively get accumulated in calandria and reduce the heat transfer coefficient that affects evaporation rate. Sometimes because of higher accumulation of incondensable gases even juice boiling is totally stopped. Continuous and efficient removal of the incondensable gases is very much essential for good performance of the evaporator. Opening of the incondensable gas cocks are to be adjusted in such a way that the incondensable gases should be removed efficiently at the same time the steam / vapour loss should be less. The temperature of the incondensable gases passing through the connection should be 1.5°C less than the temperature of the steam / vapour in the calandria. If it is less then more vapours are passing along with the incondensable gases hence vapour loss is more.

Many times the pressure of vapours going to pans is very close to atmospheric pressure, in such case venting the incondensable gas connections to atmosphere does not work and pan boiling becomes slow. If the incondensable gas connection is connected to vacuum, the working of pans is improved.

After years of use, the ammonia pipes get rusted and the rust may block the pipes. This adversely affects efficient removal of incondensable gases. Therefore, checking of ammonia piping and replacement of corroded and rusted piping in every off season and if required in the season is very much essential.

7) Juice level:

Performance of the evaporator depends on level of juice in the bodies. The juice level should be optimum, lower or higher juice level affects the working performance of the evaporator bodies.

When juice level is too low, the boiling juice cannot reach up to the top of the tube. Some heating surface remains dry and total heating surface is not utilized that adversely affects the performance.

If the level is too high, the tubes are submerged in the juice and there is no climbing film effect, hence evaporation rate is decreased. Therefore the correct level of juice boiling inside the vessel must be maintained to get higher efficiency of evaporator and to avoid entrainment

Kerr has given a graph of variation of heat transfer coefficient as a function of juice level for Robert type evaporator bodies. This graph gives a maximum value for heat transfer coefficient when the juice level is about 30-35 % of the height of the tube from the bottom. For semikestner type vessels, the optimum level value is 15-20 %.

Wang[2] has shown that the optimum level decreases in the next effects, for the first vessel 40% for second 25% for third and fourth 20%. The optimum level is lower for clean tubes but increases rapidly with duration of the run.

Thus optimum level of juice for maximum rate of evaporation varies from 15 to 40 % of the tube height depending upon the scale deposited over the heating surface. After general cleaning when the heating surfaces are very clean maximum evaporation gets only at a level of 15 -20% of the tube height. As the days pass and scale get deposited on the heating surfaces, the optimum level gradually goes on increasing. Therefore the operator depends mainly on the way of boiling seen from the sight glass.

To judge the true level of juice, 'a levelling gauge with glass tube' is fitted outside the calandria which gives true level of juice in the calandria.

When a storage tank is provided for clear juice, the clear juice feeding to the first effect can be well regulated[1]. It helps to maintain optimum juice level in the first vessel. A clear juice buffer tank must be provided when automation of the evaporator is done.

When VC + quadruple configuration is used, the exhaust pressure in the calandria and vapour pressure in vapour space of VC and first effect are usually same and the juice cannot flow from VC to first effect by difference in pressure. Therefore, the height of VC calandria is generally kept higher than the first effect so that the juice can flow by gravity.

Scale deposition or passing the scale in the pipes at the time of cleaning or any foreign material like cotton waste remained inside may block the cut line. Sometimes valves of the connecting juice pipe (cut valves of the cut line) may not work. The juice may not pass to the next body at a desired flow rate and level in the body is increased. In such case sometime use of wash out line becomes necessary however the juice levels in both the bodies are to be observed continuously and the wash valve are to be adjusted.

In factories with automation of evaporator, sometimes the actual level of juice in the body may differ from level indicated by the instrument. This may lead to improper working of the evaporator set. Therefore operator should confirm two/three times in a shift whether the instrument indicates proper level or not.

8) Leakages at calandria:

In multiple effect evaporator, in each body there is always lower pressure at the juice side than the vapour side, therefore -

- If the leak occurs at the top of the tube plate, vapours pass from the calandria into the vapour space of the evaporator of the body. This leak does not adversely affect working. The leak can even be considered as an outlet for incondensable gases.
- If leak occurs at the bottom of the tube plate, condensate pass into the juice and the juice gets diluted. The syrup brix lowers down. The condensate of the body may give sugar test.

9) Excessive scale formation:

There can be excessive scale formation for the following reasons –

- Draught affected cane gives higher scale in short duration.
- Sometimes periodical cleaning of the evaporator is postponed for some reasons and excessive heavy scale is formed.
- Scale also may be increased due poor quality of lime.
- Improper control over the clarification process parameters such as optimum lime dose, pH control etc. also cause heavy scale formation.

In such conditions, the evaporator cannot give the desire performance.

10) Steam side cleaning:

In old days when there were steam engines oil was passing along with the exhaust steam and regular cleaning of the first effect was necessary. However now with turbines in use there is no more problem of passing oil along with exhaust steam. Therefore, generally not much scale get deposited to the steam side and cleaning of steam side is a neglected matter in many factories. But entrainment of juice droplets can give scale after couple of years and if the steam side cleaning is not attended every alternative year, poor heat transfer is observed. This problem was observed at Pandurang. The second effect of the evaporator set was giving poor performance because of vapour side scale. The vapour side cleaning was attended in the off season and in the next season the working of the evaporator set was improved.

45.4 Some important points:

- The operation of evaporator set should be continuous, and steady without any fluctuations. With frequent ups and downs and stoppages the evaporation rate decreases and the possibility of entrainment also increases.
- Now in many factories FFEs have been installed. In this equipment, care has to be taken that the full heating surface remains wet every moment for which sufficient juice should be passed through each tube continuously. Therefore adequate quantity of juice circulation with proper distribution is very much essential. Failure of these results into choking of many tubes. The efficiency of the equipment and therefore of evaporator set is drastically reduced. Many factories have faced this problem. During factory operation because of mill stoppages or power turbine tripping, the juice flow may stop but steam may continue. In these conditions to avoid caramelization and choking of the tubes because of starvation, water has to be circulated in the equipment immediately, for which an automatic arrangement is needed.
- Some scale is accumulated at the strainers of the syrup pumps. These should be cleaned periodically especially in general cleaning. The suction valve of the syrup should always be kept open and flow rate of syrup should be controlled by the valve at delivery line. The vacuum equalizing connections should be cleaned in general cleaning.
- The condensates from all evaporator bodies, pans and juice heaters condensate, which gives sugar test, should not be send to boilers.
- All the savealls and external catchalls should be checked, cleaned and repaired during the off season so that the same work efficiently and possibilities of entrainment should be reduced. The drain of the savealls should not be clogged.
- Excessive higher brix of syrup should also be avoided. With higher syrup brix grain may be formed at the syrup pump. The heating surface also fouls rapidly. The SO_2 gas absorption becomes poor. The higher brix syrup may not dissolve false grain. That may also affect circulation in the pan which may result in irregular crystals formation.
- In evaporator in boiling the juice, reactions in between reducing sugars and amino acids takes place and ammonia is released. It passes along with vapours and get dissolved in condensate. If this vapour condensate is send to boiler, it turns to acidic during boiling in boiler or in boiler feed water tank and pH booster chemical is to be used for boiler feed water.

45.5 Instrumentation:

Every evaporator body is equipped with basic measurement and indicating instruments such as pressure / vacuum and temperature gauges fitted both to the calandria and vapour space of the body.

A 'level gauge glass tube' is fitted to each body to observe the juice level in the body. However we observed in most of the factories the level gauge glass tube is installed at the time of erection of the body and working in the first season. But as in most of the cases in India where the evaporator sets are operated manually, the operator (evaporator mate) generally observes the juice level on the sight glasses and not much attention is made in maintenance of the gauge glass. Therefore true level of the juice in the tubes is not observed mush accurately. The parameters to be observed and monitored in operating evaporator are given below-

- Exhaust pressure and temperature,
- Profuse vapour bleeding
- Pressure / vacuum of vapours at the vapour space and calandria of each body
- Juice level in each body and boiling of juice.
- Proper removal of incondensable gasses
- Quick withdrawal of condensate from the calandria by condensate drainpipes and pumping of the condensate.
- Brix of syrup leaving the last body.

Vacuum in the last effect: The vacuum in the last body depends on the injection water quantity and temperature, and efficiency of the condenser.

The quantity of water is always higher than required unless the strainers of the injection water pumps are choke up by some foreign materials such as algae or moss, cotton waste, jute bag pieces etc.

The injection water temperature is generally within normal limit. But if the injection water temperature rises beyond limit due to failure of spray cooling system, drop in the last body and the other bodies occurs. This affects rate of evaporation in each body and syrup brix.

Now with advanced superior designs the efficiency of condensers is improved considerably and no question arises about working of the condenser unless they are leak at joints. Thus at normal working conditions vacuum in the last body remains generally stable.

45.6 Automation of Evaporator Set Operation:

The automation and control of evaporator set works under the followings different sets of automation

Automation of exhaust pressure and temperature: This is done at pressure reducing and desuperheating (PRDS) control and recording system.

Automation of vacuum in: Generally when all the operational conditions at the injection water and condenser are normal and there are no any leakages at the evaporator set the vacuum in the last body always remains more or less constant. However now automation of condensing and cooling system is being followed in modern sugar factories mainly for optimizes of use of water at condenser keeping the vacuum at evaporator and pans steady. This reduces power consumption of the system.

Automation juice level and syrup brix control: Automation of evaporator set operation has two main parameters – the juice level in each effect and syrup leaving the evaporator set.

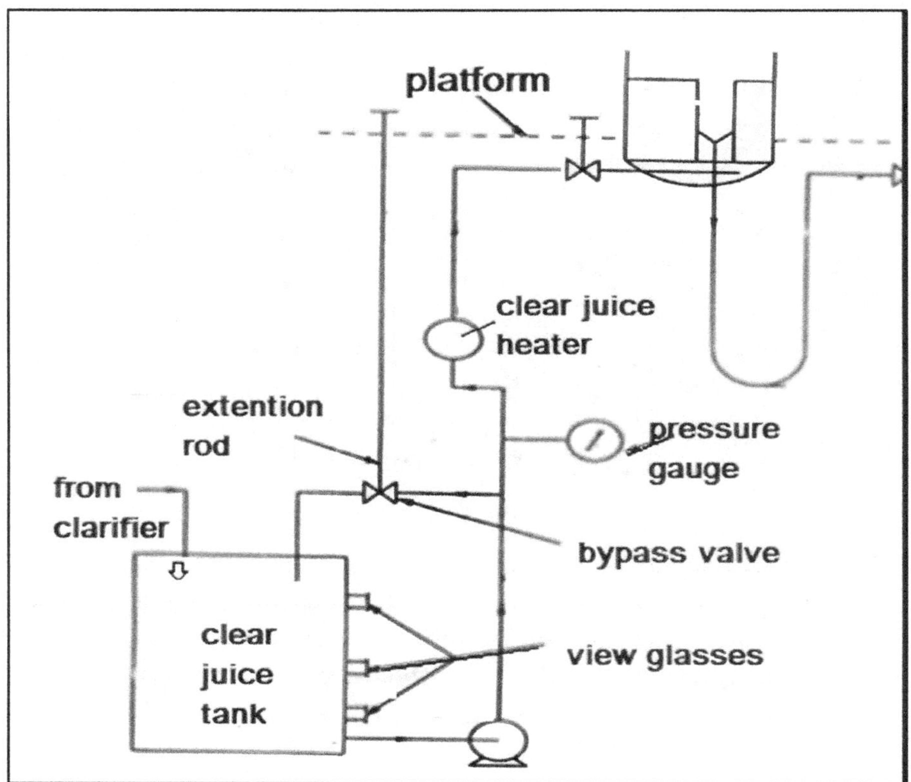

Fig. 45.1 Automation of juice inlet flow rate at evaporator

In automatic operation of evaporator set to maintain the syrup at a desired brix, accurate on line measurement of syrup brix while leaving it from the last body is very mush essential. On line brix can be measured by conductivity transducers, or nuclear density gauges. The use of nuclear density gauges is common in Australian sugar industry.

On line measurement of syrup brix based on differential pressure of a column of syrup of three meters in height between the bottom and the top is also one of the methods for measurement of syrup brix. The pressure differential is proportion to the syrup brix. It is measured and is computed to give on line brix of the syrup at 27.5°C. It also gives output of a 4-20 μA signal corresponding to the brix range selected. The syrup delivery line has a recycle connection with pneumatically operated valve. The output signal controls this valve.

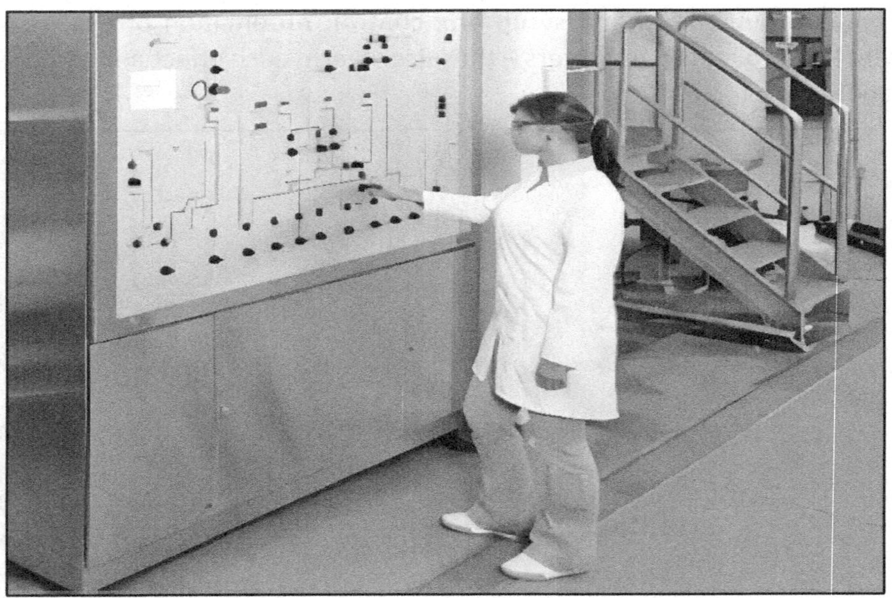

Fig.45.2 Automatic control of evaporator

In guidelines to specification of cane sugar plants for Indian Sugar Industry 'a Coriolis density sensor with dual U tube design' is recommended for online brix measurement of syrup with 4-20mA DC output for controlling and recording with alarm facility if value exceeds the desired setting. Communication ports for PC and printer are also provided.

The syrup brix value is compared with the set value in the computer and the computer transmit signal to the recycling valve. If the syrup brix becomes lower than the set point, the valve is automatically opened wide so that more syrup is recycled to the last body to concentrate it further to get the desired brix.

The cutline valves are under control of juice level regulator. A level transmitter measures the level of juice in the body and control inflow of juice in respective bodies. Set points for desired level are put in by the supervisor.

Now many factories have switched over to computerized automation system for evaporator operation. All the parameters measured and controlled are monitored and displayed on monitor screen in the control room.

The automation regulates the juice level in the bodies practically constant, which helps in giving better performance of evaporator set.

The level in each evaporator body is controlled automatically. In case with higher recycling of syrup to the last body if the level of syrup in the last body increases to higher than the set level, signal from the computer goes to control the cut inlet valve of the last body. So that the valve gets automatically reduced and level in the last body is maintained within limit. If in this way the inlet valves of preceding bodies get reduced, then a stage comes that the inlet cut valve of the first effect of the evaporator set get reduced to maintain the level in the first effect within set limits. In such case a clear juice receiving buffer tank is provided.

It should be noted that in automation the distribution of steam / vapours in the calandria, removal of incondensable gases and condensate are not considered. An operator (generally called as quadruple mate or evaporator mate) or supervisor has to monitor the removal of incondensable gases and passing of condensate. Measurement of condensate of juice heaters, evaporator set and pans can give the steam and condensate balance with quantity of vapours bleeded.

45.7 Stopping the Evaporator:

During short time mill stoppages the evaporator operation is continued. If the stoppage is going to be for more time say two hours, the evaporator operation is stopped and generally the last two bodies are liquidated. If the shutdown is for more hours, then the complete evaporator set is liquidated. The juice from the first and second body can be taken back to clarifier if there is space or in the spare evaporator body, otherwise is taken to the pan station.

When the mill is being stopped for general cleaning the power turbine is kept running till boiling house operations are finished. When evaporator set operation is to be stopped, first generally juice level in the clarifier is made lower by liquidating it. The clear juice receiving tank is made empty. Then clear juice inlet valve is stopped. The juice is taken to the next bodies so as to keep minimum level in each body. The exhaust valve is stopped. Then the last body is first liquidated. After liquidation of the last body, the juice from the preceding bodies one by one are flashed to high vacuum in the last body by wash out line. With this, some evaporation is achieved and temperature is lowered down and then the juice is liquidated. The condensate pumps are stopped. When the liquidation is completed, the injection pump is stopped.

45.8 Heat loss at evaporator:

Sandera[9] has found a loss of heat of 83 kcal/m^2/hr. when difference between the vapour from the juice and ambient air was 52°C. It means the loss of heat is approximately 1.6 Kcal/m^2/°C/hr. The temperature difference between the first effect end the ambient will be about 80-90°C and that for the last effect will be about 20-30°C. Hence heat loss at the first effect will be almost 3 to 4 times more than the last effect.

Reference:

1. Srinivasan S - STAI 1994 p M – 233 – 239
2. Wang P Y - ISJ 1956 p 72
3. Sandera –Tromp L A – Machinery and Equipment of cane sugar factory (1936) p 408

-OOOOO-

46 EFFECTS OF EVAPORATION

46.1 General:

In evaporation when juice is boiled and concentrated to syrup - because of boiling at high temperature, some inversion of sucrose, destruction of reducing sugars and transformation some nonsugars into other compounds takes place. Moreover, as concentration increases precipitation of some nonsugars of takes place. From clear juice to syrup, there is usually slight increase in purity, increase in colour and drop in pH. Normally pH of the clarified juice is dropped across the evaporator by 0.3 to 0.5 units. The RS / Pol ratio is slightly decreased from clarified juice to syrup.

46.2 Rate of inversion at different temperatures and pH values:

The percentage of sucrose inversion per hour recorded by Staddler is given in the following table

Table 46.1 percentage of inversion of sucrose per hour

Temp. °C	pH values of juice			
	7.0	6.8	6.6	6.4
120	0.11	0.18	0.28	0.44
110	0.054	0.086	0.14	0.22
100	0.021	0.034	0.053	0.084
90	0.0089	0.014	0.022	0.035
80	0.0033	0.0052	0.0083	0.013
70	0.0011	0.0018	0.0026	0.0044
60	0.00035	0.00056	0.00088	0.0014

It is seen from the above table that the percentage of inversion of sucrose increases rapidly above temperature 90°C and pH below 6.6.

46.3 Inversion loss across the evaporator:

In factory working it is not possible to collect exactly corresponding representative sample of syrup when clear juice sample is collected. Therefore, it is almost impossible to find out exact inversion loss across the evaporator. Further, the sucrose purity can also be changed because some of the volatile dissolved solids are evaporated, some dissolved solids are precipitated and some are transformed into other compounds of different densities. Thus, the exact reduction in sucrose purity due to loss sucrose by inversion cannot be found out. However, the rate of sucrose inversion at different temperatures and different pH values recorded by Staddler as given in the above table can be used and inversion loss in each body and across the evaporator set can be estimated.

46.4 Retention time of juice in evaporator body:

Let us take an example of quintuple set for a 2500 tcd plant. Considering heating surfaces recommended by Indian Standard Specifications we will calculate here retention time of juice in the third effect. The vapour bleeding arrangement and the vapours generated are taken as described in previous chapter.

Table 46.2 Assumptions

	Particulars	% on cane	Quantity per hr.
1	Crushing rate		115 MT
2	Clear juice	105	120.75 MT
3	Clear juice brix 14.30		17.267 MT
4	Throughput of brix kg/min		287.78 kg/min
5	Water evaporated in the first effect	37.66	43.31 MT
6	Water evaporated in the second effect	28.46	32.73 MT
7	Juice quantity entering the third vessel		44.71 MT
8	Brix % juice entering the third vessel		38.62
9	Water evaporated in the third effect	8.56	9.84 MT
10	Juice quantity leaving the third effect		34.87 MT
11	Brix % juice leaving the third effect		49.52
12	Average brix of juice in the third effect		44.09

Now to calculate the quantity of kg of brix in the third vessel we have to calculate the volume of juice being hold in the vessel. Considering the following specifications of the body-

Table 46.3 evaporator body specifications

1	Evaporator body third effect	Specifications
2	Diameter of the body	3.5 m
3	Bottom inlet cylindrical portion diameter	0.6 m
4	Bottom inlet cylindrical portion height	0.4 m
5	Height of conical portion	0.9 m
6	Height of collar below the bottom tube plate	0.1 m
7	Heating surface	600 m^2
8	No of brass tubes	2200
9	Tube height	2.0 m
10	Tube I D	0.042 m

Considering the juice level in the tubes as 0.6 m. the total volume of juice is to be calculated to find out weight of juice being hold by the body while in operation.

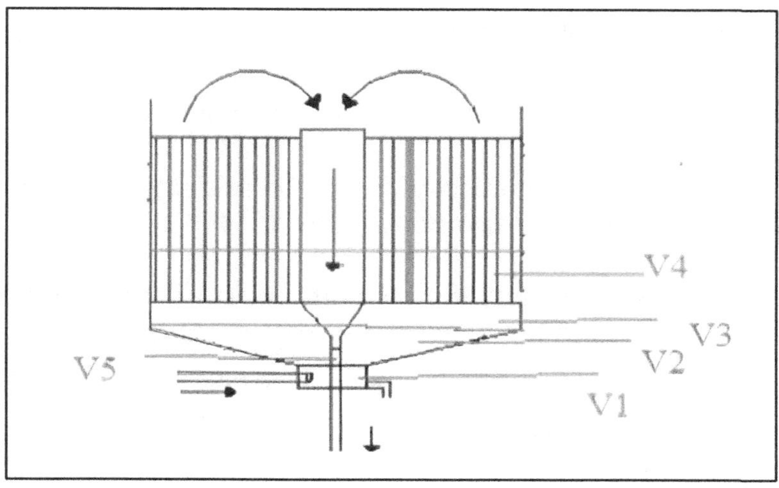

Fig.46.1 Volume of juice

V1 = Volume of juice in the inlet cylindrical portion

= $(\pi D^2 x h)/4$ = $(3.142 \times 0.6^2 x0.4)/4 = 0.113$ m^3

V2 = Volume of conical portion

= $\pi \times h (r^2 + r R + R^2) /3$

= $3.142X0.9(0.5^2+0.5X3.5+3.5^2)/3 = 13.432$ m^3

V3 = Volume bottom collar cylindrical portion

= $(\pi D^2 x h)/4$ = $(3.142x3.5^2x0.10)/4 = 0.962$ m^3

V4 = volume in the tubes

= no of tubes x $(\pi D^2 x h)/4$

= $2200(3.142x0.042^2)/4 = 3.048$ m^3

The total of volumes will be 17.555 m^3

The volume of outgoing pipe passing through the bottom portion has to be deducted through this volume which will be

V5 = $(\pi D^2 x h)/4 = (3.142x 0.150x0.150x1.4)/4 = 0.024$

Thus the total volume of juice in the body will be 17.531 m^3.

The average brix of the juice as calculated above is 44.09. The juice temperature is 99.35. The density of the juice at this brix and temperature will be 1.149. Therefore weight of the juice in the body will be 20.143 MT, or 20,143 kg and weight of brix (dissolved solids) will be 8881 kg. The throughput brix flow is 287.78 kg/min. Hence the retention time of juice in the body will be 30.86 min.

46.5 Loss of sucrose by inversion in the body:

Assuming pH of the juice as 6.8, we estimate from the table given by Staddler the inversion loss in the third effect at 100°C for 30 min will be 0.017 % of the sucrose or pol.

For standard quadruple effect with first vessel heated by steam at 112°C Classen[1] gives loss of sucrose by inversion as follows

Table 46.4 percent sucrose loss by inversion across the evaporator

Evaporator body	Percent of sucrose loss by inversion across
First body	0.020
Second body	0.015
Third body	0.010
Fourth body	0.005
total	0.050

The inversion loss across the evaporator on percent on cane can be 0.007%.

Honig says[2], *"the experience of the author at factories with different evaporator systems has amply demonstrated that the calculated loss of sucrose by inversion in the evaporators could be as high as 0.2 % on pol, but that this could not be related to the change in the glucose quotient".*

The recent development in laboratory analytical techniques such HPLC (high performance liquid chromatography) and GC (Gas Chromatography) can give more accurate figures of inversion loss across the evaporator than the conventional method of sucrose determination by double polarization.

The inversion loss across the evaporator as estimated by Schaffler[3] using gas chromatography is in the range of 0.1 to 0.2 % of sucrose. Now with profuse vapour bleeding and boiling the pans on the vapours from the second effect, temperature of the boiling juice in the first two effects is high. It is 113 to 117°C in the first effect (At Kenana it is 118°C with Robert type evaporator bodies) and 105 to 109°C in the second effect. Further heating surface of first and second effect together is 65 to 82 % of the total heating surface of the evaporator set. Therefore, the retention time at high temperature in the first two effects is increased and hence inversion loss is increased.

Low levels of glucose in juices can be measured precisely and hence inversion loss across the evaporator can be estimated by checking the glucose /sucrose or glucose / pol ratio in clear juice and syrup.

The reducing sugars are destroyed to some extent at evaporator. The destruction of reducing sugars is high at high pH and high temperature. Colour precursors[4] are formed from reducing sugars and amino acids. Addition of phosphoric acid[5] in clear juice can help to control the destruction of reducing sugars. Fructose is more sensitive to higher temperature and destruction of fructose is always more than glucose. The evaluation made by Schaffler[6] reveals that the fructose / glucose ratio across the evaporator goes on decreasing. Therefore high apparent purity rise across the evaporator station can be indicative of fructose destruction.

The measurement of low levels of inversion in evaporator is difficult. The change in glucose/sucrose ratio cannot give accurate inversion loss because some amount of glucose is destroyed during evaporation.

Shukla says[7] that *'Serious sugar loss due to inversion, in all kinds of evaporator takes place when the juice boiling temperature are above 100°C and retention time is even 25% more than the designed requirement'..*

In conclusion we can say higher inversion loss occurs across the evaporator due to following reasons

- Excessive higher heating surface in the first two effects.
- The crushing rate and hence throughput of the juice is less and retention time in the evaporator set is increased.
- Temperature of the juice in the first vessel is too high.
- pH of the clarified juice is too low (below 6.7).
- Short circuit in the vessel and part of the juice remain for too long time in the vessel.

46.6 Rise in apparent purity from clear juice to syrup:

When juice is concentrated to syrup, generally rise in purity by about 0.3 to 0.7 units is observed. This rise in apparent purity may be due to -

Some volatile compounds and gases formed due to reactions at high temperature and pass along with vapours. The effect is that the nonsugars are decreased and hence purity is increased.

- Some nonsugars from juice are precipitated and get deposited in the form of scale. Thus nonsugars are decreased and purity is increased.
- Transformation of some reducing sugars and other nonsugars into other compounds due to high temperature, which may be of different densities and may cause lowering down the brix.
- More destruction of fructose than glucose as it is more sensitive to high temperature than glucose, which may give increase in pol reading.
- Dilution of syrup for analysis may cause error resulting in higher apparent purity.

However, the total effect of all the reasons given above cannot give any satisfactory explanation of rise in purity to above 0.20 unit.

46.7 Increase in colour and turbidity:

Colour formation is mainly due to boiling of juice at high temperatures in the first two effects of multiple effect evaporator. The colour formation in the last two effects is negligible. The nature of the colour formation in evaporator is not caramelization of sugar or Maillards reaction. The intensity of colour formation shows wide variations depending on temperature and retention time. In some evaporators the colour formation may not be more than 10%, however in some cases it may be as high as 25%. In sulphitation factories colour formation across the evaporator is less than in defection factories. There is always increase in turbidity from clarified juice to syrup. Clear juice contains suspended impurities which are negligible but when the juice is concentrated to syrup these are increased by about five times. These increase turbidity of the syrup. Precipitation of certain types of nonsugars specifically silicic acid, sesquioxides and some organic non sugars that are partly precipitated as a result of the concentration increase the turbidity of the syrup. In general to control the colour formation and inversion loss, the juice boiling temperature in the first vessel should not be more than 118° C and retention time should be less than four/five minutes.

46.8 drop in pH across the evaporator:

The normal pH drop from clear juice to syrup is about 0.4 to 0.6 unit. However sometimes depending upon cane quality and process parameter, it may be as high as 0.8 unit. It is partly due to removal of ammonia however, the chemistry is not known in detail.

46.9 Scale formation in evaporator:

The scale formation primarily is the result of the concentration non sugars, specifically in the last two effects of the evaporator where some inorganic non sugar become supersaturated get precipitated and are deposited on the heating surface. This subject is discussed in detail in the next chapter.

References:

1 Prinsen Geerligs H C 1934 p.190
2 Honig P vol. III p106
3 Purchase B S SASTA 1987 p 8 – 13
4 Honig P vol. III p150
5 Gupta and Ramaiah STAI 1965 p171-178
6 Schaffler K J SASTA 1985 p 73-82
7 Shukla G K STAI 1994 p M 259 – 272

-OOOOO-

47 SCALE FORMATION IN EVAPORATOR

47.1 General:

When multiple effect evaporator is in operation, scale or incrustation is formed inside the tubes. It is developed slowly and the tubes get fouled within 2-4 weeks. In evaporator as the water goes on evaporating the concentration of impurities in the juice are increased. When they go beyond their solubilities (commonly called as passing in supersaturation phase) tends precipitate and are deposited on the heating surface of evaporator bodies in the form of scale. The scale deposition goes on increasing from the first effect to the last effect. Generally the first effect contains only 8-12% scale, however the last effect contains about 35-40% of the total scale on dry basis. The scale is generally thicker at the lower part of the tubes which portion is submerged in the juice.

The juice boiling temperature in the first effect is about 113 to 115°C, which is not reached in clarification. At such high temperature certain nonsugars such as calcium magnesium phosphate, iron phosphate and nitrogenous nonsugars which have not been removed in clarification are precipitated in the first and second effect of evaporator set. The vapours liberated from the juice also corrode vapour pipes and pump bodies.

The nature of scale depends on number of factors such as composition of juice, process of clarification, operational conditions, rate of circulation of juice in the evaporator bodies, duration of the run, and syrup brix. The scaling of organic substances such as gums, proteins and lipids is a result of coagulation effect at high temperature.

- Poor cane quality increases lime salt in clear juice.
- When the juice clarification is not proper some suspended particle may pass along with the clear juice and deposit on the tubes to form scale.
- Improper control on pH leads to higher calcium salts in clear juice that give higher scale formation in evaporator. Highly acidic or alkaline juices forming bisulphite or bicarbonate decompose at higher temperatures and the decomposition products deposit on the heating surface. Therefore to avoid extra scale formation the juice should be either neutral or slightly acidic or alkaline in the range of 6.8 to 7.1.
- When juice is over-sulphited bisulphites are formed, that decompose in evaporator forming normal sulphite and sulphurous acid. The sulphurous acid pass along with vapours and the condensate becomes acidic. Prinsen Geerligs

found sulphur in condensate. This may be due to reduction of sulphurous acid by the metal.

- Lime with higher percentage of silica and sulphate gives higher and hard scale in evaporator bodies.

The thermal conductivity of the scale is very low. In Australia tests were made[1] and thermal conductivity of scale from evaporator was found of the mean value as 0.45 kcal/m^2/°C/m. naturally it retards the rate of heat transfer and hence it must be removed to maintain efficiency of the evaporator operation. The scale formed is removed periodically by chemical and mechanical cleaning of the evaporator.

On the outside of the tubes, scale is formed when vapours / steam carry along with it juice bubbles or droplets. Rusting of the tubes also causes scale outside the tubes. The scale formed outside the tube is not much severe and can be attended every alternative year for each evaporator body.

47.2 Quantity of Scale:

First body of the evaporator contains much less and soft scale and can be easily removed whereas the last body contains thick and hard scale which is difficult to remove. The quantity of scale deposited inside the tubes per tonne of cane varies considerably from 10 to 50 g on dry basis. The average scale deposition during normal run is 100 to 800 g / m^2 of heating surface, increasing from the first effect to the last effect.

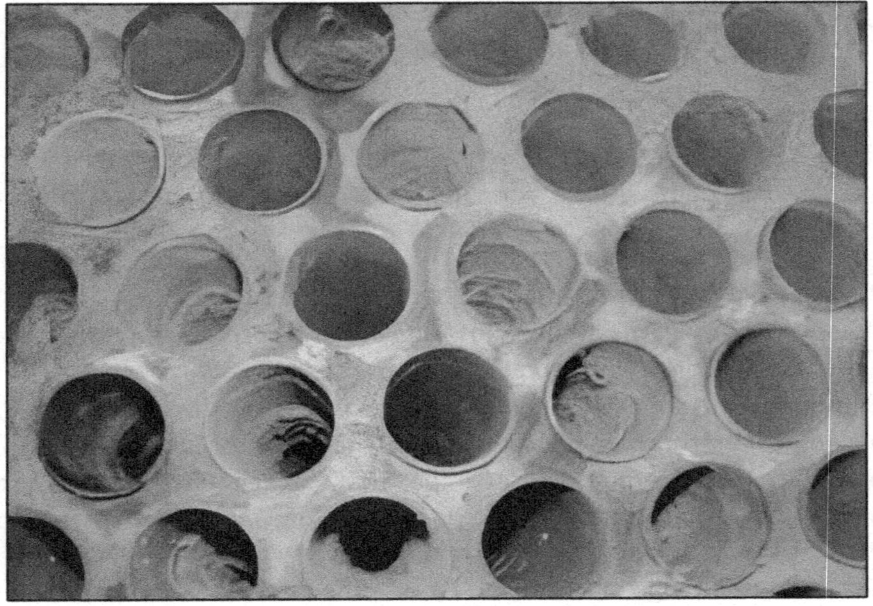

Fig 48.1 Scale formed on the tube tubes top side

- Scale is deposited more rapidly and abundantly when rate flow in the vessels is lower and there are frequent stoppages.

- High juice level increases the residence time of juice in the vessels and rate of deposition of scale increases.
- The scale becomes thick in the lower portion of the tubes where the juice is relatively stagnant.
- In Semikestner and falling film type evaporators the velocity of juice is high as compare to Robert type evaporator body. Therefore, in these types of bodies the scale formation is less.

In sulphitation factories the scales deposition is more hard and difficult to remove than in defecation and carbonation factories. In defecation factories, generally only chemical cleaning is followed in the season and chemical cleaning followed by mechanical cleaning in the off the season. However in sulphitation factories, only chemical cleaning is not sufficient and in the season mechanical cleaning is essentially followed by chemical cleaning.

47.3 Composition of Scale

The constituents of juice, which are passed in suspension form, get deposited as scale in the first effect. In sulphitation factories, the major inorganic constituents of scale are calcium, magnesium, silica, sulphate, sulphite and phosphate, while minor constituents are iron, chloride, oxalate etc. The organic constituents are not always well defined but generally are precipitated salts of organic acids.

Calcium phosphate: Precipitation of calcium phosphate which starts in the clarifier but the reactions are slow. Therefore phosphate scale is found mainly in the first and second effect. Its proportion goes on decreasing in the following effects. Magnesium and iron can substitute partially for calcium to produce a mixed phosphate. The colour of phosphate scale is whitish and it is soft in nature.

Calcium sulphite: It is one of the major scale constituents in sulphitation factories. Its concentration increases from the first effect to the last effect. It increases with drop in clear juice pH.

Calcium sulphate: It is one of the major scale constituents in sulphitation factories. Its concentration generally increases from the first effect to the last effect. It is greyish white and hard scale.

Calcium silicate: Calcium/magnesium silicate is usually precipitated in the second evaporator body onwards. Mixed silicates form hard scale.

Silica: Silica is precipitated in the later effects (and in some cases in pans). It is generally believed that it precipitates by polymerization and coagulation. Silica scale is sometimes porous and soft but may find dense and hard.

Calcium oxalate: Calcium oxalate is found usually in the last effect.

However, the composition of scale varies considerably from region to region depending upon nature of soil and cane quality. It also depends on the process parameters maintained and control over the process. Therefore, it is not possible to give range of percentage of the scale constituents within narrow range for any effect from an evaporator set.

The following table gives percentage of different constituents of scale in sulphitation sugar factories in north India and Deccan plateau of India given by technologist worked in the past.

Table 47.1 percentage of different constituent of scale

Sr. No.	Constituent	Deccan[2]	North India[3]
1	Organic matter	7-12 %	35-38 %
2	Silica as SiO_2	25-50 %	0.3-0.4%
3	Fe_2O_3, Al_2O_3	2.5-10 %	0.2-4.0 %
4	CaO	12-15 %	30-40 %
5	SO_4	1-10 %	21.30 %
6	Phosphate	1.5-10 %	1.1-10

47.4 Vapour Side scale:

After 2-3 years thin film of scale is observed on the vapour side of the calandria. As we discussed earlier the vapours from the juice contain volatile compounds from the juice and as vapours get condensed the compounds get deposited outside the tubes. These are distillation products of lipids i. e. the fatty acids from the sugar cane.

After years of use the metal of the tubes also gets corroded to some extent. Entrainment occurred occasionally also gives scale outside the tube. If the vapour side of the calandria is not cleaned for 3-4 years, the scale thus deposited reduces the heat transfer rate and fall in the performance of the evaporator is observed. Therefore this scale should be removed every year or least once in two years at in the off season. The different methods of the scale removal are discussed in the next chapter.

References:

1. ISJ 57(1955) 381
2. Bhosale R N et al Proc. DSTA 1989 p M90
3. Joshi K A et al Proc. DSTA 1951 p 170
4. Shrivastav A K et al Proc. STAI 1990 p M39.

-OOOOO-

48 CLEANING OF EVAPORATOR

48.1 General:

At the start of a 'RUN' evaporator heating surfaces are clean but as days pass scale gets deposited on the inner side of tubes of the evaporator bodies. Heat transfer coefficient of the tubes is reduced. After 15 to 30 days evaporator set fouls, syrup brix goes down and steam consumption increases. Though it is impossible to prevent formation of scale, it is necessary to avoid its formation as far as possible. Further it is also necessary to remove the scale before it becomes thick. As a routine practice, if exhaust pressure and condensate flow rate of the first vessel of evaporator is measured, it can give idea to decide the cleaning day. As days pass the exhaust pressure goes on increasing and the condensate flow rate goes on decreasing. However, in India cleaning day is generally not decided only on scale developed at evaporator.

There are many practical difficulties in deciding the cleaning day. In India factories are required to prolong cleaning for festivals like Makar Sankrati (Pongal), Maha-Shivratri, Holi, Gudhi Padwa (Ugaadi) and local festivals. On such days, generally harvesting labour takes holidays. Sufficient cane is not arrived at the factory and the factory has to run at much lower crushing rate and/or has to take stoppage for 'no cane'. Therefore, usually general cleaning of a factory is arranged on the above days. Thus, the management of the factory cannot decide the cleaning day only on the technical ground. To prolong the run, many factories now use scale inhibitors / antiscalent. Addition of phosphoric acid 10 to 20 ppm in clear juice is recommended to prolong cleaning.

The methods used for removal of scale are quite varied in frequency of cleaning, chemicals used, concentration of chemicals and time of cleaning required. In Vietnam where in most of the factories sulphitation process is followed evaporator cleaning is taken after every 12-14 days. At Kenana (Sudan) where defecation process is followed chemical cleaning of evaporator bodies for first and second effects is done after every sixth / seventh day and that for the third and fourth effects after every third or fourth day. However, in some factories from Kolhapur and Satara district of Maharashtra state (India) where scale formation is very much less and run can be prolonged for 35-40 days without use of any antiscalent.

Various methods of chemical and mechanical cleaning are adopted for softening and removal of scales.

48.2 Chemical cleaning:

The evaporator bodies are liquidated through wash out line and are made empty for the cleaning. Before taking the chemicals / chemical solution in to the bodies, the bodies are washed thoroughly. Otherwise sucrose and reducing sugar of the juice remained adhering in the vessel react with chemicals and considerable amount of chemicals are consumed by juice remained inside the vessels.

Conventional practice of chemical cleaning is to boil with caustic soda solution or acid, or caustic soda followed by acid. Generally, caustic soda is employed in most of the factories.

When the scale is thick and hard in the last two effects soda boiling is followed by acid boiling. By chemical cleaning scale becomes soft and some scale is dissolved. Some is removed and remaining becomes porous and soft.

Methods of chemical cleaning:

In conventional method of 'soak cleaning', the entire scale surface is submerged in the cleaning chemical solution. In this method the chemicals and steam requirement is more. However, in immersing the tubes in chemical solution reactions of chemicals with the scale is ensured. Further this method is simple and does not require any special techniques and therefore is followed in most of the factories.

The other method is 'Spraying of Chemicals'. In this method, the cleaning solution is sprayed into the vessel by pump through coil. 1/3rd volume of calandria level of chemical solution is sufficient. This gives saving in chemicals and steam. However, with spray system, there is risk of inadequate coverage and therefore boiling is preferable to spraying.

Soda boiling:

The caustic soda, washing soda and common salt are added to the soda tank provided near the evaporator at the ground and steam is allowed to pass into the tank to get the chemicals dissolved. The solution is recirculated for complete mixing and is kept ready to transfer it to evaporator bodies.

Generally, 10–25 % of Caustic soda concentration is used for soda boiling. Along with caustic soda, washing soda of about 5-10% of the amount of caustic soda and little quantity of common salt is always used. With addition of washing soda and common salt the effect of caustic soda is more pronounced.

In many factories instead of making soda solution in the soda tank, the bags/ drums of chemicals are directly emptied in the evaporator bodies through manhole and water is taken in the bodies. The chemicals are dissolved due to heating and boiling. The quantity of chemicals added is decided on heating surface area and severity of scale in each body.

The soda boiling is done for 4-5 hours. Generally, open boiling is followed. All the calandria are provided with a 150-mm diameter exhaust steam connection for chemical boiling. All the evaporator bodies are boiled on exhaust steam. In some factories live steam reduced at 3kg/cm^2 pressure connection is provided to the third and fourth effect. This pipe line should have main shut off valve. The first effect boiling is done on exhaust and the second effect open boiling is done by vapours from the first effect. The third and fourth bodies are boiled on live steam.

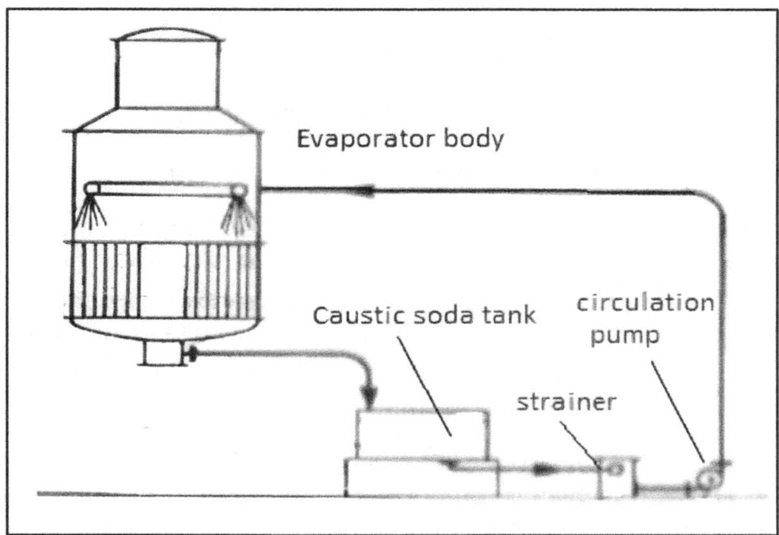

Fig.48.1 Spraying and circulation of Caustic soda solution

When steam is applied to the evaporator bodies, the ammonia valves and vacuum brake valves are closed and valve or air vent at the top of the dome is opened to release the vapours and the gases. Thus, the Soda boiling is done at 105-110°C. With live steam, the boiling starts fast, vigorous boiling takes place, and time is saved. In open boiling, as the temperature is high the chemical cleaning becomes more effective. The hard scale may get cracked off from the surface due to high temperature.

With caustic soda solution, organic constituents of the scale are hydrolysed and sodium organates are formed. Caustic soda solution disintegrates phosphate scale. The soda ash transforms the calcium sulphates and sulphites into calcium carbonates. Silica is dissolved to a very small extent by highly concentrated caustic soda solution after prolonged boiling.

$$CaSO_4 + Na_2CO_3 \rightarrow Na_2SO_4 + CaCO_3$$

$$CaSO_3 + Na_2CO_3 \rightarrow Na_2SO_3 + CaCO_3$$

$$SiO_2 + 2NaOH \rightarrow Na_2SiO3 + H_2O$$

If silica scale is present in the form of calcium silicate ($CaSiO_3$), the caustic soda at high temperatures will decompose it into sodium silicate, which is water-soluble.

$$2NaOH + CaSiO_3 \rightarrow Na_2SiO_3 + Ca(OH)_2$$

However, in the last body scale mostly the silica is present only in the form of SiO_2.

During soda boiling, the sodium hydroxide is consumed and its concentration goes down. However, NaCl used along with the soda solution regenerate part of NaOH, therefore the strength of the solution is not much decreased.

$$2NaOH + CaSO_4 \rightarrow Na_2SO_4 + Ca(OH)_2$$

$$CaSO_4 + Na_2CO_3 + H_2O \rightarrow Na_2SO_4 + Ca(OH)_2 + CO_2$$

$$2NaCl + Ca(OH)_2 \rightarrow CaCl_2 + 2NaOH$$

The used soda solution is returned to spent soda tank and the solution is brought up to the strength before using for next cleaning. Brix of spent soda / acid solution stored for next cleaning is always higher as the scale constituents are dissolved in it. Also calcium carbonate formed during soda boiling is remained in the soda solution. Therefore, the brix should not be taken as indicator of concentration and concentration of NaOH should be determined by titration only.

Sucrose and reducing sugar that remain the traces in the scales are decomposed by the action of NaOH at higher temperatures and fatty acids are formed. These fatty acids are present in spent soda solution and give heavy foaming when the solution is reused. In case of heavy foaming, it becomes necessary to make a fresh soda solution.

After soda boiling the vessels are washed with water and made empty. Then again, fill with fresh water up to 1 m above the calandria level. When the scale is relatively soft, soda boiling is sufficient, however, when the scale is thick and hard with higher silica content, the soda boiling is not sufficient. To remove the crystalline and hard scale of higher silica and sulphate content, acid boiling is followed by soda boiling.

Acid boiling:

When soda boiling is followed by acid boiling care must be taken to remove the traces of soda solutions. In soda boiling some scale gets dissolved and it becomes porous. When after boiling the soda solution is drained, some soda solution always remains in the porous scale. When soda boiling is followed by acid boiling, simply washing of the body is not sufficient. Otherwise soda solution remained in the porous scale reacts with acid as a result of which strength of the acid solution rapidly comes down and the acid boiling does not becomes effective in removing the scale. Therefore after soda boiling water boiling must be followed before acid boiling.

The most commonly used acid is commercial hydrochloric acid. By the action of hydrochloric acid, the calcium salts of simple organic acids like calcium aconitate and calcium oxalate are decomposed forming aconitic acid and oxalic acid. However, complex organic salts are not much affected by acid treatment. Only caustic soda has got the maximum access to their decomposition and removal as the reaction requires higher pH.

Hydrochloric acid generally of 0.25 – 2.0 % concentration depending upon the severity of the scale (quantity, thickness and hardness of the scale) is recommended. The strength should not be increased more than 2%. While diluting the acid it should be noted that the commercial concentrated hydrochloric acid available in market contains about 30% concentration.

Hydrochloric acid can added through the cock and cup provided to the body. However in many factories it is directly added through the manhole in to the body in which water is already taken up to top tube plate. The acid boiling should be for about 1-2 hours at 105-110°C.

Acid may react with metal and spoils the tubes, tube plates and vessel. To avoid such damage of metals due to high acidity, the acid should be analysed for its concentration and the required concentration should be carefully maintained while adding it to evaporator body to avoid metal damage.

To prevent the reaction of the acid with metals, an acid inhibitors like Rodine special 213 (0.2 – 2.0 % of acid) or formalin (6lit/100kg of HCl) is added to the solution of hydrochloric acid. Hugot recommends 5% molasses as inhibitor. Five percent molasses acts as excellent inhibitor. The acid / acid inhibitor ratio should be carefully maintained. However, with inhibitor the efficacy of the acid boiling may also be reduced.

Hydrochloric acid dissolves carbonates, sulphites, phosphates and organic acids.

$$Ca_3\,(PO_4)_2 + 6HCl \rightarrow 3CaCl_2 + 2H3PO_4$$

$$CaSO_4 + 2HCl \rightarrow CaCl_2 + H_2SO_4$$

$$CaCO_3 + 2HCl \rightarrow CaCl_2 + CO_2 + H_2O$$

The sesquioxides (Fe_2O_3, Al_2O_3) remain unaffected by soda treatment but on acid treatment they are easily decomposed and can be removed easily.

$$Fe_2O_3 + 6HCl \rightarrow 2\,FeCl_3 + 3H_2O$$

At higher temperature, HCl has a tendency to dissolve the calcium silicate into a jelly.

$$CaSiO_2 + 2HCl \rightarrow CaCl_2 \;\; + SiO_2H_2O$$

However, the calcium silicate content in the scale is quite negligible. HCl has no reaction on SiO_2 scale. Ramlingam[2] gives his experience and results obtained at Dharani Sugars. With acid boiling - using 2% HCl and 2:1 acid formalin ratio for a period of 3 hrs. at 75°C, followed by soda boiling, mechanical cleaning could be avoided in first and second effect.

During acid boiling when the boiling solution is analysed at different intervals of time it is often found that the percentage Fe_2O_3 is increasing. It may be due to dissolution of Fe_2O_3 from the scale or from the vessel surface. Practically it is impossible to find out the quantity of Fe_2O_3 that dissolved from scale and from the vessel.

During the process of acid boiling the acidic vapours produced may attack the metal of saveall, vapour piping and the metal of calandria of the next vessel. The percentage of concentration of acid inhibitor in vapours is negligible, as its flash point is always different. The disadvantage of using hydrochloric acid is that it is corrosive and difficult to handle. It corrodes the metal of vessel and tubes. With brass tubes, dezincification is effected causing pitting on the tube surface.

Further, the corrosion products are gradually accumulated in the calandria of the next vessels. In using strong acid, there is formation of H_2 with danger of explosion if the vessel is not properly vented.

Other chemicals used:

In place of hydrochloric acid, sulphamic acid or versene can also be used. Sulphamic acid is a solid inorganic acid free from corrosive properties and hence easy to handle. The acid is hydrolysed at boiling temperature with formation of sulphuric acid. Therefore, the sulphamic acid is circulated through the evaporator bodies only at about 60-70°C. It is very effective[3] in removing hard scale. The acid solution dissolves calcium salts.

The only disadvantage with this acid is its higher cost as compare to hydrochloric acid.

Versene is a powerful agent that helps to dissolve cations such as calcium, magnesium and sesquioxides in neutral solution.

Nowadays sodium or ammonium bifluoride is used along with alkali or acid to the extent of 1.5 – 3.0 % of HCl / Alkali

$$6\,NH_4F + SiO_2 \rightarrow H_2SiF_6 + 6NH_3 + 2H_2O$$

$$3\,NaHF_2 + SiO_2 \rightarrow Na_2SiF_6 + NaOH + H_2O$$

Ammonium bifluoride in aqueous solution at boiling temperature undergoes a partial dissociation into hydrofluoric acid.

$$NH_4F + H_2O \rightarrow NH_4OH + HF$$

The hydrofluoric acid thus liberated reacts with silica, the silica scale is dissolved, and the structure is disintegrated.

$$6HF + SiO_2 \rightarrow H_2SiF_6 + 2H_2O$$

This breaks up the skeleton and internal structure of the hard scale and therefore it can be removed more easily.

Scale Softeners:

Scale softeners are various blends of wetting, penetrating and solubilizing agents. Some scale softeners are used along with the caustic solution whereas others are used directly.

There is no technically reliable laboratory test available for the evaluation of scale softeners. However, the success of the use of scale softener can be determined by reduction in quantity of cleaning chemicals, time required for boiling, the softness of scale achieved as observed, and the total down time for cleaning.

Quantity of chemicals required:

The consumption of cleaning chemicals varies considerably from factory to factory depending upon severity of scale, use of antiscalent etc. however the normal range of quantity of chemicals used in the sulphitation factories is given hereunder.

Table 48.1 chemicals used for cleaning of evaporator set

Particulars	Quantity in kg/100 quintals of cane
Caustic Soda	0.1 – 0.3
Washing Soda	0.01 – 0.03
Common salt	0.005 – 0.015
Hydrochloric Acid	0.02 – 0.03
Rodine (acid inhibitor)	0.0004 –0.0006
Ammonium bifluoride	0.00 – 0.003

48.3 Steam cracking:

After the soda / acid boiling is finished, the spent solution is discharged and heavy steam cracking is done. The empty calandria is heated by the steam and then the cooled by cold water. This heating and instant cooling of the tubes loses the heavy scale.

After steam cracking, the bodies are filled with cold raw water up to 1 m above the calandria level so that the bodies are cooled. Then after draining the water the manholes are opened. Fans are provided at the manholes for further cooling of the bodies.

48.4 Mechanical cleaning:

Mechanical cleaning is carried out by Descalers that consists flexible inner shafts of 7-10 m in length, enclosed in a loose protective flexible outer sheath.

Electric motor mounted on wheel base

The inner shaft is fitted to the outer sheath at both the ends by roller bearing. Further greasing is done in between the inner shaft and out sheath so that there will be minimum friction when the inner shaft rotates at high speed and the outer sheath remains stationary. The shaft is so flexible that it can turn into arch and hence can be handled is easily in cleaning the tubes.

One end of this shaft is attached to an electric motor with power of 24 volts and speed 3000 rpm. The electric motor is fitted on a small frame mounted on wheels so that it can be shifted from one position to another or can be moved on wheels.

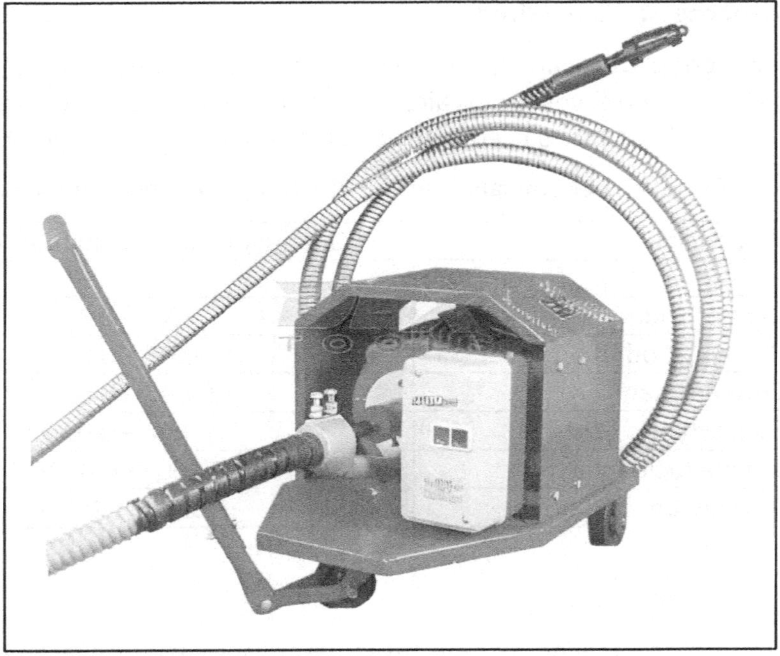

Fig.48.2 electric motor mounted on trolley

To the other end of the shaft a tool - cutter or turkhead wire brush is attached. When motor is started, the tool attached to the other end rotates at the speed of motor along with the inner. The high speed rotating tool is inserted from the top of the heating surface tube to be cleaned and is passed to the bottom end of the tube. The high speed rotating cutter removes the scale which is loosened by chemical cleaning.

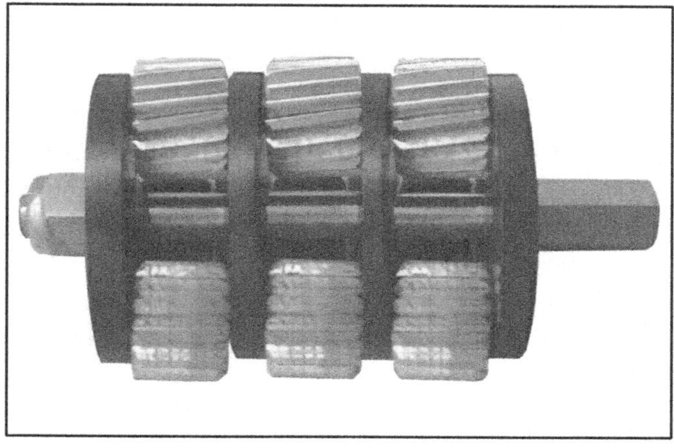

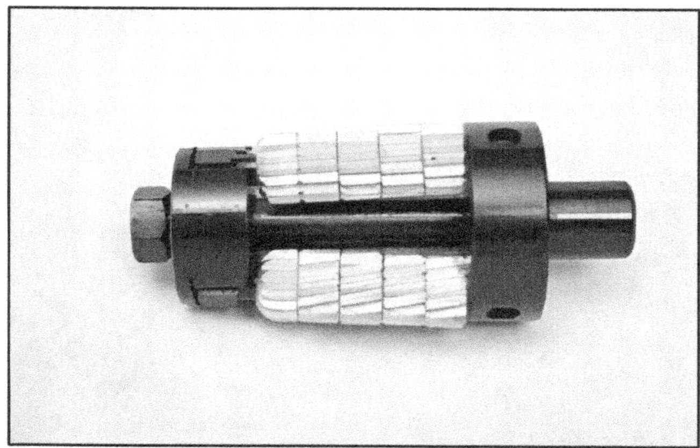

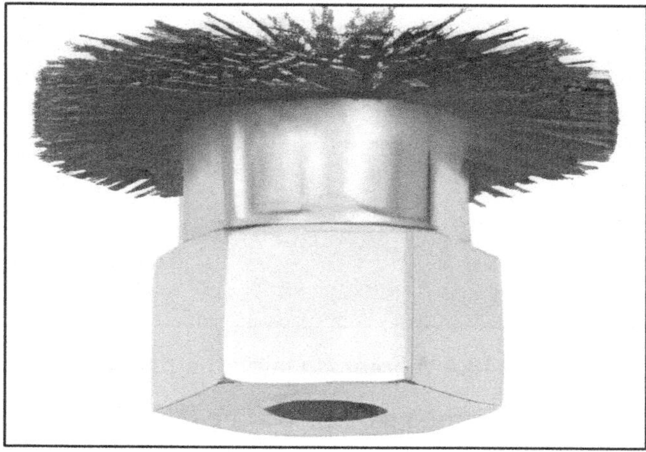

Fig. 48.3 Different types of cutters and brushes

The direction of rotation of the motor should be checked before connecting the shaft to the motor. If it is run in reverse direction the wires are unwound or cutter gets removed while cleaning the tube.

Before starting the tube cleaning, the vessel is filled with water up to top tube plate of the calandria.

While cleaning the tubes it is very much necessary that the brushes / cutters reach up to bottom of the tube. Otherwise, scale of the bottom portion is not removed properly. The shaft is clamped at a point corresponding to the height to ensure complete cleaning of the tube. Generally, two passes of the tool are sufficient to remove the scale form each tube.

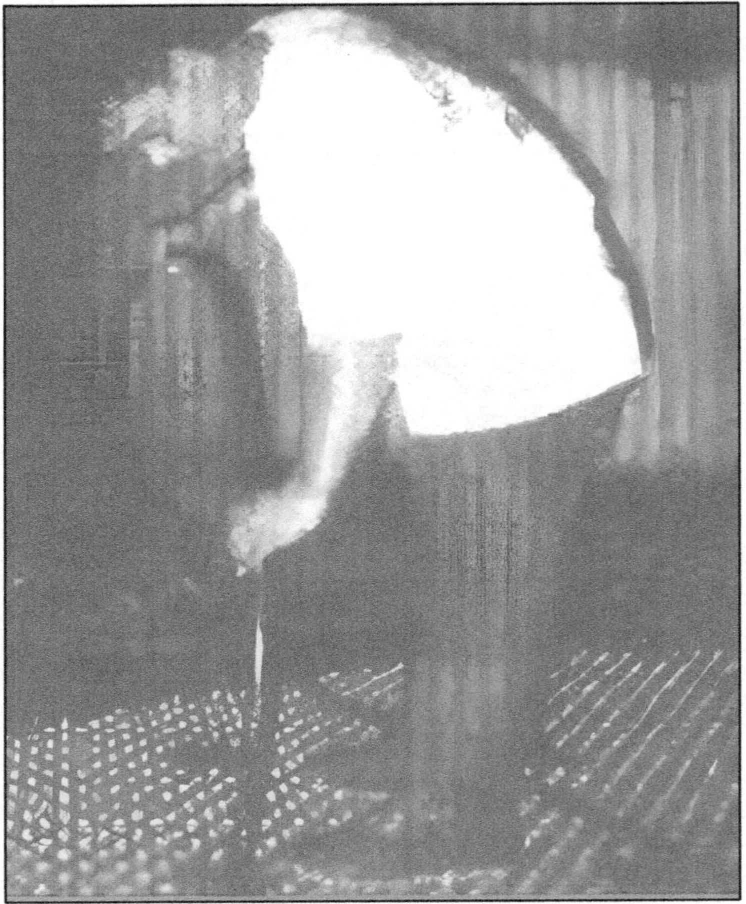

Fig. 48.4 Worker cleaning the tubes

In case of heavy scale, a tapered jet type cutting tool is preferred. The springs of the cutter should have a proper tension otherwise; a loose cutter passes easily through the tube without removing the entire scale. The cutter and the wire brushes should be replaced by new one when they start, passing easily through the tubes. Two persons working with one flexible shaft can clean about 20 to 100 tubes per hour depending upon thickness and hardness of the scale.

After the mechanical cleaning the water in the body is drained off. The bottom manhole is to be opened. The scale on the bottom saucer is to be cleaned carefully otherwise the remaining residue may pass in the pipeline and may choke it.

Then each tube should be checked by holding a hand lamp from the bottom and looking inside the tube from the top. The tubes should be very carefully observed. Sometimes particularly in the last two bodies, very hard scale is not removed but get polished by of cutters. Such polished scale reflects light and observer may feel that the tubes are clean.

Mechanical cleaning causes wear of the tubes and fine particles of metal may get detached. It is an unpleasant job and more manpower is required. Descalers operated by compressed air may also be used for which relatively powerful compressor is required. However, these are not in general use in sugar industry.

48.5 Time required for cleaning operation:

The total time required for cleaning depends on the number of evaporator bodies, their heating surfaces, number of descaling shafts available, the hardness of scale and efficiency of workers, etc. The time cycle for cleaning operation is as under –

Table 48.2 time required for cleaning

Sr. no.	Particulars	Time required in hours
1	Liquidation of evaporator	1 - 2
2	Rinsing the bodies with water	1 - 2
3	Chemical boiling	4 - 8
4	Draining and cooling of bodies	1 - 2
5	Mechanical cleaning	12 – 20
6	Testing of calandria and closing the bodies	1 – 2
7	Total cleaning hours	20 - 36

At Kothari Sugars[4] Kattur, there was a practice of taking a general cleaning at an interval of every 15 days. To reduce the cleaning time the wash out line was changed from 100 mm to 150 mm and the caustic soda line was changed from 2" to 3". Testing of calandria was done by raising the pressure by pump so that time required was reduced. The juices from the first and second effect were sent to vapour cell (stand by). With this arrangement, the load of thin juice at pan station was reduced.

48.6 Cleaning by molasses:

In the off season the chemical cleaning can be prolonged for weeks. Therefore after the closer of the season the evaporator bodies can be cleaned by diluted molasses. The bodies are filled with diluted molasses (6-8°Bx) and the molasses is allowed to ferment for three weeks. The scale is loosened. After three weeks the wash is drained out and the scale is removed by brushing.

48.7 Vapour side cleaning:

There are different methods of vapour side cleaning which includes dilute acid circulation, dilute soda circulation, diluted final molasses circulation. Methods commonly followed are given here under-

- A caustic soda solution of 0.5 % concentration containing little quantity of $KMnO_4$ is prepared in the caustic soda tank and heated up to 90°. The hot soda solution is circulated through the vapour side of the calandria for 24 hours. Then fresh water is circulated for washing and rinsing.
- In some factories after caustic soda circulation and washing, hot dilute HCL mixed with iron sulphate is circulated for one hour.
- Honig recommends use of 2% acetic acid or formic acid combined with 1-2% no foaming detergent to be circulated through the calandria.
- A 12 -15° Brix final molasses solution can be circulated through the vapour side of the calandria for 48 hours.

References:

1 Nandgopal S, IS Dec 1976 p 607 – 612
2 Ramlingam, SISSTA Aug. 1998 p 126 – 130
3 Mangal Singh, STAI 1976 p 25-34
4 Nandgopal S, SISSTA April – June 1982 p 23 – 26

-OOOOO-

49 PREVENTION OF SCALE FORMATION

49.1 General:

The scale formation at evaporators is always a headache to the sugar industry. The factory has to take stoppages of about 1-3 % of the available hours of the crop season to clean the evaporator heating surface. Lot of chemicals and labour charges has to bear to the factory. The mechanical cleaning is one of most nasty job of the factory. Therefore preventing the scale formation is always given importance.

- The scale formation in evaporator can be controlled to some extent by strict control on cane quality, lime quality and process parameters at clarification station.
- The scale formation can be reduced by using chemicals that increase solubility of calcium compounds and other salts of the juice.
- It is observed that the scale formation can also be reduced to some extent by use of magnets.
- Use of ultrasonic waves to control scale formation also have been tried. But yet it has not been proved. Further scientist oppose it because the ultrasonic waves are harmful to human ears. Higher decibels can be harmful to heart and can even cause death.

49.2 Cane quality and process parameters:

The cane quality is important in controlling the scale formation at evaporator.

- **Cane quality:** The cane should be clean, fresh, matured with minimum extraneous natter. The stale cane of low juice pH requires more lime for clarification. The lime reacts with the excess acids to form soluble salts that pass in clarified juice and form extra scale in evaporator. The clarification of juice from burnt cane is usually not so well as green cane and therefore the evaporator heating surface foul sooner.
- **Mill sanitation:** The formation of acids due to microorganisms at mills can be controlled by using proper biocides. Otherwise in clarification higher amount of soluble calcium salts are formed. These salts pass along with clear juice and get deposited in evaporator set in the form of scale.
- **Imbibition water:** Use of raw cold water for imbibition should be avoided. The raw water contains some dissolve solids that give scale in evaporator.

Shrivastava[1] found that scale formation was higher due to higher amount of SiO_2, SO_3, CaO and MgO in mixed juice and in the imbibition water used in the month of January as compare to other months.

Verma[2] has mentioned that at Ramala sugars calcium content of raw water used for imbibition was 540 ppm. After the use of soft water and condensate, the initial calcium content of clear juice was reduced from 1360 – 1380 ppm to 1280 – 1300 ppm. Since condensate is almost free from dissolved solids, it cannot increase the scale in evaporator.

- **Phosphate content:** The phosphate content in mixed juice plays an important role in clarification. The level of P_2O_5 content in mixed juice should be maintained to 280 – 300 ppm.
- **Quality of lime and dose of lime:** Bad quality of lime promotes scaling owing to the presence of silica magnesium etc. Optimum dose of lime is very important. Higher or lower dose of lime results in increase in calcium contents in clear juice that promotes scaling.
- **Clarification Process parameter:** The retention time of juice in the juice sulphiter is also equally important in getting good results of clarification. Aher[3] has mentioned that the retention time of sulphiter was below eight minutes, and hence juice clarity was not satisfactory. A new sulphiter with eight minutes retention time was installed that improved the clarification.

Optimum raw juice temperature 72 – 75° is equally important for effective precipitation of calcium sulphite. In case of juice sulphitation over gassing of juice may convert insoluble calcium sulphite into soluble calcium bisulphite that passes along with clear juice. When juice is evaporated, the bisulphite decomposes to sulphite and SO_2. The sulphite precipitates and deposits as scale. Therefore, the sulphited juice should be slightly alkaline. Addition of phosphoric acid (10 ppm P_2O_5) in sulphited juice also helps to neutralize excess of lime[4] if any.

With moist air, the formation of SO_3 in the sulphur burner is more. It forms calcium sulphate which is highly soluble, passes along with clear juice, and gives high scale in evaporator. Therefore drying of air before it enters the air compressor of the sulphur burner is essential. It is usually done by passing the air through trays containing lime. The lime should be replaced daily. Verma[4] reports that by supplying dry air the calcium content in clear juice was reduced from 1850 – 1900 ppm to 1640 ppm.

Efficient flashing of the treated and heated juice at flash tank is essential for proper mud settling. Proper venting of clarifier is also equally important. The air occluded with juice if not vented remains with the juice at 97°C for 3 hours. The calcium sulphite reacts with the occluded air and forms calcium sulphate[4], which is about 70 times more soluble than calcium sulphite. It increases calcium content in the clear juice and forms heavy scale in evaporator.

- **Lower capacity utilization and frequent stoppages:** At lower capacity utilization the residence time of juice in each evaporator body is longer with lower circulation. It helps in increasing the scale formation faster. In mill stoppages when evaporator is stopped for few hours and juice remains stagnant in evaporator bodies, scale deposition is more.

49.3 Different methods of scale prevention:

The Prevention of scale formation is followed by two different ways

i) Use of chemicals: Some chemicals increase solubility of calcium and other salts. With the addition of such chemicals in the evaporator, the salts remain in solution and the scale deposition at evaporator is reduced.

ii) Use of magnets: Ionization of inorganic nonsugars present in the juice is effected by creating weak magnetic field around juice pipe. The inorganic molecules present in the juice get ionized and remain in solution. Therefore, the scale deposition at evaporator is reduced.

49.4 Use of chemicals:

i) Addition of phosphoric acid in clear juice: Addition of diluted phosphoric acid in clear juice up to 10 to 15 ppm reduces the scale formation in evaporator bodies by increasing the solubility of calcium sulphite and calcium organic salts. This also keeps the scale soft and inhibits destruction of reducing sugars.

ii) Use of antiscalent: The antiscalent are increasingly being used the sugar industry.

Antiscalent are generally blends of low molecular weight acrylic acid polymers and organophosphates used for prevention of scaling. They contain a sequestrant and dispersant which effectively prevent scale formation. The sequestrant ties up the scale forming salt of calcium, magnesium, silica and keeps them in solution without allowing them to precipitate or delays the precipitation. Even if precipitation occurs, the chemical prevents normal crystal growth so that fine sludge is formed which remains in suspension.

Any good quality antiscalent should satisfy the following conditions.

- It should not interfere in the process.
- It should be acceptable in terms of toxicity.
- It should be economical to use.

Antiscalent dose of about 10-20 ppm on cane is used. For the first and second bodies, it is introduced in clear juice. For the third and fourth bodies, it is injected using dosing pumps and needle valves. Antiscalent should conform to FDA regulation

US: 21: CFR – 173.5 and US: 21: CFR – 172.810 and should be nontoxic[5].

```
HOOC—H2C                    CH2—COOH
        \                  /
         :N - CH2 - CH2N:
        /                  \
HOOC—H2C                    CH2—COOH
```

stucture of EDTA

Fig. 49.1 structure of EDTA molecule

```
      OH   H    OH   H    OH
      |    |    |    |    |      //O
H-O—C—C—C—C—C—C
      |    |    |    |    |      \
      H    OH   H    OH   H       O-Na
```

Fig.49.2 Sodium gluconate molecule

With the use of Antiscalent the quantity of scale is reduced but cannot be avoided completely. In antiscalent trials taken at Pondicherry Co-op. Sugar Mills[6], the factory (1250 TCD plant) could prolong the run for 70 days with an average crushing rate of 1410 TCD with imbibition 249.14 % on fibre. Advantages of antiscalent:

- The scale deposition is reduced;
- Higher evaporation rate is maintained;
- Therefore higher crushing rate at higher imbibition with higher syrup brix can be maintained;
- It increases duration of run, therefore number of cleanings in the season are reduced;
- Consumption of chemicals, steam, power, cleaning equipment such as cutters and brushes used for cleaning is reduced;
- Time required for descaling of the evaporator tubes is reduced.
- Labour cost for general cleaning can also be reduced;
- Life of the tubes is increased;
- Inversion loss during shut down is reduced;
- Deterioration of harvested cane in waiting is also controlled.

With the use of antiscalent, the salts, which do not precipitate in the form of scale in evaporator and pans, pass into final molasses. The level of inorganic constituents such as CaO does not increase in sugar. Theoretically there can be negligible increase in quantity or purity final molasses which is not detectable in weight or analysis.

In some factories, it was observed that the scale formation was reduced at evaporator but was increased at pans. Therefore, quality of antiscalent should be assessed and its dose should be decided. Antiscalent give good results in many of the factories but not much satisfactory in some factories.

Quality of an antiscalent is assessed by analysing it for sequestral value. Sequestral value of the antiscalent is defined as milligrams of calcium carbonate sequestered per ml of the given antiscalent. Kulkarni[7] gives the method for determination of sequestral value. The sequestral value of antiscalent may vary from 180 to 350 and accordingly dose is to be decided.

Although the sequestration value of an antiscalent reflects its ability to chelate the scale forming salts, it does not take into account its ability to disperse the scale forming salts. A chelating agent which has excellent sequestration values may have poor antiscalent properties. Lack of good dispersion can lead to deposition of scales in the pans, crystallizers and on centrifugal screens. Therefore in addition to the sequestration value, percent inhibition of scale forming salts and dispersion ability test are also important in assessing the quality of an antiscalent.

49.4 Use of magnetic units:

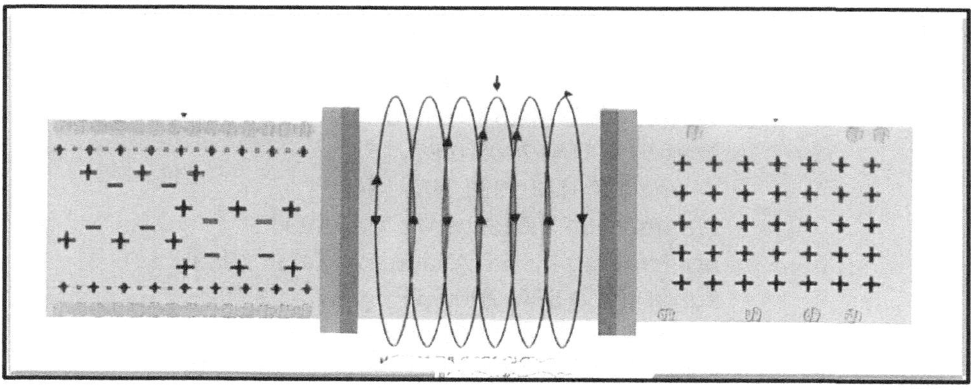

Fig.49.3 Effect of magnetic field on clear juice

When juice is passed through magnetic field, mobility of molecules of inorganic constituents from the juice is increased and they remain in ionic state. Such constituents do not precipitate and deposit on heating surface as scale. Therefore rate of scale formation is reduced and the run can be prolonged.

Some factories are using either monopole magnets or electromagnets to prevent scale formation at evaporator. Electromagnets are cheaper and can well be made by the electrical department personnel of the factories[9].

Hanumantha Rao[10] has given his experiences of using magnetizers at Shri Vaani Sugars Pungnur and claimed –

- Reduction in downtime of evaporator cleaning by 90%
- Reduction in cost of chemicals 80%
- Reduction in use of antiscalent 100%

Bangar Reddy[11] reports that by installation of monopole magnets at evaporator the factories Sarvaraya, Prudential and Nizam Zaheerabad could prolong their run maintaining higher crush rate.

Gupta[12] while using permanent magnets at evaporator found that the normal duration of the run was extended by 7-10 days. The scale deposition was less and the scale was soft. The use of antiscalent was reduced from 12 ppm to 6 ppm. Consumption of cleaning chemicals was also reduced.

49.5 Use of Ultrasonic:

Chengcan Yao[13] has described the trials of ultrasonic to inhibit scale formation at evaporator in a sulphitation factory. He give a chart of rate of scale formation with and without ultrasonic and concludes that the use of ultrasound during evaporation can give many advantages, inhibiting scale formation and even removing previous scale formed. However, he comments that an essential step in preventing scale formation is to firmly control the clarification process.

References:

1. Shrivastava M K STAI 1989 p M 31 – 40
2. Verma V K STAI 1981 p M 1 – 9
3. Aher W R STAI 2002 p E 3-18
4. Verma V K STAI 1981 p M 1 – 9
5. Marimuthu P NSI Seminar May 1994
6. Jaychandran K SISSTA January-March 1988 p M1- 4
7. Kulkarni A D STAI 2000 p mfg. 87 – 93
8. Shetty S B SISSTA 2001 p 149 – 152
9. Viswan K adham SISSTA 1999 – p 223 – 225
10. Hanumantha Rao STAI 2003, p M 61-64
11. Bangar Reddy STAI 1999 p E 50-55
12. Gupta STAI 2001, p 73-79
13. Chengcan Yao IS Feb. 2000 p 899 – 903

-OOOOO-

50 WATER BALANCE OF SUGAR FACTORY

50.1 General:

Generally sugar factories are established in the areas where there is sufficient rainfall for cane growing or round the year guaranteed canal irrigation is available. Therefore mostly abundant water is available for most of the sugar factories and hence the quantity of water being used in the factory is not generally accounted. In sugar factories generally exact water requirement for each system, water re-circulated and make up water taken is not measured.

But in some regions water shortage is observed in summer days and factories have to utilize water carefully. In some factories in summer days, arrangement of pumping of water from long distances of about 10 to 12 kms is also made. Some factories in critical situation even have arranged water tankers for supply of water. In such factories water is used very carefully and water consumption is low as necessary action towards conservation and reuses of water are taken. Therefore water consumption varies considerably from factory to factory.

The raw material sugarcane contains about 68-72% water. Imbibition water of about 30 to 40 % on cane is added at mills. Part of this water passes with bagasse and the remaining comes in boiling house. The water from the mixed juice is evaporated at evaporator and pan station.

The exhaust steam/vapours used for heating and concentrating the juice and boiling the massecuite are condensed and the hot condensate is available to use it for boilers, imbibition and process.

50.2 Water requirement of a factory:

The water requirement of a sugar factory can be broadly classified as under

- Cold water is required for cooling of equipment and for cooling of B and C massecuite,
- Hot condensate is required for boilers, imbibition and in the process;
- Cold water is required for condensers;
- Miscellaneous – water required for washing and cleaning of mills, boiling house etc. and rinsing and cleaning of vessels at the time of general cleaning. Water is also required in laboratory, at cane yard and for molasses tank cooling etc.

The water requirement and condensate produced for a 2500 tcd / 115-tch plant is considered below.

Cooling Water Requirement:

The cold water requirement is given in the following table. The cooling water requirement of a modern 2500 tcd plant is 100-120 t/hr. Water used for cooling of turbines and mills bearings is recirculated. Generally it is circulated from raw / service water tank and as the reservoir is of large capacity, no cooling of this water is done before recirculation.

Recirculation of water used for cooling of boiler feed water turbo pump, I D and F D fans, D G set, sulphur burners, air compressors, and crystallizers is not done in many factories and it goes to drain. When this is water is recirculated, the quantity of make water required is reduced. The water goes to waste / drain i.e. effluent water is treated for control of pollution and the treated water is then generally used for irrigation purpose in the nearby varietal cane development farm of the factory.

Table 50.1 Cooling Water Requirement for 2500 tcd plant

Sr. No.	Equipment	Water requirements t /h
1	Fibrizer and mill turbines, reduction gear boxes mill	35-45
2	Main power turbine (2500 kW) reduction gear	30-35
3	Boiler feed water pumps, I D F D fans D G set etc.	6-7
4	Sulphur burner and gas cooling	6-7
5	Water cooled crystallizers	11-12
6	Vacuum pumps, compressors and all other pumps	12-14
7	Total	100-120

Some water from mill bearings leak, further mill washings are contaminated with juice and oil. All this water goes as wastewater. Therefore some fresh make up water of about 3-4 % on cane may be required in the supply tank.

Condensate water requirement (for 2500 tcd/115 tch plant): Quantity of water required for boilers, imbibition and in boiling house for various purposes is given hereunder

Table 50.2 condensate water requirement

Sr. No.	Station / Equipment	Basis of estimation	Quantity required t /hr.
1	Boilers	Steam 47 %	56.35
2	Imbibition	35 % on cane	40.25
3	Milk of lime	2.80 % on cane	3.20
4	Filter cake washing	6 % on cane	6.90
5	Pan station (Molasses	4 % on cane	4.60
6	Centrifugal machines	1.5 % on cane	1.75
7	Magma making,	1.00 % on cane	1.15
8	Transient heaters	1.00 % on cane	1.15
9	Total	100.30 % on cane	115.35

Water required for condensers:

With profuse vapour bleeding, vapours going to condenser from evaporator are about 1.0 - 2.0 % on cane and from pans 18 -22 % on cane. Thus, total vapours going to condenser are 20-24 % on cane or 24 to 27 t/hr. for 115 tcd).

The quantity of vapours going to condenser may vary depending upon vapour bleeding at evaporator, syrup brix, control on use of water at centrifugals and pan station and massecuite % cane. The air leakages at vacuum vessels also add load to the condenser. Thus, the load on condenser may vary considerably. Further, the quantity of water required in kg per kg of vapours to be condensed varies from 30 kg to 55 kg depending upon the condenser water inlet temperature and approach.

The approach depends on efficiency of the condenser. The water required for air extraction also varies. Thus, for a 2500 tcd plant the injection water requirement may vary from 1500 to 3000 t/hr. All this water is recirculated.

The quantity of warm outlet water from condensers is always more than the inlet cold water quantity. It is because water of the condensed vapours is added to it.

The warm outlet water from the condensers is sprayed at spray pond to cool it and is cooled by evaporative cooling. At this time most of this added water quantity gets evaporated. Some water is lost by entrainment / drift during spraying. Water is also lost due to percolation from the spray pond and the channels. Therefore some make up water is generally required for spray pond.

In summer months at many factories, the daytime ambient temperature is above 40°C and heat released from the hot water due conduction and convection is negligible. The heat has to be released mostly by evaporative cooling. Therefore for most of the time all the spray pumps are to be operated. In such case loss of water by evaporation are high and the entrainment losses are also more and make up water requirement is increased.

In winter season the case is reverse. The water is cooled by conduction and convection, as the ambient temperature is low. Therefore, most of the time spray pumps are not required to run. As a result, much less quantity of water is lost by evaporation. The loss by entrainment is also negligible. Therefore, the water level in spray pond is maintained and no makeup water is required. Sometimes even the water level increases and the spray pond get overflowed.

As season passes, the spray pond water slowly gets contaminated by little entrainment and by vapours of sulphur dioxide (in case of sulphitation factories) and water becomes dirty. Therefore after few days some water from spray pond is discharged and fresh water is taken into the pond. In this way the spray pond water can be kept clean. The cooled excess condensate free from sugar test can be used for this purpose.

Water required for miscellaneous use:

Water is required for regular cleaning and washing of mills, fly ash arrester, juice heater cleaning, rotary vacuum filter cleaning, floor washing in boiling house, laboratory, sanitation etc. This water requirement is normally 2-3 t/hr.

50.3 Condensate available and it's uses:

Sufficient condensate is produced at juice heaters, evaporator and pans. The total condensate available from boiling house is given in the following table.

Table 50.3 condensate quantity available per hour for 115 tch plant

Sr. No.	Source	Percent on cane	Quantity of condensate t / hr. for 115 tch
1	Juice heater	17.00	19.55
2	Evaporator	80.00	92.00
3	Pans	22.90	26.35
4	Total	119.90	137.90

Thus, the condensate water production is 137.90 t/hr. (119.90 % on cane). Considering 5.0 % loss, the condensate available is 131.00 t/hr. (113.92 % on cane) and the requirement of a factory is 116.5 t/hr. (101.30 % on cane). Hence, extra condensate of 14.5 t/hr. (12.61 % on cane) can be available. Now in factories hot condensate is used for raw juice heating in one or two stages. In such factories the condensate available is reduced by 3.0 -5.5 % on cane.

Considering the makeup water required for cooling of about 5-6 t/hr., for spray pond and miscellaneous 2-3 t/h, the total water requirement is about 10-12 t/hr. The extra condensate available can be cooled down to room temperature and be used for the above purpose. However, the water should be free from heavy entrainments.

In some factories the hot excess condensate is used to heat the air entering the boiler for which tubular heat exchanger is installed before F D fan. The air is heated and the excess condensate gets cooled.

A masonry tank of 200 m^3 can be adequate for a 2500 tcd plant for storage of condensate and it can be used in mill stoppages and general cleaning so that fresh / raw water requirement can be considerably reduced.

The example of Hiranyakeshi[1] sugar factory is outstanding. Previously the water was used liberally. However, because of shortage of water near the location of the factory, the factory has to bring fresh water from a dam from long distance by tankers. It has restricted liberal use of water. The factory has taken following major step in reducing the water consumption.

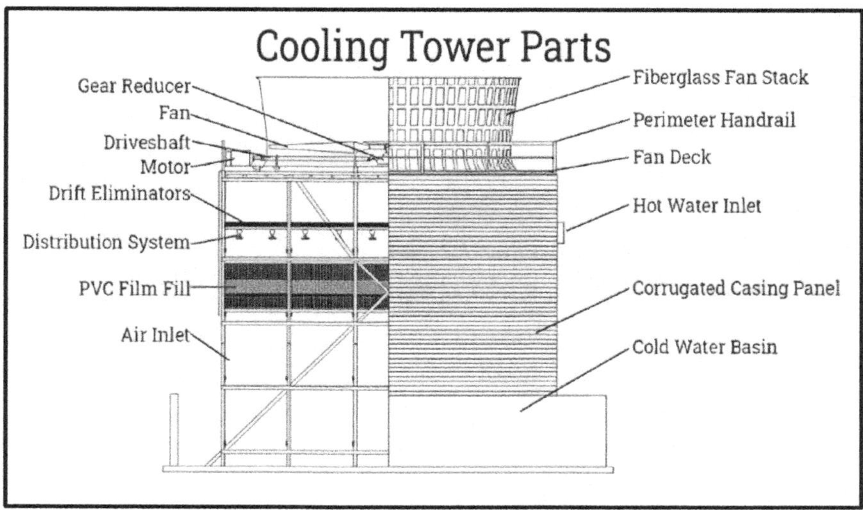

Fig.50.1 small scale cooling tower

- All the water used for cooling except for sulphur burner and water ring vacuum pump is recycled. About 90 –94 % of the cooling water used in the factory is recycled.
- Cleaning of mills is done by hot condensate water.
- Treated effluent water is used for cooling of sulphur burner.
- For juice heater brushing instead of fresh water spray pond overflow is used.
- For soda boiling of vapour cell and first body of quadruple treated effluent / spray pond water is used.
- For fly ash arrester wet scrubber treated effluent / spray pond water is used.
- Overflow of the boiler feed water tank is collected in a separate tank and the same is recycled for use it for boilers.
- Excess condensate storage tank is provided. This water is used in mill stoppages.

Thus with well-planned water management the factory could achieve zero water requirement and zero water pollution. Ponni sugars[2] has attained zero water consumption for sugar processing excluding cleaning days and laboratory use. The fresh water consumption is totally stopped. The excess wastewater generated from spray pond (about 500 – 650 m^3 /day in 2500 tcd plant) is mixed with untreated effluent (about 125 to 175 m^3 /day) and is send for treatment. Treated effluent about 400 – 500 m^3 is filtered through double stage sand filter to remove suspended solids. The outlet from filters is treated with 0.2 –0.4 ppm chlorine and is send to service water tank. From where it is used at sulphur burner, crystallizers cooling, mill oil coolers, mill turbines and power turbine, pump gland cooling, floor washing etc. The excess water is used for irrigation.

References:

1. Hiremath V C DSTA 2002 p E62-67
2. Ramegowda B STAI 2003, p M 111-120

-OOOOO-

51 COMPRESSION OF LOW PRESSURE VAPOURS

51.1 General:

The principle of vapour recompression is simple. Vapours from an evaporator body can compressed to a higher pressure. The increase in pressure of the vapours also increases the temperature of vapours. Therefore the same vapours can be used as the heating medium for the same body from which the vapours are robed. It can be achieved by two ways.

1 Low pressure steam compression by steam ejector. It is called as thermal vapour recompression.

2 Low pressure steam compression by centrifugal pump driven either by electric motor or small steam turbine is called as mechanical vapour recompression.

The oldest installation of a vapour compressor in a sugar factory is still working in Arlberg, Switzerland. In the 1980s several sugar factories in France installed vapour compressors.

51.2 Thermal Vapour Recompression (TVR):

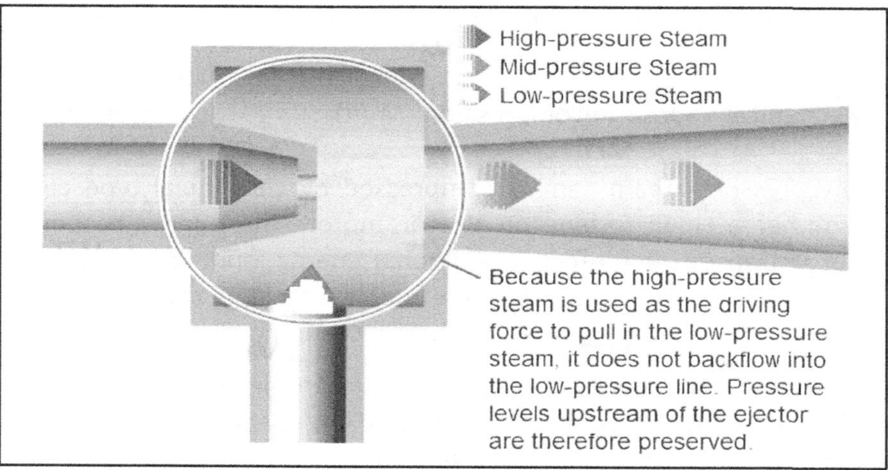

Fig 51.1 Vapour compression by steam ejector

When high pressure steam is passed through nozzles, it aspirates low pressure vapours with it. The temperature and pressure of high pressure steam gets reduced, whereas, that of low pressure vapours is boosted.

Thermo-compressor is a simple device with no moving parts. No external power is required for its functioning and it has little maintenance.

Weight of vapour aspirated per unit weight of actuating high pressure steam is known as entrainment ratio. It depends on live steam pressure, pressure of vapours and output pressure. The live steam flow can be automatically regulated. PLC controllers can used to control the system.

Three to five automatically adjustable steam nozzles can be used. In sugar factories it can be used to compress the vapours from the first effect by using live steam to fulfil the exhaust demand. However it should be noted that the output vapours are superheated and should be desuperheated before use.

Table 51.1 entrainment ratio of thermal compression

Exhaust Pressure maintained	pressure of Vapours aspirated	Pressure of Live steam ejected Kg/cm^2	
Kg/cm^2	Kg/cm^2	30.0	45.0
		Entrainment Ratio	
1.5	0.85	1.66	1.91
1.0	0.47	1.92	2.15
0.8	0.34	2.11	2.35

In India, vapour compression by steam ejectors have tried in 1980s in some factories. However it could not be successful due to inappropriate automation at that time. In deciding the use of TVR the steam / vapour balance has to be carefully worked out.

51.3 Mechanical Vapour Recompression (MVR):

Low pressure steam can be compressed by centrifugal type compressor driven by either a small turbine or electric motor. The energy supplied to the compressor is transformed into the additional energy input to vapours. It is also called as heat pump. A MVR unit will be preferable for a large unit. MVR evaporators with high evaporating capacities are common in other industries.

A compressor is an expensive machine, while an ejector is much simpler and cheap. In conclusion, MVR machines are used in large, energy-efficient units, while thermo-compression units tend to limit their use to small units. The efficiency and feasibility of this process depends on the efficiency of the compressing device i e steam ejector or compressor.

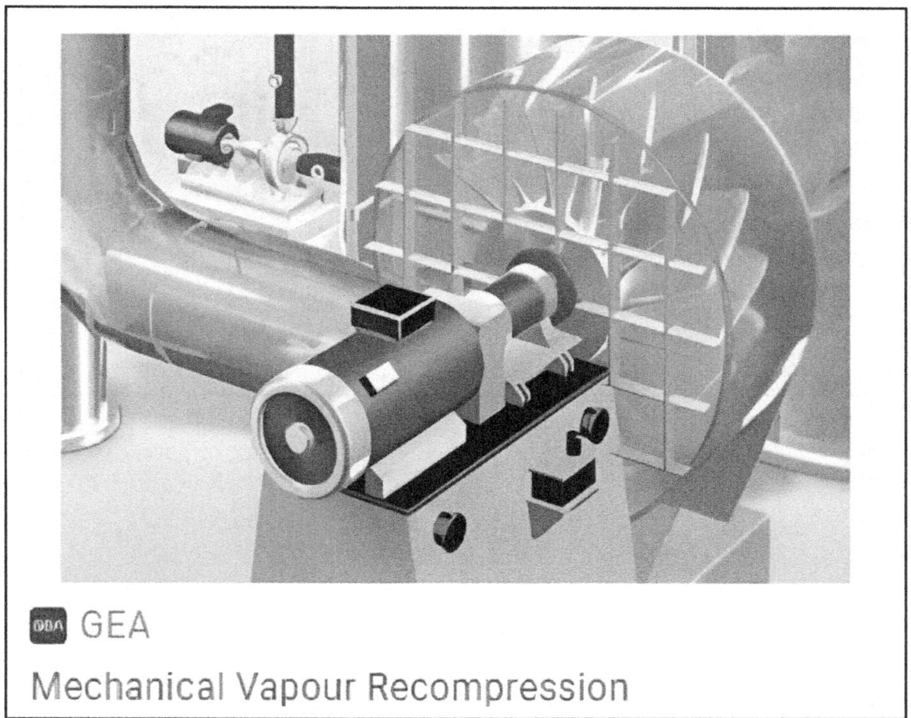

51.3 Mechanical vapour compression

LH Evaporator

MVR Falling Film Evaporator ...

-OOOOO-

52 CRYSTALLIZATION - THEORY

52.1 General:

In the sugar manufacturing process the unit operation - crystallization means crystallize out the dissolved sugar from syrup / molasses in much pure crystal form. This unit operation is the heart of the total process of sugar manufacture as the quality of marketable sugar is mainly decided in the crystallization process. It is the final stage of removing of most of the impurities contained in the raw material. The main sugar loss - loss in final molasses, is at the crystallisation and it is about 0.55 to 0.60 % of the total loses in the process of sugar manufacture. Thus crystallisation operation is important in the process of manufacture.

Sugar boiling or pan boiling is the process in which sugar from the thick juice / syrup and molasses is crystallized out by concentrating it. This is carried out in single effect evaporator called pans. The crystallisation of sugar can start at 78-80 brix. Therefore the evaporation can be carried up to 72-74 brix without any *grain formation*. However in factory working, clear juice is concentrated up to 55-65 brix. It is because the low brix syrup mixes well in the pan within short time and can dissolve fine false grain if any formed during boiling. Crystals are formed from impure syrup / molasses, but the crystals are much pure. Thus the crystallization is process that serves as a purification method. For example 'A' massecuite, from which commercial plantation white sugar of 99.8 purity is obtained, is of 86-88 purity.

Hereunder some common technical terms that are used at pan station are explained.

52.2 Common technical terms used at pan station:

- Syrup: The clarified juice after concentration up to 55-65 brix in multiple effect evaporators is called as syrup. In sulphitation factories where plantation white or mill white sugar is manufactured, the syrup is sulphited to a pH of 5.0-5.2 and it is called as sulphited syrup.
- Boiling point elevation: Difference between the temperature of a boiling sugar solution and the temperature of boiling pure water, (both measured at the same pressure) is called as elevation of boiling point or boiling point elevation (bpe). It is also called as boiling point rise (bpr).
- Saturation: A sugar solution at saturation will not dissolve any more sugar crystals at the temperature of the solution. The solution is called as saturated solution.

- Supersaturation: The degree to which the sucrose content in solution is greater than the sucrose content in a saturated solution.
- Solubility coefficient: Ratio of concentration of sucrose in impure saturated solution to the concentration in a pure sucrose solution saturated at the same temperature (with concentration expressed as sucrose/water ratio). It is referred to as saturation coefficient in the beet sugar industry.

$$\text{Solubility coefficient} = \frac{\text{Sucrose to water ration in impure saturated solution}}{\text{Sucrose to water ratio in pure saturated sucrose solution}}$$

- **Supersaturation coefficient**: it is sugar/water ratio of the supersaturated solution **to the** sugar/water ratio of a saturated solution under the same conditions (temperature and purity or nonsucrose / water ratio). It shows whether the solution is unsaturated (<1), saturated (=1) or supersaturated (>1).
- **Crystallization:** it is a process of crystallizing out dissolved solids from a solution. The process starts with nucleation and further growth of crystals takes place.
- **Pan or vacuum pan:** it is vacuum evaporative equipment used in the sugar industry to crystallize sugar from liquor, syrup or molasses.
- **Massecuite:** Mixture of crystals and mother liquor delivered from vacuum pan (at the end of each strike in case of batch pans) is called as massecuite. This can be further defined as A, B, C or D massecuites depending upon position of material in process. The massecuite from which sugar is bagged is referred as high-grade massecuites. The massecuite from which sugar separated is either used as seed or melted and the melt is used for high grade massecuite, is referred as low-grade massecuites.
- **Nucleation:** Generation and development of very fine microscopic crystals is called as nucleation. These fine crystals are further developed to grow to normal size by boiling syrup / molasses in vacuum pans.
- **Graining:** Process of initiating the formation of fine nuclei or very fine sugar crystals (which further are developed during crystallisation process) is called as graining.
- **Footing / graining material:** The material (syrup / molasses) being taken up to calandria level in the pan (concentrated just above the saturation point) to use for graining is called as footing or graining material. Generally for low-grade massecuite it is called as a graining material and for A massecuite it is called as footing.
- **Hardening:** Hardening means development of crystal in proper way by which consolidation of sucrose molecules in to crystal lattice is achieved.

- **Feeding to Pan:** During pan boiling (crystallization process) to develop Syrup / melt /molasses etc. are feed in the pan. This is generally called as feeding to the pan.
- **Movement Water:** The introduction of condensate in controlled way for movement of the boiling massecuite is called movement water. This is used mainly for
  1. To compensate rate of evaporation in boiling pan.
  2. To maintain fluidity so that circulation of boiling massecuite will be maintained.
  3. To dissolve false grain formed if any.
- **Cutting of grain / massecuite:** Dividing pan content into two pans or storing a part of grain in vacuum crystallizer or seed crystallizer is called as cutting of grain / massecuite. Mostly in factories it is loosely called as cutting of pan.
- **Strike:** Each pan with full of massecuite is called as a strike.
- **Seed Magma:** It consists of low-grade sugar (generally B seed) mingled with syrup or molasses or water and stored in crystallizer from which it is drawn in pan to start high-grade strike.
- **Conglomerates:** Two or more crystals grown together during pan boiling are called as conglomerates.
- **False grain:** Undesirable fine / small crystals formed spontaneously by secondary nucleation are called as false grain. The false rain are formed when supersaturations rises too high during boiling.
- **Mother liquor:** Liquid phase (syrup /molasses) in the massecuite during crystallization in which the crystals are growing is called as mother liquor.
- **Seed slurry:** Suspension of fine microscopic sugar crystals in alcohol or any other organic solvent in which sugar does not dissolve is called as seed slurry.
- **Seeding:** Introducing fine crystals to develop the same further in sugar boiling - crystallization process is called as seeding.
- **Strike:** Massecuite after completion of boiling when discharged from the pan into crystallizer is called as strike.
- **Drop a pan:** Discharge all of the massecuite from a pan to crystallizer. It is also referred to as striking a pan.
- **Crystallization scheme:** It defines the number and arrangement of crystallization stages involved in producing sugar.
- **Crystal content:** Proportion by mass of crystals in massecuite, often expressed as a percentage, and referred to total massecuite. (crystal % massecuite)
- **Exhaustion:** Applied to a massecuite, it represents the gram of sucrose present in crystalline form per 100gm of sucrose.
- **Exhaustibility:** Exhaustibility is the degree to which the molasses can be de-sugarised or exhausted in factory working.
- **Molasses:** Massecuite is cured or purged in centrifugal machines and mother liquor separated from crystals is called as molasses. . It may further be termed

as A, B, or C molasses according to the grade of massecuite from which it is separated. The molasses separated before washing is termed as heavy molasses and then separated using superheated hot water wash is called as light molasses.

- **Final molasses:** The molasses obtained from low grade massecuite from which no more sugar can be recovered economically under factory conditions is called as final molasses.
- **Run Off:** The mother liquor separated from refinery massecuite by centrifugal machines is called as run-off. This word is used in refineries.
- **Dissolved solids:** All the solid that are in dissolved state in solution - including sucrose, reducing sugars, organic and inorganic impurities, are called as dissolved solids.
- **Dry substance:** Total solids obtained from evaporating a solution or massecuite under vacuum to dryness. It is also referred to as total solids by drying or dry solids.

52.3 Saturation and super saturation of a solution:

The solubility of pure sucrose in water varies with temperature and increases rapidly with increase in temperature for e.g. in one kg water 2.334 kg of sucrose is dissolved at 40°C, however in one kg of water 3.703 kg of sucrose can be dissolved at 80°C. When a solution contains the total quantity of sucrose, which it can dissolve, it is said to be a saturated solution.

In practice however dissolving the sugar at slightly higher temperature and then cooling it. Otherwise by evaporating the water, the concentration can be up to or above saturation. It takes long time to make a saturated sugar solution by dissolving sugar at constant temperature.

When solution is concentrated beyond saturation point (by evaporation or by cooling), it contains more sucrose (solute) than it can dissolve at a given temperature. This state is above the saturation and even if a solution is concentrated beyond saturation, crystals do not appear immediately (or do not appear at all) in the solution. The sugar remains still in solution. This state (above the saturation) is called as supersaturation and the solution is termed as supersaturated solution.

52.4 Classen's theory of supersaturation:

Classen in 1928 put forth the process of crystallisation in scientific way. He introduced degree of concentration as supersaturation coefficient. The supersaturation coefficient is the ratio of the weight of sucrose percent water in a supersaturated solution to the weight of sucrose percent water which would present in a saturated solution having the same temperature and purity. This can be expressed by the formula

Supersaturation coefficient $= \left[\frac{S/W}{(S1/W1)}\right]_{P,\,T.}$

Where S/W = Sucrose/ Water ratio for solution of purity 'P' at temp. T
S_1/W_1 = Sucrose/ Water ratio for a saturated solution of purity P' at temp. T

The ICUMSA (International Commission for Uniform Methods of Sugar Analysis) has approved the above formulation at constant purity.
This can be written as

$$S = \frac{B`/ 100 - B`}{B / 100 - B}$$

Where B` = weight of material dissolved percent of a super saturated solution.
B = weight of material dissolved percent in saturated solution of the same purity and at the same temperature.

Table 52.1 Brix of a saturated syrup / molasses solution At various temperature and purities

Temp.→ Purity↓	62°C	66°C	70°C	74°C
90	75.9	76.6	77.5	78.3
88	76.2	77.0	77.7	78.5
86	76.5	77.2	78.0	78.8
84	76.7	77.5	78.3	79.1
76	78.0	78.7	79.5	80.2
74	78.3	79.1	79.8	80.5
72	78.7	79.4	80.4	80.9
70	79.0	79.7	80.5	81.2
62	80.4	81.1	81.7	82.5
60	80.8	81.4	82.1	82.8
58	81.1	81.7	82.4	83.1
56	81.4	82.0	82.7	83.4

For example, if syrup of purity 84 is taken in a pan up to calandria level and is boiled at 650 mm of vacuum to concentrate for graining or footing. Because of bpe the boiling point of syrup will be around 66°C and at this temperature and purity we can see from the above table that the syrup will be saturated at about 77.5° brix.

When a solution of any solute is concentrated to a required extent crystal nuclei are formed by aggregation of number of molecules and if sufficient concentration continues to remain, the nuclei thus formed will persist and grow.

Different zones of supersaturation:

On concentrating and cooling an unsaturated solution will pass the saturation line and become supersaturated. Nuclei however will not form until a considerable degree of supersaturation reached. With sucrose different zones of supersaturation are recognizable as shown in the figure below.

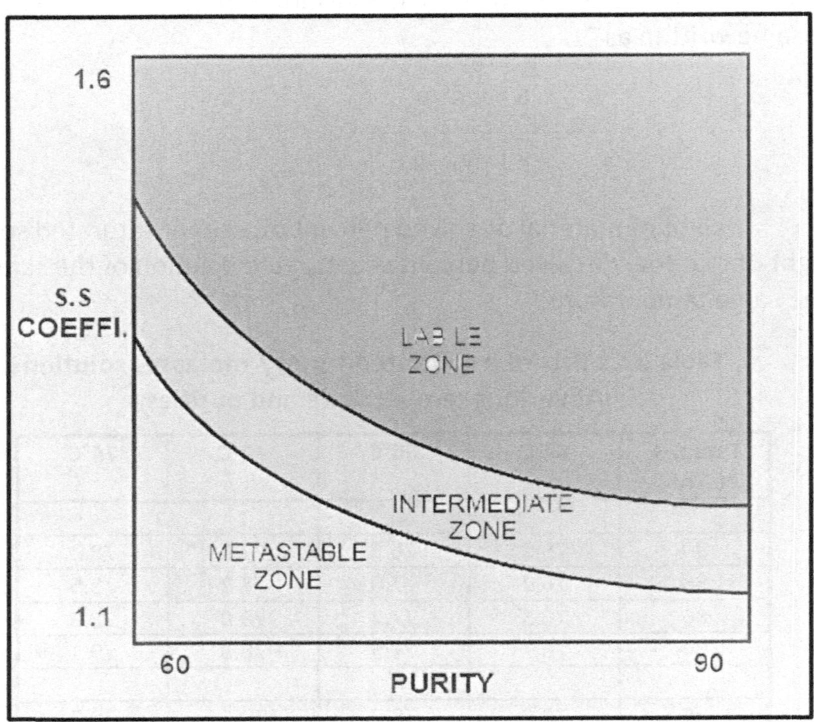

Fig. 52.1 Purity versus supersaturation coefficient and different zone of supersaturations

A—B is the saturation line. C—D and E—F lines are boundaries between different zones of supersaturation.

Let us take an example of footing taken for A massecuite in a pan. The syrup /melt taken is unsaturated and boiling under vacuum and the saturation coefficient is less than one at point **G** in the above graph. The temperature remaining constant the material is being boiled and getting concentrated the saturation coefficient goes on increasing. It crosses the line **AB** and passes in metastable zone of supersaturation. On further concentration it reaches to point **H** in intermediate zone. If boiling and concentration is still continued it will go up at point **I** in labile zone. Here an experienced panman finds the concentration has become too high so that false grain may appear and he decides to bring down the supersaturation. Now there are two ways to bring down the supersaturation 1) by feeding syrup/melt of lower brix which is unsaturated. 2) To bring down the vacuum and raise the boiling temperature so that solubility of sugar is increased and the supersaturation coefficient comes down. If the pan decides to feed a charge of syrup then on mixing the syrup charged the saturation point may come down to point **H** or lower on the same line. But if panman decides bring down the vacuum and increase the boiling temperature and bring the saturation coefficient in to lower zone, then the saturation point will go from **I** to **J** which is metastable zone.

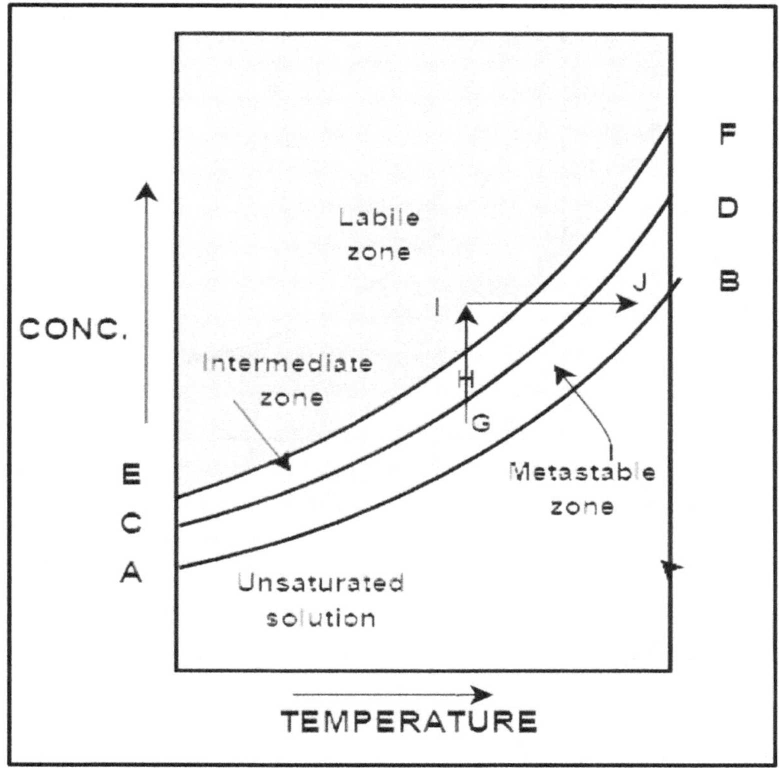

Fig. 52.2 temperature versus Different zones of supersaturation

In the first or metastable zone of supersaturation (in between A—B and C—D) the existing crystals grow in size but no new crystals are formed.

- In the second or intermediate zone (above the line C—D and below the line E—F) the existing crystal grow and new crystals are also formed. But only when crystals are already present.
- In the third labile zone (above E—F) crystals are formed spontaneously without presence of any crystals before.

To crystallize out sugar from the concentrated syrup it must be reached in supersaturation zone. In pan boiling the useful range of supersaturation is 1.25 to 1.5 depending on the grade or stage of boiling. With lower purities of solutions the degree of supersaturation has to be higher for crystallization. It can be seen in chart no. 51.1.

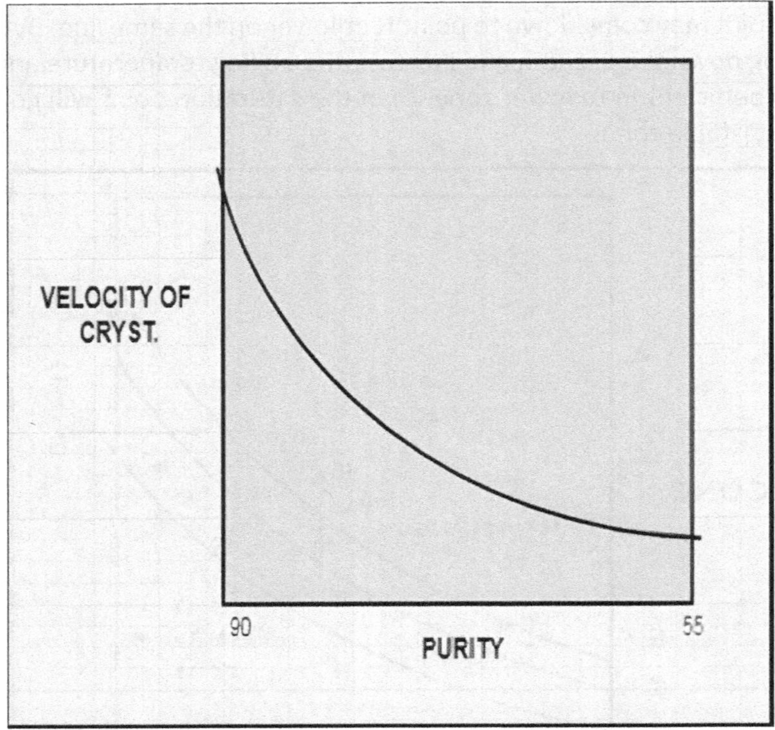

Fig. 52.3 Purity versus velocity of crystallization

Classens theory of pan boiling can be summarized as follows:

1) Depending upon the purity and the temperature of the mother liquor there is a definite coefficient of supersaturation at which crystals will appear.

2) On account of presence of non-sugars low purity liquids are less sensitive to formation of false grain.

3) The greater the number of times the syrup and molasses are boiled the more difficult the granulation becomes.

4) In order to separate sugar from impure solution by the process of crystallization driving force in the form of degree of supersaturation must be first established.

Van Hook has given the supersaturation coefficients for line marking boundaries of the respective zone for pure sucrose solutions. Values for impure solution differ depending on nature and concentration of impurities.

The width of different zones of supersaturation is sufficient so that practical use of different zones in the crystallization process is possible. Thus with proper control the necessary number of crystal nuclei per unit volume may be formed and by keeping the solution in metastable zone for rest of the charges, these crystals can be grown without formation of any false grain. Thus good control on crystal size and uniformity can be achieved.

52.5 Rate of growth of Crystal: Sugar crystal growth in a supersaturated sugar solution takes place as a result of two processes.

1. Diffusion of sucrose molecule up to crystal surface through the mother liquor,
2. Actual accommodation of sucrose molecule on the crystal surface.

Rate of diffusion of sucrose molecules:

The rate of diffusion of sucrose molecules to the crystal surface depends mainly on the following two factors;

- Degree of supersaturation of the sugar solution
- Viscosity of mother liquor

The supersaturation is the driving force which increases the rate of diffusion of sucrose molecules to the crystal surface. However the supersaturation also depends on temperature. With increase in temperature the coefficient of supersaturation decreases.

The rate of diffusion of sucrose molecules decreases with increase in viscosity. The viscosity increases as purity decreases. The viscosity also increases with decrease in temperature.

In low purity material viscosity is high and rate of diffusion is low. The lower purity decreases the rate of diffusion. In intermediate and low grade massecuites the purities of mother liquors are decreased and impurities are increased. The degree of supersaturation also has to be increased in intermediate and low grade massecuites. In the final stage of crystallisation of low grade massecuite the supersaturation coefficient is 1.5 or above.

It is not only the percentage of nonsucrose that affects the rate of crystallization but the constituents and their percentage also affects in different way.

Rate of accommodation of molecules over the crystal surface:

The process of the rate of accommodation of molecules over the crystal surface in the crystal lattice is the rate of crystal growth. As the sucrose molecules get deposited in the crystal lattice, the film of solution near the crystal surface has a lower degree of supersaturation than the solution away from the crystal. This is due to reduction in concentration of solute and also due to the effect of heat of crystallization.

52.6 Effect of impurities on rate of crystallization: T

The impurities fall into two categories.

- One group of constituents such as Inorganic constituents, carbohydrates like reducing sugars, and number of organic compounds. These retard rate of crystallization. Therefore the rate of crystallization is slower in low purity massecuites.
- The other group or those constituents like oligosaccharides, dextran, caramels, Ketoses (Raffinose in case of beet sugar boiling) these exert effect on growth rate. They have got tendency of co-crystallization and get adsorbed on different faces depending on their structures and are bound throughout the crystal.

Dextran and raffinose modify sugar crystals into the form of elongated sugar crystals. In factories when stale cane is arrived for crushing the phenomenon of elongation of sugar crystals is often noticed due to the presence of dextran formed in cane juice from stale cane. The elongation of crystals results are of adsorption of dextran on the growing sugar crystals. Needle shaped crystals are obtained in low grade boiling. Sometimes platelet formation is also observed in boiling due to retardation of growth along one axis.

52.7 False grain:

In pan boiling once seed is drawn or grain is formed in the pan, no new nucleation should occur or no new grains should be formed. Only the seed provided or grain formed by giving slurry has to be developed.

This requires the concentration of mass to be controlled and the coefficient of supersaturation to be maintained in the metastable zone. It should not shoot up to intermediate zone of further in labile zone. If in boiling the pan at certain stage the concentration of the massecuite goes above metastable zone, fresh grains appear as tiny crystals known as false grains. In pan boiling when a strike concentration overshoots and crosses the metastable zone, initially a hazy mass appears which if not dissolved quickly gives rise to innumerable of tiny crystals. Under such conditions, the pan boiler has to get rid of the false grain immediately by bringing down the saturation which is done by giving drinks of water and/or raising the temperature.

Even though this procedure is helpful in dissolving the secondary grain or crop, it also partly dissolves the original well-formed crystals in the boiling massecuite. In a pan at high massecuite level, more time is lost in dissolving the false grain. Further the crystal shapes and exhaustion of the mother liquor are also adverse affected. Thus it is very much essential to avoid formation of secondary grain in boiling.

If these false grains are allowed to grow the false grain will be an additional crop of crystals which are much smaller in size. Massecuite thus produced of different crystals sizes gives trouble in curing at centrifugals. Smaller crystals occupy the volume in between large crystals. Therefore the mother liquor / molasses cannot pass through the crystals. Hence separation of crystals and molasses is not proper. Lot of wash water is required. The sugar produced is of poor quality. Molasses purity is not controlled.

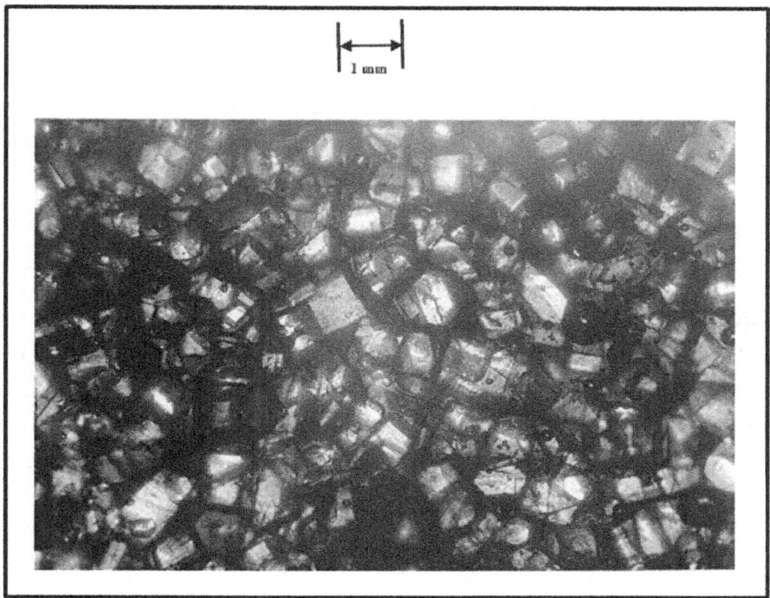

Fig. 52.4 crystals in massecuite

52.8 Conglomerates:

These are crystals joined together at some stage of boiling pan and can be seen even in the white sugar available in market. In the early stages of boiling a strike, many times two or more crystals join together and remained joined throughout the massecuite and further grow as conglomerates. These are sometimes also known as rolled grain. Apart from imparting bad appearance to the sugar crystals they are objectionable for two reasons-

- the joined grains contain impurities of mother liquor at the interface known as inclusions,
- They obstruct the proper washing in centrifugal separation and adversely affects the quality of sugar.

Conglomerates are formed mostly in high purity massecuites in the top area of metastable zone prior to false grain formation. The conglomerate formation normally is observed in pans with poor circulation during boiling.

Thus high concentration beyond a certain limit in the metastable zone and poor circulation of massecuites in some pockets induce conglomerate formation in high grade boiling. Installation of mechanical circulator in a pan improves the situation.

52.9 Crystal Size and Rate of crystallization: Rate of crystallization is directly proportional to the area of crystals in a mother liquor per unit volume and thus under identical conditions of purity, supersaturation and other boiling conditions, the massecuite with smaller crystals will give mother liquor of lower purity than the one with bigger crystal size. Greater area for deposition of sucrose is available when the crystal size is reduced. Area of crystal surface is given by the equation—

$$S = 0.00421/d$$

Where S = surface area in m^2/gm.
And d = length of crystals in mm.

In high grade massecuite (in plantation white sugar manufacture) the crystal length of about 0.80 to 1.1 mm. Whereas in low massecuite the size is brought down to 0.2-0.25 mm where maximum exhaustion of mother liquor with very high impurity content is to be achieved.

-OOOOO-

53 CONSTRUCTION OF BATCH TYPE VACUUM PAN

53.1 General:

At Pan Station massecuites are produced in vacuum pans. The vacuum Pans are used for nucleation / graining / footing and for massecuite boiling. Some times in difficulties syrup of low brix is also boiled to concentrate up to desired brix.

In most of the factories there are batch type pans. Many factories under modernization and expansion have installed continuous pans for B massecuite. Some factories have installed continuous pans for A and B massecuites. There are very few factories which have installed continuous pans far all three massecuites. Now few factories have installed vertical continuous pans.

Fig. 53.1 Pan station – batch pans

Generally pan station has about 4 to 8, batch pans of capacities_varying from 40 to 80 t and /or continuous pans of capacity 20 to 50 m per hr. are installed. In some factories 100 t to 120 t batch pans are also installed. Recently one factory in North India has installed a batch pan of 150 t capacity.

The pan station of a sugar factory is located at a height of 12-13 m. from ground floor and height of the crystallizers is about 7-8 metres above the ground floor, so that massecuite dropped in horizontal crystallizers can be taken to pugmills by gravity. There are 15-20 syrup and molasses tanks at the back side of the pans. Also seed and vacuum crystallizers are installed for seed and grains. Before 1970 some non-gravity plants were installed in India in which the pan floor was at about 7-8 m height above ground floor. In these plants massecuite is to be lifted by pumps from crystallizers to pugmills. In these the low grade massecuite has to be kept loose for lifting by pump. It was affecting exhaustion of the massecuite. Now all over the world calandria type pans are being used but up to 1930-40 coil pans were also being installed.

Coil pans: In the coil pans, a set of coils were made of 10-15 cm. dia copper tubes with wall thickness of about 2.5 - 4 mm. each coil was bent into spiral with the upper end connected to the wall of the pan while the lower end was joined to the condensate drain. The upper end of each coil was connected to steam line through valve to enable independent start or cut of any one coil. The only advantage of the coil pans was that the heating surface was gradually brought into operation as the strike level rises, but the maintenance of these pans was high. Use of low pressure vapours or exhaust was not possible. Therefore the coil pans were mostly replaced by calandria up to 1950. These types of pans are now obsolete.

Calandria pans: Calandria pan is a specially designed single effect evaporator, which operates on low-pressure exhaust / vapour. Vapours from the second or even third effect of evaporator of 0 - 0.3 kg/cm^{2}g can be used as heating media and steam economy can be achieved. In last 30 years continuous pans have been installed in most of the modern factories.

Types of calandria pans:

In calandria pans tubes are shorter and of larger diameter than in evaporators. The massecuite passes upward through the tubes. A downtake passage or a well is provided to descend the massecuite. The calandria pans can be broadly classified according to passage provided for massecuite circulation.

A) Central Downtake pans :

1) Flat calandria: - with single diameter or with enlarged body
2) Inclined plate calandria

I. inclined bottom plate,
II. Inclined both top and bottom plate.

These pans were installed in very few factories.

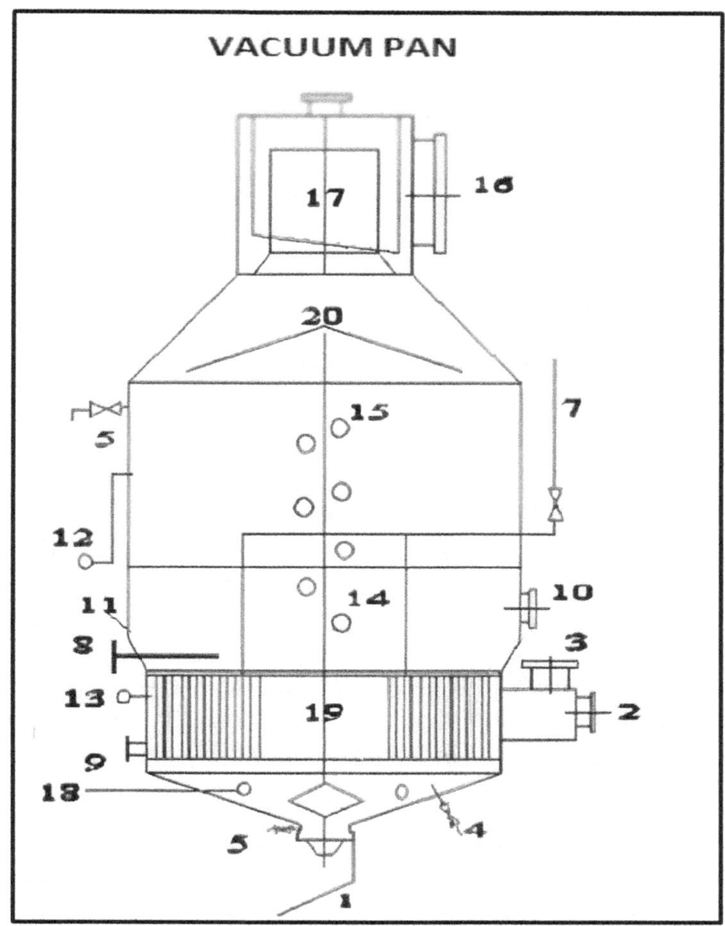

Fig. 53.2 Low head calandria pan

Vacuum pan details			
Sr. No.	**Particulars**	**Sr. No.**	**Particulars**
1	Discharge valve	11	Thermometer
2	Exhaust valve	12	Vacuum gauge
3	Vapour valve	13	Pressure gauge
4	Cutover valve	14	Sight glasses
5	Washout valve	15	Light glass
6	Vacuum brake valve	16	Vapour outlet
7	Ammonia venting	17	Catchall
8	Proofstick	18	Feed pipe
9	Condensate outlet	19	downtake
10	Manhole	20	umbrella

B) Peripheral downtake Pans:

I. Inclined bottom plate,
II. Lenticular calandria.

C) Horizontal pans

D) Continuous pans

i) with vertical tubes,
ii) Horizontal tubes.
iii) Vertical continuous pans (VCP)

The most commonly installed pans are of flat calandria and central down take with either single diameter or enlarged upper diameter.

53.2 Construction of low head calandria pan:

Here we will study a flat calandria with central downtake and enlarged body, which is called as low head pan. It is cylindrical and the body above the calandria is increased in diameter to reduce the height of the massecuite above the calandria.

Bottom of a pan:_The bottom of a pan_is conical and the discharge valve is at the centre. The angle of sides should not be less than 23° with the horizontal. A cylindrical portion (125 to 150 mm in height) is provided between the bottom soccer and calandria to expand the tubes.

Calandria: In the design of calandria, tube diameter and downtake diameter of pans are large whereas the height of tubes is less as compare to evaporator body. It is necessary for easy circulation of high brix boiling massecuite. The massecuite rises through the tubes while being heated to boil and descends down through the downtake.

The height of tubes and calandria varies from 800 mm to 1100 mm. In calandria with inclined lower plate calandria height may reach up to 1100 to 1400 mm at the centre. The interior diameter of tubes varies from 90 to 110 mm. In India according guideline of specification of sugar machinery the tube length recommended is 800 mm and external diameter of 101.6mm (4 in) in and 1.5 mm thickness in SS and 2mm in brass.

Pitch and ligament: The distance 'p' between the centres of two adjacent tubes is called as 'pitch' of the tubes. The minimum width of metal separating two neighbouring holes is termed as 'ligament'. According to standard specification the ligament should be 24 mm. however it may vary from 16 to 24 mm. The end portion of the tubes are expanded and fixed in two plates.

In case of central downtake pans, downtake diameter is about 40 to 45 % of calandria diameter. In case of floating calandria pans, calandria diameter is about 80 to 85 % of pan diameter.

Syrup / molasses drinks / feed should be distributed at the bottom under the calandria and away from the downtake. A small exhaust connection to the feed pipe gives considerable improvement.

Vapour entry: The vapour entry at the calandria should be wide or through several entries. At the entry of steam, because of impact of steam on the front tubes, there is rapid boiling of massecuite that may lead to false grain formation. To avoid such impact in many designs baffles are provided to distribute the steam.

A calibrated level board marked in ft./metres is provided adjacent to the sight glasses. The level board indicates height of massecuite above the top tube plate.

Feeding the syrup / molasses: there are two methods of feeding the syrup / molasses to the pan

1. Charging drinks time to time.
2. Continuous charging.

The second method gives better control in the pan boiling process. but with manual control the continuous charging is difficult; however with automatic control valve system the valves remains always more or less open as per degree of supersaturation of the mother liquor in the boiling massecuite. The feed velocity is below 1 m/s.

Strike level: The strike level given is normally 1.4 to 1.6 m above the top tube plate with natural circulation. With mechanical circulator a height of the strike can be 1.7 to 1.8 m without any disadvantages.

Vapour space height: Height of 2 m above strike level is provided as vapour space, above which a dome with internal save all is provided. External catchall also is provided for many pans.

Pan washing arrangement: A circular coil with exhaust steam or first effect vapours and water connection is provided at the top periphery of the pan to wash the pan after strike. Usually pan is washed after every strike.

Sight glasses: sight glasses are provided to observe the massecuite boiling in the pan. Normally 5-7 sight glasses are provided in one or two rows from the top tube plate to above the strike level on the front side of the pan. A glass with light arrangement is provided at the backside of the pan so that light can pass in to the pan and panman can see inside the pan. Wash water spray are provided to all the light and sight glasses.

Manholes: manholes of sufficient size so material like stirrer blades and entrainment packs can be taken inside and fitted or can be removed. The minimum manhole size is 450 mm in diameter.

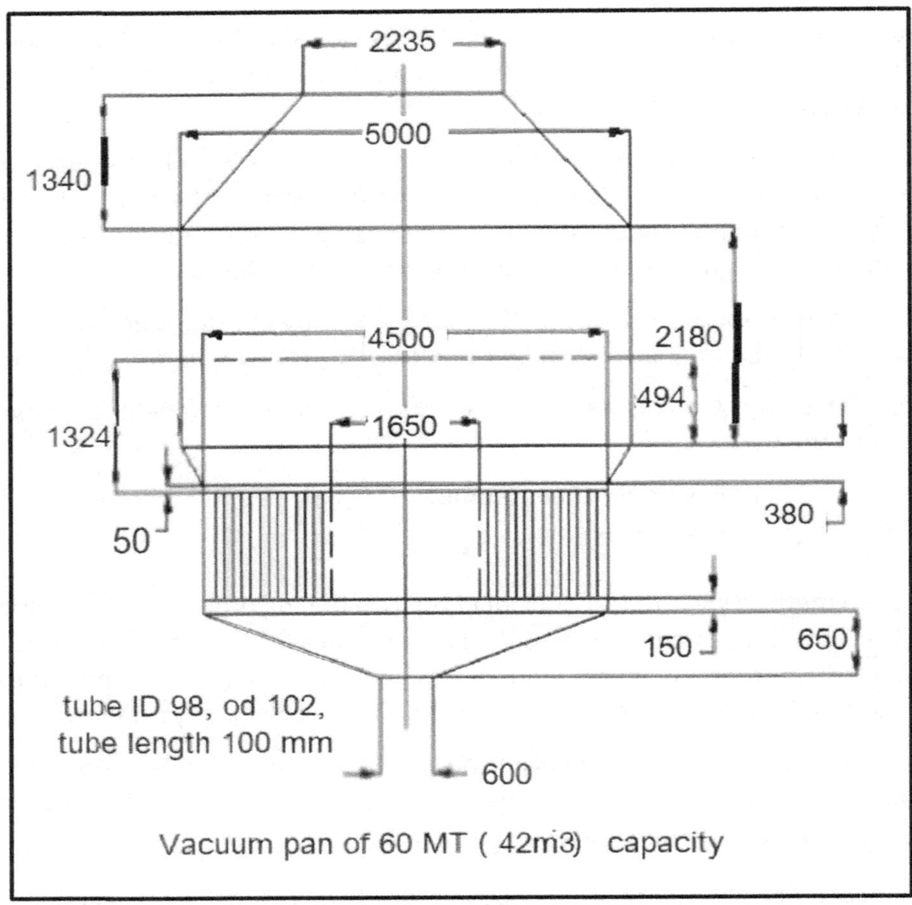

Fig 53.3 Low head calandria pan dimensions

Graining volume: The volume of up to top tube plate should be covered by footing or graining material before turning on the vapours (This volume is called graining volume). Otherwise ebullition throws syrup on the exposed portion of heating surface which is heated. This causes caramelization and increase colour of massecuite and sugar. The graining volume is generally expressed in percent of working volume of the pan. It should be as low as possible.

When graining volume is less, establishing the grain and controlling purity of massecuite becomes easy as saved volume can be used for low purity molasses feeding. From this point of view the graining volume should be about 33-35 % of the working volume. However now to reduce the hydrostatic head, - in most of the designs the tube height is reduced up to 800 mm and therefore the diameter of the calandria is increased and the graining volume has reached up to 40 %.

Heating surface to volume ratio: A very important factor in the design and performance of vacuum pans is the ratio of heating surface to its working capacity.

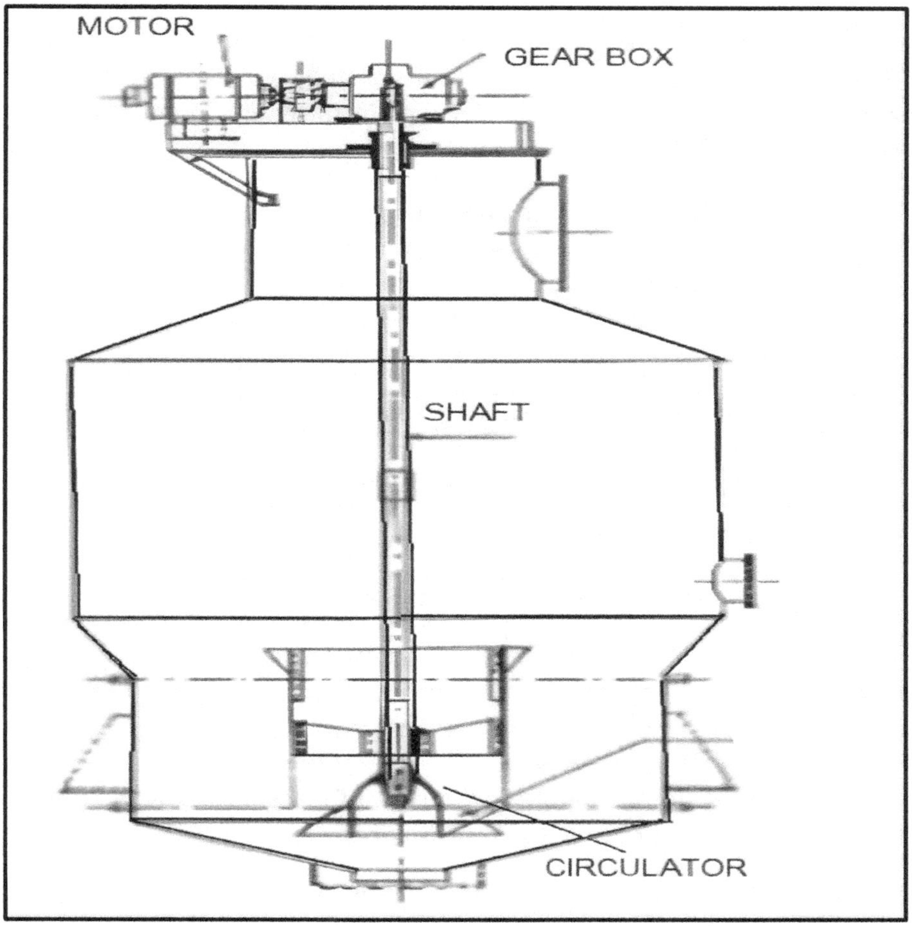

Fig 53.4 Pan with stirrer

$$\frac{S}{V} = \frac{\text{Heating Surface of pan}}{\text{Working volume of the pan}}$$

However the ratio should depend on

- Vapour pressure used for heating
- Massecuite to be boiled

Vapours: With vapours of pressure 0.3 - 0.6 kg/cm^2, the heating surface giving S/V ratio 6.0 to 6.6 m^2/m^3 can work well. If vapours are to be used at a pressure of 0.0 - 0.3 kg/cm^2, it is advisable to increase heating surface so as to raise it to 6.5-7.0

m^2/m^3.

The low grade massecuite is viscous and natural circulation of such massecuite is poor. Therefore massecuite should be heated carefully and gently, otherwise there are chances of caramelization and /or false grain formation. Hence for low grade massecuite pan the S/V ratio should be 6.0.

Table 53.1 batch pan design specifications:

Sr. no.	**particulars**	**Dimension**
1	Strike Level above tube plate	1300 - 1400 mm
2	Graining Volume % Strike Volume	35 to 40 %
3	Graining Level above tube plate	50 mm
4	Heating Surface to Volume ratio	6.0 to 7.0 m^2/m^3
5	Down Take Diameter % Pan Diameter	40 %
6	Vapour space height above strike level	2 m
7	Tube hole nominal size	4 inch, 100-102 mm
8	Annealed Tube wall thickness	1.5 mm
9	Ligament size	16 -22 mm

Heating surface, graining volume and height of massecuite: While designing of a pan following three factors have to be considered:

- Height of massecuite should be minimum to avoid high hydrostatic head and dissolution of crystals.
- Graining volume should be small to facilitate starting of the pan.
- Optimum S/V ratio depending upon massecuite to be handled.
- Optimum downtake diameter.
  It is readily seen that these factors are contradictory.
- If height is decreased - circulation will improve but graining volume is increased.
- If diameter of the centre well is increased for better circulation - the graining volume increases.

Thus for a given diameter of the tube and given downtake diameter a compromise must be reached between height of massecuite, graining volume and S/V ratio.

Heating vapours: In modern sugar factories, sugar boiling is done with bled vapours from the first or second effect of evaporator set and even with third effect of quintuple. Now in few factories vapours from the last effect of evaporator set (working under vacuum of 200-250 mm) are used for continuous pans. The boiling is obviously slightly slower. On the other hand there is less risk of decomposition of sucrose caramelization, false grain formation etc. If the boiling becomes too slow at the end of the strike a change is made to higher pressure vapours. There should be proper, effective distribution of vapour avoids dead zone to vapour side.

Condensate and Incondensable gas removal: Efficient removal of condensate and non-condensable gases withdrawals is to be taken care of; otherwise it hinders the massecuite boiling.

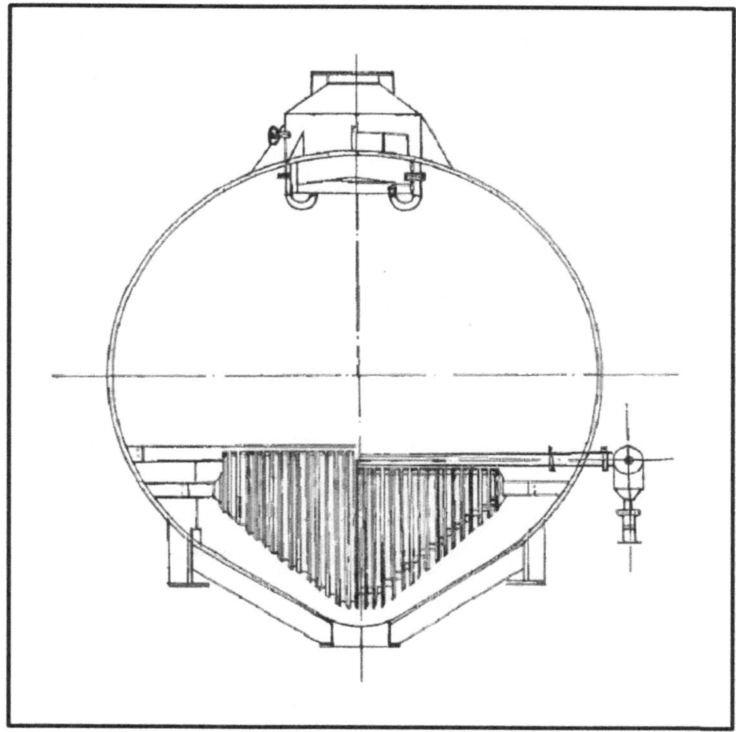

Fig 53.5 Horizontal Pan transverse section

Hydrostatic head: The boiling massecuite gives hydrostatic head of about 0.15 kg/cm^2 per m of depth. Thus at a certain depth the massecuite boiling temperature will increase corresponding to the hydrostatic pressure effected at that depth.

- This lowers the differential temperature between the massecuite and the calandria vapours this will reduce the rate of heat transfer.
- There increase increases colour formation due to high temperature.
- The supersaturation coefficient of the mother liquor decreases
- At higher temperature the viscosity of the massecuite is reduced.
- At a certain depth the massecuite boiling temperature may reach to saturation temperate. This would be critical depth below, which the boiling massecuite

reaches to a temperature at which the mother liquor becomes unsaturated and re-dissolution of the crystal may start. If redissolution takes place then crystals are partly dissolved as they reach at the bottom of the pan and hence 'soft' or rounded crystals are obtained.

Therefore

- The height of the massecuite should be restricted to a certain level.
- Vacuum should be maintained up to 635-660 mm. unnecessary high vacuum should be avoided.

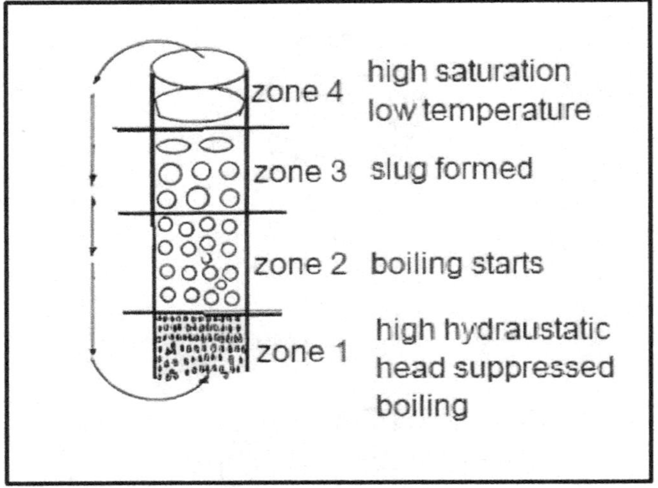

Fig. 53.6 Massecuite circulation through the tube

Massecuite circulation in the pan:

In pans natural circulation depends very much on the design of the vacuum pan. The low head pans of proper design give effective massecuite circulation for high grade massecuite.

Therefore uniform conditions for crystal growth are maintained throughout the pan. The variation in crystal size is also in narrow range. The bubbles of vapour formed in tubes promote the natural circulation in the vacuum pan. The circulation may also assisted only by the difference in specific gravity of the hot massecuite rising in the tubes and the cooler massecuite descending in the downtake. But the effect may be very little or negligible. However if heating is stopped circulation also ceases.

Initially when the footing is drawn in and boiling started, the rate of circulation is much high but as the strike level raises the circulation slows down. Even when the pan is properly designed the boiling rate of C massecuite strike slows down and circulation of the massecuite is also much reduced considerably after reaching level of 1.0 m. above the tube plate.

Studies conducted with a well-designed pan with natural circulation, showed the following circulation velocity at various levels of the C massecuite strike.

Table 53.2 speed of circulation of C massecuite against hours of boiling

Time in hour	Speed of circulation
1st	46.6 cm/Sec.
2nd	19.2 cm/Sec.
3rd	4.6 cm/Sec.
4th	0.9 cm/Sec.
5th	0.6 cm/Sec.
6th	0.3 cm/Sec.

To facilitate natural circulation Webre recommends a downtake diameter 50 % of the inside dia. of the pan. Tromp recommends 40 %.

Forced circulation by incondensable gases: Due to high viscosity the circulation is very slow for low purity massecuites at the end of the strike. In some factories the incondensable gases withdrawn from vapour side are introduced in the pan along with molasses charges. This is known as jigger steam.

Forced circulation by mechanical stirrer: With natural circulation particularly for low grade massecuite as height of massecuite increases, the movement of massecuite becomes very slow. Therefore the boiling becomes sluggish and the rate of crystallization is adversely affected. Forced circulation by mechanical circulator is a good solution for maintaining the circulation of massecuite. However the mechanical circulators were not in much use in India till 2005. But now mechanical circulators are being installed in most of the factories with new pans and also in old pans.

Mechanical circulator is like a screw pump impeller attached to a shaft that is fitted in the pan from the top. The shaft is supported in the centre of downtake by bearings and is coupled to an electric motor through a reduction gear located on top of the pan dome. The descending massecuite in the downtake is pushed downward to the bottom of the pan.

An indicating ammeter of the motor indicates the current and load variations during boiling. This helps in regulating feed to the pan.

Webre has observed following evaporation coefficients for both natural circulation and mechanical circulation of massecuite of purity 72.00.

Table 53.3 Speed of circulation with and without mechanical circulator

	Start	finish	average
Pan with natural circulation	28.74	1.03	9.27
Pan with mechanical circulation	29.65	10.30	17.25

Thus there is considerable improvement in circulation in pan with mechanical circulation over pan with natural circulation.

The load on motor goes on increasing as the massecuite level rises in the pan, reaching maximum during the final phase of tightening the strike. Speed of circulator for high grade massecuite can be 80 rpm, however in the last hour it is 60 or 50 rpm. Therefore VFD motor can be a good solution.

Advantages of mechanical circulator: The improvements in the massecuite boiling on account of mechanical circulator can be summarized as under—

- The massecuite boiling time is reduced,
- With mechanical stirrer the strike level in pan can be raised up to 30-40 cms.
- Even and uniform grain size can be maintained throughout boiling.
- In high grade massecuites conglomerate formation is prevented.
- Colour development due to local overheating is reduced.
- Low pressure vapour can be used for boiling.

D. P. Kulkarni says that *"for high grade pans unless due to faulty design the circulation and boiling rate are poor the mechanical circulator may not be essential but for final massecuite boiling pans it is certainly advantageous to install mechanical circulator".*

Discharge valve: The discharge valve should be of sturdy construction and suitable sized for quick discharge. Now for batch pans manually and pneumatically operated discharge valves are provided.

- Easy access is to be provided to Calandria for maintenance.
- Vapour space of a pan is connected to condensers by large vapour pipes.
- The various important pipe connections of pans are –
- Vapour connection to calandria,
- Condensate outlet,
- Incondensable gases connection.
- Syrup molasses and hot water connection through manifold,
- Seed intake / cutline pipe connection,
- Vapour pipe from vapour space to condenser.

Pan is provided with sampling key, basic simple gauges such as vacuum gauge in the shell, temperature gauges for shell and calandria and vapour pressure/vacuum gauge for chest of calandria. Without these gauges panman cannot understand control the pan working.

Vacuum pans are fabricated from mils steel (IS: 2062) or stainless steel (SS: 304. Standard Range of 40-120 tonnes strike capacity. Stainless steel constructed pans are provided for refinery.

A pan boiler has to control vapours entering the calandria and pan feed to cope up with the rate of evaporation and development of crystals.

Crystaloscope:

This instrument is mounted on the pan for observing crystals on the screen, throughout the process of boiling. The crystals projected on screen give an idea of the uniformity and shape of sugar crystals in the pan.

53.3 Additional equipment at pan floor:

At pan station for storage of syrup and molasses, syrup/molasses storage tanks are installed. Similarly for B seed and dry seed open crystallizers and for B and C grains vacuum crystallizers are installed. Molasses conditioners are also installed for molasses conditioning purpose.

1 Pan supply tanks:

Generally there are 14-20 or more storage tanks as per capacity of the plant for syrup, melt and molasses. The capacities are given for 4 hours storage of syrup / molasses. For example for a 2500 TCD (115 TCH) plant the syrup storage capacity given is as follows.

Table 53.4 volume of syrup for 4 hrs. storage of a 2500 tcd plant

The syrup quantity in four hours = 115 X0.25 X 4	115 MT
Volume of syrup = wt./density 115/1.28	89.84 M^3
with 30 % extra volume foam and marginal capacity	116.79 M^3, say 120 M^3

Thus for a 2500 tcd plant 8 tanks of 15 M^3 are generally provided for syrup. The total syrup and molasses tanks are given here under-

The tanks are rectangular in shape and of M.S. construction. Depending on the type of material i.e. A heavy, B heavy, C light or syrup, melt, two to four tanks are arranged in a nest or group and tanks for similar material are interconnected. This way all syrup tanks will be interconnected.

Table 53.5, number of syrup and molasses tanks Of 15 m3 capacity each

Sr. No.	material	Number of tanks
1	syrup	8
2	A light	2
3	Melt	1
4	A heavy	5

5	B heavy	3
6	C light	1
7	Total	20

Similar arrangement will be for each type of molasses like A heavy, B heavy etc. The outlets of tanks from bottom are joined to three different pipes, viz. one for melt, A light and syrup, one for A heavy and one for the last for b heavy and C light. The pipes are connected to the corresponding pans. The molasses tanks are provided with steam / vapour coils at the bottom and are thoroughly lagged. A common washout pipe is laid below for cleaning the tanks, the washings are led to raw-juice receiving tank.

2 Molasses conditioners:

Molasses received from centrifugal station usually contain fine grains of sugar. These grains must be dissolved by heating and dilution before feeding the molasses to the pans. Otherwise these grains if passed in the pan act as secondary crop and massecuite of different grain sizes is developed. Such a massecuite is not well exhausted and may give trouble in purging. Therefore the molasses has to be *conditioned* before it is fed to the pan. A tank with mechanical stirring arrangement is provided for this purpose for every massecuite at pan floor. The molasses conditioner has steam and water connections. Molasses from the pan supply tanks is taken in these tanks and only conditioned molasses is fed to the pans.

3 Seed crystallizers and vacuum crystallizers:

Generally two open crystallizers one dry seed (dust sugar from sugar grader) and one for B seed are provided at pan station. Two vacuum crystallizers are provided one for C massecuite grain and one for B massecuite grain. The vacuum crystallizers are provided light and sight glasses and with suitable connections for vacuum. The capacity of the seed crystallizers is usually half of that of A pan, while the capacity of a vacuum crystallizer is usually three-fourth the capacity of a pan. All these crystallizers are connected by 'cutline' to all the pans. Generally two molasses conditioners one for A heavy molasses and the other for B heavy and C light molasses is provided.

-OOOOO-

54 CONTINUOUS PAN

54.1 General:

Now use of continuous vacuum pans is common and in many factories continuous pan have been installed. There are two types of continuous pans

- Pans with vertical tubes as heating surface in which massecuite passes through the tubes and vapours heat the massecuite from outside the tube as in case of batch type pan.
- Vacuum pans with horizontals tubes as heating medium. The massecuite is outside the tube and the vapours pass through the tubes. These types of pans are not much installed.

There are two types of pans with vertical tubes as heating surface and massecuite passing through the tubes.

- Floating calandria with sideways down take
- Central downtake and sideways calandria

Fig 54.1 Continuous Pan

A pan is divided into 10 to 13 compartments. The size of the compartments is progressively increasing. Massecuite travels from one compartment to another compartment from the top and then from the bottom alternatively. This ensures plough flow of the massecuite and eliminates formation of stagnant pockets which may cause lumps formation. The heating surface to working volume (S /V) ratio is generally 9.5 to 10.5. The tube length is 750 to 1200 mm. Massecuite level above the top tube plate is 200 mm to 600 mm. it is highest in the first compartment and goes on decreasing to the last compartment. The massecuite flows from one compartment to the next compartment because of level difference.

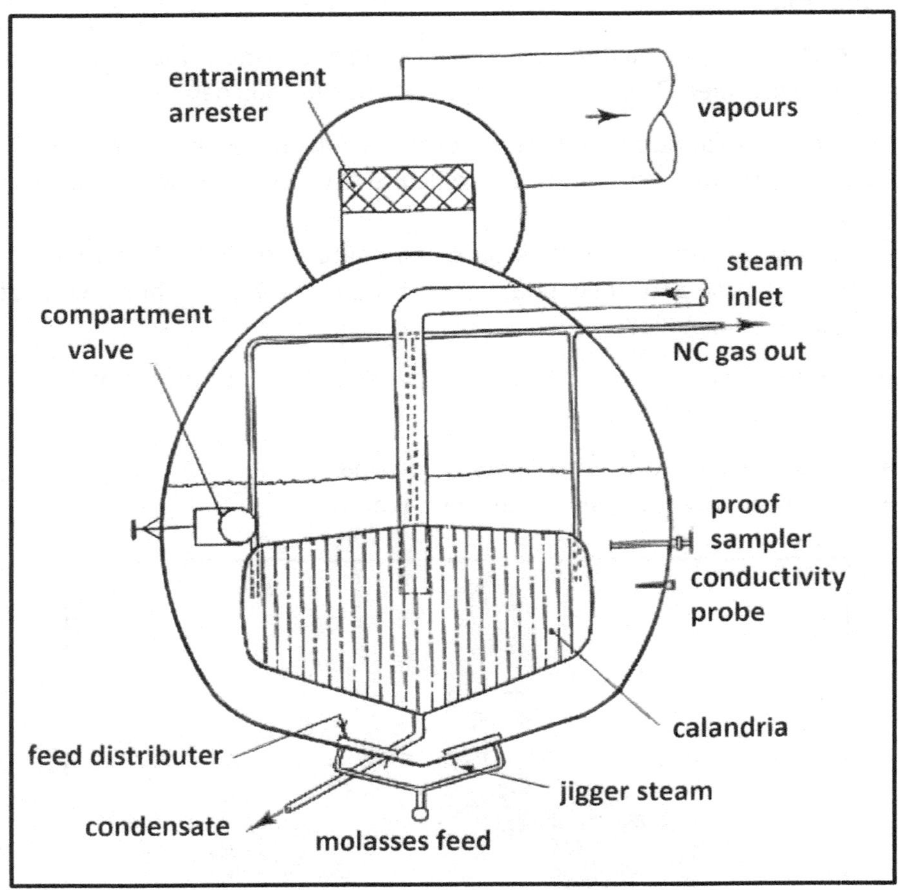

Fig 54.2 Continuous Pan

The circulation ratio – the ratio of the total cross sectional area of the tubes to the cross sectional area of the downtake is 0.6 to 1.0. The retention time for A and B massecuite boiling is about 2.5 hrs. and that for C massecuite is usually 4.5 hrs.

In most of the designs U shape flow pattern is adopted whereas in some designs straight flow pattern is adopted.

Feeding of the molasses / syrup is designed for 70-75 % of the total working volume. The remaining volume is kept for hardening (in the first compartment) and for tightening in the last compartment.

The massecuite discharging pipe is provided to the last compartment which is of about 8 m height and overflow height is about 7.5 m. In some new VCP or VKT continuous pans footing is taken directly in the pan however in most of the continuous pans at present grain or seed material is prepared in batch type pan. It is stored in vacuum crystallizer and is fed to continuous pan by metering pump at a controlled flow rate.

In continuous pans generally good circulation is observed. However, to avoid formation of stagnant pockets recirculation of incondensable gases at the bottom is usual practice in some of the designs. The pans are equipped with multi-baffle entrainment arrestor in SS construction.

The pan is liquidated during periodical shutdown for general cleaning. The percentage exhaustion of the massecuite and coefficient of variation in crystal size in continuous pan can be equal to that of batch pan if care is taken and boiling is smooth. Now continuous pans are in use for all A, B and C massecuites.

Lumps formation may occur in A massecuite pans if keen attention is not given in working of the continuous pan. There is no such problem for B and C massecuites. The continuous pan works with continuous and uniform throughput and output without any idle time. Therefore there is continuous and steady steam demand unlike batch pans. This gives better steam economy.

As the massecuite level above the top tube plate is only 200 – 600 mm, the hydrostatic head is much less as compare to batch pan. Therefore massecuite boils at lower temperature. Hence the pan works well even at lower pressure /temperature vapours. Now in modern VCP / VKT pans vapours of temperature 84-87°C are used without any trouble. Generally continuous pans of capacity 10 to 30 t/hr. are installed for low grade massecuite. For high grade massecuite the capacity ranges from 30 to 50 t/hr. The capacity of existing continuous vacuum pan can be increased by adding new compartments with mechanical circulators.

Since 1982, only continuous pans have been installed in Australian sugar industry.

54.2 Advantages of continuous pans:

Now in most of the countries continuous vacuum pans have been installed and are used for all massecuites. Continuous vacuum pans give significant advantages.

- Consistency in production of all grades of massecuite.
- Higher steam economy as pans work on low pressure vapours,
- High crystal growth rates without the formation of any fine / false grains,
- Good exhaustion of massecuites with high brix and high crystal content.

- Variation final crystal size also remains in narrow range.
- Purging at centrifugals remains normal.
- Reduced supervision because of automation.

54.3 Dual continuous pan:

Dual continuous horizontal pan can also be installed and be operated for two different grades of massecuite simultaneously. The Dual continuous pan has only one condenser. Heating steam is admitted on both ends of the pan and easy to control the steam for the respective massecuite as per requirement. This pan is having the provision to stop any one of the massecuite boiling without affecting the other massecuite boiling. The calandria will have vertical tubes with peripheral down take. The calandria is divided into multiple equal sized compartments to ensure plug flow of the boiling massecuite.

Advantages of Dual Continuous Pan

- Simultaneous boiling of two different grades of massecuite.
- Lesser initial investment compared to individual pans for two different grades for the same output of massecuite.
- Lesser space is required than the individual pans.
- Only one condenser, hence water and power saving.
- Suitable for 100% automation and man power saving.
- High exhaustion due to improved massecuite circulation.
- Flexibility to run any one massecuite.

54.4 Vertical continuous pan:

In vertical continuous pan (VCP) the batch pans (called as chambers) are arranged one above the other and utilized in series of compartments.

The seed is fed to the first compartment while the molasses is fed to all remaining compartments[1]. The heating surface to volume (S/V) ratio is generally 8 m^2 / m^3. Each chamber is provided with mechanical stirrer equipped with blades with speed running from 34.6 to 61 rpm. There is uniform and steady massecuite flow from one chamber to the next chamber. The use of stirrer makes excellent circulation.

The seed is prepared in separate batch pan[2]. The chambers are designed and erected one above the other. Therefore massecuite flows to the next chamber by gravity. Because of low massecuite level above the calandria and use of agitator for circulation of massecuite, low pressure or even vapours under vacuum can be used for VCP pans. A provision is made to bypass each chamber for cleaning. Therefore the working cycle can be prolonged for many weeks without taking break.

Now VCP continuous pans have been installed in more than 100 factories in the world. The structure of VCP pans is vertical and therefore these pans can be installed in small space and the installation is outside.

Fig. 54.4 vertical continuous pan

Advantages of vertical continuous pans (VCP)

- Cylindrical chamber is similar to batch pan
- Use of stirrer for good circulation gives higher rate of crystallization with uniformity in crystal size without any lumps formation.
- Defined flow of massecuite from chamber to chamber
- Controlled process condition in every chamber
- Cleaning of individual chambers is possible during production

54.5 Monitoring and control system for pan boiling:

Conventional pan boiling is manual and is dependent on the pan operator's skill. This results in non-uniform operation, higher energy consumption and poor exhaustion. There are different types of technologies for automation of pan boiling operation.

The control software implies with multi-parameter input module. Parameters such as resistivity/conductivity, level, temperature, viscosity are measured by sensors and accordingly control action is decided by fuzzy logic system. The control of the batch pan boiling process is accomplished by dynamic control strategy where set points are modified as pan level rises.

A new technology consists of Sensors that includes

1. Vacuum in the pan,
2. Vapour temperature passing to condenser,
3. Vapour pressure in the calandria,
4. Massecuite level above the calandria,
5. Massecuite boiling temperature above the calandria,
6. Resistivity / conductivity of the massecuite,
7. Viscosity of the massecuite,

- Transmitters
- PC and related hardware
- Fuzzy logic control software
- Feed control valve

The automatic monitoring and control gives better exhaustion of massecuites with reduced boiling time. Thus sugar losses in final molasses can be reduced at the same pan capacity is increased.

References:

1 Gil N J – ISJ Feb. 2000 p 70-74.

2 Munasamy – SASTA 1988 p 42 -44

-OOOOO-

55 GRAINING TECHNIQUE

55.1 General:

The object of crystallisation is to crystallize out maximum sugar of expected quality with minimum loss in final molasses. The crystallization process is carried under vacuum so that minimum inversion and minimum colour formation takes. Further because of reduction in boiling point the solubility of sucrose decreases and maximum sugar is crystallized in the pan boiling. The crystal size and grade to be maintained depends largely on profitability in market and crushing rate to be maintained. The bolder grain and low colour value (ICUMSA units) generally fetches more price in Asian countries. However for bolder grain production the time required for high grade massecuite boiling is increased and therefore the crushing rate has to be controlled.

55.2 Graining:

The first step in crystallisation is to take syrup / molasses of desired purity in the vacuum pan and concentrate it up to supersaturation under vacuum by boiling to make grain or fine crystals in it. In raw sugar or plantation white sugar manufacture grain is made generally for intermediate and low grade (B and C) massecuites and high grade or A massecuite is developed on seed from intermediate massecuite or dry seed. The graining volume is 35 to 40 of the finished strike. Therefore generally one-grain strike can be sufficient for two low-grade strikes. In some factories low grade massecuite pans are having graining volume only about 33% of the final strike. In such case the grain can be sufficient of three strikes. This saves much time. The graining technique decides the coefficient of variation in crystal size in the final massecuite.

To start graining the pan is closed and vacuum is raised. Graining material is taken into the pan up to calandria level and at the same time vapour is turned on. After boiling starts feeding of the graining material is continued to maintain its level above the calandria. The feeding is stopped as the material is reached the level of supersaturation.

The supersaturation is checked by thumb rule by drawing sample through proof stick /sampling key and noting its viscosity or feel or observed on the indicating instrument. The sample is taken between thumb and forefinger and length of the syrup string is noted before it breaks. If it does not break up to one and half inch length, the graining material has crossed the saturation point.

55.3 Different methods of graining:

With the use of different zones of supersaturation there are different methods graining

1. **Waiting method:** In this method the graining material is concentrated up to labile zone as a result of which grain appears spontaneously in the boiling graining material without any addition of sugar dust or slurry in it.
2. **Shock seeding method:** In this method the grain is formed in intermediate zone. When the boiling material reaches the intermediate zone sugar dust is feed into it. It cats as a 'shock' to the saturated boiling material and grain appears.
3. **True seeding method:** In this method the graining material is concentrated to reach up to metastable zone of supersaturation. As soon as this condition is reached, the required numbers of fine microscopic sugar crystals in slurry form are introduced into the boiling graining material.

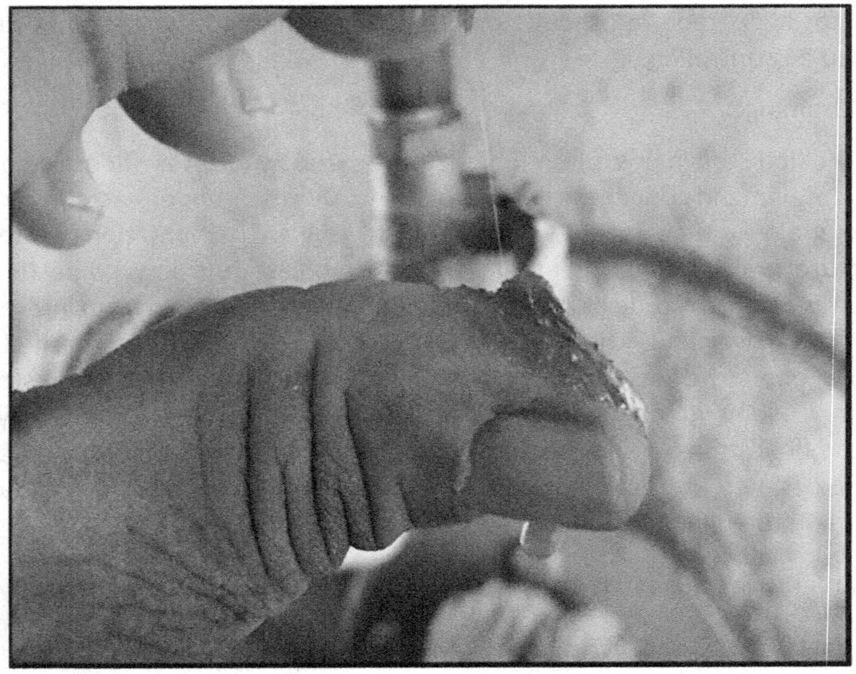

Fig. 55.1 checking the supersaturation level by length of string

Waiting Method:

In waiting method the graining material is concentrated till labile zone of supersaturation is reached. The coefficient of supersaturation is determined by instrument like panometer or cuitometer or by difference in temperature of the boiling material and the temperature of the vapours going to condenser (bpr).

When instrument is not available it is gauged by thumb rule. Sample of graining material boiling in the pan is taken by proof stick or sampling key. Experienced panman takes the material in thumb and fore finger of the same hand and forms a string of that material. In labile zone the string will not break till the length reaches more than eight centimetres. Grain will appear spontaneously.

Consider graining material of 80 brix and 72 purity is boiling in vacuum pan at vacuum of 630 mm (25") of Hg. Now at this vacuum water will boil at 55°C. The bpr of this material will be about 8°C. As the graining material is just above the calandria level considering hydrostatic head of about 0.2 m, the increase in boiling point due to hydrostatic head will be about 3°C. Thus the boiling temperature of the graining material will be about 66°C. Now at purity 72 and temperature 66°C the material crosses the saturation line at about 79.4 brix. As the material will be concentrated further and the brix will increase the supersaturation and the bpr also will increase.

At about 88.9 brix the bpr will be about 16°C and coefficient of supersaturation will be more than 1.45. On further concentration the material crosses the line between intermediate zone and labile zone. As the material will boil further in labile zone, at this concentration fine microscopic crystals will be formed spontaneously without the presence of any crystals in the boiling material. The quantity of grain desired is determined by observing sample spread on a piece of glass under magnifying lens. The sugar particles sparkle on the glass.

When sufficient grain is formed further grain formation is stopped by lowering the degree of super saturation up to metastable phase. This is done by taking a charge of water and decreasing the vacuum by 2" to 3." serves this purpose, the super saturation comes down to 1.25. The bpr will be approximately 13-14°C.

Disadvantages of waiting method:

- Time taken for formation of grain is more.
- Control of number of nuclei and their size is very difficult as different crops of crystals appear at different intervals.
- Grain hardening also becomes time consuming.
- Thumb rule type manual judgement is used in which there are chances of error in determining the exact required grain.

Shock seeding method:

Shock seeding method was introduced in beet sugar industry by Zitkowskey and rapidly replaced old waiting method. Graining material is concentrated to above supersaturation. By thumb rule it is determined by string length of about 1.5" of a sample tested between thumb and forefinger. Then about 500 g of sugar dust is fed into the pan. The feeding of sugar dust gives a kind of shock to the supersaturated boiling graining material, and grain appears spontaneously.

The shock should be applied as soon as saturation is exceeded. With instrument control the powder should be fed in just beyond the saturation point which is at approximately 9-10°C bpr or supersaturation 1.1 with purity 72.

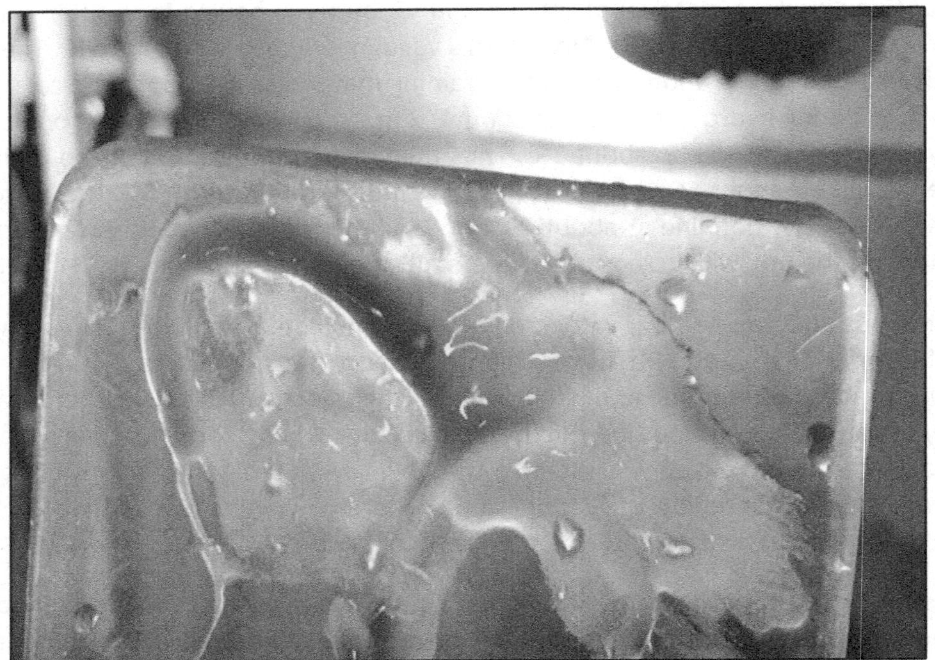

Fig. 55.2 uniformity and right number of grains assessed by observing the grain

Grain will not appear immediately upon introduction of powdered sugar. Late shock may result in the formation of conglomerates because of excessive concentration. Air admitted in the pan when sugar is introduced in the pan should be minimum so that the temperature equilibrium will not be disturbed.

When grain begins to form, after a few minutes the further grain formation is to be arrested. It means the supersaturation of graining material is to be lowered down from intermediate phase to metastable phase. This is achieved by feeding condensate/syrup in to the pan. Vacuum is maintained constant during entire process of establishing the grain.

Advantages over waiting method:

- By shock seeding method, relatively more uniform grains are formed than by waiting method.
- Time is saved in appearing the grain and to check it.
- Easy to check the grain.

True seeding method:

True seeding is now general practice in all factories. In this method the microscopic crystals of 3-5 micron size in slurry form in organic solvent are used as seed. The number crystals in the final strike are fed in sugar slurry form of 3-5 micron size for development of sugar crystals. Ideally there will not be any increase or decrease in the number of crystals. Neither fine crystals fed in the form of slurry get dissolved, nor is any fine of secondary crop formed. The concentration is hold in the metastable or crystals growing zone. Fine microscopic crystals in slurry form in rectified spirit are introduced as soon as saturation point is reached as observed by the thumb rule or indicated by the instruments.

The proper time to seed is when saturation is exceeded. When graining material boiling in the pan is of 72 purity and its temperature is 68°C, the bpr at saturation is 9-10°C. Allowing 1-2°C as margin, proper seeding point would be at bpr 10-11°C which is at supersaturation 1.15.

To determine the supersaturation without instrument 1 g of coarse sugar is placed in the notch of the proof stick. It is wetted with the graining material in the pan for about 30 seconds. Then the coarse sugar withdrawn and observed with magnifying lens. If the corners of the crystals are rounded saturation has not yet reached. If corners remain sharp and square, saturation has been exceeded. Now it is the time of introducing sugar slurry in the pan.

After introduction of the slurry into the pan supersaturation continue to increase for 3-5 minutes. As soon as the time has elapsed it is necessary to start the use of movement water in order to maintain the supersaturation in metastable phase. Use of movement water should be continued till the fine crystals are developed up to 25-30 micron in size. Then first molasses charge is fed and movement water flow rate is slowly reduced. With the right amount and size of seed slurry, massecuite with regular sugar crystals of required size, free from conglomerates are produced. The variation in crystal size remains within narrow range. This method is used universally.

When the grain becomes visible to the naked eye, panman observes on watch glass whether there is sufficient grain or not. If the grain is too much he dissolves a portion of it by reducing the supersaturation below saturation. Unnecessary time is wasted in dissolving the excess grain and when factory working is in full swing and there is load of molasses it gives trouble.

If there is insufficient grain the panman has to be careful that no false grain appears while boiling the pan. Further the expected crystal size in final massecuite is reached before the strike level is reached. Therefore either pan is to be discharged at lower level or the crystals will be large in the final massecuite when it is discharged at normal level. Large size final crystals will lead to unsatisfactory exhaustion in the low-grade massecuite. To avoid having an irregular grain it is necessary to take care to keep the material in metastable zone of supersaturation.

Advantages of true seeding:

- If there is no mistake in establishing the grain, there is always sufficient grain.
- Time is saved in establishing grain and hardening it.
- No false grain or conglomerates form in pan if it is further boiled carefully at steady conditions keeping the material in metastable zone.
- Uniform crystals of predetermined size are produced. Therefore higher exhaustion and better purity control is achieved.

55.4 Establishing the grain:

While starting the pan - at first all the valves of the pan are closed then the exhaust /vapour valve to the calandria is opened slightly and vapours are introduced. The non-condensable gases valves are opened. The steam valve to the body of the pan is opened slightly and to heat the pan from inside. Then the injection water valves of the condenser are opened. The steam valve to the body is closed and vacuum is created. The graining material is taken in the pan from the pan supply tanks. The vapour valve is opened. The graining material is boiled to concentrate it to supersaturation.

In establishing the grain the following precautions are to be taken for best results.

- Graining should be done by true precipitated slurry method. The size of fine crystals being 4-5 microns.
- The slurry dose should be calculated and the exact dose should be given.
- Purity of graining material should be 66-67 for C massecuite and around 72-73 for B massecuite grain.
- Graining by seeding takes place in the metastable zone at about coefficient of supersaturation 1.10 - 1.20. Maintain the supersaturation preferably by instruments.
- Control the steam within two minutes when grain is appeared and starts movement water so that the graining material will remain in metastable zone.

55.5 Hardening of grain:

Once grain is established, the next step is to make the grain hard. Hardening of grain means to dissolve fines if any present in the grain and make the grain size uniform as far as possible. In grain hardening it is also claimed that the molecules are rearranged in proper crystal lattice so that the crystals of definite shapes are developed. The grain is developed slowly by boiling the material without feeding any molasses.

When the grain is appeared, the crystals are fine and widely separated. There exists much less surface area of the crystals to take up sucrose molecules from the molasses. Further the distance between the fine crystals is also more. Therefore critical condition exists.

- In hardening of grain the vacuum is lowered down by about 2-3 inch (50-75 mm) of Hg and boiling temperature is raised by 7-10°C. This will maintain the supersaturation within the metastable zone. It reduces the rate of evaporation. The vacuum should be hold steady.
- If the rate of evaporation is high, the supersaturation rises and goes in intermediate zone and false grain are formed. To avoid this, movement water feeding is started. Usually the condensate water from the same pan - instead of leave it to condensate receiving tank, is taken into the pan so that whatever water gets evaporated at any moment almost the same quantity of water is introduced in the grain. With circulation of hot condensate water in pan the level of grain in the pan remains almost constant.
- In grain hardening no syrup / molasses is fed as it increases the quantity of mother liquor and the fine crystals are more widely separated. This may increase the supersaturation and danger of false grain formation while concentrating the material.
- As the crystals develop and crystal surface area increases the amount of movement water is slowly reduced and finally closed off.
- The fine grains are developed in proper crystal lattice. The coefficient of super saturation of the mother liquor slowly comes down to very close to 1.00 and crystal growth is almost stopped but the rearrangement of sucrose molecules continues and the fine grains are developed in proper crystal lattice.
- It should be noted that to check the evaporation rate, the steam pressure can be reduced but then vacuum increases and circulation in the pan decreases. This may lead to increase in viscosity because of lower temperature and insufficient circulation may lead to give localized higher supersaturation causing formation of false grain and conglomerates. Therefore panman has to be very careful in maintaining steady vapour pressure and vacuum.

The grain is slowly developed in proper crystal lattice and becomes hard and well-shaped uniform crystals are formed. The hardening of grain further helps in giving massecuite of uniform crystals. The graining time may vary from 30 minutes to 1 hr. or even more depending upon purity and viscosity of the graining material

After hardening of the grain, the pan is fed frequently of continuously with AH/CL molasses and the grain is fill up to strike level. The C massecuite grain purity at strike level is maintained to about 61-65. It takes generally 3-4 hour to boil up to strike level.

Generally grain is prepared for 2 or 3 strikes. Grain of 2 strikes or of one strike is transferred to vacuum crystallizer or another pan. The remaining grain is boiled for one C massecuite strike.

Continuous pan

-OOOOO-

56 SUGAR BOILING TECHNIQUES

General:

Crystallization or sugar boiling process can be divided into following phases:

1. Concentration of footing (in case of A massecuite) or graining material (in case of B and C massecuite) up to saturation.
2. In case of A massecuite taking required seed and washing it keeping required seed for further development. In case of B and C massecuite establishing grain, making it hard and develop up to certain level and size.
3. The third stage is dividing seed (in case of A massecuite) or grain (in case of B and C massecuite) for two/three strikes. This is done by transferring (cutting) the seed partly to another A pan. In case of B and C massecuite the grain is partly send to vacuum crystallizer. A seed or grain for singe strike is kept in the pan for further development.
4. The next stage is giving the charges or drinks to the pan to building up the grain or seed up to desired level of pan.
5. Final tightening of massecuite. And discharging the strike.
6. Washing the pan by steam and hot water before starting for next strike.

We have discussed the concentration of graining material and establishing the grain and develop it for two/three strikes in last chapter. In plantation white sugar manufacturing, the object is to produce white marketable sugar from syrup of 78 to 85 purity. In this process the different grades of sugars of crystal size 0.6 mm to 2.0 mm are produced. The loss of sugar in final molasses also has to be controlled. In maintaining final molasses purity to a minimum level, the final massecuite purity has to be kept low.

1) Operating conditions and control:

The pan operation remains steady when the vacuum in the pan and vapour pressure at the calandria remains steady. Vacuum of 630 to 650 mm (25 - 25.5") of Hg is always recommended for better boiling of massecuite and good exhaustion of mother liquor of sugar in the pans. The feeding of syrup / molasses (continuous in case of automation or frequent small charges in case of manual operation) should be steady.

The pans must be equipped with instruments and gauges which indicate temperature of massecuite, vapour pressure and temperature at the calandria. Conductivity and degree of supersaturation indicators should be installed. This is helpful in control of pan boiling.

2) 'A' massecuite seed preparation:

A mixture of syrup, 'A' light molasses and melt is taken as footing material for 'A' massecuite while producing plantation white sugar. Seed for 'A' pans must be as white as possible. Now usually dry seed (sugar dust from grader) which is fairly white is used in most of the factories.

However in some factories B seed is used. While using B seed, the B massecuite should be cured in batch centrifugals, so that in the seed there will not be broken crystals. In centrifugals the B massecuite should be well washed after separation of heavy molasses. B light molasses should be separated by syrup separator.

Generally panman after concentrating the footing material takes some excess of seed and washes it with syrup / water. When the seed is washed the fine crystals from the seed are dissolved. The remaining seed is more or less of equal size. The exact required seed is to be kept. It means the right numbers of crystals are to be kept so that the required final crystal size will be obtained. Generally footing for two or some time three strikes is taken at a time. This is highly skilled technique and only expert panman can do it. No automation has so far been perfectly developed for washing of the seed. Establishing grain we have studied in the last chapter.

3) Developing the seed / grain:

Once grain is made hard or seed is washed, the next work is to develop the grain / seed. In developing the grain /seed care has to be taken that conglomerates or false grain are not formed.

By experience sugar boiler maintains required feeding rate of syrup/ molasses in the pan. Generally small frequent charges /drinks are given. These small charges compensate the quantity of water evaporated and maintain the supersaturation coefficient in desired limits. As feeding is continued the fluidity of the massecuite is also maintained and the massecuite level / volume slowly and progressively increase.

For high grade massecuite if molasses charging is delayed the massecuite becomes over tight. If giving drinks /charge further delayed there is possibility that the mother liquor may pass to intermediate zone and false grain formation may occur.

4) Molasses Conditioning:

The molasses being fed to the massecuite may contain fine sugar crystals that are to be dissolved before feeding the molasses to the pan. Further molasses temperature also should be either equal to or more than the massecuite boiling temperature in the pan. Therefore the molasses should be *conditioned* before feeding. It is done in molasses conditioners. The molasses is diluted to 70-73 brix and heated up to 72 -75°C and well stirred in molasses conditioners. In conditioning the molasses the fine sugar grains if any present in molasses are dissolved. When the temperature of molasses is higher than that of boiling massecuite in the pan, it does not bring down the temperature of the massecuite. Further it gets well mixed immediately with the massecuite in pan.

5) Massecuites Brix:

The brixes of massecuites at the time of discharging should be high but at the same time sufficient fluidity has to be maintained so that the massecuite can be discharged from the pan. At the beginning of strike when the strike level is low the brix is not high but as the grain size increases and the strikes builds up the brix increases. It reaches to maximum level at the time of dropping. The highest brix at the time of discharging gives maximum crystal percentage in the massecuite. The purity drop from massecuite to heavy molasses increases. The brix should be kept high right from the being of the strike. However in maintaining high brix care has to be taken that no conglomerate or false grains are formed.

6) Purity Control:

In white sugar manufacture as the syrup purities vary from 78 to 85 from factory to factory. The syrup purity is also different for different periods of crushing season. The A massecuite purity also varies from 85-90 and therefore the B and C massecuite purities also vary to some extent. However to maintain the loss of sugar in final molasses to a minimum level the C massecuite purity has to be maintained at a minimum level. Therefore as season passes and purity of syrup and A massecuite increases, the chief chemist / process manager of a factory has to think of adopting three and half or four massecuite boiling scheme.

7) Recirculation of sugars and nonsugars:

Sucrose and reducing sugars in solution are sensitive to heat and undergo transformation due to heating. The extent of decomposition depends on retention time and temperature. Therefore the more we boil the syrups and molasses the more losses of sugar are obtained.

The organic nonsugars also undergo undesirable transformation as a result of prolonged heating. Recirculation of sugar and nonsugars should be minimum. The massecuites must be concentrated to a high degree to achieve high exhaustion of mother liquor of sugar. Further the separation of molasses from sugar crystals at centrifugals has to be efficient.

8) Size of sugar crystals:

The standards for size of sugar in India are - large (L) medium (M) and small (S) and super small (SS). The bolder grain sugar of size 1.5 – 2.0 mm fetches more prices in Indian market. In producing bolder grain regularly, three and half or four massecuite boiling scheme instead of conventional three massecuite boiling scheme is adopted. Capacity at the pan station hampers further loss in final molasses also may be increased. Therefore in adopting such scheme, capacity utilization, capacities at pans station, receiving crystallizers and centrifugals etc. have to be taken into consideration.

9) 'C' massecuite boiling:

In 'C' massecuite boiling as viscosity of the massecuite is high, there should be maximum possible crystal surface area for easy deposition of sucrose from the mother liquor. The final crystal size should be maintained to 170-200 micron for giving higher surface area. Accordingly the slurry dose is calculated while feeding it for graining. Graining material for 'C' massecuite should be at around 63-65 to bring down the final purity of 'C' massecuite. In many factories the C massecuite purity is maintained up to 50-52 or even lower.

The smaller the crystal size the higher will be the surface area offered for the deposition of sucrose from the mother liquor. Higher surface area available helps in rapid desugarisation of the mother liquor and higher exhaustion of mother liquor.

Fig. 56.2 Computerisation of pan boiling

The boiling time taken for C massecuite is generally 4½ to 6 hrs. The longer the duration of boiling the better is the exhaustion of mother liquor in the pan itself. Therefore the boiling time of 'C' massecuite should not be unnecessarily brought down below 4½ hours.

10) Tightening of the massecuite:

During the last phase of the strike, the massecuite brix increases and massecuite becomes viscous. Concentration is continued to tighten the material to maximum practicable limit. In this last phase, many times little movement water is also introduced to facilitate the circulation of the massecuite. The movement water also helps in maintaining supersaturation at an optimum level so the risk of formation of fine false grain is reduced.

Table 56.1 C massecuite Grain size at various levels:

Sr. No.	Grain size	Size in microns
1	Slurry size	3-5
2	After hardening	65 – 75
3	At the time of cutting the grain	110 –125
4	C massecuite hot discharged	170 – 200

When the pan is full and tightened, in case of high purity massecuite the crystals occupy the maximum space and that mother liquors occupies only the voids in between the crystals. Hence further tightening of the massecuite may cause lumps formation in the massecuite. It is the maximum limit and strike is to be discharged. In case of low-grade massecuite it is not the voids but the viscosity becomes the limiting factor for tightening of the massecuite.

Attention has to be given to the maximum exhaustion of final molasses in the pan itself. Tight boiling should be practiced right from the lower level of the strike and C massecuite should be dropped at highest possible brix say 101-102.

11) Discharging and washing of pan:

When the strike is ready for dropping the vapour valve to the calandria is closed simultaneously the condenser water is stopped and vacuum breaking valve opened. It gives a whistle. Then discharge valve is opened. The massecuite flows into a gutter positioned below the pan and connected to the different crystallizers. After the strike is discharged and pan emptied it is necessary to wash the pan from the inside of the pan to remove the massecuite adhering to the inside of the pan before starting of the pan for next strike.

The method of washing the pan is to spray of high pressure hot water at 75-90°C followed by steam for 5-10 minutes. A circular pipe, with number of openings towards wall of the pan is located inside the top part of the pan for washing purpose. High pressure hot water is sprayed through this pipe. Then exhaust steam or vapours from the first effect of evaporator set are admitted through this pipe. The washings are let out through the discharge valve of the pan. Then the discharge valve is closed and condenser water is started to take vacuum in the pan for next strike.

12) Disposal of washings:

Usually the washing of pan is passed into the same crystallizer in which the massecuite is dropped. The washings dilute the massecuite to some extent and dissolution of crystals definitely occurs. It increases heavy molasses purity. In the case of C massecuite this is certainly not desirable. The pan washing contains more colour than the massecuite discharged. In white sugar manufacture it is almost 60 to

100 % more than the massecuite. It is high in A massecuite pans. However increase in colour due to pan washing is much less for raw-sugar production.

Due to pan washings the colour of heavy molasses also increases. To overcome this problem in some factories an arrangement is provided to collect the washing separately and use the same with C light molasses.

-OOOOO-

57 MASSECUITE BOILING SCHEMES

57.1 General:

It is not possible to exhaust the syrup of the sugar to a maximum practicable limit in one operation. Therefore the operation is carried out in two to four stages. Different purities of massecuites are boiled so as to maintain minimum sugar loss in final molasses. It is known as massecuite boiling system or scheme.

There are five/six different systems of massecuite boiling are followed for production of raw or plantation white sugar.

- Two massecuite boiling scheme for Raw sugar manufacture
- Two massecuite boiling scheme for plantation white sugar manufacture
- Two and half massecuite boiling scheme
- Three massecuite boiling scheme
- Three and half massecuite boiling scheme
- Four massecuite boiling scheme

When factory is erected the number of equipment and machinery and their capacities at pan section and centrifugals section is decided according to the massecuite boiling scheme to be followed. Factory adopts a system, which is more suitable, depending upon syrup purity and quality sugar to be produced. Generally whichever may be the boiling scheme, it is taken in to consideration that the loss of sugar in final molasses should be as low as possible.

However in Brazil and now in India sugar recovery is controlled and higher purity molasses (B heavy) is passed to distillery for production of ethanol that is mixed with petrol and used as fuel.

In selecting the massecuite boiling scheme it is also seen that the recirculation of sugar as well as non-sugars should be maintained to a minimum level and the total massecuite quantity would be minimum. This can be attained by exhaustion of mother liquor up to utmost practicable limit during pan boiling.

57.2 Two massecuite boiling scheme for raw sugar manufacture:

In many Latin American countries raw sugar is manufactured by two massecuite boiling scheme - generally called as A-C massecuite boiling scheme. The essential points in the two massecuite boiling system adopted are

- The A massecuite purity is kept fixed (82).

- The A massecuite is cooled down up to 48-50°C to get maximum exhaustion as a result of which 62 A heavy purity can be achieved.
- C massecuite graining and further boiling is carried out on A molasses and C massecuite purity is maintained to 62.

Seed magma of the C massecuite is taken as footing for A massecuite and the massecuite is boiled on syrup. If required A molasses is re-circulated to A massecuite (back boiling) to maintain the A massecuite purity 82.

This boiling scheme can be followed well where raw sugar of standard quality is to be produced and syrup purity is around 79 to 81. When syrup purity is up to 81.0, no recirculation of A heavy molasses to A massecuite is generally needed. But when syrup purity increases above 81.0, to maintain the purity of A massecuite 82.00, recirculation of A heavy molasses to A massecuite is required. Further the recirculation is increased to such a high level that two and half or three massecuite boiling system has to be followed.

57.3 Two and half massecuite-boiling scheme for raw sugar manufacture:

In this system the intermediate massecuite is made up of low-grade massecuite seed, syrup, and molasses of high-grade massecuite. The low-grade massecuite or final massecuite is boiled on molasses from intermediate massecuite. The grain is developed on intermediate molasses and if required syrup is taken. The sugar from high-grade massecuite and intermediate massecuite is mixed together for bagging. This boiling scheme is definitely convenient when normal raw sugar is to be manufactured.

However now in International market mostly high pol raw or VHP (very high pol) raw or VLC (Very low colour) raw are preferred, these two boiling schemes described above not much followed.

57.4 Three massecuite boiling scheme:

This boiling scheme is simple, straight and commonly adopted in most of the factories for raw as well as plantation white sugar manufacture. In this scheme sugar is crystallized out in three stages as shown in the flow sheet.

'A' massecuite: 'A' massecuite is essentially boiled on syrup, melt and 'A' light molasses. B seed or dry seed is taken as a seed for 'A' massecuite boiling. The seed footing is taken for 2/2.5 or sometimes for three strikes. Massecuite purity is in the range of 85 to 89. 'A' massecuite is concentrated to maximum practicable limit (Bx 91-92).

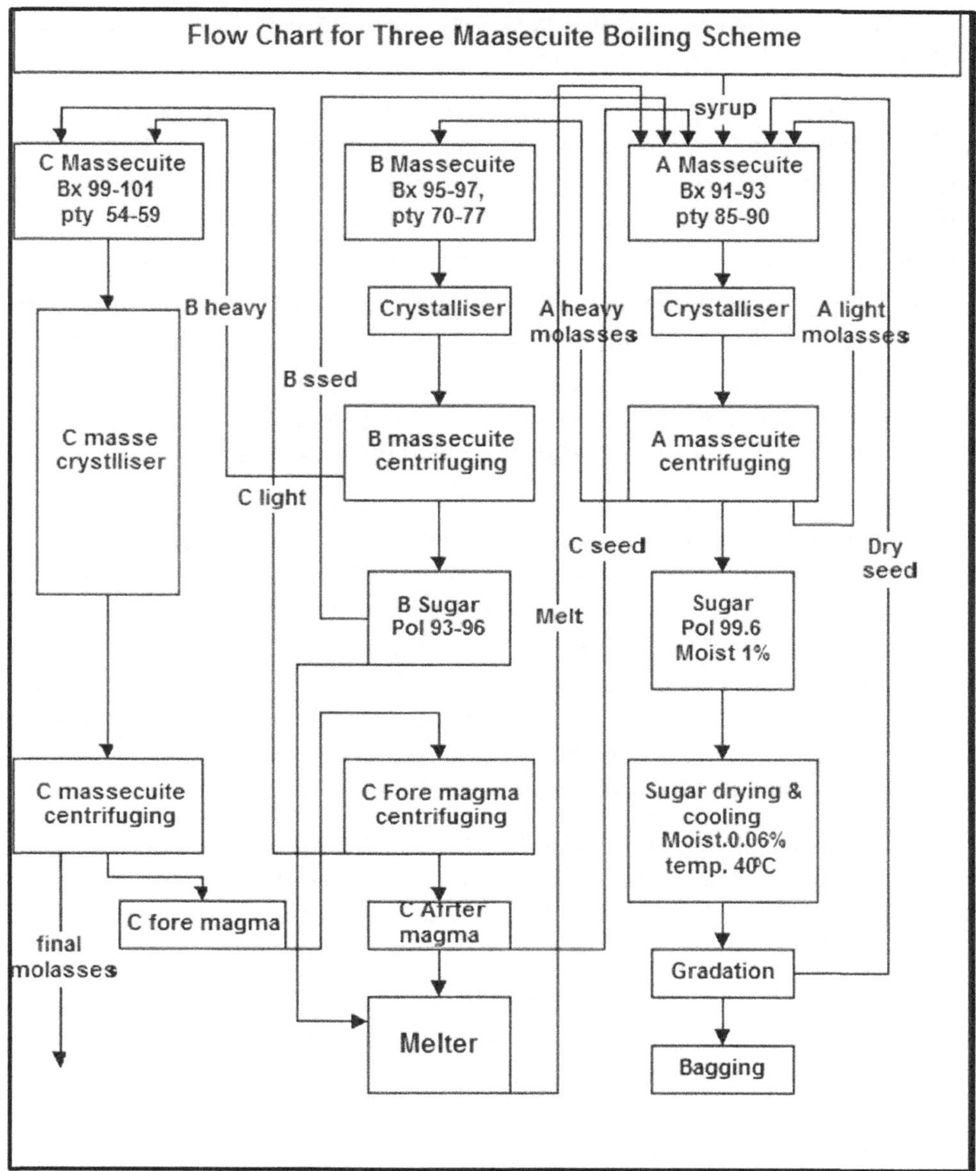

Fig. 57.1 Three massecuite boiling scheme

When optimum size required for bagging is reached and pan becomes full, it is discharged in air cooled crystallizers. The hot massecuite is then passed to pug mill and from pugmill to 'A' massecuite centrifugals. 'A' massecuite is generally single cured. But the molasses before superheated water wash and after water wash are separated by a chute called syrup separator.

A heavy molasses of 70 - 75 purity is used for 'B', 'C' grain and B massecuite boiling. While molasses separated after washed i.e. A light molasses (purity 90 and above) is used for 'A' massecuite boiling. Sugar is discharged on hopper from where it is conveyed to elevator. While conveyed it is dried with hot air blower & the cooled down to 45°C by cold air blower.

B massecuite: Now generally separate grain is developed for 'B' and 'C' massecuite. Normally B massecuite grain is developed for two strikes, half portion of which is cut to vacuum crystallizer for the next strike. The remaining grain in the pan is developed with 'A' heavy molasses. The factories in which double curing is followed, 'B' light molasses is used for 'B' massecuite.

In some factories to maintain high quality plantation white sugar, CFW sugar is heavily washed hence the 'C' light molasses obtained is of purity 67-70. A part of this molasses is also used for last charge of 'B' massecuite to control purity of B massecuite. B massecuite which is boiled on A heavy molasses has purity 72-76 and final brix around 95-96 is discharged in water cooled crystallizers.

The massecuite is cooled up to 45 -50°C in 6-10 hrs. And then passed to centrifugals for curing. B massecuite is generally single cured in continuous centrifugals. However in most of the factories part of B massecuite is cured in batch machine seed of which is mostly without crystal breakage and therefore is used as seed for A massecuite. The purity drop from B massecuite to B heavy molasses is about 22-24 units. Thus the B heavy molasses obtained is of purity 48-52 which is used for 'C' massecuite boiling. 'B' sugar that is collected in magma mixer is of 94-96 purity. It is mixed with melt/syrup/water to give B magma which is lifted by magma pump to melting tank or to B seed crystallizer at pan floor.

The factories where purities are high curing of B massecuite in batch machines with use of syrup separator can be helpful in controlling the B heavy purity. The B light molasses being separated after wash is used for B massecuite. The quality of B seed is also improved.

C massecuite: Grain for C massecuite is prepared usually for two strikes from syrup/A heavy/C light molasses. Then the C massecuite is developed on B heavy molasses. 'C' massecuite purity will be about 54-58, brix will be more than 100. The C massecuite is cured in continuous centrifugals. Grain size of 'C' massecuite and its uniformity is important for exhaustion of the mother liquor in 'C' massecuite, so that the final molasses purity can be controlled.

'C' massecuite is cooled down up to 38-40°C in 24 hours. The cooled massecuite is passed to pugmill, then from pugmill through transient heaters to continuous centrifugals. The massecuite is reheated in transient heaters to saturation temperature (50-52°C) of the mother liquor. This is necessary to reduce viscosity of the massecuite. 'C' massecuite is double cured. In the first curing diluted molasses or warm water (2% on massecuite) is used as a lubricant. Final molasses is separated from 'C' massecuite in the first curing. The final molasses purity is usually 28-34. The C single cured sugar or 'C' fore worker sugar (CFW) of purity 78-80 is mingled with C light/water to a magma. The magma is lifted and send to pugmill provided to C after worker (CAW) continuous centrifugals. The CFW magma is cured in CAW centrifugals in which 'C' light molasses is separated. The 'C' light of molasses of purity will be around 60-65. Little quantity of C light molasses is used for CFW magma and the remaining is send to pan floor for 'C' massecuite boiling. The 'C' seed sugar or 'C' after worker sugar (CAW) of 92-94 purity is send for melting. The melt is used for 'A' massecuite boiling.

57.5 Three and half massecuite boiling scheme:

C massecuite purity is normally kept 52 to 58 to keep the final molasses purity to a minimum practicable level for which the B heavy purity should be around 45 to 50. However when syrup purity remains above 83 and/or factory has to produce more bold sugar, all massecuites and molasses purities increase and C massecuite purity cannot be brought down to 58 or below with conventional three massecuite boiling system. In such case three and half boiling system may be adopted. The intermediate massecuite may in between A and B massecuite or in between B and C massecuite, whichever process manager/chief chemist finds suitable depending upon equipment at pan station and centrifugals. The intermediate massecuite is called as A1, B1 or C1 massecuite. In such case heavy molasses of a high massecuite is used for low grade two massecuites. For e. g. in A, A_1, B, and C massecuite boiling scheme, A heavy molasses is used as per requirement for A_1 massecuite and the remaining is used for B massecuites. The intermediate massecuite percentage may be about 2-4 % on cane. In this boiling system as complete AH molasses not used for A_1 or B massecuite, the system is called as three and half massecuite boiling system. With this boiling system the B and C massecuite purity can be maintained at a desired level.

57.6 Four massecuite boiling scheme:

If the juice purities are high and/or factory has to produce superior quality bold grain plantation sugar then regular four massecuite boiling scheme can be adopted. In 'four *massecuite boiling scheme'* heavy molasses of high grade massecuite goes to immediate next low massecuite.

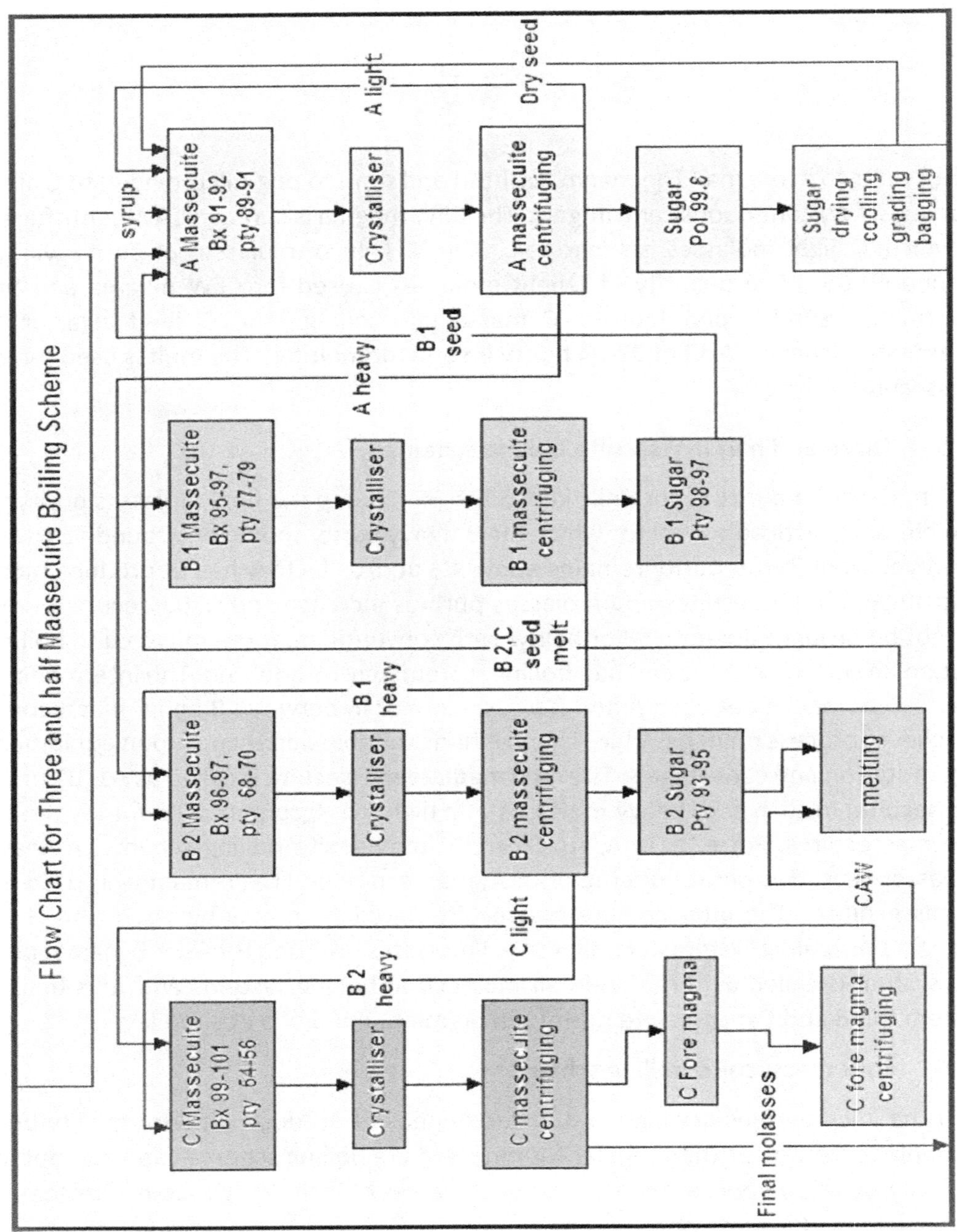

Fig. 57.2 Three and half massecuite boiling scheme

-OOOOO-

58 PAN STATION MATERIAL BALANCE

General:

In this chapter we will calculate the theoretical material balance at pan station based on three massecuite boiling scheme which is followed in most of the factories for manufacture of plantation white sugar. While calculating the material balance the normal purities of all materials and purity drops that can be attained are assumed.

Assumptions:

- Cane – 100 t
- Syrup - purity 82.00, solids 15.50 t, sugar (pol) -12.71 t, nonsugars 2.79 t,
- Sugar – purity 99.8
- Final molasses – brix 90.00, purity 30.00,
- C massecuite is double cured
- A and B massecuite are single cured.

Table 58.1 assumptions for material balance at pan station

Sr. No.	Material	Brix	Pol	Purity
1	Syrup	62.00	51.15	82.50
2	A massecuite	92.50	81.40	88.00
3	Sugar	99.90	99.70	99.80
4	A heavy molasses	82.00	60,68	74.00
5	A light molasses	73.00	66.43	91.00
6	B Massecuite	96.00	71.04	74.00
7	B seed	96.00	92.16	96.00
8	B heavy molasses	84.00	42.00	50.00
9	C massecuite	100.00	56.00	56.00
10	C single cured sugar (CSCS) or C fore worker sugar (CFW)	94.00	75.20	80.00
11	Final molasses	90.00	27.00	30.00
12	C light molasses	73.00	46.00	63.00
13	C double cured sugar (CDCS) or C after worker sugar (CAW)	95.00	89.39	94.00

1 Cobenze's diagram:

This diagram is used in boiling schemes for the calculation of the relative quantities of two products to give a mixture of the required purity.

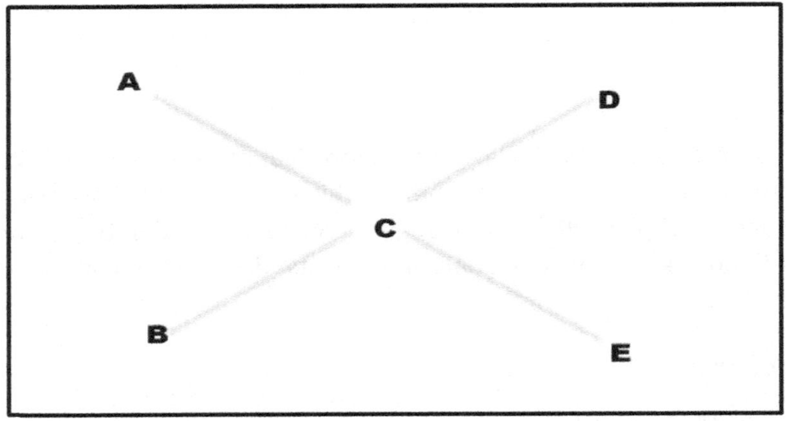

Fig. 58.1 Cobenze's diagram

Where

A = purity of product 1(sugar separated from massecuite)
B = purity of product 2 (molasses separated from massecuite)
C = purity of mixture (massecuite being cured)
D = quantity of product 1
E = quantity of product 2
C - B = D
A - C = E

This formula is used to calculate the material balance at the pan station.

2 Sugar recovery and final molasses % cane:

From 100 t of cane syrup containing 15.0 t of solids is obtained. In pan boiling process water from the syrup gets evaporated and the solids pass either in sugar or final molasses. Then we can calculate quantity of solids pass in sugar and in final molasses by Cobenze's diagram.

Table 58.2 calculations of solids in sugar and final molasses

Cobenze's diagram			Solids in		
			Sugar	+ Final molasses	= Syrup
99.80		52.50			
			52.50	+ 17.30	= 69.80
	82.50				
			11.658	3.842	= 15.50
30.00		17.30			

Thus from the calculations we get	
Sugar in sugar recovery % cane	11.658
Solids pass final molasses	3.842
Final molasses % cane	4.268

3 Quantity of C massecuite:

When C massecuite is cured final molasses is separated and C single cured sugar or C fore worker (CFW) sugar is obtained.

Cobenze's diagram			Solids in		
			CFW sugar	**+ Final molasses**	**= C massecuite**
80.00		26.00			
			26.00	+ 24.00	= 50.00
	56.00				
			4.162	+ 3.842	8.004
30.00		24.00			

Therefore

Solids in C massecuite are 8.004 t

C massecuite % cane is 8.004 t

Solids in CFW 4.162 t

$$\text{Final molasses \% cane} = \frac{\text{solids in final molasses \% cane}}{\text{brix \% final molasses}} \text{ X } 100$$

$$\text{Final molasses\% cane} = \frac{3.842}{90} x100 = 4.268$$

4 Quantity of C seed:

A magma of CFW sugar is made in magma mixture and it is pumped to pugmill for second curing to get C double cured sugar (CDCS) or C after worker sugar (CAW) while C light molasses is separated.

Table 58.3 calculations – C seed quantity

Cobenze's diagram			Solids in		
			C after worker	**+ C light molasses**	**= C fore worker**
94.00		17.00			
			17.00	+ 14.00	= 31.00
	80.00				
			2.282	1.880	4.162
63.00		14.00			

Solids in CAW or C seed 2.282 t

Solids in C light molasses 1.880 t

The C seed is melted and the melt is used for A massecuite boiling.

5 Purity of imaginary mixture of A heavy and B heavy molasses:

The C massecuite is boiled on B heavy molasses and C light molasses. Some A heavy molasses is often taken for C massecuite graining. To maintain expected purity of C massecuite, calculated quantity of A heavy is to be used for C massecuite graining. To calculate the quantity of A heavy molasses an imaginary mixture of AH and BH molasses is assumed and its purity is calculated.

Table 58.4 purity AH + BH molasses of imaginary mixture

	C massecuite	C light molasses	A H + B molasses
Solids	8.004	= 1.880	+ 6.146
Pol	4.482	= 1.184	+ 3.298
Purity	56.00	63.00	53.65

Thus the mixture of AH and BH molasses contains 6.146 t of solids and 3.34 t of sugar and purity is 53.65.

6 Quantity of AH and BH used for C massecuite:

From the quantity of solids of the AH + BH imaginary mixture and its purity, solids in AH used for C massecuite and solids in BH can be calculated.

Table 58.5 quantity of AH and BH molasses in imaginary mixture

Cobenze's diagram			Solids in		
			A H molasses	B H molasses	AH + BH molasses mixture
74.00		3.65			
			3.65	+ 20.35	= 24.00
	53.65				
			0.935	+ 5.211	= 6.146
50.00		20.35			

Thus the solids in AH used for C massecuite are 0.935 t,
And the solids in BH are 5.211 t,

7 Quantity of solids in B massecuite and B seed:

The BH molasses is obtained from B massecuite. Therefore solids in B massecuite and BFW can be calculated from the solid of B heavy molasses.

Table 58.6 calculations for solids in B seed and B massecuite

Cobenze's diagram			Solids in		
			B seed	B H molasses	B massecuite
96.00		24.00			
			24.00	+ 22.00	= 46.00
	74.00				
			5.684	+ 5.211	= 10.895
50.00		22.00			

The solids in B massecuite 10.895 t
B massecuite % cane 11.350 t
Solids in BFW sugar 5.684 t

8 Total solids in A heavy molasses:

Solid in A heavy molasses used for C massecuite 0.935
Solids in A heavy molasses used for B massecuite 10.895

Total solids in A heavy molasses 11.830

9

Table 58.7 Quantity of solids in A massecuite and A fore worker

Cobenze's diagram			Solids in		
			A fore worker	A H molasses	A massecuite
99.00		14.00			
			14.00	+ 11	= 25
	88.00				
			15.056	+ 11.830	= 26.886
74.00		11.00			

Thus
The solids in A massecuite are 26.886 t
And solids in A fore worker 15.056 t
Solids in A light molasses 11.830
The A massecuite % cane 29.224 t

10

Table 58.8 Quantity solids of A after worker and A light molasses:

Cobenze's diagram			Solids in		
			A after worker	A light molasses	A fore worker
99.80		8.00			
			8.00	+ 0.80	= 8.80
	99.00				
			13.687	+ 1.369	15.056
91.00		0.80			

Thus the solids in A after worker that discharged in hopper are 13.687 t

Solids in A light molasses 1.369 t

The solids in sugar bagged are 11.670 t

Therefore solids in rory, dry seed and dust are 2.017 t

Sugar recovery % cane 11.682

11 Total massecuite % cane

A massecuite % cane	29.224
B massecuite % cane	11.350
C massecuite % cane	8.004
Total massecuite % cane	**49.578**

Table 58.9 A massecuite counter balance:

Sr. No.	Material	Tonnes of solids	Purity	Tonnes of pol
1	Syrup	15.50	82.00	12.710
2	C seed	2.282	94.00	2.145
3	B seed	5.684	96.00	5.457
4	A light	1.369	91.00	1.246
5	Dry seed	2.017	99.80	2.013
	Total	26.852	87.78	23.571

-OOOOO-

59 CRYSTALLIZERS

59.1 General:

The massecuite from vacuum pan is discharged and stored before curing in vessels called crystallizers. In crystallizers the massecuite is gradually stirred so that there is relative motion of crystals and mother liquor. Therefore the crystals always come in contact with fresh supersaturated mother liquor. While stirring the sucrose in the molasses crystallizes out and deposit on the existing crystals. The existing crystals grow but no nuclei are formed. Because of stirring the temperature of the massecuite also remains same throughout the crystallizer. As time passes the temperature of the massecuite goes down. It decreases solubility of sucrose in the mother liquor and crystallization is continued.

Fig. 59.1 Air cooled crystallizer

59.2 Crystallizers:

Simple horizontal air cooled crystallisers are U shaped vessels with stirrer in the form of a helical ribbons or arms mounted on a longitudinal central horizontal shaft. The shaft is rotated at about 0.3-0.7 rpm. The walls of the crystallizers and the surface of the massecuite act as cooing surface. In these crystallizers cooling of massecuite is very slow. Water cooling crystallizers are installed for rapid cooling. Various types of crystallizers using water cooling elements have been introduced since 1930.

Plants erected from 1972 are gravity plants. In these plants crystalizers are mounted at a platform approximately 6.5 to 7.5 m above the ground level. Therefore massecuites can slowly flow from crystallisers to pugmill by gravity. Before 1972 some plants erected are non-gravity plants. In which the crystallisers are mounted on ground foundation just one meter above the ground level of the plant. In such plants the massecuites are drawn to pugmills by pumping. In such plants crystal breakage may occur to some extent while lifting the massecuites.

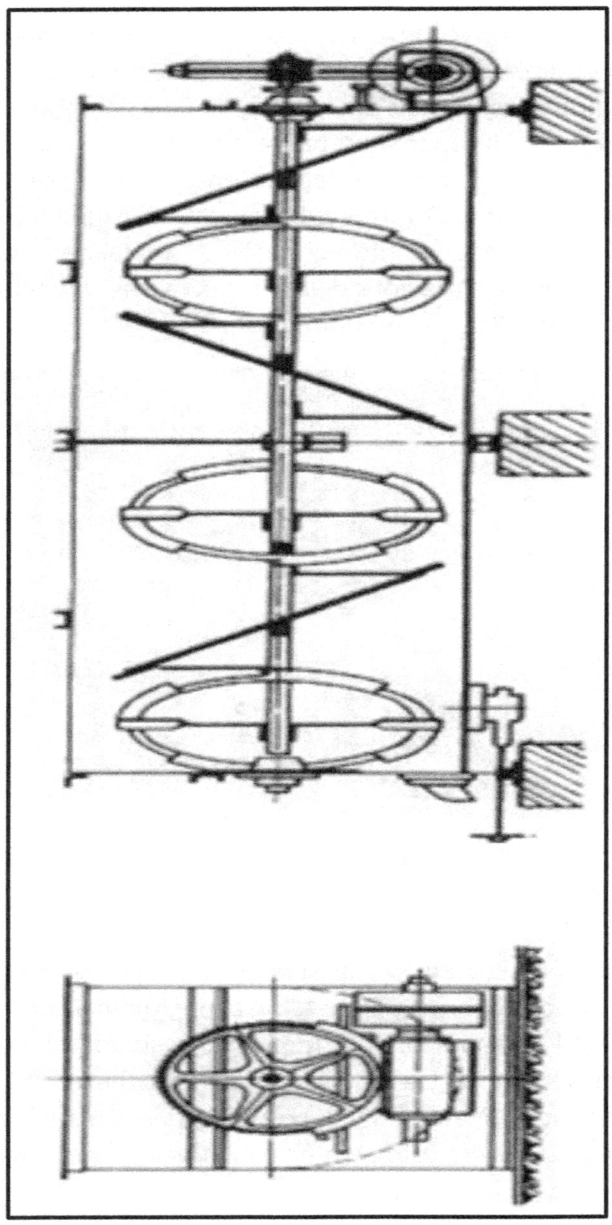

Fig. 59.2 Horizontal Crystallizer

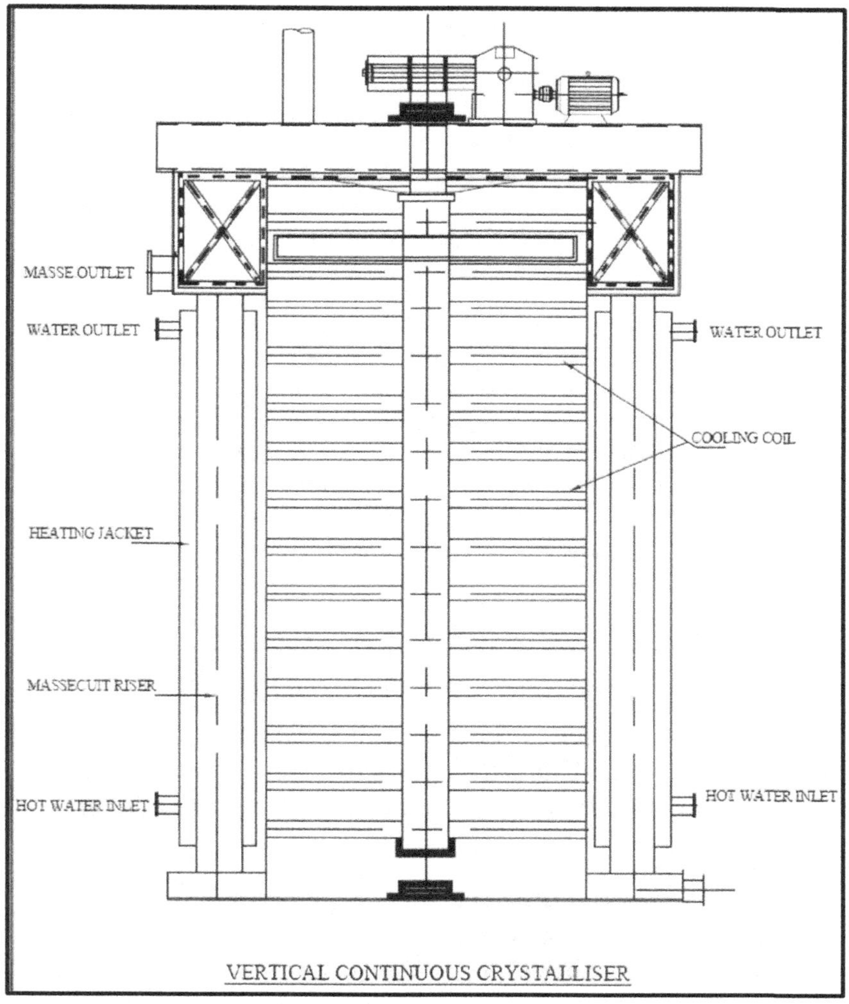

Fig. 59.3 Design of vertical water-cooled crystallizer

59.3 Water-cooled crystallisers:

There are many different types of water cooled crystallizers in the industry out of which Werkspoor and Fletcher-Blanchard crystallizers are more common. In Werkspoor crystallizers water cooling elements are provided in the form of thin hollow discs with a gap in the form of a sector of about 45°. The discs are mounted at right angles to the horizontal hollow centre shaft.

The Blanchard crystallizers are provided with a hallow shaft of about 300 mm in diameter with rows of radial pipes that are closed at the ends works as cooling elements. Now in new designs curved pipes are used to give increased cooling surface which are commonly used. Generally air cooled crystallizers are used for A massecuite and water cooled crystallizers are used for B and C massecuite.

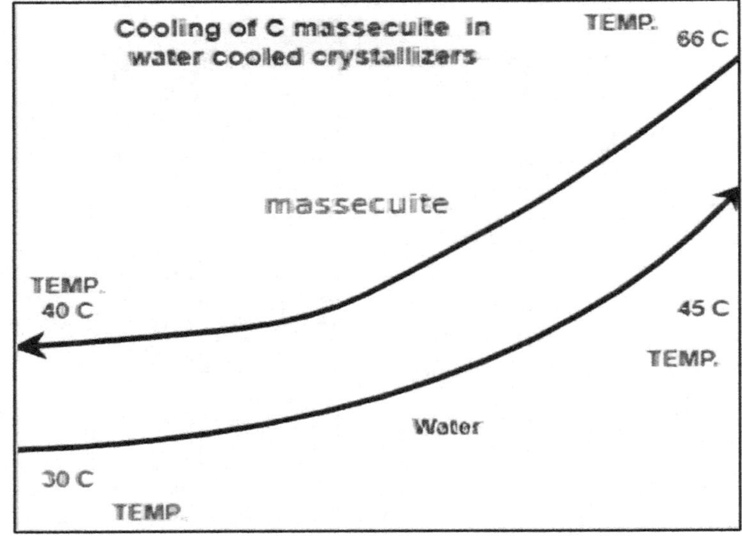

59.4 A graph showing cooing of massecuite

In low grade massecuites the rate of crystallisation is very low due high concentration of non-sugars. Therefore all possible steps have to be taken to ensure maximum exhaustion of mother liquor. To achieve higher exhaustion the following steps are taken.

- Higher surface area is provided with smaller crystal size,
- Optimum cooling of massecuite,
- Adequate time in crystallizers for exhaustion.

After dropping the low-grade massecuite it is cooled up to 38-40º C in water-cooled crystallisers. After cooling down the C massecuite up to 38-40º C it should be retained at that temperature for 3-4 hrs., so that the mother liquor gets exhausted of sucrose to practically possible limit on industrial scale. For a semi-gravity plant the massecuite has to be pumped and in such cases if cooled massecuite is to be pumped, pumping problem are encountered and hence higher exhaustion may not be possible.

Corrosion of cooling surface:

The cold raw or fresh water used for cooling of massecuite contains dissolved oxygen that causes for corrosion of the cooling elements which are made of steel. When corrosion occurs heat transfer coefficient of the metal is considerably reduced, the metal becomes weak. Therefore, the cooling of massecuite up to desired temperature is not achieved. Leakages may also occur. When leakages are developed the water percolates into the massecuite dilute it and crystal dissolution starts.

59.4 Reheating of C massecuite:

The C massecuite, which is cooled up to 38 to 40°C, is to be reheated to saturation point (at about 51-53°C) of mother liquor to reduce the viscosity of the massecuite so that the massecuite can be cured easily at centrifugals. The saturation temperature of massecuite can be found out in laboratory by using saturoscope.

The viscosity of the massecuite falls by about 50% when temperature is increased by 5°C. Thus C massecuite cooled up to 38°C and having viscosity 1200 poise when heated to 52°C (to saturation point) the viscosity of the massecuite will come down to about 250-300 poise. Thus purgeability of the massecuite will be increased. Transient heaters are used to reheat the C massecuite.

59.5 Transient heaters at C centrifugals

Diluted final molasses of 15 to 25 brix and temperature 48-50°C is used as a lubricant at the continuous centrifugal machine. This further reduces the viscosity of the massecuite to 30-50 poise at which the continuous centrifugal machines can work well.

When C massecuite purity is about 55-56 or lower - which is normal now in sugar industry, and the grain size is about 180 micron, cooling of C massecuite can be started immediately after dropping. No false grain formation occurs due to immediate starting of cooling because of higher surface area available for deposition of the sucrose that is coming out of mother liquor. Air cooling for first 3-4 hours is not required. Therefore receiving crystallizers provided are also water-cooled crystallizers. In continuous horizontal or vertical crystallizers it should be confirmed that the massecuite flow is stream lined or plug flow and there should not be any short-circuiting. The crystallizer to pugmill gutter should be well wide and deep with an angle of min 24° to have sufficient flow of viscous cold massecuite.

Leaking of coils of water cooled crystallizers commonly occurs in old factories.

Therefore all the cooling coils of the crystallizers should be well tested and made leak-proof in the off season. If water leaks from the cooling elements to the massecuite, dissolution of the crystals starts. In such case cooling water is to be stopped and cooling of 'C' massecuite cannot be attended and therefore maximum possible exhaustion cannot be achieved.

The C massecuite should not be diluted in the crystallizer or pugmill either by water or by diluted final molasses as it increases the purity of final molasses by re-dissolution of crystals.

Reheating In crystallizer and/or in the pugmill takes almost two hours and there are all chances of localized over heating of the massecuite causing redissolution of crystal which ultimately increases the purity of final molasses. Therefore now days this method is not used. Instead use of transient heaters is now common in almost all sugar factories. The transient heater is installed in between pugmill and centrifugal machines. The massecuite is heated in very short time and then immediately is fed to the centrifugal machine. Hence the possibility of redissolution of sugar crystal is almost eliminated.

The transient heater is a tube heat exchanger. The massecuite flows from outside of the tubes and hot water flows through the tubes. The hot water temperature should be 57-60 C so that there will not be overheating of the massecuite. The hot water velocity should be 1.5-2.0 m/sec. The overall heat transfer coefficient (OHTC) can be taken as 200 Kcal/m^2/h/°C. The specific heat of massecuite can be taken as 0.5 KCal/kg.

The massecuite from the crystallizer is sent to pugmill a U shape small crystalliser mounted right over the centrifugal machines. Like crystalliser the pugmills are also provided with revolving arms or helical ribbon to prevent settling of crystals. Pugmill is also useful for double curing. Pugmills equipped with heating elements for hot water circulation to reheat the low grade massecuite were in use. But in recent years with the development of and wide acceptance of transient heater, the practice of reheating massecuite in pugmill has been obsolete. From the bottom of the pugmill individual piping with valve and charging chute for each centrifugal machine is provided. The massecuite flows from the pugmill to the centrifugal by gravity.

-OOOOO-

60 EXHAUSTIBILITY OF MOLASSES

Effect of nonsugars on exhaustibility of final molasses:

Final molasses purity varies considerably from region to region, factory to factory and also in different periods of the same season in a factory. A final molasses of purity 29.0 may not have been be reached to maximum exhaustion of sucrose. On the other hand a final molasses of purity 34.0 may be well exhausted.

Many sugar technologists in different part of the world have studied composition of the final molasses – particularly percentage of reducing sugars and ash, to evaluate the exhaustibility of molasses.

Experiments made by technologists shows that the minimum purity that can be attained depends on the composition of the impurities contained in the juice and consequently in the molasses.

- The entire reducing sugars pass in final molasses and when reducing sugars are higher, the purity of the final molasses will be lower.
- Inorganic solutes in the molasses - particularly the chlorides of potassium and sodium at high temperature, generally increase the solubility of sucrose. Therefore when ash proportion is higher the final molasses purity increases.

Many technologists have put forth different formulae to estimate minimum final molasses purity - that can be achieved, based on the percentage of reducing sugars and ash content in the molasses.

We have given hereunder two formulae for targeted purity or the minimum purity that can be reached with at its best exhaustion.

1 The Douwes Dekker formula for targeted purity:

$P = 36 - 0.08r + 0.26c$

Where

P = true purity attainable with normal exhaustion of the final molasses, (target purity)

r = reducing sugars % of non-sucrose in the molasses

c = sulphited ash % non-sucrose in the molasses

Non-sucrose = dry substance – sucrose

The Douwes Dekker formula applied only to defecation factories is as follows

$P = 33 - 0.055r + 0.31c$

2 Reunion formula for targeted purity:

$$P = 40 - 4\ R/C$$

While analysing final molasses in laboratory it is difficult to determine dry substances accurately. The value for dry substances varies greatly depending on the method of analysis. However, the refractometric brix approaches the true figure more closely than the gravimetric figure.

All the formulae given by different technologist give only approximate idea of the final molasses purity that can be achieved. In practice no formula gives the exact figure of minimum final molasses purity that can be attained. Therefore we can say that there is no standard to indicate, whether the final molasses obtained is really exhausted or not.

In our opinion a more accurate method - to find out whether the final molasses is well exhausted or not, is to find out total reducing sugars in it. If TRS are equal to 50 per 100 brix or less, then we can say the molasses is well exhausted.

-OOOOO-

61 CENTRIFUGALS - THEORY

61.1 General:

When mother liquor of massecuite is exhausted in the pan and in the crystallizers to the practically possible limit, the next unit operation is to separate the sugar crystals from the mother liquor. This operation is basically a solid (sugar crystals) liquid (molasses) separation and it is carried out in centrifugal machines. Centrifugal force is used for separation of sugar crystals and molasses. The operation is called as curing, purging or centrifuging.

The curing operation carried is important from the view of quality of sugar, re-circulation of sugars and non-sugars and loss of sugar in final molasses. Centrifugal machine consists of a perforated basket (drum) which revolves at high speed. When massecuite is charged in the basket and machine is revolved because of centrifugal force the molasses is thrown off through perforation whereas sugar crystals remain inside perforated lining.

61.2 Theory:

Centrifugal force is defined as that force which tends to impel thing outward from the centre of rotation. This force is

1. Directly proportional to the mass of the rotating body,
2. Directly proportional to the square of the speed of rotation,
3. Inversely proportional to the radius of rotation.

$$f = \frac{mv^2}{r} = \frac{wv^2}{g\,r}$$

Where f - centrifugal force
m – Mass of the material being rotated
v - Speed of rotation in radians per second
r – Radius of gyration
w – Weight of the material being revolved
g – Gravitational force

Therefore,

$$f = \frac{w}{g} \times \frac{(2\pi rN / 60)^2}{r}$$

Where,

N – No. of revolutions per min. of the basket
$v = 2\pi rN / 60$

$$\text{or} \quad f = \frac{w\,D\,N^2}{1789} \quad \text{or say} \quad \frac{w\,D\,N^2}{1800}$$

Where D = 2r and g = 9. 81

Thus for unit weight of purging material the centrifugal force is directly proportion to the diameter and rpm of the basket of a centrifugal machine.

61.3 Gravity Factor:

Effect of various centrifugal machines is compared by a term gravity factor. It is the ratio of centrifugal force exerted by the machine at spinning speed to the gravitational force.

$$\text{Gravity factor } G = \frac{\text{Centrifugal force}}{\text{Gravitational force}} \quad \text{or} \quad \frac{f}{mg}$$

$$G = \frac{WDN^2}{1800} \times \frac{1}{mg}$$

$$G = \frac{DN^2}{1800}$$

The gravity factor increases with increase in diameter and speed of the basket. Thus higher basket diameter increases capacity as well as gravity factor of the machine. The higher gravity factor reduces time cycle with better separation of molasses. Modern trend in the world sugar industry is to use high capacity and fully automatic batch centrifugals.

Mean Equivalent Radius:

The centrifugal force on sugar crystals against gauze is obviously greater than that on the crystals at the interior surface of sugar layer. To calculate the mean centrifugal force on the whole mass of massecuite 'Mean Equivalent Radius' is taken which is calculated as follows,

Consider the cylinder as above having

The interior radius at the gauze	r_1
The radius at inner wall of sugar layer.	r_2

Hence thickness of sugar wall is ($r_1 - r_2$).

The mean equivalent radius is calculated as,

$$r_m = \frac{\mathbf{0.67\,[r_1^3 - r_2^3]}}{\mathbf{[r_1^3 - r_2^3]}}$$

Example

- Diameter of basket 1.500 m
- Interior radius at the gauge (r) 0.750 m
- thickness of sugar layer 0.170 m
- Radius corresponding to inner wall of sugar layer 0..580 m

Hence, mean equivalent radius

$$r_m = \frac{0.67[r_1^3 - r_2^3]}{[r_1^2 - r_2^2]}$$

$$r_m = \frac{0.67\,\{(0.750)^3 - (0.580)^3\}}{(0.750)^2 - (0.580)^2}$$

r_m = 0.679 m. or D = 1.357 m

Suppose N = 1090 rpm, then

$$G = \frac{D\,N^2}{1800} = \frac{1.357 \text{ X } 1090^2}{1800}$$

G = 896

Batch Centrifugals

-OOOOO-

62 BATCH CENTRIFUGALS

62.1 General:

Modern trend is to use batch centrifugals for A massecuite and for some percentage of B massecuite (which is cured to serve as seed for A massecuite). The remaining B massecuite (seed of which goes for melting) and C massecuite are cured by continuous centrifugals. The batch centrifugals needs 2.5 – 4.0 times more power than continuous centrifugals,

Continuous centrifugals yet are not being used for A massecuite curing because in continuous centrifugals, there is considerable breakage of sugar crystals (12–16 %). Moreover sugar discharged from continuous machine contains higher moisture and lumps are formed.

However in 1993 STG Eng. (Australia) constructed a proto type continuous centrifugal machine which is being used for raw sugar A and B massecuites as well as for refinery massecuites (at Manildra Harwood Sugar refinery in Australia) since 1997. The sugar produced by the continuous centrifugal is equivalent in purity, colour, percentage of fines and moisture as compare to batch centrifugal sugar. The day is not far away when continuous centrifugals will be used for A massecuite curing in sugar factories as well as in refineries.

62.2 Details batch type centrifugal machine:

Batch type centrifugal machine consists of a cylindrical basket that is hanging on a vertical shaft or spindle, which is driven from upper end by electric motor. The centrifugal basket rotates about its centre of gravity because of flexible suspension.

Basket is a hollow vertical cylinder made up of special quality stainless steel. It is provided with optimum number of holes to pass molasses through it and is attached at the bottom to the shaft by a spider.

It is open at the top and at the bottom so that massecuite can be feed into it and sugar can be discharged through it. There is gap in the spider for discharge of sugar. The basket thickness depends on tensile strength of the metal. The baskets are fabricated of special steel of higher strength. The holes of the basket are of 5-7mm diameter and spaced at about 22-25 mm between the centres.

The shaft or spindle is made of high grade forged steel. The shaft is hanging on a floating bearing assembly.

A thin metal sheet cone is provided to close the discharge opening when the machine is running.

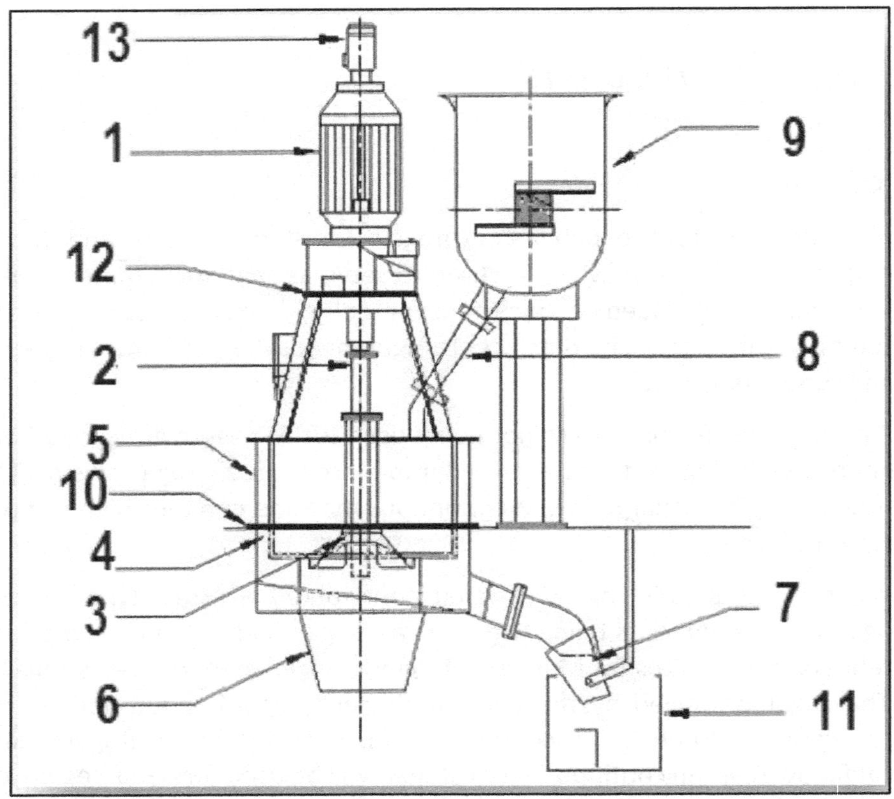

1 elec. motor	2 shaft
3 lifting hood	4 basket
5 monitor casing	6 discharge chute
7 syrup separator	8 massecuite feed pipe
9 pug mill	10 platform
11 syrup gutter	12 mounting frame
13 fan	

Fig. 62.1 Batch centrifugal

The basket is covered from sides by monitor casing which is open from the top and bottom. The molasses separated from the massecuite is collected inside the monitor casing. The top opening of the monitor casing can be closed by means of two halve covers provided on it. The covers are provided with a matching hole through which shaft is passed.

The basket is lined with a backing screen and working screen. Sugar crystals are retained inside the working screen and molasses is passed through it. A water / steam pipe with specially designed nozzles is fitted inside the basket from the top. Molasses is washed off from the massecuite by spraying water / steam through the nozzles.

In operating cycle when massecuite curing is done, the speed of centrifugal machine at first is brought down by regenerating braking and then by means of mechanical brakes. The brakes are consist of a drum and brake shoes fitted on the shaft. The brake shoes are provided with special rubber pads.

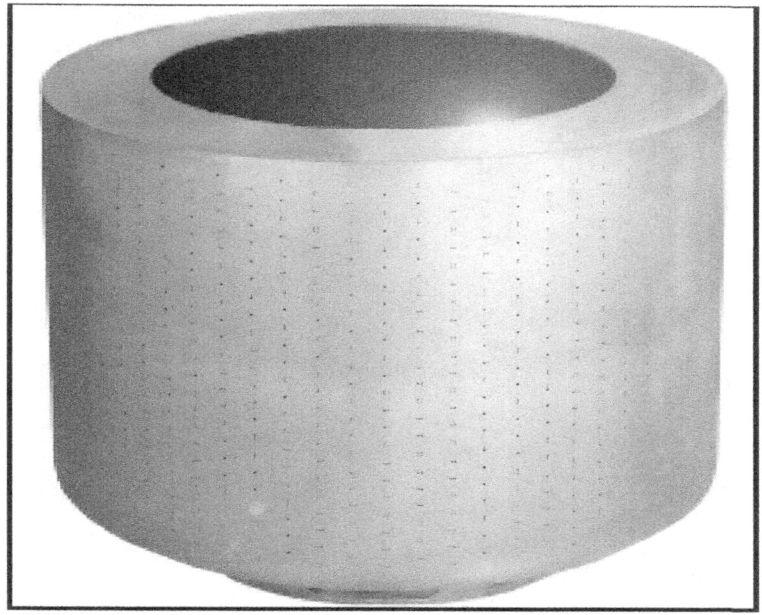

Fig. 62.2 basket of a batch centrifugal

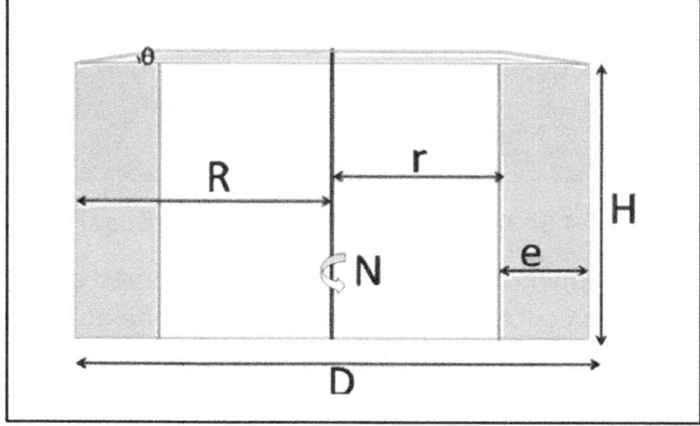

Fig. 62.3 working volume of a batch centrifugal

Fig. 62.4 Shaft and hood of batch centrifugal machine

When the brake is applied the brake shoes are tightened on the brake drum and the machine stops. The centrifugal is fed with massecuite at a speed between 100-300 rpm and the sugar is discharged at a speed between 30-60 rpm. For flat bottom machines pneumatically operated plough is provided for discharging the sugar. Its blade remains outside the casing during charging and spinning. In modern high capacity machines now washing arrangement to wash monitor casing from inside is made.

Backing screen:

If a single perforated working screen is used for purging it cannot work because the great portion of the liner lies against the smooth side-wall of the basket and it does not allow the molasses to pass through. Therefore the basket is furnished with brass or copper wire backing screen. The backing screen separates the working screen from the wall of the basket and gives passage to molasses. In some factories two backing screens are used. Tromp recommends a 7 mesh backing screen in between 4 mesh backing screen and working screen.

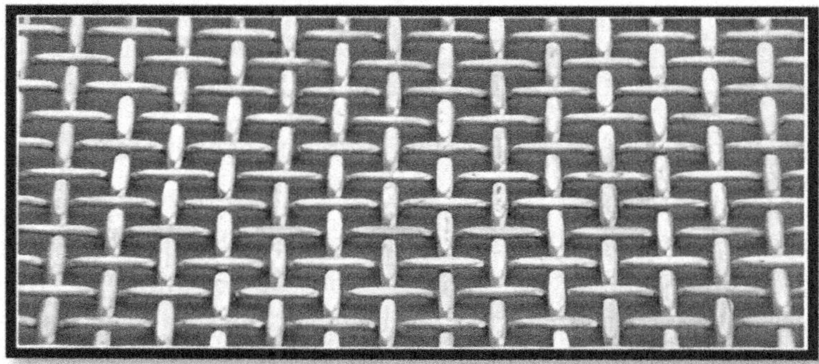

Fig. 62.5 backing screen for batch centrifugals

Working screen or liner:

The working screen or liner is a perforated sheet of pure copper, stainless steel or special alloy. They are provided with horizontal slots of size 0.35 X 4 mm arranged either in columns or in staggered formation. The pitch of these perforation in the vertical direction is 1 mm. The opening area is about 24- 26 % of the total area. The width 0.35 mm is for interior face of the screen. The perforations are pyramidal to avoid jamming of exact size of crystals in the slots. Liner with round holes of 0.4 mm is also used.

When slot size is increased, the opening area of the working screen is increased. This results in increase in capacity. However increase in slot size results in passing of fine crystals. It increases molasses purity. Therefore screens with optimum slot size has to be used.

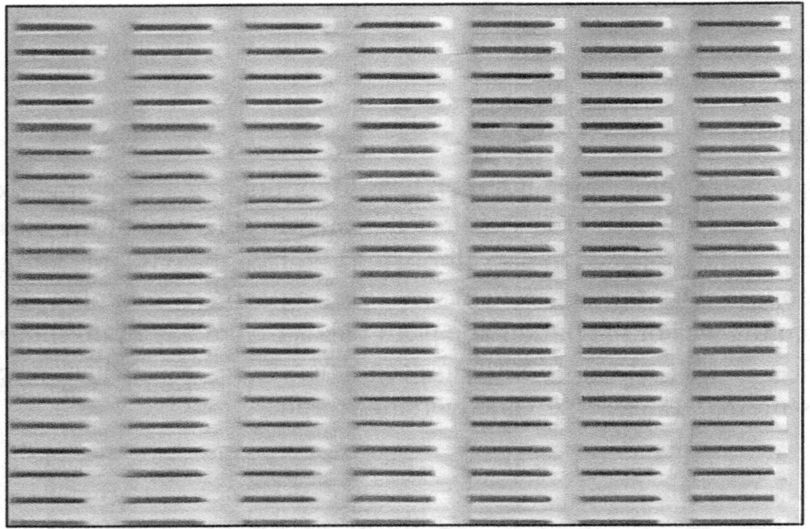

Fig. 62.6 working screen of batch centrifugals

Thickness of the liners varies from 0.45 to 0.8 mm often 0.5 mm stainless steel and 0.7 mm brass. Thinner liner has shorter life but give better operation. The liners are fitted in such a way the overlapping is about 50- 60 mm. The overlapping is arranged in such a way that while the basket is rotating, the resistance of the air tends to hold the free ends against each other.

Electric Motor for Centrifugal Drive:

A vertical electrical motor drives centrifugal machines. The shaft of the electric motor is in line with shaft of the centrifugal machine and is fitted with it by direct coupling

There are two means of electric drive for centrifugals.

i) Three phase induction motor generally of squirrel cage type.
ii) D.C. motor supplied through thyristor from AC.

Both types of motors are in use. However now inclination is to use centrifugals of higher capacities and that are with DC motors. The advantages of D.C. motor are as under.

- Choice of speed is independent of the frequency of the A.C supply with precise control on the speed
- Specific power consumption is 0.4 – 0.6 of that of the induction motor. 3-4 kWh per tonne of sugar as against 8-10 kWh per tonne.
- Power of D.C. motor is constant hence no peak current demand as in case of AC motor.
- Less heating of the motor.

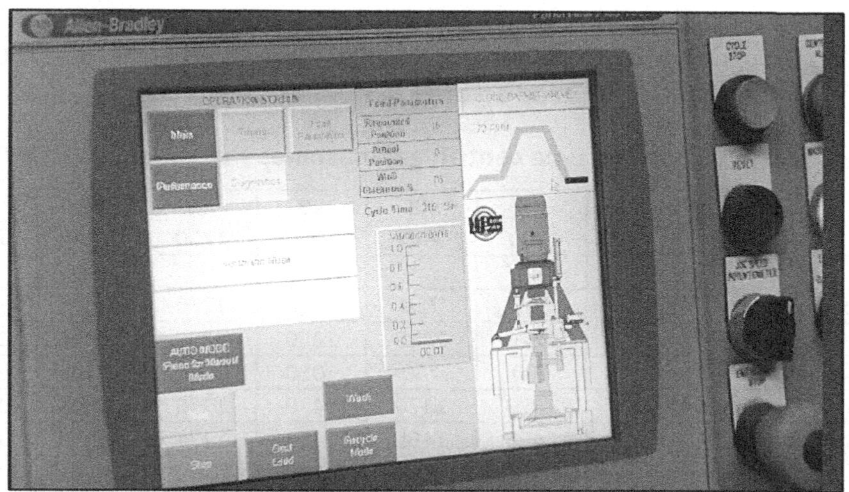

Fig. 62.7 Monitor for batch centrifugal machine

Regenerating braking:

The interesting characteristic of electric centrifugal machines is regenerative braking. In this system the energy consumed in accelerating and running the centrifugal is partly regained.

Efficiency of electrical motors driving centrifugals is low since it operates in acceleration spinning and braking during each cycle. The power factor increases to 0.8-0.9 during acceleration and falls down to 0.4-0.5 at running speed.

Arrangement of spray nozzles:

The number of wash nozzles and their arrangement is such that water is sprayed uniformly throughout the surface of massecuite. Uniform distribution of wash water in fine spray is very important for efficient washing with optimum quantity of superheated water. There should not be any area which is comparatively less washed or highly washed. Otherwise it may affect either purity of molasses or quality of sugar. The washing is usually less at the upper of the basket. Therefore the spray nozzles are fitted slightly in upward direction.

There is trend in recent years to increase the capacity of centrifugals and reduce the number of centrifugal machines. Batch centrifugals with capacity up to 2.0-2.2 t/charge are available. Modern variable frequency drives operate at high efficiency and make full use of regenerative braking. Variation in harmonic and peak currents are minimised by using the DC bus technology with electronics systems.

The machines operate at 2% on massecuite of total wash including chute and screen wash. The reduction in wash water gives considerable saving in steam at the pans.

Plough design and performance have also been improved reducing the amount of sugar left on the screen. This minimized the recirculation of sugar. The high gravity factor has reduced the moisture percent in sugar discharged from the centrifugals up to 0.5 % this minimizes the load on drier or hopper.

Table 62.1 Batch type centrifugal machines sizes and capacities

Machine type	KB 500	KB 750	KB 1000	KB 1250	KB 1500	KB 1750
Capacity kg per charge	500	750	1000	1250	1500	1750
Basket inside dia. (mm)	1220	1220	1220	1260	1650	1430
Basket inside height (mm)	600	750	1050	1050	1100	1150
Spinning speed (rpm)	1470	1470	1470	1350	1250	1200
Gravity factor	1465	1465	1465	1276	1172	1144
No. of charges per hr.	18-20	18-20	20-22	20-22	20-22	20-22
Lubrication	Grease	Grease	Oil Mist	Oil Mist	Oil Mist	Oil Mist

Now the batch centrifugal machines are fully automated. All the actions are put under the control of time switches or time delay relays. The actuating medium is generally compressed air. Pneumatic operation can thus be applied to

- massecuite feed valve
- wash water
- syrup separator
- mechanical braking
- Plough discharging
- closing of basket bottom
- cover the monitor casing

The batch machines of one battery are co-ordinate by electronic system to regulate sequence of commencing the cycles of different machines. So that simultaneous starting or stopping can be avoided. This helps in maintaining steady power consumption at centrifugal station.

-OOOOO-

63 OPERATION OF BATCH CENTRIFUGALS

The operation of the batch centrifugal starts with water wash applied under a pressure of 3.0-3.5 kg/cm^2 (45-50psig) by a pipe with 6 to 9 nozzles each delivering 8-10 l / min. The wash applied may be adjusted between 3-10 seconds.

The mother liquor or molasses of the massecuite can be distinguished into two parts

- The molasses which fills the voids in between the crystals and any excess amount which can be removed easily.
- The molasses film adhering to the crystal which is difficult to remove.

At first the excess molasses - molasses filling the voids is removed. This molasses of high brix and lower purity is called heavy molasses. When this portion of molasses has been removed hot water or superheated water in a fine spray is applied to wash the partially cured massecuite. The water washes the molasses film and removes it. At the same time a small amount of sugar gets dissolves. Therefore the molasses separated from the crystals with the application of wash is of higher purity and lower brix. This molasses is called as light molasses.

Table 63.1 Working cycle timing

Sr. No.	Particular	Min. sec.
1	Charging and acceleration	0: 50
2	Spinning	0 : 50
3	Electrical braking	0: 15
4	Mechanical braking	0: 15
5	Discharging	0:30
6	Total	2:40

The hot water wash is usually followed by saturated steam wash and it is continued till spinning is complete. The steam is generally of 4-5 kg/cm^2 pressure. Recent trend is to replace water and steam wash by super-heated water wash. Water under pressure is heated to 115-120°C, the pressure being regulated to about 3.0-3.5 kg/cm^2 at the nozzles. The water as it is superheated is broken into very fine spray, which gives an effective and uniform distribution on the wall of massecuite so that water quantity required is reduced. The time of applying wash and its duration is also important and is to be decided by trials.

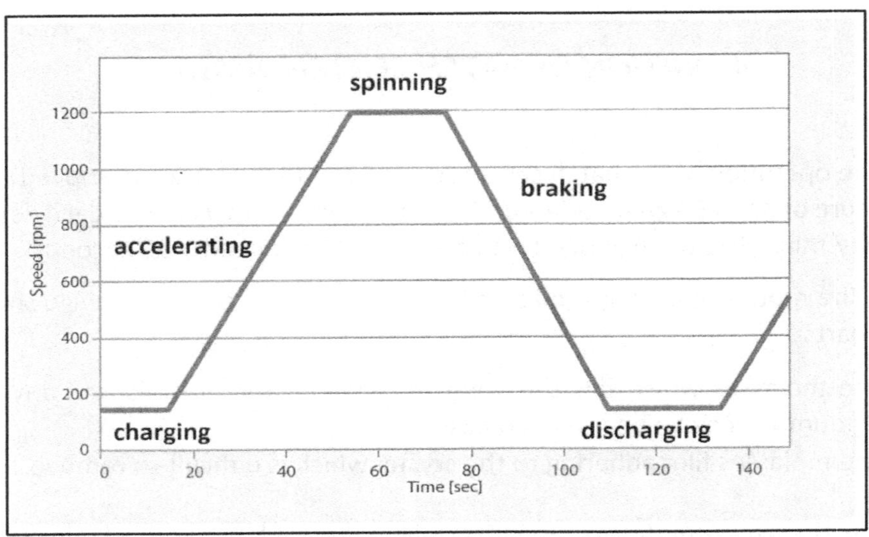

Fig. 63.1 working cycle – time versus speed of a batch centrifugal

Generally the wash water is applied two times by an interval of 10-20 seconds so that washing becomes more effective. The best moment for applying wash is when 75 -80 % of the molasses has been separated. The washing is effected 3-12 seconds about 40 - 50 seconds after charging or 30 seconds before the end of running at high speed. The quantity of wash water required is usually 1.5-2.5 % on massecuite.

Spinning time depends on the centrifugal force developed at the time of spinning, which ultimately depends on

- The basket diameter
- Machine rpm.
- Viscosity of the massecuite / molasses - which mainly depends on temperature and purity of the massecuite.

However, sometimes when the juices of over matured and stale cane do not clarify well, the massecuites are stickier. In such case the spinning time is to be increased. The capacity of the machine is generally expressed in weight of massecuite per charge. Centrifugals are designed to receive a layer of massecuite proportional to the diameter. The machines are designed for a massecuite layer of maximum thickness equal to 14% of the diameter of the basket.

Auto Mode of Operations:

1. Main Motor "ON"
2. Monitor Case Open (feeding chute up)
3. Massecuite Thickness Controller "ON"
4. Screen Wash (Start/Stop)
5. Syrup Separator to Heavy

6. Pugmill gate open (PGO)
7. Pugmill Gate Close (PGC)
8. Water Wash for Tilting Chute (Start/Stop)
9. Acceleration from Charging Speed to Spinning Speed
10. Monitor Casing Close (Feeding Chute Down)
11. Spinning
12. First Water Wash
13. First Curb Wash
14. Second Water Wash
15. Syrup Separation Heavy To light
16. Second Curb Wash
17. Braking
18. Steam Wash
19. Ploughing

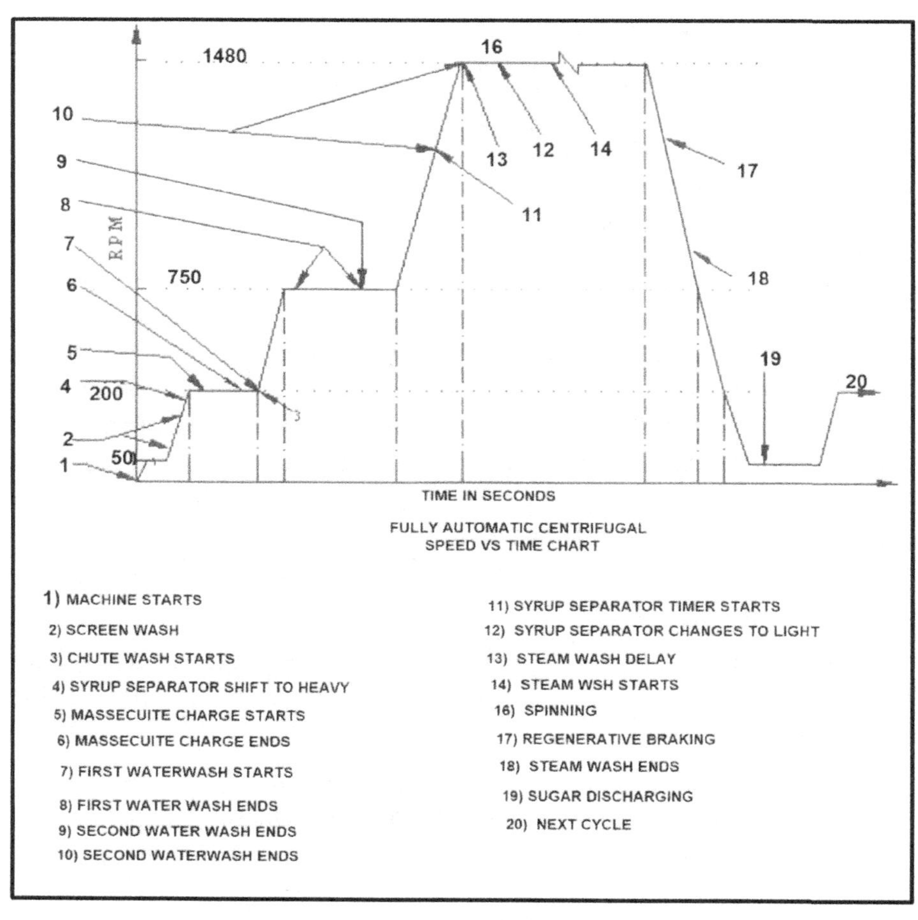

Fig. 63.2 Automation of batch centrifugals

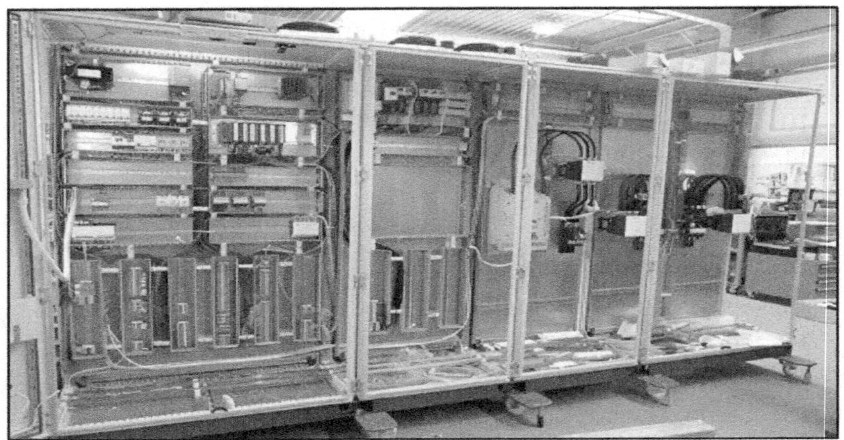

Fig. 63.3 batch machine automation system

Performance of batch centrifugals:

The performance of centrifugals depend mainly on three following factors

- maintenance of the machines
- massecuite characteristic
- operating parameters maintained,

Massecuite characteristics:

The brix, purity, crystal content, viscosity of the massecuite as well as crystal size, crystal uniformity, presence of false grains and conglomerates all these parameters give the characteristics of massecuites. The capacity of the machine is greatly influenced by viscosity and also crystal size and uniformity. If crystals of the massecuite are uniform in size and the massecuite is free from false grain and conglomerates, the machine can be loaded to its full capacity. With smaller and comparatively uneven crystals with fines and conglomerates curing becomes difficult.

At such time batch centrifugals cannot be loaded to its full capacities. The spinning time also has to be increased. Hence capacity of the machines is considerably decreased.

'A' massecuite is hot cured whereas 'B' massecuite generally is cooled up to 50-52 C. cooling of massecuites gives good results. There is higher exhaustion of mother liquor upon cooling and therefore higher drop in purity from massecuite to heavy molasses is obtained.

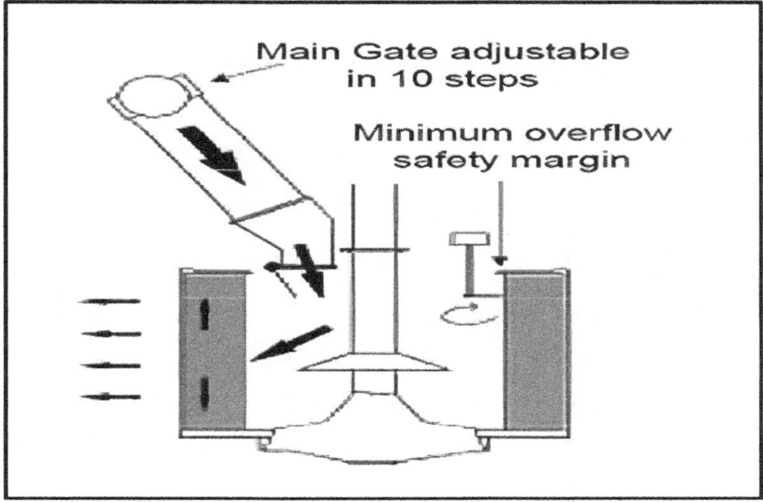

Fig. 63.4 charging of batch centrifugal

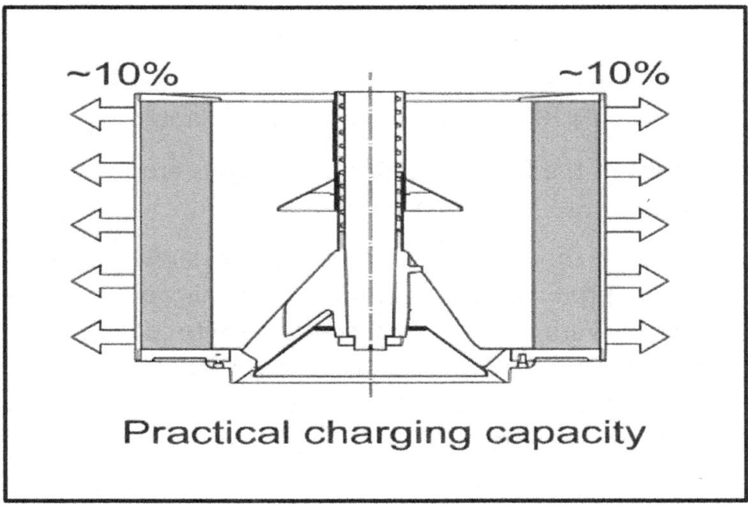

Fig.63.5 charging capacity of batch centrifugal

A-heavy and A-light separation:

In single curing of A massecuite the heavy and light molasses are separated in molasses discharge gutters/pipes by means of syrup separator. The 'A heavy' molasses is used for B massecuite boiling whereas 'A light' molasses is used for A massecuite. If A- Heavy is mixed with A light molasses sugar colour of the next A massecuite is affected. On the other hand if A light is mixed with A-heavy, B massecuite purity is increased. B massecuite percentage is also increased and then control of final molasses purity becomes difficult. To avoid this, proper separation of A heavy and A light is very much essential. However there are many factories where the syrup separators are not working satisfactorily and the separation is not proper.

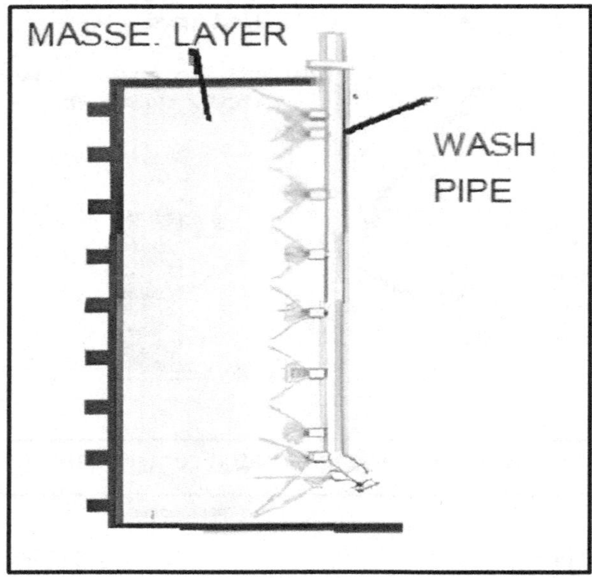

Fig. 63.6 level sensor and wash pipe

Nowadays in most of the factories batch centrifugals are made fully automatic, however proper preventive maintenance is necessary.

The quality of B seed should be superior, the purity should be + 97 to maintain better quality of sugar production. However in maintaining purity of B-seed, in case of single curing B-heavy purity may increase. B-Massecuite cooling and double curing can control B-heavy purity. The B seed required for A massecuite is normally taken by curing B massecuite in batch centrifugals to avoid irregular seed because of crystal breakage in continuous centrifugals.

-OOOOO-

64 CONTINUOUS CENTRIFUGALS

64.1 Construction of Continuous Centrifugals:

Now continuous centrifugal machines having conical baskets with vertical axis are used for intermediate and low grade massecuites all over the world. The construction of the machine is described hereunder.

Conical basket: A conical basket of a vertical continuous machine is made up of stainless steel (ASAI 316 / 321/347 standards) capable of withstanding stress, corrosion, and cracking. Holes are provided to the basket for discharge of molasses. Besides molasses discharge is effected through annular space provided at the top of the basket. The thickness of the basket is about 2.4-2.8 mm and has a semi-cone angle of 30° to 35°. The diameter of the basket is 1000 to 1500 mm. The basket rotates at a speed of 1800-2600 rpm, maximum gravity factor being 2000-2600. At the top a sugar transfer ring is provided to transfer the sugar in sugar compartment inside monitor casing. The basket is perfectly balanced both statically and dynamically.

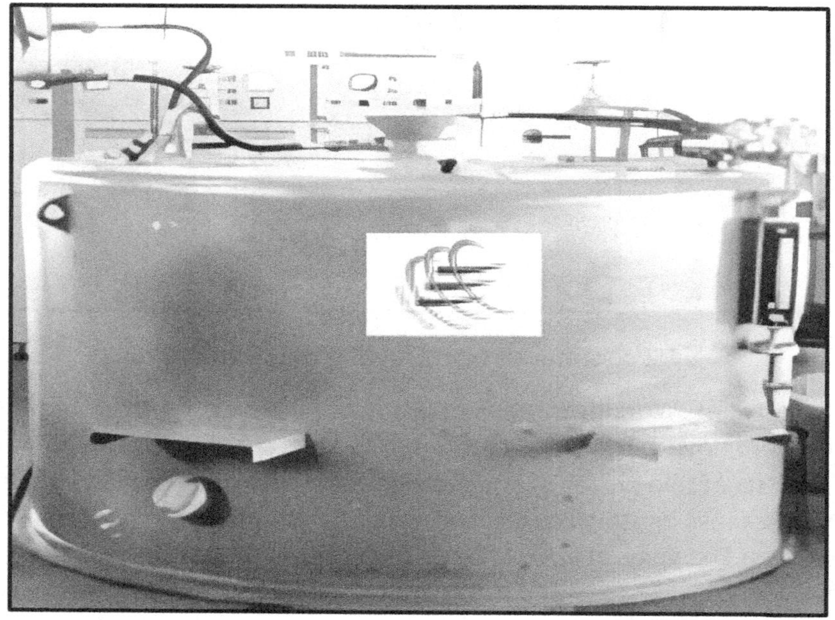

Fig. 64.1 continuous centrifugal

Central shaft with Bearing: A dynamically balanced forged shaft of IS 1875 standard with bearing housing which supported on inner housing is provided as a driving assembly.

The whole assembly of basket and bearing housing is supported on rubber buffers which take care of vibrations.

Driving unit: A motor mounted behind the machine is provided to drive the machines it is connected to the machine at the bottom by pulley with belts. Synthetic neoprene V belts are used. The power required is 60-90 kW.

Backing screen: A backing screen used is made up of stainless steel (ASAI 304/316/316Ti) wire of 1.6-mm dia. and is of 5 mesh.

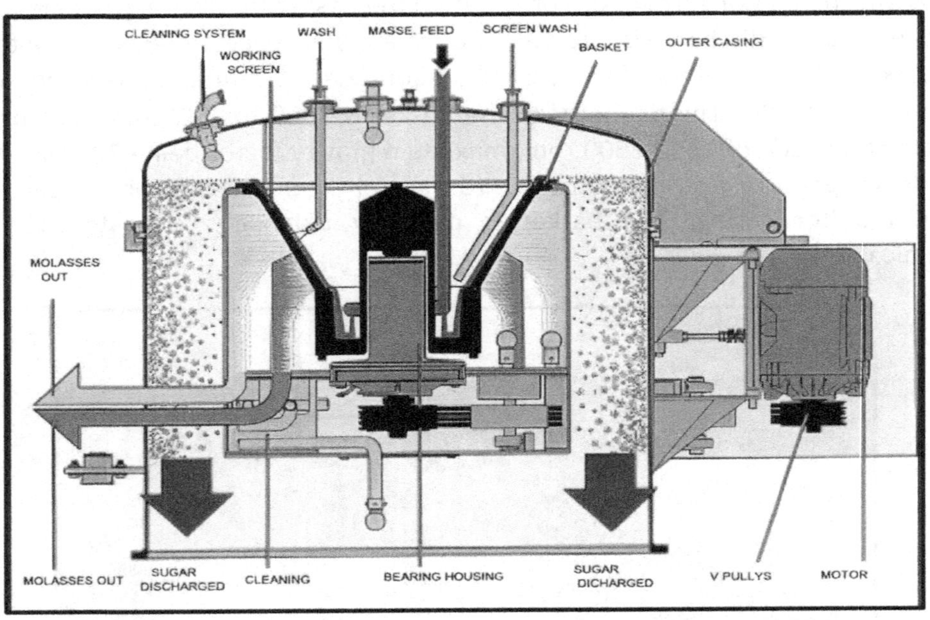

64.2 continuous centrifugal schematic diagram

Working screen: Working screens made up of pure nickel with chromium plating are generally employed. The nickel chromium screens are of 0.25-0.30 mm thickness. Each set consists of two pieces of equal weight with tolerance of ± 5 g. Brass screens of 0.7-mm thickness or stainless steel screens of 0.5 mm thickness are also used in some countries. The general aperture size of the screens is as follow:

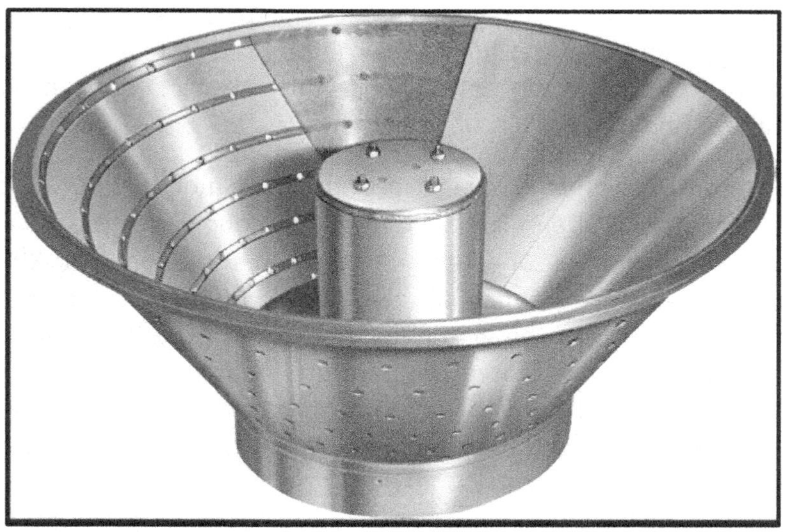

Fig. 64.3 basket of continuous centrifugal

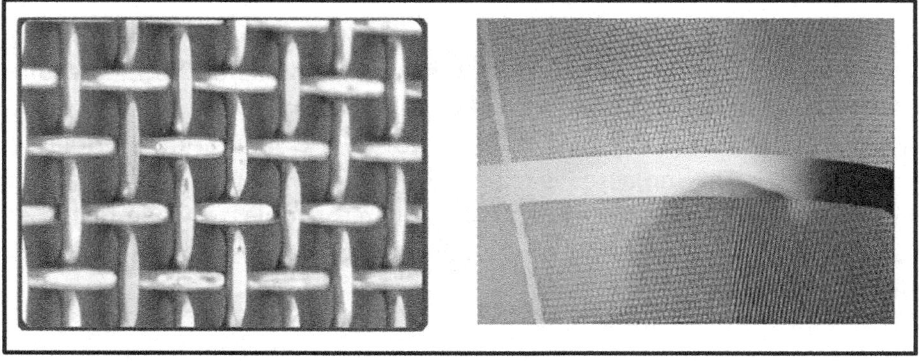

Fig. 64.4 backing and working screens of continuous centrifugals

Table 64.1 aperture sizes used

0.05 x 1.66 mm	for c-massecuite
0.06 x 1.66 mm	for c-massecuite
0.09 x 1.66 mm	for b-massecuite

A positive screen clamping arrangement is provided to the basket for proper fitting of working and backing screen.

Monitor casing: The monitor casing or outer casing made from MS plates is of large diameter. There is enough space in basket and casing so that there would not be any jamming of sugar. An inspection door and sampling device is provided to the casing.

Massecuite feeding: The massecuite feeding pipe passes at the centre bottom of the basket. It is jacketed so that hot water can be used to heat the massecuite. A disintegrator is provided at the bottom end which restricts entry of any foreign material or lumps on the screen.

Molasses outlet: The inner housing is provided with molasses outlet for removal of molasses. The massecuite of 50°-52°C is fed continuously at the middle bottom of the basket. Some hot water or diluted final molasses solution of 20-30°brix of 50-60°C is fed along with massecuite flow to work as a lubricant and to reduce the viscosity of the massecuite. The water quantity to be used is about 1.5-2.5% on weight basis. However the water temperature quantity is decided upon the massecuite quality and purity of final molasses. If the massecuite is of low purity and high viscosity temperature of the water can be increased up to 60-65°C. Steam can also applied at the feed point.

Wash water /diluted molasses can also be applied by a pipe provided with nozzles (1.2 mm size) inside the machine parallel to the screen surface about 50 mm from the screen.

Massecuite when reaches at the lower end of the feed pipe it is disintegrated evenly in the feed cup and passes over the working screen. The massecuite gets accelerated in the lower part and then flows upward over the surface of the working screen towards the rim of the basket. As it moves, the separation of molasses and sugar takes place. The molasses is expelled through the apertures of the working screen. The expelled out molasses gets collected in the molasses chamber. The sugar particles travel to the upper end of the basket and are thrown out into the sugar chamber.

In contrast to the batch machines continuous machines work with thin layer of massecuite. The thickness of layer of massecuite that is travelling over the working screen is only 3-5 mm at the lower end of the conical basket and is hardly 200-400 micron at the discharge end. The molasses faces least resistance while passing through this layer.

The massecuite feed can be controlled automatically according to the power consumption of the motor.

64.2 Different designs of continuous machines:

Magma mixing type centrifugal:

Sugar crystals from cont. C/F are usually mixed with syrup to produce magma for further processing for magma preparation, magma mix is used. In magma mixing C/F, syrup is pumped through jets (similar to water wash jets) in the top of centrifugal casing, where it mixes with sugar being discharged. Mixing process is helped by high speed motion of sugar as it leaves the lip of basket. Unlike standard crystal discharge continuous C/F machine, the sugar is discharged via pipe as magma. Advantages are saving in space and capital equipment, separate mixture, tanks, pumps.

Melting type centrifugal:

Further development of magma mixing C/F machine. In melting C/F sugar is totally dissolved inside the outer casing of C/F by addition of water/syrup for sugar to melt fully, it has to remain in contact with water syrup for suitable length of time. This is achieved by using internal trough in which magma is collected. Suitable retention is given in casing to dissolve sugar & the dissolved syrup sugar overflow out. Additional controls loops are provided to match syrup addition to the through put or brix of discharged syrup.

Double purged type continuous centrifugal machine.

Double purging is technique used to reduce circulation of impurities back to higher grade with minimizing loss of sugar in molasses. The design shows two different basket one above other. The first stage operates as magma mixing type while discharge magma is directly feeding to the lower second stage basket via pipe work inside the casing. High grade continuous centrifugals (HGCC) with low speed spins, steep basket angles and curved deflectors are developed to prevent crystal damage.

Table 64.2 Different Models Continuous Centrifugals:

Particulars	Centrifugal machines			
Dia. of basket (mm)	1100	1350	1500	1750
Basket inclination deg.	30/34	30/34	30/34	30/34
Basket speed (rpm)	1800/2000	1500/1800	1500/1680	1500/1680
Max. Gravity factor (G)	2100/2445	2050/2430	1605/ 1680	
Motor rating (kw)	55	90	100	120
Centre distance between centrifugals(mm)	2000	2500	2500	2700
Capacity t/hr. for low grade massecuite	6-8	10-12	12-15	15-18

64.3 Pros and cons continuous centrifugals:

Continuous operation is obviously ideal from the mechanical point of view i.e. machine runs at constant speed without stopping, is fed continuous stream of material and produces a constant output.

- Easy operation.
- no power peak loads

- Can work with high brix and low purity massecuite satisfactorily.
- Small size sugar crystals can be kept.
- Can tolerate little false grain.
- Maintenance cost low.
- Labour saving.
- No additional accessories required.
- less power consumption

Crystal breakage:

The main drawback of the continuous machine is that the percentage of crystal breakage is considerably high. This is because

- The crystals scratch while travelling over the working screen.
- The crystals when thrown out of the cone at high speed strike heavily on the inner surface of the monitor casing. To avoid this manufacturers have increased the diameter of the monitor casing. So that the crystals will have to travel more distance before striking and air resistance can bring down their speed so that crystal breakage will be reduced.
- Sugar drying in this machine is not up to the mark. so sometimes these machine held lumps

The working screens exhaust rapidly. The opening size increases from 60 microns to 80-100 microns in a run and the final molasses purity can be increased by 1-3 units. Therefore it is necessary to change the screen when the openings are increased.

Crystal take-off speed at upper edge of basket reaches up to 110-140 m/sec or 400 to 500 km/hr. Medium and large sugar crystals of size 0.8 - 1 mm with 20 - 30 m/sec velocity may break unto 20 - 30%, whereas smaller crystals of size 0.4 - 0.6 mm the danger of crystal breakage is not so severe.

The crystal breakage occurs in continuous centrifugals due to its very high peripheral speed due to high rpm. So continuous machines are not preferred for high grade massecuite as crystal size and shape is the important aspects of sugar in market.

-OOOOO-

65 SUGAR DRYING AND COOLING

65.1 Drying of sugar:

Moisture percentage of sugar is considered as one of the important characteristics of sugar quality. Sugar with higher moisture content on storage in godowns undergoes rapid degradation and colour is developed. The degradation is proportional to the moisture percentage and the bagging temperature of sugar. Therefore, drying and cooling of sugar before gradation and bagging is essential.

Drying of sugar is done to some extent in centrifugals and further on grass hopper conveyers /fluidized bed hopper or in rotary dryer.

Fig. 65.1 conveying and drying of sugar by grass hopper

The sugar is dried to some extent in the centrifugals by either steam wash or superheated water wash. With steam wash the sugar becomes somewhat hard and cake formation is observed. Therefore now in most of the factories superheated water wash is employed. With superheated water hardening of sugar does not observe and temperature is also lower.

Sugar discharged from centrifugals is of 65-80°C temperature containing 0.4 - 0.8% moisture. Because of high moisture lumps are formed. Sometimes dried lumps are very hard and difficult to break. Further higher the moisture percentage higher the growth of microbial organisms which leads to deterioration of sugar. Hence the sugar should be dried to a safe limit i.e. up to 0.04 % moisture.

The hot sugar is hygroscope in nature. If this hot sugar bagged directly it absorbs moisture from atmosphere during cooling operation and form lumps. Further as days pass sugar becomes pale yellow and then blackish. To avoid these draw backs sugar has to be cooled down to ambient temperature (35 - 40°C) before bagging.

65.2 Conveyers with drying and cooling arrangement:

The conveying, drying and cooling is done by various means as follow:

Multi-tray grass hoppers

- Fluidized bed hoppers
- Rotary dryers

Multi-tray Grass hopper:

It conveys the sugar from centrifugals to elevator and while conveying the sugar is dried by hot and cold air blowers. Thus the function of a grass hopper conveyor is twofold. The conveyor should be adequate width and length so that a thin layer of sugar is maintained for effective drying.

Usually there are three hoppers of 10-12m in length and 1.0-2.0 m width depending upon capacity of the factory.

- The first hopper placed under the centrifugals to receive the discharged sugar is single tray.
- The second hopper is a multi-tray hopper with an arrangement of hot air bower. Hot air (80-90°C) is blown through the trays of second hopper to dry the sugar up to 0.04% moisture.
- The third hopper is also a multi-tray hopper with cold air blower arrangement.

The trays are supported on flexible strips at 60° to the horizontal. Each grass hopper is driven by a 12-15 BHP TEFC electric motor. The hopper eccentricity is about 25-35 mm. The vibration is effected by means of an eccentric motion given by 250-300 rpm. Each hopper is provided with two connecting rods. The portion of the hopper where connecting rods are fixed is strengthened by a 10 mm thick and 1 meter wide MS plate. Because of eccentric motion the sugar is pushed forward and travels to the next hopper. While travelling the sugar is dried with hot air on the second hopper and is cooled on the third hopper. The sugar from the third hopper is fed to sugar elevator by chutes provided to the hopper at the end. The grass hoppers are generally made of galvanized trays.

The hoppers can handle sugar up to 0.4 MT/m2/hr. For a 5000 TCD plant three hoppers of sizes 12 m length X 2.0 m width are recommended. The hoppers cause minimum damage to sugar crystals. There is no erosion and lustre sugar is maintained.

Fluidized bed hoppers:

In some sugar factories fluidized bed hoppers are in use. The basic advantages of fluidized bed hoppers are efficient cooling and drying of sugar. Space is also saved.

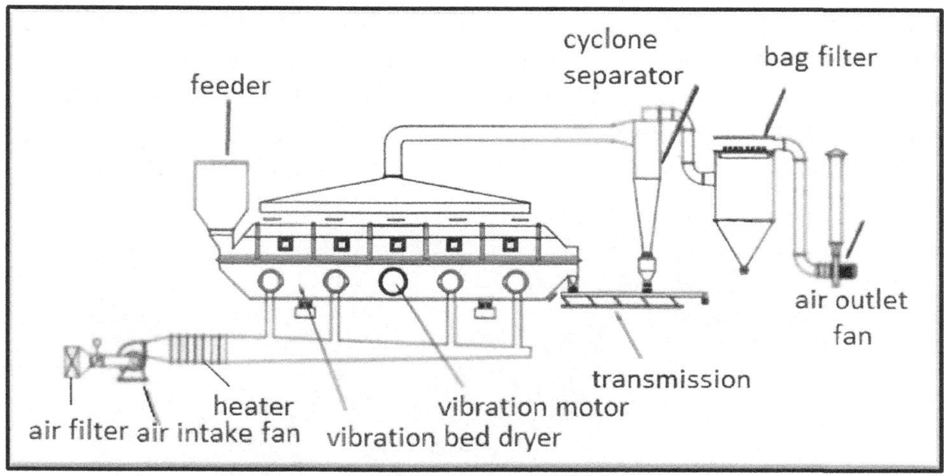

Fig. 65.2 fluidized bed hopper

In fluidized bed hoppers air is passed from below trough a bed of sugar resting on a distributor plate. The sugar crystals are fluidized and made to hover turbulently. Technical specifications and working conditions for fluidized bed hopper installed in a sugar mill are as follows

Table 65.1 fluidized bed hopper details

SR. NO.	PARTICULARS	
1	Overall size of the fluidized bed hopper	4600 mm (2) x 2000 mm (W) x 5225 mm (H)
2	Input sugar	Moist. % - not more than 1 %
		Temp. - not more than 85°C
		Crystal size - not less than 0.8 mm
		dust % - not more than 1
3	Output sugar	Moist. % - not more than 0.06%
		Temp. - 5-7°C above ambient
4	Drying media	Hot air generated by steam radiators With maximum temperature 110°C
5	Capacity	20 MT/hr.
6	Steam pressure used for radiator	5 kg /cm^2

However in many installations it is observed that grain breakage is increased while using fluidized bed hopper. Therefore in fluidized bed hoppers efficient dust catcher system is also to be provided.

Rotary dryers:

A typical rotary dryer consists of an inclined rotary drum of 2-3 m dia. and 7-8 m in length through which a current of hot air is passed. Inside the periphery 12-24 ladles sloping at an angle of 45° with the radius are fitted. The dryer revolves at 6-10 rph. The sugar enters at the upper end. The revolving motion of the drum causes the sugar to fall from one ladle to another until it reaches the discharge end of the drum. Sugar travels through the dryer in about 10-20 minutes. The sugar dust is carried out by air current and is collected in dust catcher of suitable design.

Fig. 65.3 Rotary dryer

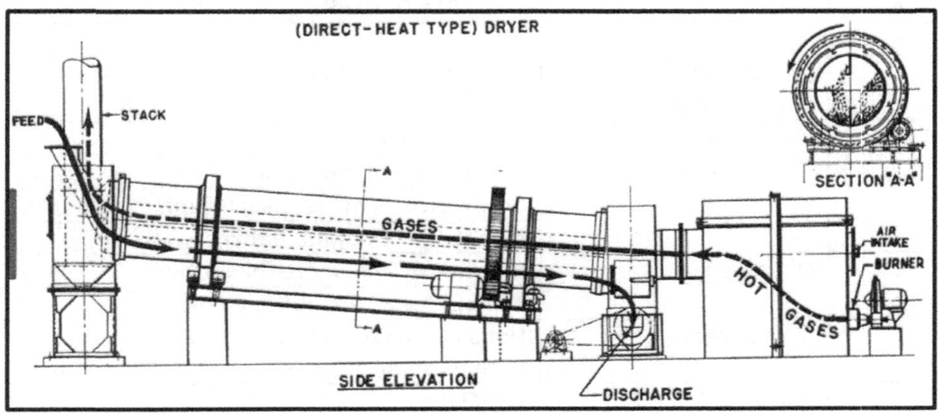

Fig. 65.4 rotary dryer

When the air is in counter current direction the sugar dust being carried by the hot air current may adhere to the moist sugar crystals that are entering the rotary dryer. This adversely affects the gloss of sugar. The hot air temperature should be controlled. Too hot air may heat the discharging sugar abnormally high and this may affect the lustre of sugar and also it delays cooling of sugar. Sugar of higher temperature may enter the silo or bagged which may cause caking problem.

65.3 Sugar Elevator:

Cooled and dried sugar is elevated by the sugar elevator from hopper to grader. The grader is consist of number of buckets fitted to at two parallel endless chains that are running on pulleys one at the bottom and one at the top. The top pulley is driven by a motor and reduction gear box. Distance between the buckets on the chain is 25 to30 Cm. Speed of chain is kept in between 15 to 24 m/min.

Fig. 65.5 sugar elevator

Fig. 65.6 bucket of sugar elevator

65.4 Sugar dust collection system:

Generally the dust collectors can be classified into two types:

- Cyclone separator (dry collection system)
- Particulate scrubbers (wet scrubbers)

Cyclone Separator:

The most widely used type of dust collection system is cyclone separator. This is probably the simplest and least expensive type of dust collector. In this system the dust laden air enters a conical chamber tangentially and leaves through a central opening. The sugar dust by virtue of inertia, move towards outside wall. Then the dust is lead into a receiver. A cyclone is essentially a settling chamber. It should have adequate retention time so that the sugar dust should reach the settling velocity and thus get separated from the air.

The limitation of cyclone separator is that it is less efficient with very small dust particles particularly with sizes below 30 micron.

Wet Scrubbers:

The wet scrubbers is a device in which a liquid generally water is used to assist the collection of dust particles. In wet scrubber the predominant mechanism is inertial deposition only.

Generally in a wet scrubber a film of liquid flow on the surface of the liquid level to avoid re-entrainment of particles and also help flushing away the deposited particles. However for dust particles like sugar this problem practically does not arise. Hence wet scrubbers are ideally situated for sugar this problem practically does not arise. Hence wet scrubbers are ideally situated for sugar dust collection.

Fig. 65.7 Dust collector

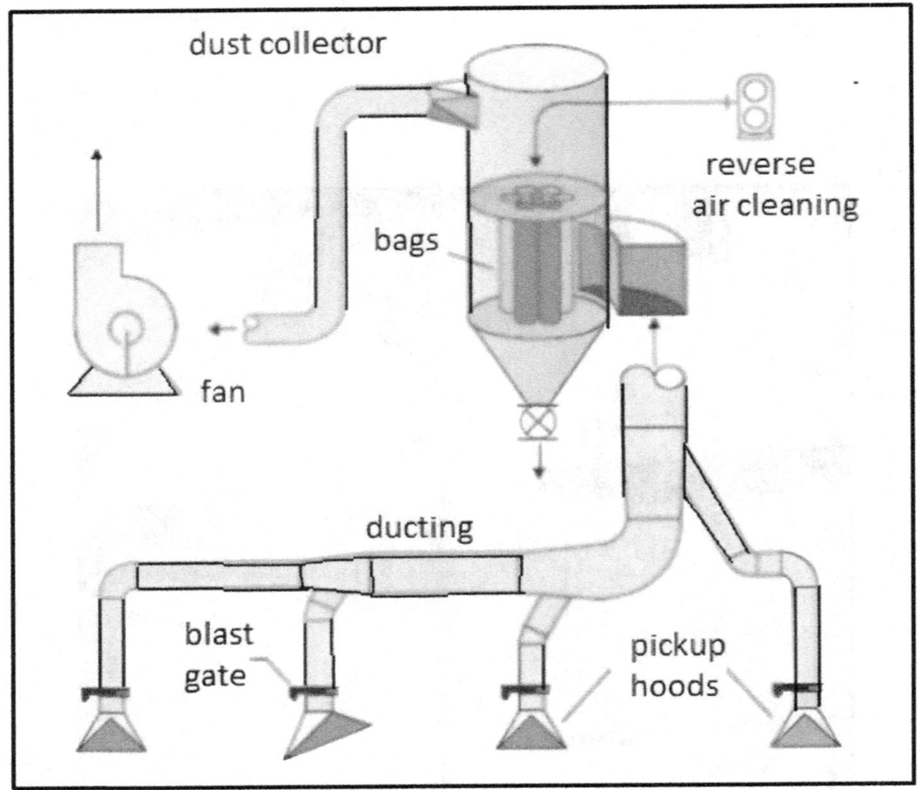

Fig. 65.8 Dust collector

-OOOOO-

66 SUGAR GRADATION AND BAGGING

66.1 Sugar Graders:

The sugar from grass hoppers is of varied in crystal sizes and needs to be well sieved and graded. The classification is achieved by using wire mesh through which sugar crystals smaller than screen aperture are mostly passed and larger crystals are retained on the surface. Vibrating type inclined graders are commonly used. Screens of different mesh size are fixed in frames arranged one over the other. The frames are vibrated at high speed by means of eccentric movement having a small throw 0.5 - 1.0 inch and are belt driven by a motor.

Fig 66.1 sugar grader

The vibrations prevent choking of screen and also keep holes in the mesh open. Angle of inclination to horizontal varies from 37-44. The screens exhausted torn out screens can be changed easily within short time

Table 66.1 Screen mesh sizes fitted to grader

mesh no.	Suitable for 80 % M, 20 % S	Suitable for 80 % S, 20 % M
1	8	8
2	10	14
3	14	20
4	20	24

Generally four decks of mesh/screens with chutes for collecting retained sugar are provided. Below the last lowest deck a chute is provided for collecting sugar dust. The rori and lumps are separated in first chute while the powder or dust sugar is collected in the last chute. In 2nd and 3rd and 4th chute different grade sugar is collected. Usually following combination of mesh size is used.

66.2 Sugar Silos:

Now in many sugar factories indoor silos - in which 12 hours sugar production can be stored, are installed for different grades of sugar in the sugar house. Sugar from the graders is lifted to silos. Weighing, packing and conveying of the sugar bags to godowns is generally carried in two shifts.

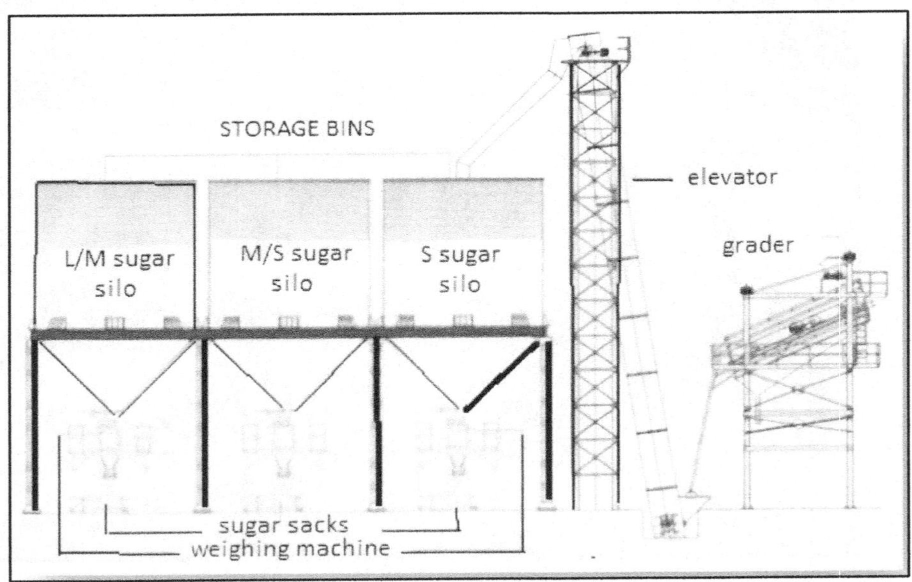

Fig. 66.2 Indoor sugar silos

-OOOOO-

67 COLOUR IN PROCESSING

67.1 Colour in the juice and colours developed during process:

The Coloured compound in the process can be broadly categorized into two main types:

1. Natural colorants in the cane plant that are extracted in juice extraction process. These Colorants are grouped as - Chlorophyll Anthocyanin Saccaretin, tannin, Flavanoids, Melanin, Xanthrophylls and Carotene etc.

2. During the process reactions takes place and highly coloured compounds are formed. The compound or groups of compounds are Melanoldins, Maillard Reaction, Caramels, HADPs (Hexoses Alkaline Degradation Product) etc.

The deterioration of cane starts from cutting of the cane. It causes degradation of sugars with formation of dextran. Process of enzymatic browning also starts cutting and it continues till the juice is heated and the enzymes are denatured due to heat. The formation of coloured substances is bound to happen as the material passes through the process. The following table shows the colour predominantly generated at different operations in the process.

Various soluble green coloured polyphenols derivatives like tannin are present in the tops and buds of the cane. These react with iron and give dark colour to the juice.

During the liming stage at high temperature and high pH flavonoids and saccaretin associated with fibre gives brewing effect and yellow colour is developed. Some quantity of bagacillo always comes along with the raw juice. However it becomes colourless in neutral or acidic medium. Therefore maximum removal of bagacillo from mixed juice by screening is good for process. Iron coloured complexes are also generated.

During Clarification sugar factories and in Carbonatation process refineries, when juice/melt are treated milk of lime. In highly alkaline conditions invert sugars, present in juice / melt, form highly coloured group of compounds which are called as HADP (Hexoses Alkaline Degradation Products). HADP are polymers of carboxylic acids, of brownish yellow colour. These compounds and melanoidins, caramels and phenolic compounds increase colour of the intermediate products in the process.

Some colouring compounds such as chlorophyll anthocyanin are removed in clarification and do not affect the process.

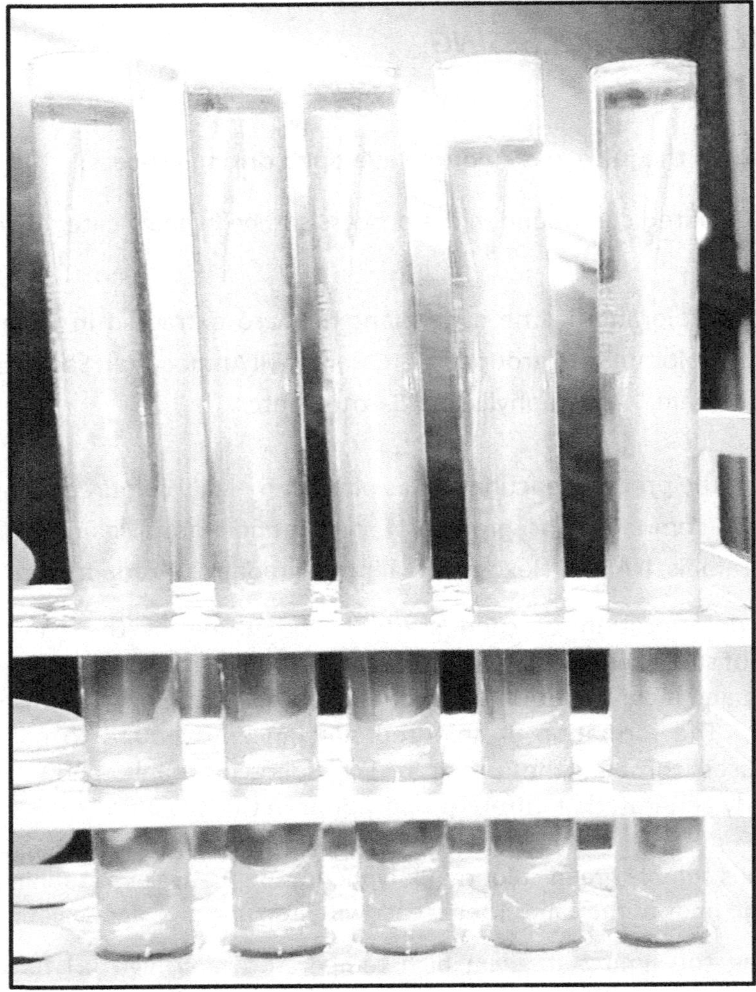

Fig. 67.1 clarified juice

Table 67.1 main gourps of compounds developing colours in process

stage	temperature	Main group of compounds generated that develop colour
Extraction	25-75°C	Enzymatic browning
Purification	30-100°C	Alkaline degradation of reducing sugars
Evaporation	80-118°C	Maillad's reaction and caramelization
Crystallization	40-70°C	Maillards reaction

During the evaporation the colour development can be attributed to the caramelization as a result of the high temperatures. This causes sucrose loss and the colouring matters developed pass into the final sugar crystals. Therefore residence time of juice in the first effect should be minimum and the temperature of juice should not be too high. Honing recommends maximum temperature 118°C, with a condition that the juice will not remain for more than two minutes at this temperature.

In clarification of cane syrup by Phosphaflotation mainly turbidity is removed to a large extent. But there is no significant removal of from colour of syrup.

Crystallization is the process where 40 to 70% percent of the total colour in the sugar is generated primarily by the Maillard reactions, the sugar colour is determined by the amount of syrup the crystals dragged, and this can be corrected in some extent during centrifugation and washing.

The most important practice for reducing the generation of colour in the vacuum pan is related to the efficiency of massecuites circulation that avoids overheating. The reactions can be minimized by boiling the vacuum pan at lower temperatures. Particularly with the C massecuites, temperatures should be maintained below 68°C.

At last, it is important to emphasize that throughout the process, the juices syrup and massecuites come in contact with iron equipment. The rusted iron causes increase in colour of the products.

Fig. 67.2 sugar

67.2 Colour values of products in process:

Colour measurement at various stations gives guidelines to control the process. The average range of colour values of boiling house products in different factories following double sulphitation process is given in the following table. This data is from the factories in Deccan plateau in India. The colours are by ICUMSA method no GS1-7. However the raw juice colour may vary considerably depending upon region, soil, variety maturity etc.

There is about 45-50 % reduction in colour from mixed juice to clear juice. Therefore the clarification process is of prime importance in the production of sugar of low colour value. Ten minutes retention time in juice sulphiter gives higher colour removal in clarification.

With higher retention in clarifier at lower crushing rate the percentage of colour removal is decreased. Hence retention time in clarifier is also equally important for efficient removal of colour at clarification station.

Table 67.2 average range of colour values in boiling house product

Sr. no.	Material	Average range of colour values I U
1	Mixed juice	18000-22000
2	Clear juice	8000-11000
3	Unsulphured syrup	9000-12000
4	sulphited syrup	8000-11000
5	A massecuite	6500-8500
6	Melt	4000-6000
7	A heavy	15000-18000
8	A light	6000-8000
9	Dry seed	140-180
10	B massecuite	16000-20000
11	B seed	1000-1200
12	C massecuite	25000-30000
13	CAW sugar	3000-3500

There is increase in colour by about 7- 10 % from clear juice to unsulphured syrup. With sulphitation up to 5.2 pH, the colour of syrup is decreased by about 15 – 20%.

The massecuite boiling time is also important in manufacturing sugar of low colour value. The boiling time should be less. With due precautions and control over the process, in sulphitation factories sugar of colour of 45 - 70 I U can be produced.

-OOOOO-

68 SPECIAL CARE IN RAW SUGAR MANUFATURE

While manufacturing raw sugar the process has to be handled carefully. Care has to be taken for cane quality and mill sanitation. Care also has to be taken in the process of clarification, and crystallisation and purging.

The stale cane may result in dextran formation due to bacteria. The higher dextran content in the raw sugar melt is responsible for lowering the rate of crystallization and also for elongation of crystals in refined sugar manufacture. Cane should be fresh and free from tops. Nitrogen compounds are higher in tops that pass with clear juice and get embedded in crystal and affects keeping quality of raw sugar.

The cane wax, starch and gums as well as dextran that pass in raw sugar affect the rate of filtration in refineries. However these are well within limits in Indian raw sugars and hence can easily be acceptable in International market.

Every unit at mills is kept clean to prevent the bacterial infection. Mill sanitation practices with optimum dose of quality chemicals should be followed properly to control the growth of microorganism and consequent formation of dextran.

Clarification:

Defecation process is followed for raw sugar manufacture. Screening of raw juice should be efficient. Cush-cush particles may pass up to pan boiling and work as nucleus for sugar crystal formation or remains in the adhering molasses as a place for growth of microorganisms.

Phosphoric acid addition should be adjusted to get P_2O_5 content of mixed juice around 300-350 ppm for good clarification results.

The liming of juice is done under perfect control. Generally hot liming is preferred. The pH of the limed should be so adjusted that clear juice pH should be about 6.9 to 7.2.

If required flocculating agent should be used so as to maintain the clear juice free from suspended matter and turbidity.

Pan boiling:

Three boiling scheme can be followed for raw sugar manufacture. In case B heavy is to be diverted for ethanol production then two massecuite boiling also can be followed. Little quantity of A Heavy molasses can be recirculated to A massecuite.

The size of raw sugar crystal should be between 0.6 mm to 1.0 mm, preferably

0.8 mm. All crystals should be uniform in shape and hard.

The purities of low grade should not be very low. Otherwise viscosity increases and that affects the quality of raw sugar because of poor centrifugation.

Enzymes like amylase can be used at syrup and B-Heavy molasses to control microorganism.

Curing:

Moisture in raw sugar is maintained by adjusting the curing time and application of superheated water wash so that free flowing washed raw sugar is produced. Now with high speed machines, desired moisture percentage of raw sugar can be easily maintained.

Fig. 68.1 Raw sugar

While curing A massecuite the drying of sugar should be appropriate so that it should be free flowing. Time cycle and application of superheated wash water timing are important in controlling pol and moisture of the raw sugar. Longer time cycle results stickiness of raw sugar. It may cause problems in bagging. Shorter time cycle will give high temperature of raw sugar which after cooling may absorb moisture and forms lumps.

Storage in godowns:

During storage raw sugar is affected mainly due to presence of microorganisms and their activity in the molasses film on the crystals. Raw sugar during storage develop colour very fast. It is better to keep the godown closed at all times.

Fig. 68.2 Heavy duty plastic jumbo bags of 1.0-1.5 t capacity for raw sugar

Raw Sugar standards:

Now-a-days generally high pol raw sugar is traded in international market. The range of specifications followed in International market are given here under.

Table 68.1 variation of raw sugar standards in international market

Particulars	Standard
Pol %	min. 98.3, max. 99.3
Moisture	0.10-0.20 % max
R.S. %	0.5 Max.
conductivity Ash %	0.1-0.3 Max.
Crystal size	0.6 to 1.2 mm
C.V.%	40 (Max.)
Safety factor max.	0.3
Starch	200 ppm (Max.)
Dextran	100 ppm (Max.)
ICUMSA	600 -1200 IU
Conductivity Ash %	0.01% max

Parameters of Raw sugar as per revised BIS 2020 Standards (5975:2020) are given here under

Table 68.2 raw sugar standard
Given by Bureau of Indian Standards (BIS)

Sr. No.	Characteristics	Requirements		
		LP	VHP	VVHP*
1	Polarization, °Z, Min	96.5	98.0	99.0
2	Reducing sugars % by mass Max	1.0	0.8	0.6
3	Sulphated ash % by mass Max	0.8	0.6	0.5
4	Safety factor, Min	0.3	0.3	0.3
5	Crystal size MA, Min	0.8	0.8	0.8
	CV Max	40	40	40
6	Colour (IU) Max	3000	1500	650
7	Starch mg/kg, Max	200	150	100
8	Dextran mg/kg, Max	150	125	75
9	Sulphur dioxide mg/kg, Max	10	10	10

Now there are many refineries having no affixation station and such refineries buy only very high quality (VHQ), very high pol (VHP) or very low colour (VLC) raw.

-OOOOO-

69 RIFINING OF RAW SUGAR

69.1 General:

Demand for food products that contain minimum chemicals (as per FDA norms) is increasing worldwide. The plantation white sugar manufacturing involves double sulphitation and sugar produced by this process contains 30-60 ppm of SO_2 whereas refined sugar is sulphur-free white sugar. Now consumers in most of the developed and developing countries are becoming health conscious and do not want sugar with sulphur dioxide. Refined sugar is most pure white sugar. It is low in ash and has a better keeping quality.

In manufacturing of refined sugar there are many advantages over plantation white sugar manufacture by double sulphitation process. There is 100% saving in sulphur cost and 35% saving in lime cost. However cost of phosphoric acid and colour precipitant is increased. The total cost of chemicals is not much increased. There is much reduction in scale formation at juice heaters, evaporators and pans. The scale is much soft and cleaning of the equipment becomes very easy. Reduction of Corrosion of equipment is considerably reduced. Therefore reduction of maintenance cost of equipment is reduced. There is no pH drop of injection water. Therefore maintaining pH of injection water pH by addition of lime is not needed.

There are two types of cane sugar refineries

i) Refineries operating in conjunction with raw sugar factory
ii) Independent refineries, which may purchase raw sugar

Many sugar mills only operate during the harvest season, whereas refineries may work round the year. The raw sugar is stored in large warehouses and then transported into the sugar refinery by means of conveyor or transport belts.

Refined sugar plants are ranging in capacity of refined sugar output from 100 TPD 2500 TPD output

There are various operations in the refined sugar manufacture the flow chart is given hereunder.

Refineries can also be classified according to types of refined sugar they produce.

1. Refineries that produce granulated sugar.
2. Refineries that produce granulated sugar as well as liquid sugar.
3. Refineries that produce liquid sugar only.

The process may differ from refinery to refinery. One or the other step is either added or omitted. Liquid sugars contain various proportions of sucrose and invert sugars. 'Medium invert' liquid sugar which is most widely traded in USA contains 76-78 % solids with 50 % sucrose and 50 % invert sugars. In the refineries which produce only liquid sugar there is no pan boiling station.

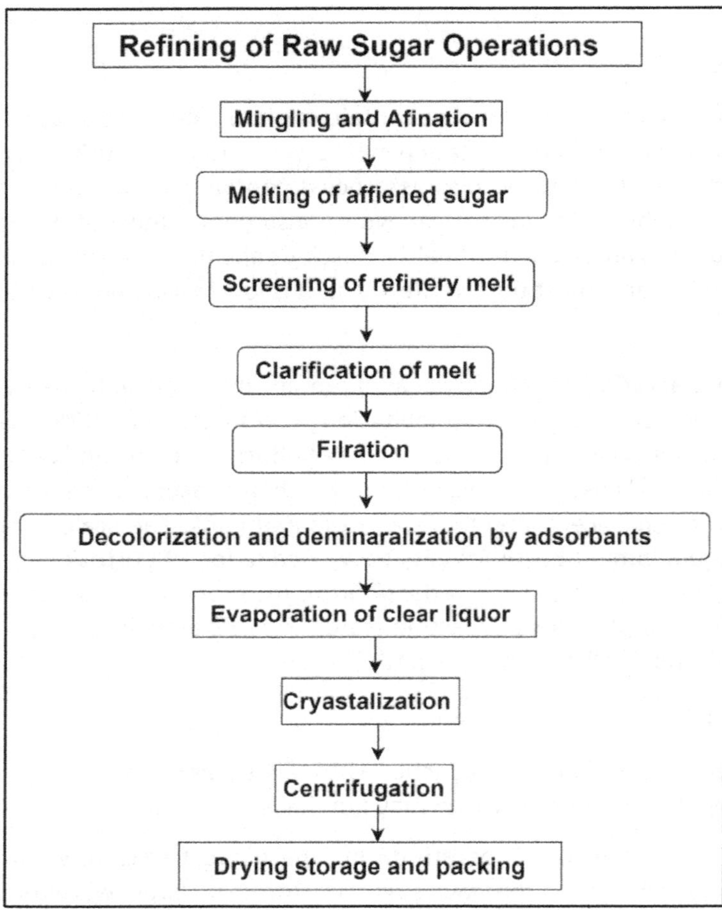

Fig. 69.1 Refinery flow chart

69.2 Mingling and Affination:

In a standalone sugar refinery, raw sugar will arrive in bulk by barge, rail or truck for unloading into the raw sugar warehouse. The raw sugar will undergo metal and debris removal and large lumps will be crushed.

The raw sugar is first weighed through the Duplex weighing machine before going for affination or melting. Some refineries process raw sugar of about 97.5 pol and colour of 1500-3500 ICUMSA units. In such refineries affination is the first step in the process. Magma of raw sugar is made and it is purged in centrifugals. This process is called as 'mingling and affination'. The main aim of affination process is to obtain affined sugar over 99% purity containing minimum colour and ash.

The sugar crystals are covered with a thin film of molasses has an average purity of about 63-65 and contain higher impurities and colouring matter. This film of molasses is removed from the surface of raw sugar crystal by purging.

Raw sugar is mingled with runoff from R3 / R4 massecuite (skips) or sweet water from filtrates in a mingler to form magma. The runoff /sweet water temperature is maintained at about 72° C before using it for mingling. The temperature of the affination magma is maintained to 42-43°C and the brix is maintained close to 93°. So that maximum colour and non-sugars can be removed in purging. The run-off softens the adhering molasses.

Raw sugar of larger and uniform size of crystals can be easily affined. However if raw sugar is of small crystals, large amount of molasses is adhered to them and separation of adhering molasses becomes little difficult.

Affination centrifugals are fully automatic batch centrifugals. When molasses coating of the sugar crystal is removed in affination about 50-65 % of the colour is removed from the original raw sugar and the purity of affined raw sugar is increased to about 99.0 – 99.5. Thus, in affination process non sugars are removed by mechanical means at a cheaper cost.

Table 69.1 Effect of Affination:

Sr. No.	Constituent	% in raw sugar	% in affined sugar
1	Sucrose	97.5	99.2
2	Invert sugars	0.50	0.17
3	Ash	0.45	0.15
4	Organic matter	0.6	0.16
5	Colour	1500-3500	700-1200

The brix of runoff / affined green syrup is maintained to 75, however when it is recirculated and when the purity goes below 84-85, it is discarded from affination and is send to raw sugar house for recrystallization. Even though the spent affined syrup has relatively high purity, its crystallising power is low because it carries large amount of Coloured impurities as it is boiled previously several times.

One method of using this spent green syrup is to keep this in separate storage tanks in raw house and use it for topping 'A' raw massecuite. A second method the green syrup can be treated with lime and phosphoric acid and clarified by air floatation, filtered and boiled to produce soft sugar or bura sugar.

In the refineries where VHQ (very high quality), VHP (very high pol), or VLC (very low colour) raw sugar is used, affination is not required. Many raw sugar factories, which refine their own raw sugar have abandoned affination and wash their raw sugar in their centrifugals in raw house to get purity above 99.25.

69.3 Melting and Screening of refinery melt:

The affined raw sugar or high quality raw sugar (VHQ, VHP or VHQ) of about 99.0 to 99.60 pol is melted in a sugar melter to 60-62 brix in condensate /sweet water. The temperature of the melt is maintained to about 75°C. Retention time of melt in melter is 25 to 30 minutes. Now continuous horizontal melters with multiple compartments have been installed in many modern refineries.

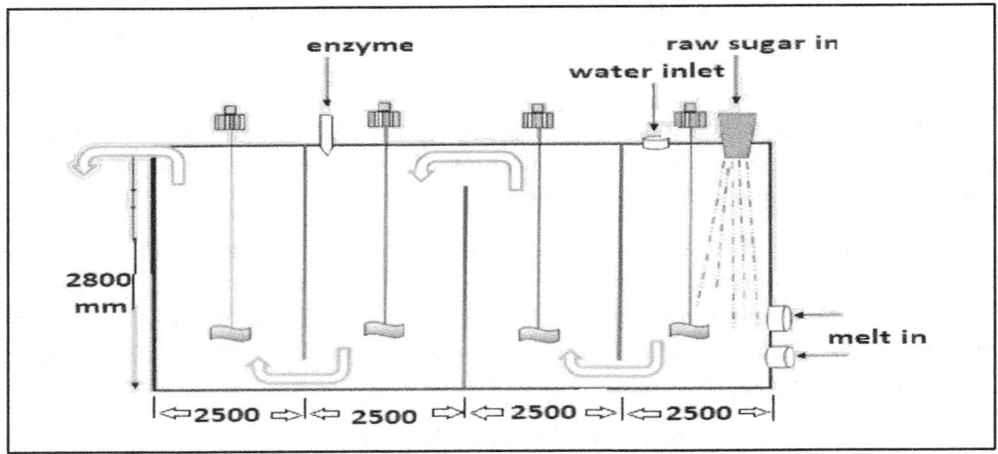

Fig.69.2 Sugar melter

Sometimes enzymes like Dextranase and /or amylase are added for degradation of dextran and/or starch in this sugar melter.

The sugar liquid / melt is then screened through a rectangular vibrating nickel metal 80-mesh screen to remove strings, jute twin, lint and other coarse suspended material. Screening of melt decreases load on 'Continuous Flotation Clarifier'. The screening of melt also reduces requirement chemicals for the clarification process. The Dorr-Oliver 300-DSM screen is also used for fine separations of solids.

69.4 Melt Clarification:

After screening the refinery melt is clarified. In clarification process some impurities and coloured compounds or colorants of the melt are removed.

Colorants in melt:

Sugar colorants are a very complex mixture of organic compounds. They can be broadly categorized into two main types;

1. Natural colorants resulting from the cane plant - Natural Colorants are - Flavanoids, Melanin, Chlorophylls, Xanthrophylls and Carotene.
2. Colours developed during juice processing - Melanoldins, Maillard Reaction, Caramels, HADPs (invert alkaline degradation product).

The melt is clarified either by phosphatation process or Carbonatation process.

Phosphatation process:

The phosphaflotation or phosphatation process works on the principle of forming low density aggregates of precipitated particles adhering to air bubbles that float on the melt.

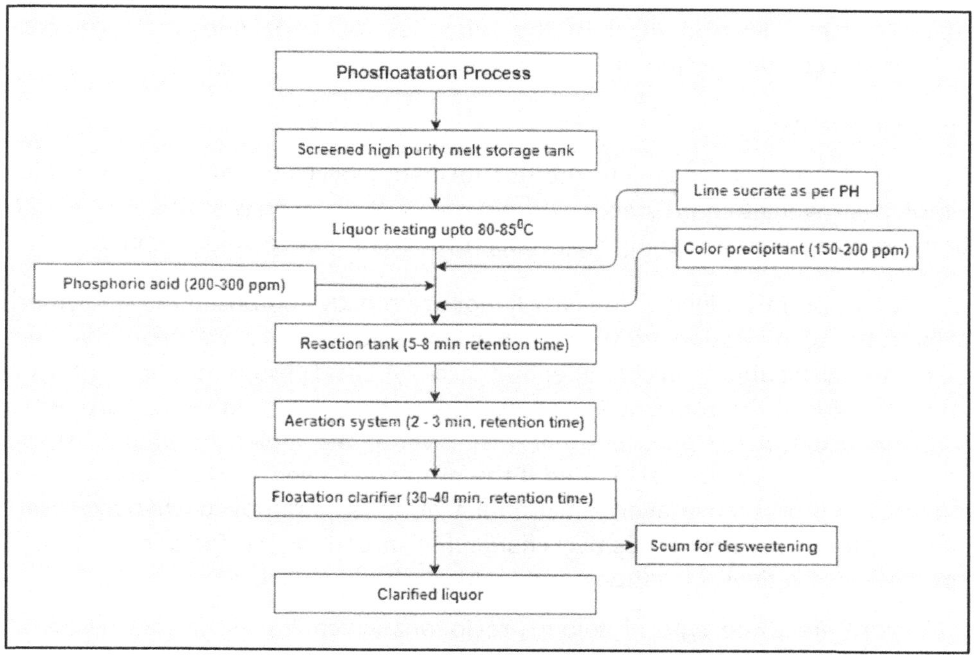

69.3 Phosphatation process of clarification of refinery melt

The screened melt of 70°C is treated with phosphoric acid. About 150-200 ppm of P_2O_5 on solids is required. The phosphoric acid is well mixed with melt in 7 to 10 minutes. Then lime sucrate is added to the melt to get a pH of about 7.2-7.4. Flocculent is added for rapid floc formation. Nowadays colour precipitants are also added.

After the lime addition liquor is heated to 85-90°C in tubular heater or direct contact heater and then is send to reaction tank. 20-30 minutes are given for floc formation. The phosphate lime treatment with heat first neutralises the organic acids present in the melt. Reaction between lime and phosphoric acid takes place and tricalcium phosphate is formed that occludes the colloidal impurities.

$$3Ca(OH)_2 + 2H_3PO_4 = Ca_3(PO_4)_2 + 6H_2O$$

The polyphenols of iron compounds which impart brown green colour are precipitated by phosphatation treatment. The majority of colouring matters and colloids by nature are negatively charged and the addition of cations such as calcium neutralizes these charges. This results in precipitation, flocculation and coagulation of organic non sugars. The anionic colour bodies are therefore effectively removed in phosphatation process. Some soluble coloured compounds are absorbed by the tri-calcium phosphate.

Heavy flocculent participate is formed that entraps suspended material from the liquor together with particles of bagacillo, clay etc., that are escaped through the screening. The precipitate occludes most of colloidal matter and some amount of colouring matter. Now in most of the refineries following phosphatation process flocculants and colour precipitants are being used.

Colour precipitants:

The precipitants coagulate the floc formed. Flocculants are polyacrylamide of high molecular weight co-polymers of sodium acrylate and acryl amide. It is very long linear molecular chain containing high concentration of mobile anionic charges.

The colorant molecules in sugar liquor are mostly composed of depolymerised carbohydrate fragments and are strongly hydrated and extremely water-soluble. Almost all the colorants are anionic in character and carboxylic acid groups may be located down the entire length of the colorant molecules. The layer of hydration water around each colorant molecule makes absorption and removal of the molecule difficult. However colorant molecules can be precipitated if the protective layer of hydration water could be removed. This protective layer of hydration water can be removed when hydrocarbon derivative carrying a strong positive charge (i.e. a cationic surfactant) is used. The hydrocarbon derivatives are cationic and hydrophobic in nature.

There are three type of colour precipitant which is used in melt clarification system - Dimethyl ditallow ammonium chloride (Talofloc), a waxy material developed by Bennett at Tate & Lyle Ltd. Talofloc is a semisolid at ambient temperatures and a solid under cold conditions. *Hence,* it is required to heat for liquefied the chemical and then to use it.

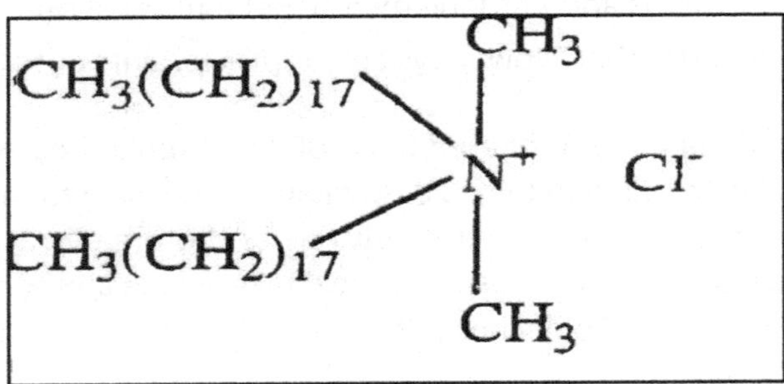

Fig. 69.4 Chemical formula of Talofloc precipitant

The 2nd colour precipitant is developed by American Cyanamid is an epichlorohydrin-methylamine Copolymer. This substance is prepared as 50 % aqueous solution and hence it is more easily handled than Talofloc.

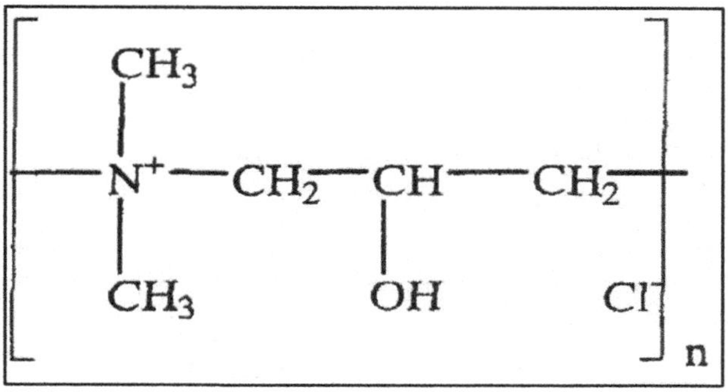

$$\left[-\overset{CH_3}{\underset{CH_3}{N^{+}}}-CH_2-\underset{OH}{CH}-CH_2- \quad Cl^{-} \right]_n$$

Fig. 69.5 Chemical formula of a copolymer precipitant

The 3rd colour precipitant is a cyclic polymer and is known as Dialkyldimethyl ammonium chloride. The trade name is Talomel.

Now some other manufacturers have also developed similar chemicals and are marketed under different trade names.

Aeration of melt:

The heated and treated melt is then send to air absorption tank in which air is dissolved under compression at 7-10 kg/cm^2. In some refineries only some portion of the melt is send though air saturation and then it is send the main feed inlet of the clarifier.

The micro air bubbles occlude in the calcium phosphate and the flocks become lighter density. Maximum possible aeration is needed. The melt then is send to floatation clarifier where floc is carried upward by the buoyant air bubbles and forms a layer of scum on the top of the sugar liquor. Retention time of treated melt in a continuous flotation clarifier is generally 45 minutes. Residual phosphate in the clarified liquor is measured to determine the completion of reaction.

Clarifier:

Construction of clarifier and features adopted in the design are unique. This ensures uniform and undisturbed melt flow in the clarifier. Clarifier is a rectangular or cylindrical steam jacketed or heavily lagged tank.

The treated melt is fed to the clarifier from the bottom. A portion of the treated melt is send though a specially designed air saturation tank, wherein the liquor is thoroughly mixed with compressed air and then it is send to the main feed inlet of the clarifier.

69.6 Refinery clarifier for phosphatation process

Clarifier has slow moving skimming blades or scrapping paddles at the top level of the liquor. In clarifier scum is formed in the liquor and is carried upward by expanding air. The upper part of the clarifier is open and a blanket of scum on the top of the liquor acts as a heat insulator and also delays to escape air into the atmosphere. The volume of scum is 5-10% of the volume of the liquor. The scum is scrapped by continuously moving paddles and. discharged into a gutter provided on the periphery of the clarifier.

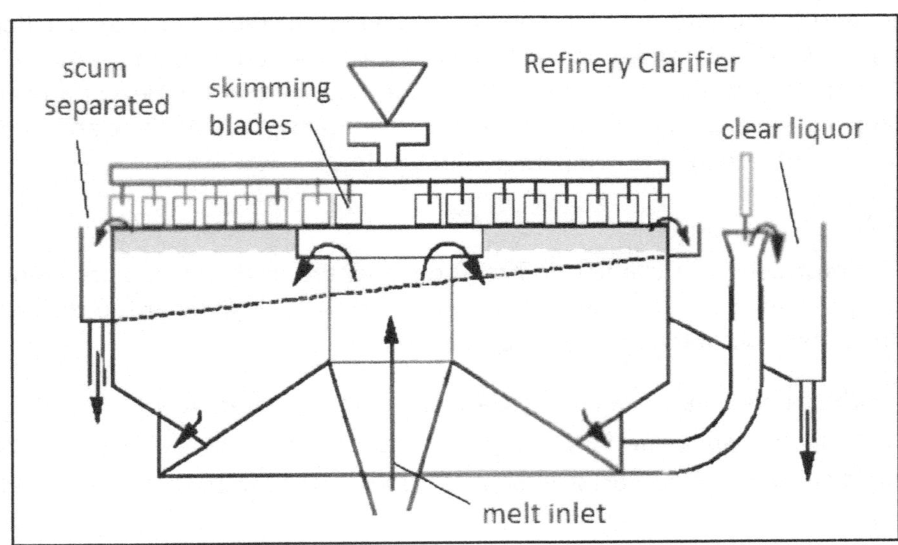

69.7 Design of refinery clarifier for phosphatation process

Some precipitated fine particles do not float above in the clarifier and pass along with clarified liquor. These fine suspended impurities are to be removed by filtration.

The clear liquor underflow from the clarifier is led to one or two filtration processes where any carry over are removed. The clarified liquor is discharged in a weir box and is send for further process. The melt /scum level in the clarifier can be controlled in weir box.

The precipitate that floats above is scraped / skimmed off as a scum. The scum is de-sweetened in several ways, the most popular of which is a series of two or three counter current clarifiers.

In clarification colour removal is about 30-50 %. The melt clarification system is provided with complete automation to regulate pH, temperature, melt flow and dosing proportional to melt solids. The melt clarifier plays an important role to ensure the throughput of the refinery plant and also in downstream process.

Melt filtration:

The clarified liquor from the melt clarification system still contains little suspended solids which may blind the pores of de-colorization resins. Deep bed filters are used remove the fine suspended solids. The deep bed filtration system removes suspended solids from clarified melt. Then liquor is send through ion exchange columns. Periodically the trapped solids from the DBF are flushed out to clean the filter media.

De-sweetening of scum

This is optional for back-end refinery, but is required for stand-alone refinery. In case of standalone refinery, the scum from scum tank is sent to de-sweetening system. It is diluted to 20° Brix in a tank is reheated and then settled in another continuous clarifier. Then it is filtered with addition of filter aid. Bowl type centrifugals also can be used for desweetening the scum. The low brix filtrate is used for affination of raw sugar, diluting refinery run off and /or for preparation of lime saccharate.

Tate and Lyle Company has improved the phosphatation process by optimum use of flocculants, colour precipitants and proper automation of the process. This improved process is being called as Talofloc process.

Carbonatation Process:

Carbonatation process for clarification of refinery melt is being used for about 140 years. This is best clarification process. It is a robust and can take variations in process parameters, however plant investment is more.

In this process lime and Carbon Dioxide are added in the sugar liquor.

$$Ca(OH)_2 + CO_2 = CaCO_3 + H_2O$$

A voluminous and gelatinous precipitate of calcium carbonate is formed by reaction of calcium hydroxide and carbon dioxide. The process of crystallization of calcium carbonate is also equally important.

The calcium carbonate precipitates adsorb the impurities, polysaccharides, colouring matters and ash present in the sugar liquor. A mass of calcium carbonate crystals of the right size and optimum size distribution facilitate the filtration.

Less soluble salts such as sulphates, starch, wax, gums and many other impurities are entrapped by the calcium carbonate. The impurities adsorbed and entrapped are then removed in the succeeding filtration. The precipitated carbonate crystals act as a filter aid at filtration process. The pH and temperature of the melt are important for efficient precipitation and removal of inorganic constituents. Destruction of reducing sugars and filterability of the carbonate cake are also affected by pH and temp conditions.

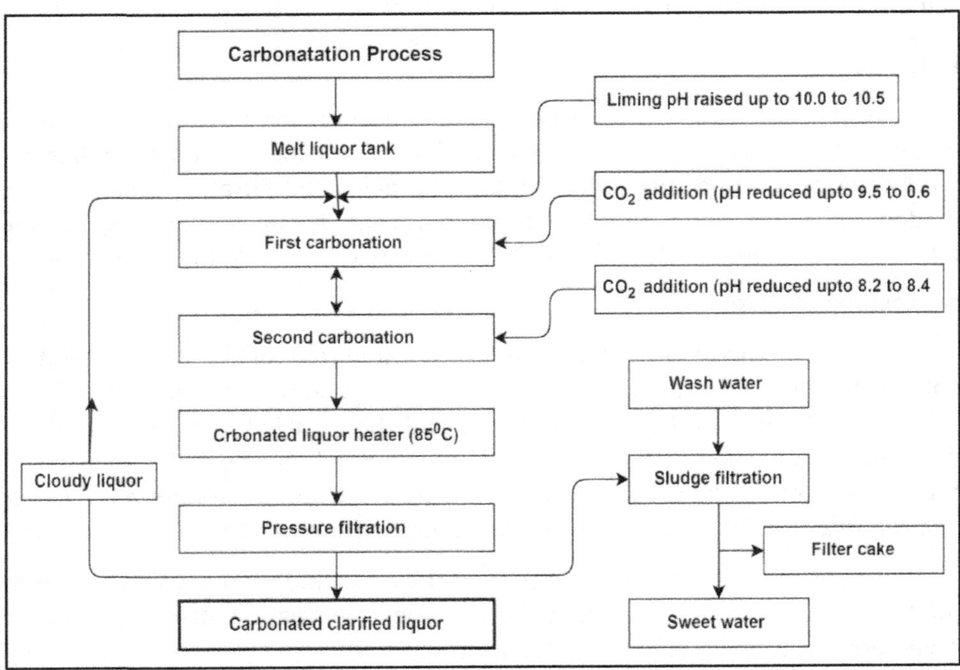

Fig. 69.8 Carbonatation process

The screened melt of 65-67 brix is pumped at a controlled flow rate through a heat exchanger to maintain the temperature of about 70° C. Then it is sent to a mixing tank where the liquor is mixed with hot milk of lime (or sucrate) of 20 Brix. The addition of milk of lime is about 0.5-1.2 % CaO on liquor solids. The lime must be well mixed with the liquor. The retention time of the liquor in the tank is only one to two minutes to avoid destruction of reducing sugars. Then the limed liquor is passed to the first saturation tank in which the temperature is raised to 80-85°C and carbon dioxide gas is introduced.

Fig. 69.9 Carbonatation plant

The CO_2 is evenly distributed across the cross section of the tank through pipe containing a large number of holes. Proper and effective distribution of CO_2 gas is important for better precipitation.

In first carbonator pH drops from 10.6 to 9.6 and more than 90 % of the added lime is precipitated as Calcium carbonate.

Then the liquor is passed into the second saturator, which operates at pH 8.0-8.3. Here the remaining lime is precipitated. A voluminous crystalline calcium carbonate precipitate entraps much of the insoluble and semi-colloidal matter as well as some ash and coloured compounds. Carbonation can be followed in two or three stages. Soluble bicarbonates are formed below pH 8.0, which not only increase the ash but is also detrimental to the filtration. Retention time in all saturators is one to one hour fifteen minutes.

Boiler flue gas is used as source of carbon dioxide required for the process. The gas drawn from the boiler chimney base is washed and cooled with water in scrubber. Where fuel containing sulphur is used the gas is washed second time with soda to remove corrosive sulphur compounds and dust. The concentration of carbon dioxide gas in the scrubbed flue gas is about 12-14% and temperature is about 70-80° C. The gas is then compressed and pumped to saturation tanks. In standalone refineries with gas fired boilers quality of flue gas is good and purification process is not required. It can be directly injected to the carbonators.

Table 69.2 Removal of Impurities during Carbonation

Sr. No.	Particulars	Removal %
1	Colour	55-60
2	Turbidity	85-95
3	Starch	90-95
4	Gums	25-35
5	Sulphates	85-90
6	Phosphates	95-100
7	Magnesium	65-70

The carbonated liquor from the second saturation tank passes to a receiving tank equipped with stirrer from where it is pumped to Carbonatation filters.

In carbonation colour reduction is about 40 % to 60%. The ash content also is removed by about 20-25%.

The major disadvantage of this process is that the high pH levels lead to a certain degree of destruction of invert sugars (monosaccharides). The breakdown products are generally organic acids, which require more lime in neutralization and increase the ash in the final molasses, thereby increasing the loss of sugar in molasses.

Table 69.3 Percentage Removal of Nonsugars By phosphatation and Carbonatation Processes

Sr. No.	Non sugars removal percentage	Phosphatation	Carbonation
1	Colour	30-40	50-60
2	Turbidity	85-90	90-97
3	Starch	90-95	90-95
4	Polysaccharides	60-65	88-94
5	Sulphate	25-30	84-88
6	Phosphate	85-90	95-100
7	Magnesium	30-40	65-70

69.5 Filtration:

In phosphatation process the clarified liquor is filtered as well the scum from the clarifier is filtered. In case of Carbonatation process the whole treated liquor is filtered through filters. The precipitate is retained on the filter cloth and clear liquor passes through. There are different types of filters used in industry. The more common are horizontal / vertical leaf filters, candle filters, multi-bed filters.

Leaf filter consists of number of filter elements (leaves), each composed of a metal frame with a backing screen over which nylon, polypropylene, stainless steel or Monel metal cloth is fitted. Leaves generally have 3.8 – 5.1 mm spacing. The filter elements are fitted in a shell that can work up to 3.5 to 4.5 kg/cms pressure. In carbonation process precipitated calcium carbonate itself acts as filter aid.

The candle filters are also used for phosphatation process. In phosphatation process filter aid like diatomaceous earth (Diatomaceous Earth Powder is also known as D.E, diatomite, or Kieselguhr) or other fibrous material is essential for pre-coating of the filtering surface.

The filter aid is used at filter station in two step-

- First, a thin layer of filter aid, called a pre-coat, is built up on the filter septum by recirculation of filter aid slurry through the filter.
- Second, a small amount of filter aid is regularly added to the treated liquid to be filtered. In filtration the filter aid from the unfiltered liquid, is deposited on the pre-coat. The minute filter aid particles entrap suspended impurities but allow clear liquid to pass through without clogging.

The liquor is mixed with filter aid. The temperature of liquor is maintained at about 82 – 85°C for better filterability. Carbonated liquor is pumped to each filter and forced under pump pressure to pass through the cloth which retains the precipitates and allows the clear liquor to flow through. At first cloudy filtrate is received. It is sent back to the lime melt. The flow rate remains about 30 -50 m^3/hr. Filterability of liquor depends on crystals growth of calcium carbonate in the precipitate.

When more cake thickness is developed on the filter elements, the operating pressure is reached from 2 kg/cms to 4.2 kg/cms and filtration rate is dropped. Then the filter is stopped for cleaning the cake formed over the leaves is discharged from the filter by sluicing the cake with water. Nozzles in between the leaves above each leaf are provided for this purpose. The cake is diluted and again filtered through filter presses or another filter for sweetening off. In some refineries *in situ* sweetening off procedure is followed. The total operation cycle of the filters is generally of two and half hours.

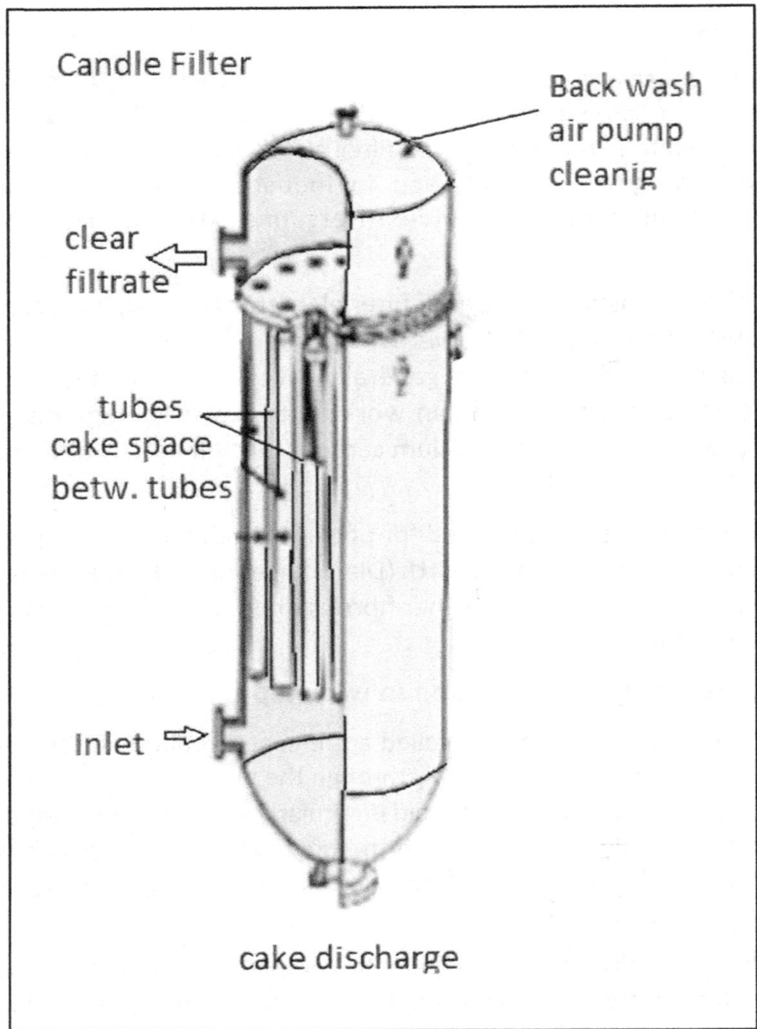

Fig. 69.10 candle filter

Periodic cleaning of cloth with sulphamic acid with inhibitor is generally followed.

In case of refineries attached to raw sugar factories the scum is sent to raw juice receiving tank.

The sweet-water obtained from sweetening-off filter cake is stored in a sweet water tank and is used for melting of washed sugar and for diluting refinery run-offs.

Installed capacity of filters is designed so as to handle worst quality of raw sugar liquor. Considering the minimum filterability of raw sugar, filter area requirement of 15

m^2 filtering area per tonne of melt per hour is considered to be standard. On volumetric basis, filtration rate of 0.1 m^3/ m^2 / hour is acceptable.

In candle filter high throughput rate is maintained. The operation can be fully automated by PLC system. The candle / filter cloth can be exchanged faster. No additional energy is required for cleaning of the filters.

Plate and frame type Membrane filter press:

Now modern plate and frame type membrane filters are being used for filtration of carbonated liquor and desweetening of carbonated cake. These filter presses do not require any pre-coat. The filter press operates at pressure of about 5.0 to 6.0 kg/cm^2 and good filtration is achieved even for high viscous liquor at low temperature. The filtration rate maintained is also high in comparison with conventional filters. The dry cake discharged contains about 30 % moisture.

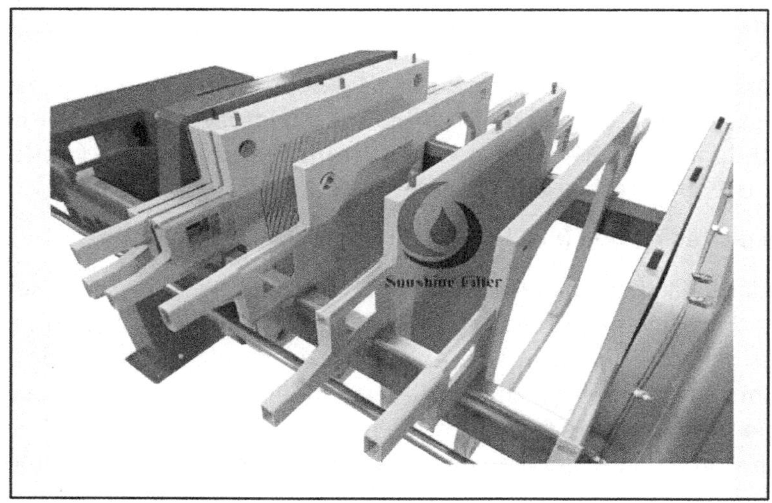

69.11 Filter Press

69.6 Decolourisation and demineralisation by adsorbents:

The clarified liquor, which still contains some colour, is treated with either of the following adsorbents.

Bone char:

In the past a material called 'bone char' was used extensively to remove colour from melt in the refining process. Modern technology has largely replaced bone char decolourization but it is still used in a few refineries. Where bone char is still used, it is prepared by incinerating animal bones to leave activated carbon. The bone can remove colouring matter as well as ash. Bone char can be regenerated 100 times.

Granular Activated Carbon (GAC):

Granular Activated Carbon (GAC) is technically preferred as the adsorbent for removing colour after carbonation. The highly porous structure of granular activated carbon gives it excellent binding capabilities and bind different impurities. GAC is highly effective for decolourization. GAC can also remove odour of the liquor. GAC also has the added benefit of reacting with certain compounds to reduce toxicity.

However it is unable to remove ash and auxiliary ash removal system becomes necessary. In GAC system the main disadvantage is the high sucrose loss associated with the process. The sucrose loss will be around 0.04%.

GAC is generally used when liquid sugar is directly produced from decolourised liquor without crystallisation.

In GAC system the main disadvantage is the high sucrose loss associated with the process. The sucrose loss will be around 0.04% on an operating day and in shutdown periods.

A GAC system is known to be capital cost is in higher side as compared to an IER system. GAC can be derived from a variety of organic materials. It can be regenerated for about 25 times. A GAC system is known to be capital cost is in higher side as compared to an IER system.

Powdered activated carbon (PAC):

Powdered activated carbon (PAC) is generally used in small scale and /or seasonal refineries. In this process colouring matters are absorbed on the surface of activated carbon. These adsorbents used in single stage or double stage, depending on the colour loading and effectiveness of other process stages. Filtration is a main limiting factor in this process. The quantity required per unit of refined sugar production is less however it can be regenerated only 1-2 times.

Ion Exchange Resins (IER):

With the development of polymer chemistry synthetic ion exchange resins or adsorbent resins have been in use in sugar refinery since 1970 for decolourization of refinery melt and are now common.

The IER system involves ions from polymer holds temporarily high molecular weight impurities from liquor. The Mechanism of colour removal in this process consists of

- Ion exchange reaction in between negatively charged colorants,
- Hydrophobic interaction between the non-polar colorants,
- Adsorption of colorants inside the pores of the resin bed.

- When in contact with a sugar melt solution to the ion-exchange resin reacts according to a simple acid-based chemical reaction. Most of the sugar colorants are anionic in nature, being negatively charged at high pH. Therefore strong base anionic resins are efficient decolorizes for sugar melt. In IER up to 65% - 70% of input colour is removed.
- The ion exchange process in the refinery is a continuous process whereby sugar liquor is passed through a resin bed and colour is trapped within the bed. The liquor needs sufficient contact time with the resin for effective colour removal. Flow rate in relation to the amount of resin present is important. The lead resin bed obviously always takes the major colour impact. The trail bed always takes a lighter load.

Fig.69.12 Ion Exchange Resins plant

The impurities are then released when the polymer is regenerated.

When a bed of resin in operation becomes saturated with colour after some time, it is isolated and standby bed is brought on line. The saturated bed then is washed, cleaned and regenerated.

One advantage of ion exchange resins as sugar decolorizer is the economy. The costs are considered to be lower than traditional granular carbon or bone char methods.

IER requires a less bulky apparatus. Also shorter retention times reduce sucrose degradation during decolourization. As regeneration can be performed inside resin columns, no special handling systems or other equipment for regeneration are needed.

The resin process is easily automated, and the liquor and adsorbent are always contained inside a closed vessel, so the process is more hygienic than other decolourization processes.

To produce refined sugar with 45-50 ICUMSA colour (with back end refinery) the second step of decolourization and de-ashing by PAC or GAC or IER is not required. Whereas, to produce refined sugar equivalent to EEC1 and EEC2 grade sugar this secondary it is required.

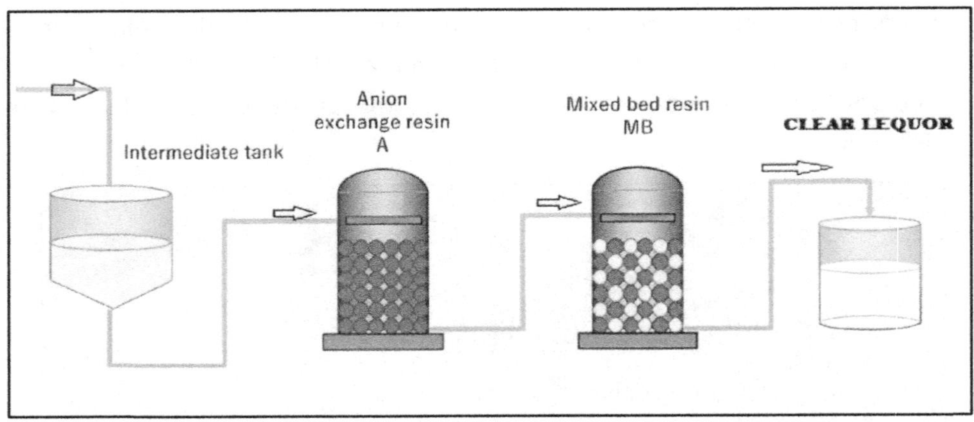

Fig.69.13 Ion Exchange Resins

69.7 Evaporation:

After clarification, decolourisation and demineralisation the clear white liquor is sent to double effect or triple effect evaporator to concentrate it to 72-76° brix. Then the concentrated fine liquor is pumped to the pan supply tanks for feed to pans.

69.8 Crystallisation / Sugar Boiling:

The clear water white liquor directly or concentrated in evaporator, is boiled at pan station for crystallisation. Any one of the following systems is used for crystallisation. Many refineries have been constructed to use a traditional strait four boiling scheme to produce refined sugar, while there are several smaller refineries that operate with single white strike boiling schemes. Other refineries using continuous pans or vacuum crystallizers have adapted their boiling schemes to suit these particular processes.

Straight boiling system:

In this system first run off from R_1 massecuite is used to boil R_2 massecuite and so on. The run-offs, before they are pumped to storage tanks at pan station, are diluted to 70-72° Brix to dissolve all the crystals remained in them. The liquor drawn into the pan is evaporated to the supersaturation point and seeded with fondant sugar or powdered sugar slurry in alcohol. All the straight strikes are boiled using the same procedure. For refined massecuite boiling, vacuum pan should be of low head with fast natural circulation. The massecuites are discharged from vacuum pans to enclosed receiver and then to refined sugar centrifugal distributor. Colour increase during boiling of massecuite is about 5%.

The runoff storage tanks required are more as enough run-offs for consecutive strikes is to be stored. The run off of R4 massecuite is sent to affination station for mingling purpose. The sugar from each strike is off different colour. The sugar from the second strike has well enough colour so that it can be packed separately but the third and the fourth strikes sugar must be blended with sugar from the first strike. Therefore sugars of all strikes are stored in separate bins and mixed proportionately and dried together.

Four boiling scheme:

A typical simplified 4 boiling mass balance based on 50% yield on dry solids at each stage and sugar colour for each massecuite is given in the following table. Colour elimination factor is the ratio of colour of massecuite to colour of sugar produced from the massecuite. It is lower for low massecuite colour and increases with increase in massecuite colour. It varies from 15 to 35.

Boiling Scheme

- Yield 50%
- Colour elimination factor is assumed,
- Feed colour 340
- Feed purity 99.4

Table 69.4 Traditional four Boiling Mass Balance

Strike	Massecuite		Colour elimination factor	Sugar		Runoff	
	Solids	Colour		Solids	Colour	Purity	Colour
R1	100	340	15	50	23	98.8	399
R2	50	680	20	25	34	97.6	796
R3	25	1360	25	12.5	54	95.2	1588
R4	12.5	2720	30	6.25	90	90.4	3168
Blended sugar output				93.75	32.10	--	--

Table starts from 340 ICUMSA fine liquor and the calculations show that the blended sugar is 32.10 is only. The maximum colour requirement for EC2 sugar is 45 IU.

In some refineries three massecuite boiling scheme is followed along with back boiling. Refined sugar discharged from R1, R2 and R3 massecuites are having different ICUMSA colour and are stored in different silos after drying and gradation. According to requirement of final blending of all different R1 R2 R3 sugars is done and packaging is done,

Now in some refineries sugar of M grade is manufactured. R2 and R3 sugar taken as footing material of R1 massecuite boiling and Dry seed taken as footing material of R2 and R3 massecuite boiling.

In back end refineries clear condensate and sweet water from filter station is used for raw sugar melting. And the R3 or R4 massecuite runoff are send to raw house to use them in A massecuite boiling.

In standalone refineries the heavy molasses separated at affination station is recycled and when the purity comes down below 85 it is send to raw house. A separate raw recovery house is maintained in standalone refinery. The excess molasses / 4th run off / sweet waters are send to raw recovery house where three massecuite boiling system is followed for sugar recovery and final molasses is produced which is send to distillery. The raw sugar from A massecuite is send refinery for melting.

Back boiling scheme:

The single strike boiling is particularly suitable for small or medium size refineries (up to 200 TPD RSO). In many Russian factories single strike boiling scheme is followed.

In Dubai a standalone refinery operates at capacity 1700 t/d RSO. In this refinery VHP raw sugar up to 1200 IU is processed. Carbonatation followed by PAC process is followed. The liquor is then concentrated in double effect evaporator. The target color of the fine liquor remains 340 IU. The pan and centrifugal station machinery installed and back boiling scheme followed at Dubai refinery is discussed hereunder.

Machinery at pan and centrifugal station in Dubai:

There are total 6 pans

R1 massecuite

- Four pans of capacity 90 t pans,
- 140 t strike receiver or crystallizer
- Four batch centrifugals of capacity 1850 kg/charge.

R2 massecuite

- One pan of 90 t capacity
- 140 t strike receiver
- Two batch centrifugals of 1250 kg /charge capacity.

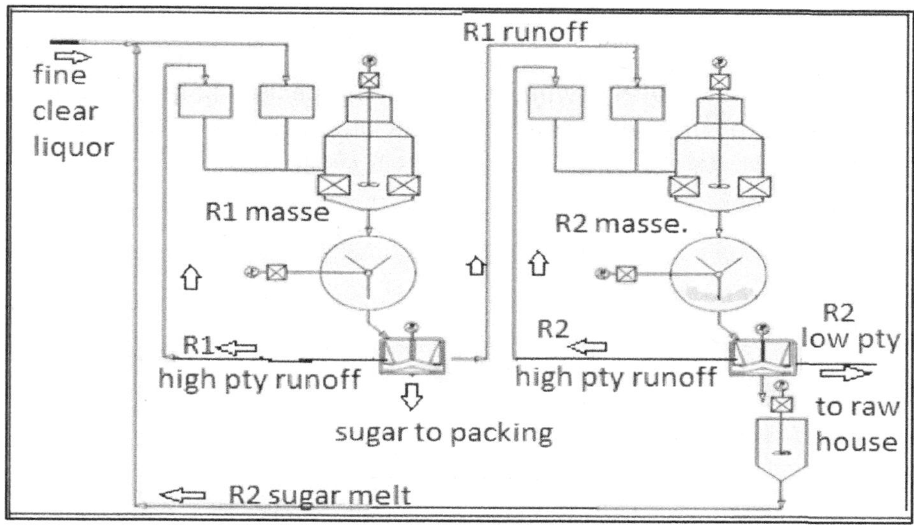

Fig. 69.14 Refinery back boiling scheme

Raw house

- One 50 t recovery pan for both B and C massecuite boiling.
- Three 60 t receiving crystallizers.
- One for B and two for C massecuite.
- Three BMA K1300 continuous centrifugals machines

At centrifugals runoff separators are used to separate low and high purity runoffs called LP runoff and HP runoff. About 75 % run off is HP runoff and it is returned to R1 massecuite for back boiling. R2 massecuite is boiled on the LP R1 runoff. Sugar from R2 massecuite is melted and the melt is mixed with clear liquor to feed to R1 massecuite. The HP run off from R2 massecuite is returned to R2 massecuite for back boiling and the LP R2 run off is send to raw house. In raw house there will only two massecuite boiling is followed.

Thus refined sugar crystal size 0.6 mm is produced only from the R1 massecuite. With this scheme the refinery can produce 100% refined sugar to bottlers' standards. The boiling time of the strike is about 2 hours. For R2 massecuite 55 % of the runoff is returned to back boiling and 45 % of the run of goes to recovery house.

To achieve good performances, the pan operator has to optimize the recirculation of the run-off to R1 massecuite. The sugar produced is of 35 IU. There are runoff separators of Heavy and light run off.

69.9 Refined sugar drying, cooling, storage and packing:

Wet sugar from centrifugal is conveyed through rotary sugar drier or fluidized bed dryers' hoppers for sugar drying and cooling. In drier moisture is brought down to less than 0.04% and temperature to 40°C. Then the sugar is conveyed to a vibratory screen for lump separation and then to sugar grader for gradation.

Sugar from bins automatically weighed and filled in bags. The bags are sewed in automatically with sewing machine. The packed bags are transported by belt conveyor to sugar warehouse and stacked automatically / manually.

-OOOOO-

70 WHITE SUGAR STANDARDS

70.1 General:

A common person recognizes sugar quality by clear white colour crystals of even size. Some prefer sugar of bolder grain. More scientific way to recognize better quality sugar is higher pol percent and lower colour value. Sugar may contain impurities like gum, organic acids, starch, dextran, coloured and colloidal complex organic compounds in traces. These are present in the mother liquor get included in the sugar crystal while crystals are growing. Sometimes even mother liquor gets included in the sugar crystal when the crystal growth rate is high and circulation in the pan is poor. The major constituents as regards to quality of sugar are,

- Colour value
- Grain size
- Moisture %
- Pol %
- Level of different impurities

In case of refined sugar colour of sugar solution is important rather than crystal size and there are strict limits for impurities. The sugar quality in international sugar market is evaluated by colour, SO_2 content, invert sugars, ash, and insoluble matters.

70.2 Keeping quality of sugar:

The keeping quality of sugar depends on impurities, moisture percentage, process followed for clarification, process parameters maintained etc. Moisture percentage of plantation white sugar generally varies from 0.04 to 0.08 percentages. Higher moisture percentage is detrimental to keeping quality of sugar. Sugar is hygroscopic in nature and absorbs atmospheric moisture if relative humidity is higher. There exists equilibrium between moisture percentage of sugar and surrounding atmospheric relative humidity at about 60-65%. Increase in relative humidity increases moisture content of sugar and sugar deteriorates rapidly.

70.3 Different standard specifications of white sugar:

In the international trade, sugar is purchased and sold by many standards and varies from country to country which are fixed by different organizations as follows:

1. Sugar Standards of the Codex Alimentarius Commission
2. Indian Standards for plantation white sugar
3. Indian standard specification for refined sugar

4. EEC sugar grades
5. Brunswick Sugar grades
6. Paris sugar grades

1 Codex Alimentarius Commission (CAC):

The commission was established in 1962 at Rome to implement the food standard programme of Food and Agricultural Organization (FAO) and World Health Organization (WHO). The CAC was of the opinion that there should be two separate standards for white sugars one for refined sugar and other for plantation white sugar the commission therefore introduced two standards in October 1965. The two standards known as specification "A" for refined sugar and specification "B" for plantation white sugar. However sugar standards for special sugars such as used in food industry, soft drink industry, pharmaceutical industry or liquid sugars are not given by CAC.

Table 70.1 CAC Specification for white sugar

Sr. No.	Particulars	'A' sugar	'B' sugar
I)	Polarisation (min.)	99.7%	99.5%
Ii)	Invert sugar (max.)	0.04%	0.1%
Iii)	Conductivity ash (max.)	0.04%	0.1%
Iv)	Sulphated ash	-	-
V)	Moisture (max.)	0.1%	0.1%
Vi)	Colour (max. ICUMSA units)	60	150
	Food additives		
I)	Sulphur dioxide (max. Mg per kg)	20	70
	Contaminants		
I)	Arsenic (as) (max. Mg per kg)	1	1
Ii)	Lead (pb) (max. Mg per kg)	2	2
Iii)	Copper (cu) (max. Mg per kg)	2	2

A: specifications for refined sugar

B: specifications for plantation white sugar

Table 70.2 Specification of Refined Sugar As per European Economic Community (EEC) – sugar grade

Sr. No.	Characteristic	EEC1	EEC2	EEC3
1	Pol % (min)	99.7	99.7	99.7
2	Moisture % (max)	0.06	0.06	0.06
3	RS % (max)	0.04	0.04	0.04
4	Ash % (max)	0.011	0.02	NS
5	Colour % (max)in ICUMSA	22.5	45	NS

Table 70.3 Refined Sugar Specifications (II REVISION)
INDIAN STANDARD IS1151:2003

Sr. No.	Characteristic	Requirement
1	Loss on drying %by mass (Max)	0.05%
2	Polarization (Min.)	99.7
3	RS% by mass (Max)	0.04
4	Conductivity Ash% by mass (Max)	0.04
5	Colour ICUMSA unit (Max)	60 IU
6	SO_2 Content mg/kg (Max)	15 PPM
7	Lead mg/kg (Max.)	0.5
8	Chromium mg/kg (Max)	20

Fig.71.1 Refined sugar

2 Indian Standards for plantation white sugar:

In India, grades of plantation white sugar made on the basis of crystal size and colour of the sugar. The bureau of Indian Standard (BIS) has given three grade for sugar crystal size L (Large), M (Medium) and S (Small) (IS 498-1985)

Further all the grades are classified according to surface colour of sugar crystal. The colour code numbers are given as 31, 30 and 29. Physical standards are produced every year by National Sugar Institute Kanpur. These standards consists of typical qualities of sugar filled and sealed in square glass bottles. The colourless square glass bottles used for sugar standards are of the design and dimensions given in IS: 11102 - 1984.

The standard sugar samples filled in bottles are termed as Indian Sugar Standards (ISS) and are denoted by arbitory numbers or symbols to indicate their grain size and colour. Thus at present there are nine standards namely L31, L30, L29, then M31, M30, M29 and S31, S30, S29 and SS31.

For categorizing and grading purposes the sugar to be graded for its colour it is filled in a square bottle similar to that for preparation of ISS sample and then compared visually with suitable standard grade bottle.

The following table shows the requirements for colour of crystal sugar.

31	Equal to or whiter than the sealed samples of colour 31
30	Equal to or whiter than the sealed samples of colour 30 but inferior to colour 31
29	Equal to or whiter than the sealed samples of colour 29 but inferior to colour 30

Thus the quality is judged in terms of its visual colour in solid state. Since the sugar deteriorates the physical sugar standards are not useful in the next season and as such are prepared by NSI every year.

Determination of Sugar colour:

The colour of the sugar is determined in solid phase by photoelectric reflectance measurement (R) and grain size of the sugar is determined by sieves. Then the modulated reflectance value (MR value) is calculated. The sugar having particular MR values are selected and the standard bottles made by National Sugar Institute. These standards are sent to sugar factories in India for adoption. The **Modulated Reflectance (MR) Value** is given as (IS 7424-1987)

$$\mathbf{MR = R \times G}$$

Where

R = Mean reflectance value of 4 surfaces of sugar bottle

G = Grain size of sugar in mm.

These samples are used for visual comparison in determining the colour of sugar.

Table 70.4 Grain sizes and MR values for different grades of sugar:

<table>
<tr><th>Grain size</th><th colspan="2">Retained on</th><th rowspan="2">Cumulative percent retained by mass (min.)</th><th colspan="3">Mr values (min)</th></tr>
<tr><th>Design ation</th><th>IS sieve</th><th>Tyler sieve</th><th>31 Grade</th><th>30 grade</th><th>29 grade</th></tr>
<tr><td rowspan="3">L</td><td>1.70-mm</td><td>10</td><td>60</td><td rowspan="3">94</td><td rowspan="3">81.8</td><td rowspan="3">71.4</td></tr>
<tr><td>850-micron</td><td>20</td><td>95</td></tr>
<tr><td>600-micron</td><td>28</td><td>99</td></tr>
<tr><td rowspan="3">M</td><td>1.18-mm</td><td>14</td><td>60</td><td rowspan="3">66.6</td><td rowspan="3">60.8</td><td rowspan="3">52.8</td></tr>
<tr><td>600-micron</td><td>28</td><td>95</td></tr>
<tr><td>425-micron</td><td>35</td><td>99</td></tr>
<tr><td rowspan="3">S</td><td>600-micron</td><td>28</td><td>60</td><td rowspan="3">39.9</td><td rowspan="3">37.0</td><td rowspan="3">34.5</td></tr>
<tr><td>300-micron</td><td>48</td><td>95</td></tr>
<tr><td>212-micron</td><td>65</td><td>99</td></tr>
<tr><td>SS</td><td>212- micron</td><td>65</td><td></td><td></td><td></td><td></td></tr>
</table>

The Bureau of Indian Standards has also given another additional standard for other characteristics of plantation white sugar. (IS: 5982 - 1970). The essential requirements for plantation white sugar are as under-

Table 70.5 BIS standard (IS5982-1970) for plantation white sugar

Sr. No.	Characteristics	Requirement
i	Moisture percent by weight max.	0.08
ii	Pol, percent min.	99.5
iii	Reducing sugars, percent by weight max.	0.1
iv	Specific conductivity 5% soln. X 10^6 mhos/cm	100
v	Sulphur dioxide, ppm, max.	70
vi	Calcium oxide (CaO, mg/100g, max.	30
Vii	Turbidity, percent by weight, max.	15
viii	Water insoluble, percent by weight, max	0.1

Fig.70.2 plantation white sugar by sulphitation process

-ooooo-

71 CONDENSERS

71.1 General:

Vapours generated from the last body of the evaporator and vacuum pans pass along with incondensable gases and air through a vapour pipe to equipment called condenser. In the condenser the vapours come in contact with cold water and are condensed. Cold water is continuously pumped into the condenser to condense the vapours. The water formed from vapours occupies very little volume as compare to that of vapours and therefore vacuum is created and maintained in the closed vessels. The cold water used for condenser is commonly called as injection water as it is injected through specially designed nozzles into the condenser.

The air and incondensable gases are also removed either by a separate air extraction system provided with the condenser or by water jets provided in the condenser.

After condensing the vapours, the cold injection water becomes hot. In case of barometric condensers the hot water flows down by gravity through a tail pipe provided to the condenser. In case of short condensers a pump is provided to remove the water from the condenser. The short condensers are not used in sugar industry and therefore not discussed here.

The hot water going out from the condenser is cooled down and recirculated to the condenser. A special cooling system is installed to cool the hot water going out from the condensers.

71.2 Types of condensers: There are two types of condensers:

- Surface condensers
- Direct contact condensers

i) **Surface condensers:** In surface condensers cooling water passes through a number of tubes and vapours pass over the outer surface of the tubes or vice versa. With the use of surface condenser, the condensate remains separate from the cooling water. Therefore, the condensate formed remains pure and can be reused. However surface condensers are costly and therefore are only used where the condensate is required for reuse.

In sugar factories where condensing turbines are used for cogeneration of electricity, surface condensers are used with the condensing turbine and the pure condensate is to send it to boilers for reuse.

However, in sugar factories there is no so much scarcity of water and water formed from vapours from last effect of evaporator or pans are not required for reuse. Therefore, these types of condensers are not in used for evaporator and pans. However, as an exceptional example at Al Khaleej sugar refinery Dubai, where water is scarcity surface condensers are in used for vacuum pans.

ii) Direct contact or jet condensers: In these types of condensers the vapours directly come in contact with cooling water and are condensed. These type or condensers are simple in designs, cheap to manufacture and give good performance. High vacuum can easily be attained. Therefore these types of condensers are used in sugar industry. We are studying here only direct contact condensers. There are two types of direct contact condensers or jet condensers,

- Short condensers
- Long or Barometric condensers

Short condensers: In short condensers the hot water leaving the condenser is extracted by pump. However, such types of condensers are not used in sugar industry. Therefore these type of condensers are not discussed here.

Barometric condensers: In barometric condensers, the hot water leaving the condenser flows out by gravity through a vertical tail pipe. Bottom end of the tail pipe is dipped below water level in a channel or well.

The water column in the tail pipe works as barometric column which is dipped into the water that is open to atmosphere. Thus the condenser becomes barometric equipment.

The barometric column for mercury is 76 cm, density of which is 13.6. However, for a barometric condenser it is water column density of which is 1.0. Hence the height of the water column required will be

$$0.76 \text{ m} \times 13.6 = 10.34 \text{ m}$$

In the factory working the highest vacuum is 690 mm of mercury which corresponds to 9.38 m water column. About 1.0-m head is required to maintain downward velocity of water. Considering 0.5 m margin for safety, the height of the water column comes out to be 10.88 m or say 11.00 m. Thus, the tail pipe height above the water level in the hot water channel should be 11.00 m. Considering 1 m length dipped in the channel, the length of the tail pipe should be 12 m.

The vapours going to condenser also contains some air. The incondensable gases are evolved from the juice / syrup. Some air is introduced because of leakages in evaporator bodies / pans and piping. The removal of air along with vapours is most essential to maintain the vacuum, otherwise air would get accumulated and vacuum would fall in spite of efficient condensation of vapours. To remove the air either a separate air extraction system is provided or high velocity jets of water that entrap air are provided in the condensers. Thus there are two major types of barometric condensers.

i) **Condensers with separate air extraction system or dry air condensers:** In this system, a separate air pump or water / steam jet ejector is provided outside the condenser to remove the air. These types of condensers are generally called as barometric condensers in the industry. Now these type of condensers are not used in sugar industry. Therefore these have not been discussed here.

ii) **Condensers with combined vapour and air extraction system or wet air condensers:** The wet air condensers - generally called as multijet condensers. These are used in sugar industry,

71.3 Multijet condensers:

In this type of condensers, no separate air extraction system is provided. The water is pumped in the condenser through spray and jet nozzles. The vapours are condensed by water sprayed by spray nozzles and air is extracted by water jets that are ejected through jet nozzles.

Noel Deerr gives the cross sectional area of condenser as 0.16 m^2 per tonne vapours per hour to be condensed. The minimum volume recommended by sugar engineers is 0.3 m^3 per tonne of vapours per hour going to condenser. In factory working the vapour line juice heater may be kept idle in case of tube leakages. Therefore while designing condenser for evaporator set the capacity of the condenser of evaporator is to be calculated on the basis of vapours produced from the last effect and not on the quantity vapours going condenser when vapour line juice heater is in operation. The efficiency of condenser greatly depends on design features of the condenser.

Foreign material that passes along with water in the condenser may choke the perforated baffle. Therefore screening of the cold water before entering it into the condenser is necessary to avoid or minimize the choking of nozzles. For this purpose a conical bucket type strainer is fitted at the water inlet of the condenser. The inlet water channel for condensers is provided with screens to arrest foreign material which can choke either the pumps or the condenser.

The condensers are provided with two types of nozzles of specific designs called spray nozzles and jet nozzles and are fitted in specific settings. Vapours are condensed by the cold water sprayed by spray nozzles. The water ejected from the jets nozzles entrains the air from the condenser and passes into the water in the tail pipe.

When velocity of water in the barometric column is maintained 2 –2.5 m/s, the air bubbles entrained by water jets do not escape and rise into the condenser again but pass along with water.

Fig. 71.1 Condensers at a Factory

The main injection water pipe inlet is divided into two branches for spray nozzles and jet nozzles. An annular jacket is provided at the top periphery of the condenser in which spray nozzles are fitted. Water entering the jacket is injected through the spray nozzles. There are two separate valves are provided to regulate the spray and jet water. However now with single water entry condensers there are no separate branches of injection water for spray and jet nozzles.

In the multijet condensers the condensation takes place mainly on the walls of the condenser. Therefore the condensers are designed in such a way that the water sprayed from the spray nozzles discharges mainly on the walls and flows down. Water pressure at the jet is around 0.40 kg. Now there are some multijet condensers with new designs, which work at almost zero pressure at the jets. Generally a pressure gauge is provided at the water inlet of the condenser to observe the water pressure entering the condenser.

71.4 Quantity of water required for vapour condensation:

The quantity of water required per kg of vapours to be condensed is calculated as follows:

H — total heat in KCal/kg of vapours
t_1 —temperature of inlet water °C
t_2 –temperature of outlet water °C
W —kg of water required per kg of vapours condensed

Then

Heat taken by the water = Heat given by the vapours

$$W(t_2 - t_1) = H - t_2$$

Or
$$W = \frac{H - t_2}{(t_2 - t_1)}$$

Example:

Consider

Vacuum at the pan/last body of evaporator	660 mm
Total heat of vapours per kg	620.6 Kcal
Inlet temperature of water	32°C
Outlet temperature of water	46°C

Then,

$$W = \frac{(620.6 - 46)}{(46 - 32)}$$

Or
$$W = 41.04 \text{ kg}$$

Thus, 41.04 kg of water will be required to condense one kg of vapours when inlet temperature is 32°C and the outlet temperature is 46°C.

When the temperature difference in inlet and out temperature is higher, the quantity of water required is considerably reduced. On the other hand when the difference is decreased, the required water quantity is increased. This difference depends on inlet water temperature and the "Approach". The difference in between vapour temperature and the outgoing temperature is called as Approach. It gives efficiency of the condenser. The efficiency of the condenser operation is evaluated on the basis of minimum water used to condense unit weight of vapours. It is possible when approach temperature is minimum.

Considering temperature of the vapours to be condensed as 52°C and the condenser water outlet temperature as 46°C (hence the approach 6°C) the water requirement kg per kg of vapours to be condensed is given in the following table and graph.

Table 71.1 water requirement in kg per kg of vapours to be condensed

Condenser water inlet temperature	Water quantity require to condense one kg of vapours
°C	kg
24	26.12
26	28.73
28	31.92
30	35.91
32	41.04
34	47.88
36	57.46
38	71.83

It can be seen from the above table that the water requirement increases rapidly above and decreases slowly below 30-32°C.

For wet air or multijet condensers extra water is required for air extraction.

The bottom cone of the condenser has a slope of 70° to the horizontal to flow the water fast into the tail pipe. With less angle there can a fear of filling the condenser with injection water and even pass it to the last body of the evaporator or vacuum pan.

The injection water pumps are of 22 m head. Now in most of the factories all the injection pumps delivery pipes are joined to a suitable common header from which branches pass to each condenser. Each pump is provided with valve and non-return valve in the delivery line.

The injection water and spray pumps are of centrifugal type bronze impeller and fittings and directly coupled to SPDP slip ring induction motor. The injection and spray water pumps have 950 rpm running speed. To start the injection pump lifting the water the suction pipe of the pump is to evacuated and water is to be sucked up to the pump for which a priming pump is provided.

71.5 injection water pumps required

Table 71.2 Capacity and number of Injection water pumps recommended For multijet condensers:

Plant capacity TCD	Pump capacity	No of pumps (one as stand by)
1250	850 m^3/hr.	3 pumps
2500	850 m^3/hr.	3 pumps
	plus 1700 m^3/hr.	1 pump
5000	3000 m^3/hr.	3 pumps
7500	3000 m^3/hr.	4 pumps

1. Nozzles of condensers: The design and proper size of nozzles plays an important role in saving power in multijet condensers. Higher power is required for under size nozzles.

2. Velocity of water in the tail pipe: The velocity of water in the tail pipe should be 2 to 2.5 meter. If it exceeds 3 m the water may not pass at required flow rate and water column may rise to submerge the throat of the condenser. In that case, the incondensable gases cannot be removed effectively. That results into low vacuum at evaporator and pans. Sometimes water even may come from the condenser into the last body of the evaporator or into the pan. Chinnaswami[1] has discussed such case.

71.6 Single water entry multijet condensers:

Now in newly coming up factories specially designed single entry stainless steel multijet condensers are being installed. As well in most of the old factories old condensers have been replaced by new single entry condensers. The condensers are made of stainless steel 6 mm plate thickness up to 1200 mm dia and 8 mm plate thickness above 1200 mm dia. with hoop rings and stainless steel tail pipe of 4 mm thick. In single entry condenser only one water distributing box is used for spray and jet nozzles. Nearly 80 % water is used for spray nozzles and 20% for jet nozzles. The jet and spray nozzles are of gun metal or S S 304 or PVC. Vapours are condensed by fine mist enforced by hydro dynamically designed spray nozzles with minimum requirement of injection water[2]. This reduces quantity of water used in the condenser.

Vacuum is created only in 1- 2 minutes and remains steady therefore downtime of pans is reduced. The water vapour ratio is least. The difference between vapour temperature and condenser tail pipe water temperature (approach temperature) does not exceed 6°C. The pumping head and power consumption is low. Singe entry condensers are made up of stainless steel (AISI 304) which is resistant to acidic or basic environment. It is fully resistant to corrosion, erosion, and abrasion. The spray nozzles are made up of thermoplastic to absorb abrasive forces. A perforated strainer is provided at the inlet of the spray head to prevent clogging of nozzles. It can be removed easily for inspection and cleaning.

Ravishankar[2] has mentioned that by installation of single entry condenser at Ugar factory a remarkable achievement was made. The water requirement of the condenser has been considerably reduced. Consequently, the power consumption, which was previously 5.79 kW/t of cane, was reduced to 2.88 kW/t of cane.

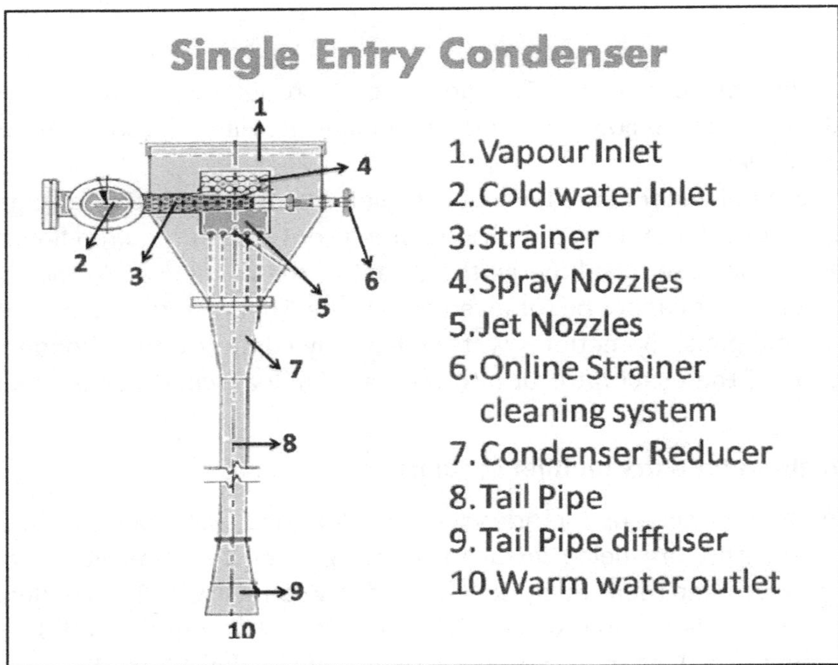

Fig, 71.2 Single Entry Condensers

71.7 Precautions to improve performance of condensers:

The performance of condenser depends on operating conditions such as quantity and temperature of vapours, injection water inlet temperature and volume of incondensable gases etc.

Profuse vapour bleeding at evaporator, higher syrup brix and proper control at the pan station can reduce the load on condensers.

In case of vacuum pans at the beginning of the strike the evaporation rate is high on the other hand at the end of the strike it is much low. Therefore the load of vapours on the condenser varies considerably.

The load of air on condenser should also be minimum. The main sources for air are leakages at vacuum vessels and condensate piping. The leakages at different vessels are to be tested by water testing at about 3.0 kg/cm^2 pressure. The air leakages at condensate pumps should be properly taken care of by use of mechanical seals. The globe type valves at pans are to be fitted in such a way that gland-section will not be toward vacuum side.

Because of reduced vapour / air load on condensers, water requirement of the condensers is also reduced.

71.8 Condenser Automation:

Generally when there is no air leakages optimum vacuum required at last body of the evaporator or at pans is maintained steady with no difficulty without automation. It is because capacities of the pumps provided for injection water are for maximum water required at condensers. Therefore the main purpose of condensers automation is not maintaining optimum and steady vacuum but it is power saving, because the power requirement at condensers is large.

Power consumed by the injection pumps is directly related to the quantity of water used in condensers. The cold water requirement of the condenser varies according to the load on the condenser and cold water temperature. The vapour load on the condenser at evaporator remains more or less constant, however at pans it fluctuates with type of massecuite, level in the pans and vapour pressure at calandria of pans. The load of incondensable gases at evaporator remains more or less steady. The load of air also remains less and steady if there are no leakages.

In setting automation at condenser optimum quantity of water is to be injected in the condenser as per load on the condenser and injection water temperature.

The conventional method of condenser automation is to control the water flow going into the condenser by throttling pneumatically operated control valves installed at the water inlet pipes of the condenser. But with this system decrease in water quantity at nozzles adversely affects spray formation and force of jets due to loss of pressure. Therefore efficiency of the condenser is reduced. As efficiency of the condenser is reduced the vacuum at the evaporator / pan is reduced and to maintain the desired vacuum the control valve gradually gets opened. Thus the basic purpose of energy saving by reducing the water flow through nozzles becomes ineffective. In throttling the valve at 50% flow rate almost 75% energy is lost.

Spray Engineering Devices (SED) has made an innovative idea of opening and closing of number of spray and jet nozzles. So that the differential pressure at the nozzles remains always same. All the open nozzles transfer the entire pressure energy into the condenser giving good performance. Therefore even if few nozzles are open and only 15% of water injected into the condenser the condenser efficiency does not affect. In this case there is no loss of energy as in throttling of valve.

SED has developed a control system based on SCADA (Supervisory control and Data Acquisition) software that includes analog inputs (4-20mA, pressure and temperature transmitters), digital input (24 V DC) and digital outputs (24 V DC).

The nozzles are divided into sets to vary the quantity of water flow based on set vacuum. SCADA system[12] controls the spray and jet nozzles, with pressurized (1.2-2.0 kg/cm2) clear water. The default mode of these nozzles is in open condition in case of any failure of auto system.

The following are the sensing and calculated parameters for control of quantity of water for the condenser i) Vacuum ii) Vapour temperature iii) injection water inlet temperature iv) injection water outlet temperature v) dry bulb temperature vi) wet bulb temperature. vii) Total water consumption viii) total vapour condensation ix) water / vapour ratio.

Additional water ejector is provided for air evacuation in case of higher air leakages. The automation of the dry air condensers is comparatively simple and the injection water quantity can be controlled as per vapour load on the condenser.

Cooling and condensing system with conventional multijet condensers without any automation consumes 4 to 6 kWh/t of cane. However, with single entry condensers and complete automation, the energy consumption can be reduced up to 2.0 kWh/t of cane in many factories.

References:

- Chinnaswami A P SISSTA July - Sept. 1976 p 26-28
- Ravishankar T STAI 2002 p E 116-126
- Jain P F STAI 1991 p E 15 - 17
- Patil C B STAI 2000 p M 62 -71
- K K Menon STAI 1975 E 5 -10.
- Perk C G M ISSCT 1956 vol. II p 82-97
- Mukharji J P STAI 1978 p E 1-4
- Verma H L STAI 1978 p E 5 – 16
- Ramlingam M STAI 1992 p M 29 – 36
- Ravishankar T SISSTA 2002 p 185 – 191
- Suresh Babu D G SISSTA 2002 p 199 – 203

-OOOOO-

72 CONDENSER WATER COOLING SYSTEMS

72.1 General:

The factories, which have ample supply of water, can allow hot water from the condenser to use it for irrigation purpose. However, for most of the factories, the hot water is cooled in spray ponds or cooling towers and the cooled water is recirculated to the condenser.

The spray pond or cooling tower operates on principle of removing heat from the hot water by evaporating some portion of the hot water. The latent heat required for vaporization is taken from the hot water. Each one kg of water that is evaporated takes about 530 KCal of heat from the hot water and the water is cooled.

In spray-pond or cooling towers, the hot water is pumped and sprayed in the air by nozzles. The nozzles split the water into fine droplets that are exposed to air with increased surface area of water and heat from the water is lost by following two ways

- Some water gets evaporated. The latent heat required for vaporization is taken from the remaining water and it is cooled. This is called as evaporative cooling. This raises relative humidity of air surrounding the spray pond.
- Some heat is conducted from warm water to the surrounding air.

The amount of heat to be removed from the hot water is called as heat load. The size of spray pond and number of nozzles or size of cooling tower is decided on the maximum quantity of water to be circulated and the maximum heat load on the cooling system. Out of the heat lost by warm water, about 75 to 90 % of the heat of warm water is taken by water evaporated and the remaining heat is removed by convection by the air. Centrifugal type pumps of 15 m head are used to lift the hot water are generally called as spray pumps. A priming pump is provided to suck the warm water in the suction pipe of the spray pump.

72.2 Spray ponds:

Spray pond is a large masonry built pit dig in ground. To get good performance a spray pond should be rectangular preferably long and narrow crossing to wind direction so that advantage of wind can be taken more effectively. A spray pond of 3600 to 4000 m^2 area is sufficient for a 100 tch plant. More area may not be more advantageous because in hot summer days water is being heated by sun rays. The depth of the pond is 1.2 to 1.6 m. however it does not give any effect on cooling.

The water distribution piping at the spray pond are mounted and fixed on RCC pillars. Now the use of stainless steel fabricated common header and main pipes and use of thermoplastic for branch pipes and nozzles has reduced the corrosion and maintenance. The durability also has been improved. The distance between two adjacent spray nozzles is generally kept 4 to 5 m. So that the sprays do not overlap each other and air can pass through them. Then the sprays give good results.

Fig. 72.1 Spray pond

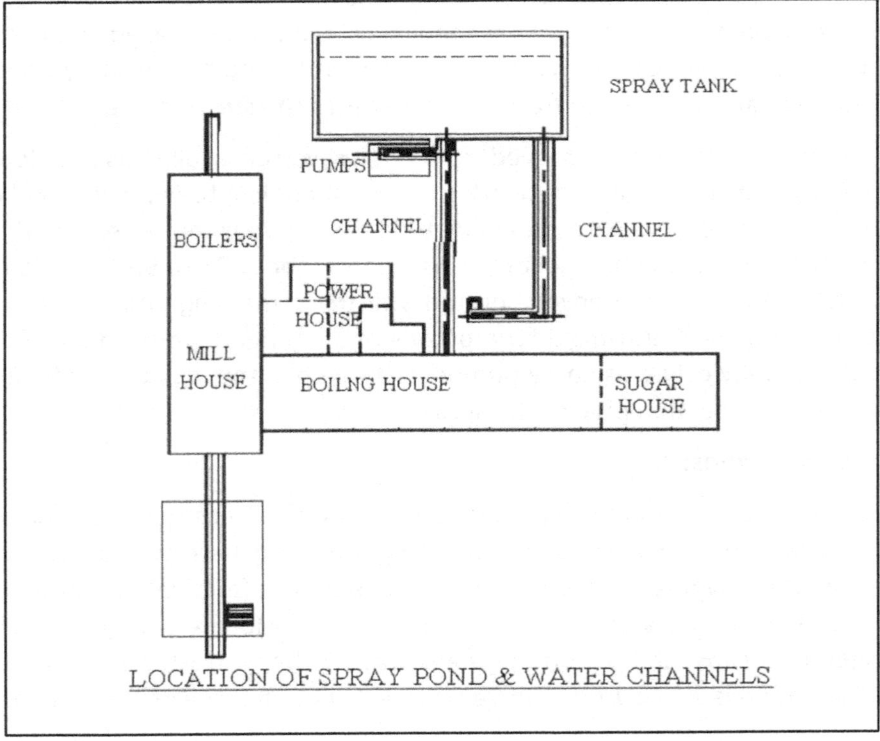

Fig. 72.2 Location of spray pond

In arranging the spray nozzles each spray pump is connected to different set of nozzles. Therefore when one pump is working, water is supplied to only set of nozzles connected to that pump. So that water pressure at the nozzles is maintained. However each set of nozzles are distributed throughout the spray pond area. So that for any spray pump working the whole spray pond area is covered and cooling efficiency is maintained.

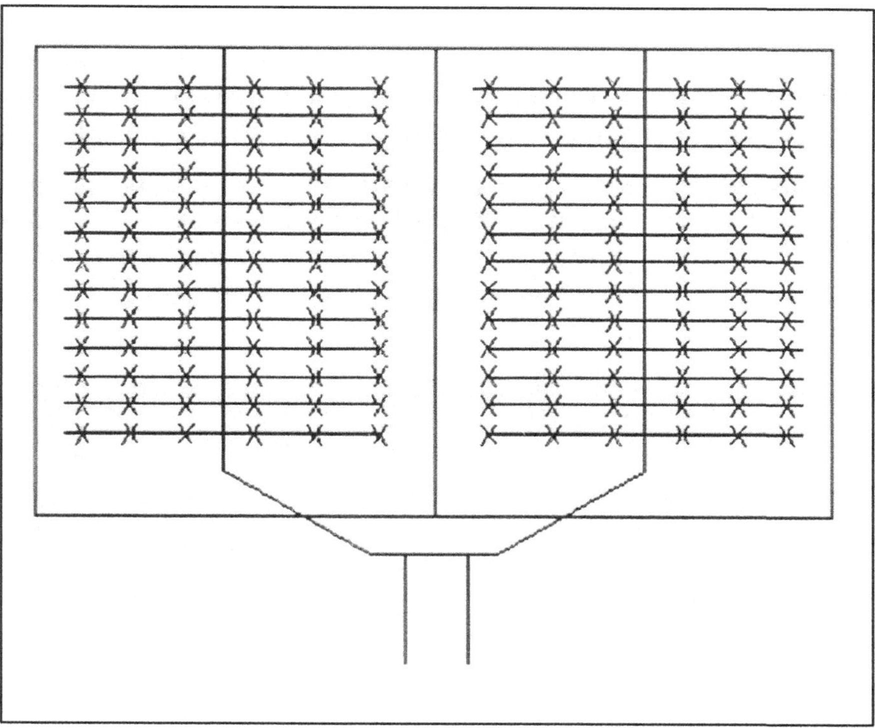

Fig. 72.3 Arrangement of spray nozzles

The nozzles are generally arranged in staggered arrangement. They are arranged on branch piping with single nozzle arrangement. There are many types of nozzles – conical jet vortex and volutes types. The nozzles are of PVC with stainless steel insert at throat of nozzle. Now PVC nozzles that can be readily fitted and dismantled are more popular. The water is sprayed through the nozzles with high velocity in all directions. The small droplets with maximum surface area come in contact with air. This gives better evaporation and cooling.

Foreign bodies can choke up the nozzles. To avoid this, there must be fine grid /strainer at the spray water pump inlet. Further in case of any nozzles get choked it should be easy to replace the same.

The water level in the inlet and outlet channels should be sufficient. The strainers of the suction pipes of injection pumps are dipped and the tail pipes are well sealed. According to Tromp[4] the loss of water by entrainment from the spray pond is 3 %. However in winter when spray pump are not in operation it is almost negligible. Generally in layout of the factory, provision of space for additional pumps for expansion of the plant capacity is kept. The spray pond and masonry water channels are suitably designed so that additional piping and nozzles can be installed in expansion stage.

Jain[2] has described the zero level integrated concept, in which spray are installed separately from the pond. The bottom surface of the spray area is inclined towards the inlet channel and therefore water flows directly to the inlet channel.

Thus mixing of low temperature spray water with high temperature pond water is avoided. As a result of which low temperature spray water is always available at injection pit. Jain says that with this system, the cooling water efficiency was increased to almost double in comparison to conventional spray pond system.

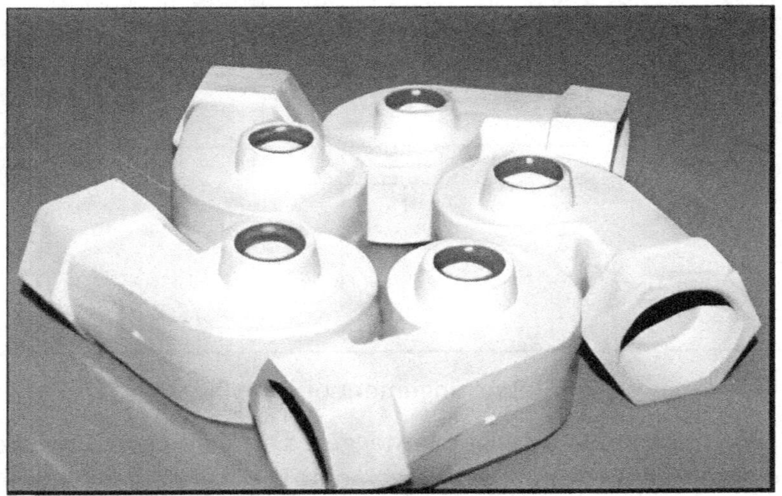

Fig, 72.4 Spray nozzles

72.3 Cooling tower:

Now cooling towers are being used in sugar factories. There are two types of cooling towers –

1. Natural draft cooling towers – in these types of cooling towers air circulation through the tower is achieved naturally. These cooling towers are generally not used in sugar industry. Author has observed such type of cooling towers installed in some sugar factories in India, however the results are not much encouraging.

2. Mechanical draft cooling towers – in these cooling towers air is circulated by fans.

There are two types of mechanical draft towers - induced draft towers and forced draft towers.

In induced draft towers fan is fitted at the top at the discharge and it pulls the air up through the tower.

In forced draft towers the fan forces the air into the tower.

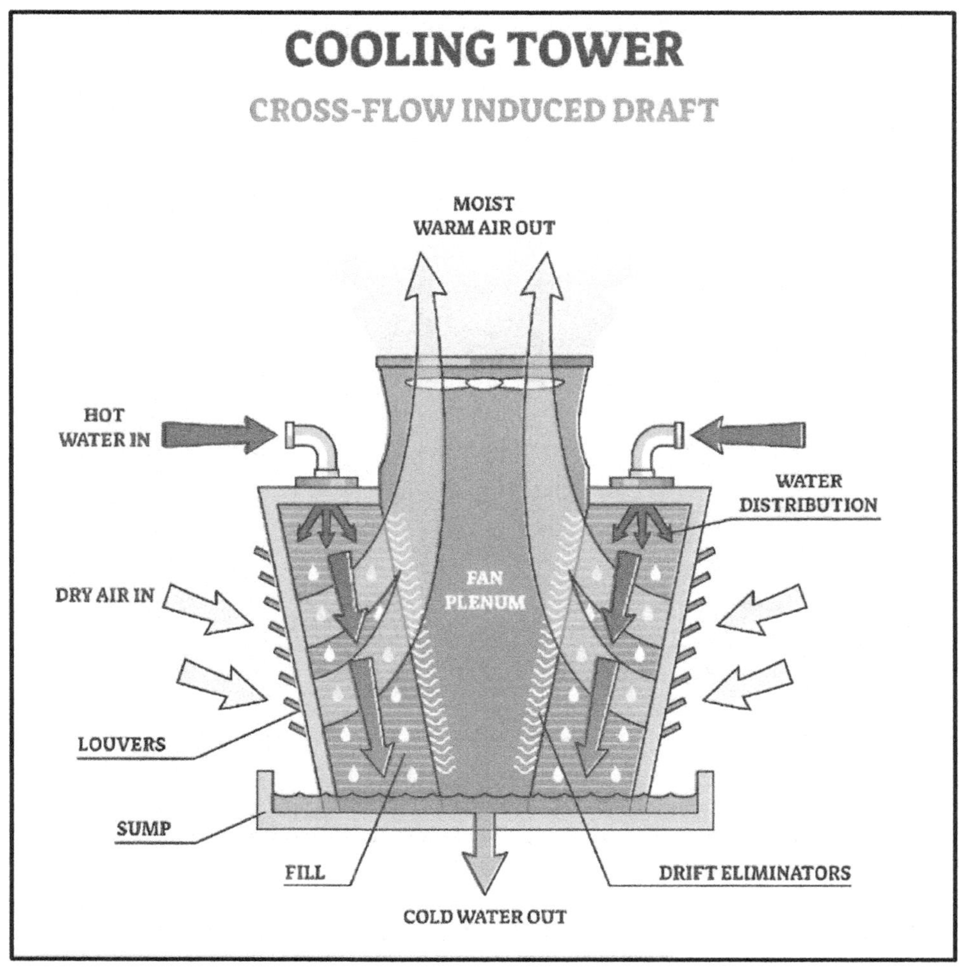

Fig. 72.5 Cooling Tower

The induced draft towers are considered as more efficient than that of the forced draft towers. The flow arrangement can be counter flow or cross flow. Thermodynamically a counter flow induced mechanical draft fan is better. However from practical consideration, cross flow type design find more efficient and compact.

Inside the tower, fills are provided to increase contact surface as well as contact time between air and water to get better heat transfer. Wooden or plastic fills (stacked in horizontal rows) are used. Water is pumped and distributed over the fills through nozzles.

Towers of Fibre Reinforce plastic (FRP): A most suitable structural material for small and medium towers is Fibre Reinforced Plastic. It is highly corrosion resistant. The cost is low and maintenance is also less as compare to reinforced concrete structure. Therefore now small and medium cooling towers are being made of FRP structures.

Structure is made by FRP lined heavy M S angles Tees and channels. Louvers are made up of moulded FRP and are fixed with fibre filled nylon nut bolts. Nozzles of nylon 6 are fixed to G I pipes distribution header.

Fig 72.6 FRP Cooling Tower

In India, in about 100 sugar factories cooling tower system is in use.

72.4 Makeup water: There is loss of water being circulated through the system due to evaporation entrainment / drift and percolation.

The TDS also goes on increasing as days pass. With higher TDS scale formation corrosion and biological growth may be higher. Therefore to maintain the TDS of the water, some water must be drained off. Therefore fresh water must be added periodically to the circulating water system.

If entrainment is occurred from the last body of evaporator or to pans to condenser, then the syrup/molasses is mixed with injection water. The spray pond water becomes dirty with dark colour and heavy large foam is formed at the outlet channel of the injection water. The pH of the water also is lowered down and to maintain the pH generally lime or pH booster is added in the spray pond water.

If to maintain the pH of the injection water alkaline lime is continuously added to it instead of replacing the water, then TDS of the water goes on increasing and condensing capacity of the water gets reduced. Vacuum at the pans and evaporator is adversely affected.

72.5 Efficiency of cooling System:

It is assumed that when the warm water is cooled efficiently it can be cooled down very close to wet bulb temperature. The difference between the cooled water temperature and the wet bulb temperature is called 'Approach' (approaches to wet bulb temperature). With spray pond cooling system a drop in temperature of about 10-13°C or up to 7°C of wet bulb temperature can be achieved.

The efficiency of the system is calculated[1] as follows

$$\text{Efficiency} = (t_1 - t_2) / (t_1 - t_o)$$

Where

t_1 = temperature of warn water entering the system, °C
t_2 = temperature of cooled water leaving the system, °C
t_o = temperature of wet bulb, °C

This should be in between 50% and 70%. In India[2] the cooling efficiency of spray pond ranges from 35 – 48 % which is 20% lower than the figures reported by Hugot. Observation made by V.K. Jain[3] shows that when the water is sprayed the temperature of water droplet becomes even lower than the wet bulb temperature. However, the temperature of these droplets is immediately increased as it is mixed with spray pond water as the temperature of pond water is generally on higher side than ambient temperature; the cooled water temperature again increases.

According to Jain, the efficiency as calculated by the above formula is higher than 100% when the spray water temperature is measured before it is mixed with the pond water. He applied empirical constant "C" to the formula and the formula becomes

$$\text{Efficiency} = C (t_1 - t_2) / (t_1 - t_o)$$

Based on the experiments conducted with water cooling system in the factories he gives the value of empirical constant as 0.814

72.6 Comparison in spray pond and cooling tower:

Bhojraj[5] gives comparative data of various cooling systems. The floor area needed for spray pond for a 2500 TCD plant is 85 x 70 m, while in case of mist cooling it is 50 x 50 m and minimum in case of cooling tower 18 x 18 m. (Mist cooling system is in fact an improved version of spray pond involving less area of spray pond). Generally, the approach temperature for spray pond is 10-12°C, for mist cooling 0-5°C and for cooling towers 0-5°C.

In spray pond cooling system, in case of spray pumps failure, temperature of the spray pond water does not increase immediately as the volume of water is large. On the other hand in case of cooling tower failure of pumps and / or fans immediately increases the temperature of inlet water channel. It adversely affects working of condenser and consequently evaporator and pan stations. Thus cooling tower system is more sensitive in case of failure of spray pumps.

The water entrained in the air flow and discharged out of spray pond or cooling tower is called as drift. The drift loss (entrainment) due to wind in case of spray pond is more as compare to cooling tower because spray pond has large area. In spray pond the average drift loss can be 3% whereas for cooling towers the drift losses[6] can only 0.2%. It is generally observed that in many factories proper care or maintenance of cooling system is not taken which affects efficiency of condensing system.

References:

- Kulkarni D P 'Cane sugar manufacture in India' p 404
- Jain V K IS Nov. 1997 p 601 – 604.
- Jain V K IS Feb. 1999 p 899 – 902
- Tromp L A p 491
- Bhojraj S K STAI - DSTA Joint 1994 p M 225-230
- Menon K K STAI 1978 p E 25 – 31

-OOOOO-

73 *STEAM GENERATION*

73.1 General:

It is the fortune of sugar industry that it is self-sufficient in electrical power and heat energy. In sugar mills steam is generated in boilers using bagasse as fuel and it is used for power generation and in process operations. Steam at high pressure of 21 kg/cm^2 to 124 kg/cm^2, of temperature 340° to 500 °C is generated. This high pressure steam is used to generate electricity is generated using steam turbine and alternator. High pressure steam is also used at prime movers at Fibrizer / shredder, mills, boiler feed water pumps etc. The exhaust steam (low pressure steam) from these turbines /prime movers is utilized for process operations. The exhaust generated is about 75-90% of the requirement of the process.

Fig.73.1 high pressure boiler

The saturated steam may contain some moisture. It means it is wet steam. Such steam cannot be used for turbines and prime movers. Therefore the steam is superheated in the boilers. Superheating the steam means, heating the steam above saturation temperature of steam (corresponding to the pressure). for e.g. at 32 kg/cm^{2}abs pressure the saturation temperature of steam is 236.368°C, but in boiler it is heated up to 380°C. It means the steam is superheated. Such a steam is completely dry and has higher heat content. Such a supersaturated steam is absolutely essential for turbines and prime movers.

The exhaust steam from the prime movers is of about 0.7 to 1.5 Kg/cm^2 and is entirely used in the boiling house. However the exhaust is also superheated and such exhaust is bad conductor of heat, therefore the exhaust has to be desuperheated. As the exhaust generation is always less than the requirement of the boiling house some high pressure live steam is bleeded in the exhaust steam and desuperheated to meet the demand.

The bagasse percentage on cane generally varies from 27 to 30 % on cane. The falling from the last mill contains 48-49 % moisture 2.5 - 3.0 % brix and 48 – 49 % fibre. The fibre constitutes the insoluble cellulosic matter of sugar cane.

73.2 Boiler:

The function of boilers is to convert water into steam of a desired pressure and temperature. Heat produced by combustion of fuel is transferred into water. The boiler is a pressure vessel and water is pumped into the boiler at operating pressure. Water gets evaporated and steam is formed of desired pressure and temperature.

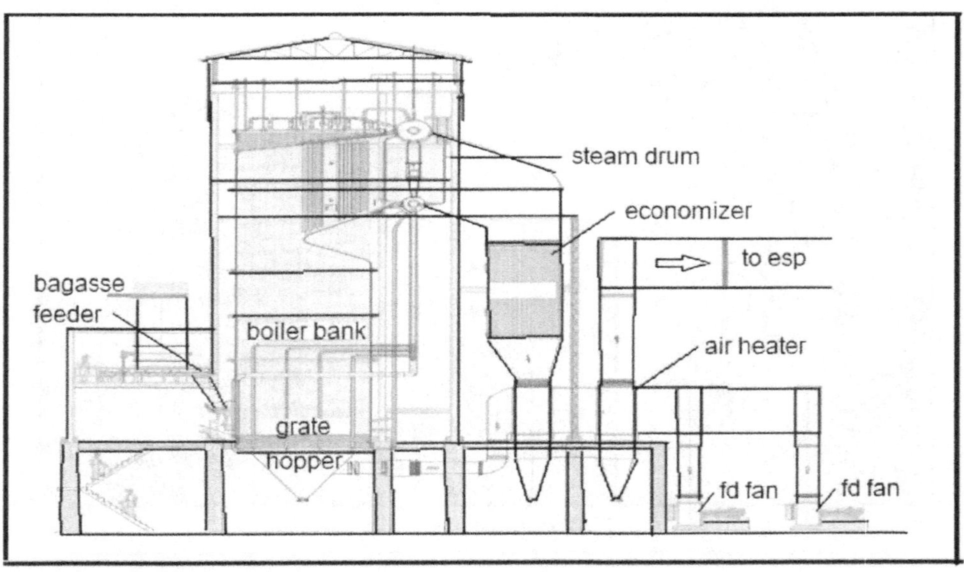

Fig.73.2 schematic drawing of a typical boiler

In the last three decades the concept of exploiting the potential of bagasse as a renewable source of energy has gained wide acceptance. The bagasse is now being used for generating electric power for public use. Major development in this area is installation of high pressure boilers for surplus power generation.

Bagasse fired boilers of 45 Kg/cm^2 (g), 425°C, 66 kg/cm^2(g), 86 kg/cm^2(g)

110kg/cm^{2}g from 40 TPH to 180 TPH are being used in sugar industry. Bagasse and biomass fired Boilers of capacity up to 300 TPH with pressure up to 160 kg/cm^2 (g) and temperature up to 560°C now are available. Such boilers are specially designed tall furnaces of higher overall efficiency. Now bagasse fired boilers are commonly being installed outdoor. Boiler is consists of pressure parts such as

- Heating surface with attached drums or shells for storage of water and steam.
- Super heater surface to super heat the steam up to desired temperature.

The heating surface is mainly seamless steel tubes of standard specifications. Drums are cylinders with spherical ends, of welded steel plate construction. Furnace area is provided for combustion of bagasse. The pressure parts are properly connected to produce some desired pattern of flow of water. The drums and tubes are supported with adequate steel structure. The whole unit is surrounded by suitably shaped casing or walls. This directs flow of combustion gases and insulates the boiler interior from outside atmosphere.

Mainly there are two modern water tube types' boilers

- Straight tube boilers and
- bent tube boilers,

In these boilers, banks of tubes constitute heating surface. The tubes are connected to two or more drums. The main pressure parts of boiler are economizer, drums, modules banks and superheater tubes.

1 **Steam drum:** At the top of the boiler, a steam/water drum, provided with suitable dished ends with pressed steel manhole doors is mounted. All parts are of boiler quality constructed in accordance with IBR. The steam drum of the boiler is provided with suitable internal fittings comprising cyclones and scrubbers to ensure steam purity with silica content not exceeding 0.02 ppm.

Boiler feed water is pumped into the lower section of the drum. The boiler water is treated using chemicals in the drum.

Steam comes in the upper section of the drum. Fine water droplets coming along with steam are separated in cyclone separator and dry steam goes out. The water droplets are drained to the water in the bottom section.

2 **Boiler Bank tubes:** The steam drum and mud drum are connected by a set of tubes are called bank tubes. Heat generated by combustion of bagasse is transferred from the flame / gases to the water and it is transformed into steam. The arrangement of bank tubes mostly is in line.

Generally two or three rows of these bank tubes are called as down comers. The low temperature feed water entering the top drum is passed to the lower drum through these tubes. The steam rises through the remaining bank tubes into upper section of the top drum. Thus natural circulation of water is established through the boiler bank.

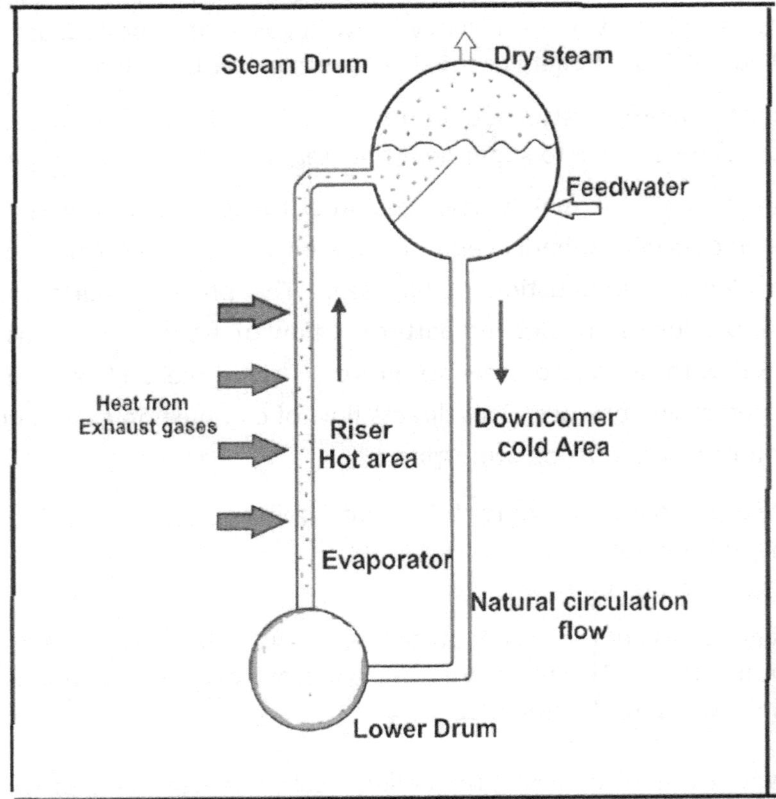

Fig.73.3 water circulation and heating in boiler

3 **Mud drum:** The mud drum is located at the bottom of a water tube boiler and is connected to the steam drum via boiler tubes. The mud drum remains full of water. In this drum impurities in the water are settle down and sediments are removed from the boiler by blow downs.

4 **Furnace:** Furnace / combustion chamber of bagasse fired boiler is of special design with sufficient volume and height for efficient mixing of incoming bagasse and air. In the combustion chamber / grate bagasse drying and burning takes place. Boiler design is fuel specific keeping optimum exhaust gas temperature with respect to ash fusion temperature of the specified fuels, thereby eliminating fouling and deposition. Furnace height is selected to provide adequate residence time for efficient combustion of fuel with minimum unburnt carbon loss.

Boiler Water Wall Panels are a series of tubes welded together tangentially or with membrane bar between them to form the walls of the boilers combustion chamber.

5 Super heater: For efficient operation of prime movers like turbines it is essential to use superheated steam. To superheat the saturated or slightly wet steam from boiler drum a heat exchanger called superheater is installed.

Superheater is a bank of tubes suitably located in the path of hot gases. Superheater is made-up of smaller diameter tubes with several U bends interposed between two headers.

Saturated or slightly wet steam from boiler steam drum passes in superheater. The flue gas coming from the boiler flow across the superheater tubes where the heat transfer takes places. The saturated steam is heated in it by the flue gases. Water droplets entrained with wet steam are completely evaporated in a superheater and steam becomes superheated dry steam.

In superheater the saturated steam is heated up to 100° to 200°C above the steam saturation temperature at boiler working pressure. This depends on boiler working pressure and also on the requirement of electricity and power of prime movers. The superheated steam is sent out from superheater into the live steam mainline.

6 Attemperators: An attemperator controls steam temperature by injecting a spray of very fine water droplets into the superheated steam flow. These droplets evaporate immediately. The evaporating water reduces the steam temperature to a desired set point.

Attemperator is usually located at the exit superheater or in between two stages of superheater at an intermediate position. Usually, boiler feed water or water from economizer is used for attemperation.

7 Economiser: A heat exchanger called economizer consisting of tubes is installed in path of flue gases passing from boiler to chimney. The Economizer recovers sensible heat from flue gases to raise the temperature of feed water. Water from feed water tank of temperature of 90-97°C is passed through the economiser to heat up to 220°C before entering into boiler.

Economizer is made from tubes generally have fins, through which the water circulates in series. Fin tubes are arranged in groups, and water passes from one tube to the other by means of U bend. The tubes are made of carbon steel. Economizer increases the efficiency of boiler.

8 Air preheater: A heat exchanger is provided in the duct after the economizer for further recovery of sensible heat from the flue gases during their passage to chimney, special equipment is installed in which air used for combustion of bagasse in the furnace is heated. This equipment is known as air pre heater. There are two type of air heaters is of

- tubular
- plate types

Air heaters are located in the flue passage after the economizer and unlike the economizer operate at atmospheric pressure. Preheated air improves the combustion of bagasse.

Ambient air is forced by a fan through ducting at one end of the pre-heater tubes and at the other end the heated air from inside of the tubes emerges into another set of ducting which carries it to the boiler furnace for combustion.

Recovery of heat from the gases going to chimney is accomplished by both economizer and air preheater which have found important place in the present day high pressure or even medium pressure boiler installations.

The dew point of the combustion gases is around 60° - 65°C and the fear of condensation on the exterior of air heater does not exist. In practice it is possible to bring down the temperature of flue gases to 150° - 170°C in boilers equipped with economizer and air preheater.

9 Bagasse Feeding: The bagasse handling system consists of main bagasse carrier (MBC) and return bagasse carrier (RBC) of suitable sizes. MBC will take-off the mill bagasse from last mill to the boiler through bagasse elevator. Further the MBC and RBC conveys the excess bagasse to yard. When mill operation is disturbed, the system supplies the bagasse from the yard to the boiler.

Bagasse is feed from carrier through sliding gates to the storage silos. The sliding gates are mounted just below MBC above the storage silos and control the bagasse feed to Silo. Bagasse feeding system - depending upon boiler capacity, consists of 2 to7 bagasse silos. The silos have total bagasse holding capacity of 10-15 minutes requirement of boiler.

The modern high pressure boilers use either dumping or travelling grate with pneumatic distribution for bagasse firing which ensures efficient burning. However the quantity of bagasse available on the grate inside the furnace is limited. To maintain the steam pressure and temperature in the boiler continuous bagasse feeding in the furnace without any interruption is very much essential. Therefore an arrangement of silos are made that can hold up to 10-15 minutes bagasse required.

In mill stoppages the feed of bagasse to silos is stopped. At that time bagasse is taken from the bagasse yard is taken by return bagasse carrier. Bagasse baling machines as well as bagasse de-bailers or briquette crushers are provided in the bagasse yard.

The bagasse extractor / feeder extracts the stored bagasse. The extractor is of rotary drum with spikes on the drum, the drum is rotated by variable speed drive arrangement. A screw feeder feeds the bagasse from the outlet of the extractor and conveys it to the mouth of the mechanical / pneumatic distributor for bagasse. The screw feeder is rotated at constant RPM with drive arrangement.

10 Supporting Structure: The boiler is top supported and is provided with suitable steel supporting structure of bolted/welded construction from rolled steel sections and designed with adequate strength for the loads imposed by the boiler and associated equipment. The supporting structure is provided with monorail beams for floors.

11 **Chimney:** The waste gases are discharged through a chimney of R.C.C. or steel construction. The height of chimney is usually about 30 m to prevent the fly-ash from spreading in the immediate vicinity of the boiler installation.

12 **Chemical Dosing System:** Chemicals dosing is followed to the boiler feed water circuit at appropriate points for control of boiler water quality. Two types of dosing systems are present, high pressure and low pressure. Both the systems consist of tanks for solution preparation, pumps and agitators with motors.

13 **De-aerator:** Dissolved gases particularly oxygen and carbon dioxide - if passed along with feed water into the boiler, attack the boiler tubes at the high temperatures. Therefore the feed water has to be treated to remove dissolved gases. For this purpose, the feed water is passed into a de-aerator where it is atomized in steam under low pressure of about 0.2-0.3kg/cm^2 g.

The deaerator consists of *a de-aeration section*, a storage tank, and a vent. In the deaeration section, steam bubbles through the feed water. The steam heats the water and agitates and gets condensed. The feed water gets heated up to saturation temperature corresponding to the steam pressure in the deaerator. Non condensable gases and some steam are released through the vent.

The feed water is also treated with suitable reagents. The pressure in the deaerator remains constant. The make-up water is taken from the demoralization plant.

14 **Fly ash arresters / Electro Static Precipitator:** In sugar mills the unburnt solid particles carried along with flue gases settle down in the neighboring areas and cause environmental pollution. The factories have to install fly-ash arresters to meet this problem. This equipment installed in the flue passage operates on the principle of centrifugal action or change of direction of the gases.

The flue gases leaving the boiler are passed through an ESP to remove the dust particles. The ESP unit consists of transformers, rectifiers and controls. In ESP, fly ash is collected in hoppers. The fly ash collected is in dry form to ease the process of transportation by means of dense phase conveying system to Ash Silo of adequate capacity.

15 **Boiler Feed Pumps:** Boiler feed pumps are multistage centrifugal pumps with a higher head and reasonable discharge capacity. As these pumps are mainly designed to supply water to the boiler during steam generation, its discharge head should be sufficient to feed the boiler by overcoming its internal pressure.

Reserve feed water pumps are necessary, in case of any possible mechanical and electrical failure. In some cases where an alternative power source need is absent may arise to have a turbo-feed water pump for the plant to use in case of power failure.

16 **Induced draft fan:** Supplying the required air for the combustion process and further withdrawing it is called as draught. This includes supplying fresh air to

the combustion chamber by forced draught and the evacuating of flue gases by induced draught.

A fan or blower is located at or near the base of the chimney. *The pressure over the fuel bed is reduced below that of the atmosphere. The gases are drawn by creating a partial vacuum in the furnace and the flue gasses are passed up to the chimney.*

17 **Forced draft Fan:** A draught is produced by a fan. In a forced draught system, a blower or a fan is installed near or at the base of the boiler to force the air through the boiler combustion bed and other passages through the furnace, super heater economizer etc.

18 **Secondary air fan:** It is a secondary air fan. In this system, a fan is located at or near the sides of the Furnaces. The discharge is connected to the boiler's body with multiple nozzles that direct air to the inside combustion chamber to create turbulence in the flue gas. Hence ensures the perfect mixing of fuels which puts the combustion under suspension.

19 **Pneumatic spreader fan:** A pneumatic spreader fan is in which the discharged air is a motive force in spreading the bagasse from the spreader to the boiler furnace.

20 **Ash Handling Equipment:** Ash handling equipment is the most important type of equipment that need frequent follow-up, which otherwise disturbs the working atmosphere badly and significantly reduces the efficiency of the steam generating unit. It is also appropriate to respect the country's rules and regulations regarding air pollution.

21 **Water Tanks:** There are different tanks in the boiler plant used for handling feed and boiler water.

- Feed Water reservoir tank
- Deaerator tank
- Deaerator water storage tank
- Blowdowns tank

22 **Boiler Mountings**: These are fittings primarily intended for the boiler's safety and control of the steam generation process. These are -

1. Pressure Gauges
2. Safety valves
3. Water level indicators
4. Feed Check Valves
5. Steam Stop valve
6. Fusible Plug
7. 7 Blow off Cock
8. Manholes & Mud holes

73.3 Air required for combustion:

Air supplied for combustion of bagasse contains 23.15% Oxygen and 76.85% nitrogen and other inert gases by weight. On volume basis the corresponding percentages are 20.84 and 79.16 respectively. Combustion of bagasse involves following reactions

For calculating air requirement for combustion bagasse, its composition essentially should be known. The composition of dry bagasse is as under

- Carbon 47 %
- Hydrogen 6.5 %
- Oxygen 44 %
- Ash 2.5 %

Amount of oxygen required for complete burning of dry bagasse is as under

Oxygen required for burning of carbon

$$C + O_2 = CO_2$$
$$12 + 32 = 44$$
$$0.47 + 1.253 = 1.723$$

- Oxygen required for burning of hydrogen

$$2H_2 + O_2 = 2H_2O$$
$$4 + 32 = 36$$
$$0.065 + 0.52 = 0.585$$

Hence total oxygen required for complete burning of 1 kg of dry bagasse is 1.253 + 0.52 = 1.773 kg. Out of 1.773 kg of oxygen required, 0.44 kg of oxygen is available in the bagasse itself. Therefore oxygen required to be provided from air is 1.773 – 0.44 = 1.333 kg. However air contains 23.15 % oxygen by weight. Hence to provide 1.333 kg of oxygen, theoretically required air will be 5.76 kg. But in actual practice bagasse is wet.

Let w be the weight of moisture per unit weight of bagasse. Therefore the actual air required for burning of wet bagasse will be 5.76(1- w) kg per kg of wet bagasse.

In practice some excess air than theoretically required for complete burning of bagasse is to be supplied. Generally 35 to 45 % excess air is supplied. With little excess air incomplete combustions of carbon may take place and carbon monoxide will be formed. The carbon monoxide formation gives only 2361.0 Kcal heat as against 8000 KCal per Kg of carbon when carbon dioxide is produced.

73.4 Combustion of bagasse:

Heat evolved by combustion of bagasse increases gases temperature. The temperature of the furnace in efficient boiler ranges between 1100°C to 1200°C. This combustion temperature decreases with higher excess air and increases with higher temperature of the preheated air. Moisture in bagasse has adverse influence on the temperature of combustion.

Major portion of the heat is transmitted through the boilers tubes to the boiling water and steam. The remaining heat passes along with the flue gas. Some heat is lost by radiation and boiler blow downs. Heat loss in the flue gases increases with excess percentage of air supplied to the furnace. Higher moisture content in bagasse also causes loss of heat.

The flue gas from the boiler chimney contains carbon dioxide, water vapors, nitrogen and oxygen from excess air. In modern boilers optimum efficiencies are obtained with CO_2 content of the gases of 12-15% and excess air varying from 35-40%.

Boiler efficiency is decreased due to incomplete combustion of bagasse and formation of carbon monoxide. The unburnt bagasse goes along with ash...

In practice out of the total heat obtained in combustion of bagasse in furnace, the **losses take place as under—**

- The sensible heat loss in flue gas ranges from 8 to 15 %.
- Loss in unburnt bagasse that passes along with ash.
- Loss due to incomplete combustion of bagasse. This results in formation of CO instead of CO_2
- Radiation loss, leakages, blowdowns etc. These losses ranges between 6 to 12 %.

Heat loss in flue gases increases with—

- percentage of excess air
- moisture of bagasse
- temperature of flue

Taking all the above losses into account the heat transferred from fuel to steam is around 65-68% of the heat theoretically obtained under certain standard conditions of temperature and pressure. In practical terms this is stated as overall efficiency.

To reduce the losses and attain high efficiency by following features are incorporated in modern designs of boilers

- To reduce the moisture content of bagasse at mills itself so that calorific value of bagasse is increased,
- Supply of preheated air,
- Raising the temperature of boiler feed water
- Thorough mixing of air and bagasse in the furnace so that there will be minimum unburned bagasse left out and there will minimum formation of carbon monoxide.

Generally boilers of higher pressure and temperatures of steam are more efficient. The most important sections in boilers is the furnace.

73.5 Calorific value of bagasse:

The bagasse percent cane is about 26-29%. The bagasse contains 48-49% moisture, 2.0 - 3.0 % dissolved solids or brix and 48 - 50% fibre. The fibre constitutes the insoluble cellulosic matter of the sugar cane.

Calorific value i.e. heat generated by combustion of a unit weight of bagasse is expressed in two ways

i) Gross calorific value (GCV) and
ii) Net calorific value. (NCV)

The GCV or higher CV is the heat generated by combustion of one Kg. of bagasse and the products of combustion being reduced to the conditions of 0°C under 760 mm of mercury. The water vapour formed as a result of combustion of hydrogen as also the water present in bagasse is condensed under these conditions.
The net calorific value or the lower calorific value is based on the condition that water generated due to presence of hydrogen or the water accompanying bagasse is presumed to be in vapour state.

The gross CV gives the amount of heat theoretically liberated but the net CV indicates the heat available under practical conditions. The formulae for the CV of bagasse taking into account the major components are as under—

For dry bagasse

GCV = 4600 KCal/Kg and NCV. = 4250 KCal/Kg.

For wet bagasse

GCV = 4600 - 12S - 46 W (KCal/Kg)
NCV = 4250 - 12S - 48.5W N.C.V. (KCal/Kg.)

Where S = Sucrose % bagasse W = Moisture % of bagasse.

The dissolved solids and moisture content in bagasse reduce the heat value of bagasse. The moisture content has greater influence on the heat value.

73.6 Boiler efficiency (η):

Boiler efficiency can be calculated by two methods

A) Direct method – when all the instruments like flow meters, pressure gauges, temp. gauges, are in working conditions and fuel burnt is weighed

$$\eta = \frac{\text{Heat output in Kcal}}{\text{Heat input in Kcal}} \times 100$$

$$\text{Or } \eta = \frac{\text{Steam generated X Heat in steam above feed water temp.}}{\text{Fuel (bagasse) X GCV}}$$

Heat gained by the steam above feed water temperature is calculated by using steam table or heat entropy chart.

B) Indirect method:

In sugar factory it is not possible to weigh the bagasse that is burned in the boilers and most of the instruments like flow meter etc. are not in working conditions. With some instruments like pressure gauges, temp. gauges, and readings Co_2 by gas analyser the boiler efficiency can be calculated.

$$\text{Boiler efficiency } (\eta) = \frac{\text{Heat output}}{\text{Heat input}} \times 100$$

$$\therefore \eta = \frac{\text{Heat input} - \text{heat losses}}{\text{Heat input}} \times 100$$

Heat losses:

1) Condensation loss % $= \frac{GCV - NCV}{GCV} \times 100$

(Condensation loss ranges from 15 to 18 %)

2) Sensible heat loss in flue gases %

$$= \frac{(T_g - T_r)\left[(1-w)(1.4m - 0.13) + 0.5\right]}{NCV}$$

Where

- T_g – flue gas temperature
- T_r - room temperature
- w – moisture per unit weight of wet bagasse
- m ratio of actual air to theoretical air

(The sensible heat loss in flue gas ranges from 8 to 15 %)

3) **Unknown heat loss:** the loss of Leakages, radiation, blow downs, unburnt bagasse, or loss due to CO formation, ranges between 6 to 12 %.

Evaporation ratio or Seam/Fuel Ratio:

The quantity of steam generated from one kg of bagasse

$$= \frac{\text{Steam generated (kg)}}{\text{Fuel (bagasse) burnt (kg) during the same period}}$$

The evaporation ratio or steam/fuel ratio varies from 2.0 to 2.7 kg steam /kg of bagasse and can be reached as high as 2.9 kg steam /kg of bagasse in case of modern efficient boilers with large combustion chambers and large air heaters. The overall efficiency of modern boilers vary from 60 TO 68 % on GCV or 78 to 83 % on NCV.

73.6 Bagasse drying

Moisture in bagasse brings down the heat value of the bagasse as shown by the formula

$$\text{N.C.V.} = 4250 - 12S - 48.5W$$

Where S and W represent sugar and water percentages respectively while N.C.V. (net calorific value) stands for net calorific value of bagasse in KCal/Kg.

Flue gases after the economiser and preheater can be used for drying of bagasse to recover more heat from flue gases. The temperature can be reduced up to 100°C i.e. well above their dew point of 60° - 70°C. It can be seen from the above table that the drying of bagasse gives higher production of steam per unit weight of bagasse. Bagasse drying is carried out in some factories in the world up to 33 to 40 % moisture content in bagasse. At Waiialu in Hawaii bagasse was being used and moisture in bagasse was being brought down from 47.8%to 33.50%.

Table 73.3 reduction in bagasse moisture % and percentage increase in NCV

Moisture % in bagasse	G.C.V.	N.C.V.	% increase in N.C.V.	Weight of dry bagasse	corrected N.C.V.	% increase in corrected N.C.V.
50	2270	1800	0	100	1800	0
40	2730	2300	+27.8	83.3%	1915	+6.4
35	2960	2525	+40.3	77%	1944	+8.0

In Brazil separate drying of bagasse is followed to each furnace with hot gases after the air heaters. The moisture of bagasse was dropped from 52 to 40%. However in most of the factories in the world it is not being followed because of high cost of dryer.

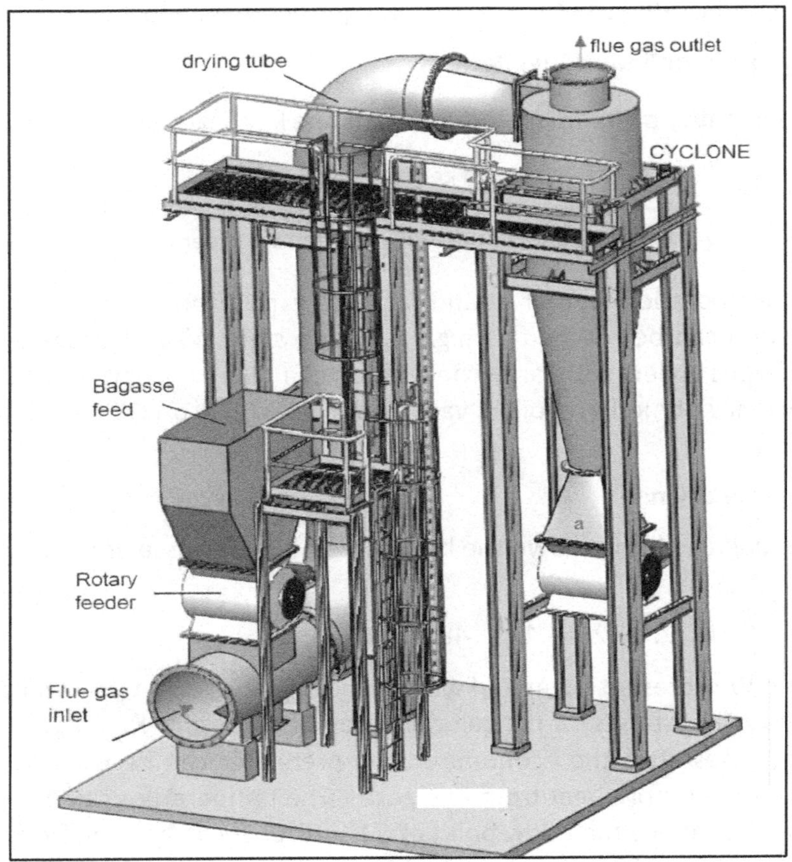

Fig.73.4 Bagasse dryer a pilot project proposed at Colombia University

A pneumatic bagasse drying system is proposed to reduce the moisture content of bagasse from 48% to 30%. This work provides bagasse dryer design parameters, including specifications for dryer system components, such as feeders, fan, drying tube, and cyclone. The total bagasse drying system proposed can be fitted within a 6 x 6 x 25 m space to dry 60 tph of bagasse, reducing the moisture content from 48% to 30%.

-OOOOO-

74 BOILER WATER TREATMENT

74.1 General:

Water to be feed to boilers should be demineralized and made soft with standard limitations of dissolved impurities. In sugar factory the exhaust steam get condensed in heat exchanger in boiling house is most pure water and if not contaminated by sugar, is used as boiler feed water with minimum treatment. The vapour condensate also has to be used as makeup water. Now in modern sugar factories mostly all exhaust steam is used in the first effect of evaporator and juice heating and pan boiling is done on vapours from the evaporator bodies. All the exhaust condensate of first effect is taken to the boiler feed water tank.

In some factories still exhaust steam is used for sulphited juice second heating and clear juice heating. However this exhaust condensate should be avoided to use as feed water for boilers. It is because in juice heaters there is high pressure to juice side and comparatively lower pressure to steam side and if tube gets leak, mostly heavy leakage of juice into condensate occurs and such water should not be allowed to go to boiler.

The total exhaust condensate sent to boiler is not sufficient for boilers because some steam is directly used in the boiling house and some steam is lost due to leakages in drains and lost to atmosphere. Further some water from the boiler passes out as blowdowns. Therefore some make water is always required and part of condensate of second effect of evaporator is usually send to boiler as make up water.

The condensate from vapours generally contains some traces of volatile matter distilled off from juice. Further if entrainment occurs the condensate gets contaminated with sugar traces. Therefore the vapour condensate going to boiler has to be regularly tested for sugar test pH etc.

At the start of the season when condensate is not available or due to some unavoidable circumstances the boiler operators have use raw water that has to be treated carefully and demineralise. Sometime in stoppages - like general cleaning or any other prolonged stoppage, factory can use the excess condensate stored is separate tank.

At any time if condensate gets contaminated with juice and passes to the boiler feed tank and then in boiler, then the pH of boiler water goes down due to decomposition of the sugars and other organic matters, moreover the carbonaceous material gets deposited inside the boiler tubes. In such event the factory operation is disrupted until the boiler water quality becomes normal after the necessary blow down and replacement of contaminated water by pure good quality feed water.

In Sugar Industry mostly boilers of pressure 32kg/cm^2 however in many old sugar factories still 21 kg/cm^2 boilers are used. In recent years, several sugar factories have installed high-pressure boilers of 45, 67 and 87/110 kg/cm^2 pressure for co-generation of power. In old sugar factories having 21 kg/cm^2 pressure boilers due attention is not being paid to the monitoring and control of boiler feed water as well as boiler water quality. On the other hand many sugar factories are successfully and regularly controlling boiler water for 45/67/86kg/cm^2 pressure boilers and turbines.

Boiler water purity is very important for efficient and reliable operation of medium (20 to 60 kg/cm^2) and high-pressure (61kg/cm^2 and above) boilers. For high-pressure boilers stringent quality parameters are to be maintained for boiler feed water. Boiler water and boiler feed water quality norms have been specified for different pressures of boilers operation and it is absolutely essential to rigidly follow the quality standards of water for smooth boiler operation.

Therefore, treatment and control of boiler feed water has become more important. Steam purity is also important for safety and reliability of turbines. The problems encountered on account of deviation from the quality standard, in so far as they disrupt the smooth functioning of boiler station in cane sugar factories are as under—

74.2 Scale formation on boiler pressure parts:

During the course of evaporation, water is evaporated as steam and impurities that remain accumulated in boiler water get concentrated. The salts become insoluble and get deposited as scale on the boiler tubes. The sodium disilicate, calcium carbonate, calcium sulphate and calcium silicate exhibit a retrospective solubility, being more soluble in cold water than hot. Hence at high temperature in boiler drum, these salts are precipitated and get deposited as hard scale.

The iron and copper ions present in boiler feed water also get segregated in the boiler drum and deposit as scale on boiler tubes. The scale deposits in boiler tubes have much lower thermal conductivity and decreases the overall coefficient of heat transfer of the tubes. The steam generating capacity is decreased. The loss in efficiency due to scale formation can be as follows.

Table 74.1 thickness of scale and loss in efficiency

Thickness of scale (mm)	Loss in efficiency %
1.6	9
3.2	18
6.3	38

- The steam/fuel ratio decreases and this affects fuel economy. More bagasse is consumed for the same steam production.
- If the scale is too thick the tube metal temperature rises which may lead to distortion or even bursting of the tubes resulting in costly boiler breakdown.

Therefore it is paramount important to keep the heating surface of the boiler tubes clean and to protect the boiler tubes and drums from scaling and corrosion to maintain high boiler efficiency. This is possible by feeding water that contains impurities below certain standard norms, to the boiler for which the boiler feed water should be suitably treated.

74.3 Foaming:

Foaming is due to the chemical composition of the water. Pure water does not foam and in a boiler steam bubbles are large, rising quickly to the surface and bursting quickly. In presence of certain dissolved or suspended substances the surface tension is altered and the steam bubbles remain small. Therefore they do not rose quickly and in effect expand the water foam over at the steam take off. Thus excessive level of surface-active chemicals in boiler water can cause foaming and priming.

Operating the boiler at excessive less than the design pressure increase the volume of steam produced although steam generation rate is same. For example a reduction in pressure from (21 kg/cm^2 to 18 kg/cm^2) increases the volume by 17% and enhances tendency of foaming. As the layer of foam thickness and approaches the steam take-off, drops of water from the bursting bubbles or from the foam itself may pass over with the steam.

Factors causing foaming:

- High suspended solids in boiler water

- High alkalinity of boiler water
- High dissolved solids in boiler water
- Contamination of boiler water with oil and /or detergents.

Table 74.2 Requirements of feed water and boiler water and condensate (Permissible parameters)

	Particulars	21 to 40 Kg/cm^2	41 to 60 Kg/cm^2	61 to 80 Kg/cm^2	81 to 100 Kg/cm^2
Boiler Feed Water					
i	Total hardness as CaO mg/l, (max.)	1.0	0.5	Nil	Nil
ii	pH value	8.5-9.5	8.5-9.5	8.5-9.5	8.5-9.5
iii	Oxygen as O_2 mg/l, (max.)	0.02	0.02	0.01	0.005
iv	Iron and copper mg/l, (max.)	5	0.5	0.02	0.01
v	Silica (SiO_2) mg/l, (max.)	-	-	0.05	0.02
vi	Oil mg/l, (max.)	-	-	Nil	Nil
vii	Residual hydrazine (as N_2H_4) mg/l (Max)	-	-	0.05	0.05
viii	Conductivity at 25°C microsiemens/cm. max	-	-	0.5	0.3
viii	Oxygen consumed in 4 hrs. mg/l Max	-	-	Nil	Nil
Boiler water					
i	Phosphate as PO_4 mg/l	15-30	5-20	5.20	5.20
ii	Caustic alkalinity ($CaCo_3$) mg/l (Max.)	200	60	20	7
iii	Total alkalinity ($CaCo_3$) mg/l (Max.)	500	300	50	30
iv	Silica (as SiO_2) mg/l (Max.)	80	24	10	5
v	Total hardness (as $CaCo_3$) mg/l (Max.)	Nil	Nil	Nil	Nil
vi	Residual sodium sulphite as Na_2SO_3 mg/l	20-30	-	-	-
vii	Residual hydrazine (as N_2H_4) mg/l	0.1-0.5	0.05-0.3	0.05-0.1	< 0.01
viii	Total Dissolved Solids mg/l (Max.)	2500	1500	120	100
ix	Chlorides mg/l (as NaCl) Max.	-	-	6	4
x	pH value	11.0-12.0	10.5-11.0	10-11	10-11
Condensate					
i	pH value	-	-	8.5-.9.0	8.5-9.0
ii	Hardness (as $CaCo_3$) Max.	-	-	Nil	Nil
iii	Oil mg/l, (Max.)	-	-	Nil	Nil
Iv	Silica (as SiO_2) mg/l (Max.)	-	-	0.02	0.02
v	Iron and copper mg/l, (Max.)	-	-	0.01	0.005
vi	Ammonia (as NH3) mg/l, (Max.)	-	-	0.5	0.5
vii	Sodium (as Na)) mg/l, (Max.)	-	-	0.06	0.03
viii	Dissolved oxygen as O_2 mg/l, (max.)	-	-	0.03	0.03

The water film around each bubble is made more stable by an increase in suspended or dissolved solids in the boiler water. Oil, detergent or organic matter can also cause foaming. The suspended solids act as nuclei for bubble formation and thus increase foaming.

Reducing sugars if enter in the boiler, - due to high-pressure and temperature conditions, decompose to organic acids and increase foaming.

74.4 Problems associated with impurities in steam:

In general steam is free from impurities the water might have contained. However sometimes entraining of boiler water along with steam may occur which is termed as priming or carry-over.

Carry Over or Priming:

Carry-over or priming is contamination of steam by boiler water and contained solids. The carry over sludge and silica in the steam get deposited inside the turbine inlet valves, jamming its operation, which may lead to over-speeding the turbine, which is dangerous. Scale may also deposit on turbine blades causing imbalance and vibrations. The priming causes heavy erosion of turbine blades.

Factors causing carry over:

- Excessive dissolved solids in boiler water
- Excessive suspended solids in boiler water
- High boiler water level
- Fluctuations in steam load
- Over loading of boilers
- Sudden increase in steam demand
- Mechanical defects in the boiler including faulty design

If the concentration of dissolved and suspended solids are allowed to rise beyond a certain limit carry-over can occur but this can be controlled or minimized by blowing down the boiler water to limit the concentration of dissolved solids and suspended solids in it.

Sudden addition of excess dose of chemical like sodium carbonate may lead to increase in sodium content and violent evolution of CO_2 also may lead to violent ebullition leading to carry-over. When boilers are continuously over loaded sudden variation in load leads to leakages in steam drum internals that can cause carryover.

The problems of carry over due to mechanical factors and operating parameters such as overloading of boilers or high water levels cannot be controlled by chemical treatment.

Silica carryover:

At high temperature the dissolved silica in boiler water volatilise and pass along with the steam in the vapour phase and when the steam temperature is lowered forms a scale on turbine blades. This phenomenon commences at about 40 kg/cm^2ab and becomes increasingly serious as the pressure increases. If the silica content of the steam is kept below 0.02 ppm there is usually no problem and this can be assured by limiting the boiler water silica content, the value depending on the pressure and alkalinity.

74.5 Corrosion:

The presence of oxygen in the feed water is the primary cause of corrosion. Low pH, scale deposits and carbon dioxide also induce and accelerate corrosion. Corrosion can take place at economizer boiler drums and tubes and steam piping.

Oxygen accelerates the reaction of iron and water. It can react with iron hydroxide to form a hydrated ferric oxide or haematite. This action is generally localised and forms a pit in the metal. When the pit becomes progressively anodic severe corrosion results. At the cathode surface oxygen react with hydrogen and depolarises the surface permitting more iron to dissolve the anode creating a pit.

74.6 Raw Water Characteristics:

The raw water characteristics vary considerably in composition from place to place in respect of pH, total hardness, magnesium, chlorides, sulphates silica and total dissolved solids.

All natural water contains various impurities suspended and dissolved. The impurities are dissolved from the atmosphere while raining and while flowing over the land and percolating through soil. These impurities include suspended, bacteriological as well as dissolved inorganic and organic substances. Ca^{+2} Mg^{+2} Na^{+} K^{+} Fe^{+2} Al^{+3} traces of Zn^{+2} and Cu^{+2} are the main cationic impurities while (HCO_3^-) CO_3^{-2} Cl^- $SO4^{-2}$ and NO_3^- are the general anionic one. Bicarbonates, carbonates and hydroxides cause the alkaline hardness. The bicarbonates associate with Ca^{+2} and Mg^{+2} contribute to temporary hardness. The raw water does not contain any caustic alkalinity and is incorporated during treatment say the soda lime process.

74.7 Condensate:

The steam is generated in the boilers, supplied to the prime movers and the exhaust steam from the prime movers is utilized in boiling house for juice heating, evaporation & pan boiling. The condensate is returned back to the boiler. Thus the steam and the condensate form a close circuit. Steam used for miscellaneous purposes and blown off, which is normally 5-8 % of the total steam cannot be returned back to boiler. However vapour condensate is used for makeup. Sugar Industry is perhaps the only industry in which condensate obtained in boiling house is more than boiler feed water requirement. In routine sugar factory working almost 100% condensate water is used for boilers. Therefore need of treated raw water for boilers is limited only at-

- The starting of the season
- Restarting the factory after general cleaning
- Emergency need if factory working is disturbed.

Exhaust condensate except that of juice heaters is totally sent to boiler feed water tank. In case of juice heaters the juice side pressure is always higher than the steam side pressure. Therefore in juice heater if any tube leakage occurs large quantity of juice is injected to steam side and by the time it is noticed heavily contaminated condensate may pass to boiler feed water tank and into the boiler, which becomes dangerous for boilers. Therefore even exhaust condensate from juice heaters is not taken to boiler feed water tank. On the other hand at evaporators the juice side pressure is always lower than that of steam side. Therefore even if tube leakage occurs little juice can pass in condensate till it is noticed and action can be taken to divert the condensate to sweet water tank / hot water overhead tank without much trouble at the boiler station. Therefore vapour condensate (from the second effect of evaporator or sometimes pan condensate) is usually taken as make up water.

Vapour condensate usually contains mineral impurities ranging from few ppm to 100 ppm. Sometimes oil contamination and dissolved gases like ammonia are also found in condensate.

74.8 Nature and source of impurities: The nature and source of impurities can broadly be classified as under.

Organic contamination:

Sometimes pH of vapour condensate is found high which is due to ammonia and ammonium compounds but as the water enters boiler the ammonium compound decompose at high temperature and ammonia and corresponding acids are formed. Ammonia gets evaporated and acids reduce the boiler water pH.

Sometimes due to entrainment or tube leakage at evaporator juice passes along with condensate to the boiler. The sugar in the boiler feed water when enters the boiler due to high alkalinity and high temperature decompose in the boiler and corrosive acidic compounds which are detrimental to the boiler tubes and shells are formed. Sugar contamination leads to dark colouration to the boiler water, which then gives characteristic odour. In this condition giving blow downs and replacing the boiler water with quality clean water is necessary.

Iron and Copper ions:

These contaminants are picked up by the condensate from heating and boiling equipment from boiling house in which steam /vapour are condensed.

Inorganic Impurities:

When untreated or improperly treated raw water is taken to boiler feed water tank, the inorganic impurities such as calcium, magnesium sulphates; calcium silicates or sodium silicates pass into boiler water.

The vapour condensate may content SO_2 that is used for juice and syrup sulphitation. This may cause sudden drop in alkalinity of feed and boiler water.

Dissolved oxygen

The treated raw water as well as condensate picks up oxygen from atmosphere. Under practical conditions dissolved oxygen is usually found between 0.2 - 0.3 ppm. Dissolved oxygen even at trace level and presence of acidic ions cause corrosion and pitting in boilers.

Table 74.3 boiler water temperature and dissolve oxygen in ppm

Temperature °C	Dissolved oxygen Content in ppm
50	8
80	2
90	1.0
100	0.3
105	0.007

Oxygen gets released immediately when feed water enters the economizer/boiler tubes. Oxygen reacts with iron to give ferric oxide (Fe_2O_3) which will not protect metal from further attack and metal is continuously lost. The attack can be localised and result in deep pitting. Therefore it is necessary to remove dissolved oxygen from feed water.

73.9 Boiler feed water control:

Boiler water conditioning may be divided into two parts - External treatment and internal treatment. External treatment means conditioning of the boiler feed water before feeding it to the boiler and internal treatment means conditioning of the boiler water inside the boiler drum. The external treatment used for removal of suspended impurities as well as removal or reduction of calcium magnesium silica and other dissolved impurities. The internal treatment is done by adding chemicals in the boiler feed water to prevent scaling, corrosion, foaming and priming in the boiler.

External treatment - Boiler Feed Water Conditioning:

Fig. 74.1 boiler feed water treatment plant

Number of processes such as sedimentation, filtration, chemical, thermal etc. are recommended for the treatment of raw water. The primary objective in selection of external treatment of raw water is to enable it to produce a feed water of desired characteristic with minimum impurities so that it will minimize the operational problems at the lowest possible cost of chemicals. Typical examples of external treatment are

- Sedimentation
- Cold lime / soda softening
- Hot lime soda / zeolite softening
- Cat ion exchange softening

Natural river waters or lake waters contain less dissolved impurities but contain high percentage of suspended solids. The suspended impurities are difficult to remove therefore generally well water is preferred for boilers.

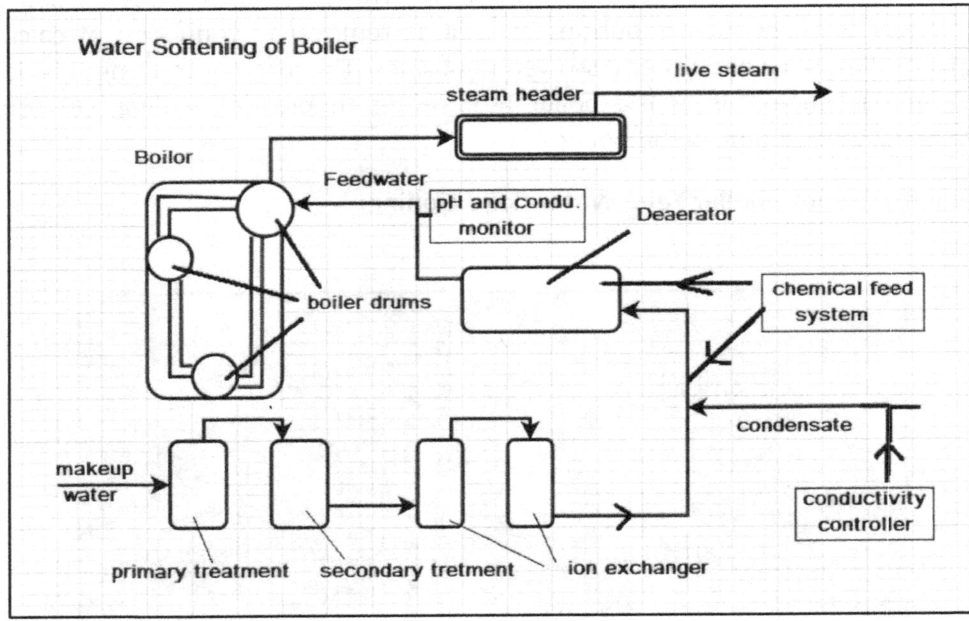

Fig.74.2 boiler feed makeup water softening system

Suspended impurities in water are usually removed by sedimentation. Addition flocculating agents such as Alum, sodium aluminates, ferrous or ferric sulphates etc. are used to expedite coagulation and accelerate the sedimentation process. Mechanical filters like gravity type pressure filters are sometimes used to remove suspended impurities. Microorganism impurities are removed by the use of potassium permanganate.

Lime softening

It involves removal of impurities by chemical reaction with lime. Carbonate hardness of calcium and magnesium are precipitated as calcium carbonate and magnesium hydroxide. Silica is removed as an insoluble magnesium complex. The reaction generally proceeds at ambient temperature.

Concentration of silica in high-pressure boilers can be controlled by the method of lime softening of raw water. If cold lime softening process is used ferric sulphate can be added to form floc of hydrous-ferric –oxide which absorbs silica.

Lime soda softening

Lime softening does not remove non-carbonate hardness; soda ash is supplemented with the lime for the removal of non-carbonate hardness from water. It takes too long time for completion of the reaction. Calcium carbonate and magnesium carbonate both are slightly soluble. Therefore the hardness in water is never reduced to value below 30 ppm.

Ion Exchange Processes:

Base Exchange softening:

This is the simplest ion exchanger process commonly used for softening the raw water in which calcium and magnesium in the water are removed and replaced by sodium. Nothing else is changed so that the solids remain the same.

An ion exchange plant consists of a vertical cylindrical vessel having a coarse and fine gravel layer at the base. Above the gravel layer, a layer of a strong cat ion exchange resins are filled. The resin is initially in Na^+ form, which is replaceable from resin matrix by the hardness ions (Ca^{++}, Ma^{++}) from raw water and provide softened water.

When the resin is exhausted it is regenerated. First the column is back washed for removing suspended matter arrested in the gravel layer. Then the resin is regenerated to the sodium form by passing saturated brine solution through the column. Calcium and magnesium are discharged to waste. The cycle is of 6-10 hours.

Dealkalisation

In dealkalisation a week acid resin is used. It is capable of exchanging alkaline hardness into the corresponding acid (carbonic acid) but it leaves non-alkaline hardness partially undisturbed. The carbonic acid produced is blown in a degassing tower and the water is then adjusted for alkalinity with caustic soda. In this process the total solids are thus reduce by the amount of alkaline hardness. The regeneration of the unit is carried out by sulphuric acid or hydrochloric acid.

Split stream dealkalisation

In this process we have a parallel arrangement of ion exchange resins in sodium and hydrogen form through which water flows and combine again after treatment. The treated water is practically soft and of low in alkalinity. Alkalinity and hardness are to be adjusted in the treated water. Common salt and acid are used for regeneration of sodium and hydrogen cation units respectively.

Demineralisation

This process uses two types of resins either in separate units or as a mixture in a single unit. One resin removes all cations (Calcium, magnesium, sodium etc.) and the other removes all the anions (sulphate, chloride bicarbonate silica etc.) the treated water is almost free from dissolved solids.

74.10 Water Analysis:

Water from treated water reservoir, boiler feed water tank and boiler drum should be tested for pH, TDS, dissolved silica, hardness sulphates etc. So that corrective steps can be taken in case the level of impurities rises above the permissible limit. These tests are recommended for all boilers having working pressure of 21 kg/cm^2 and above.

The impurities in treated feed water should be in specified permissible limits. The total organic carbon should also be continuously checked in boiler water in respect of working pressure of 67 kg/cm^2 and above to know the level of organic contamination.

74.11 Deaeration

Dissolved oxygen should be removed before the feed water enters the economizer. In low-pressure boilers all deaeration is accomplished by use of chemicals only but in high-pressure boilers the combination of mechanical and chemical deaeration is necessary

Boilers operating at pressure 32 kg/cm^2 and above should be provided with de-aerators. Solubility of oxygen in water at temp above 105°C is 0.007 ppm. Therefore to remove dissolve oxygen the boiler feed water before it is feed to boiler is passed through a de-aeration tank in which it is heated upto105°C (at 0.5 kg/cm^2) by using exhaust steam.

74.12 Internal treatment:

The boiler steel reacts with water to form Fe_3O_4 (Magnetite). The magnetite forms a very thin, dense and adherent layer on the metal protecting it from further corrosion. Corrosion will take place if this thin is removed or penetrated. The magnetite film is most stable at 10.5 to 11.5 pH. Therefore pH of boiler water should be maintained in this range. Some impurities always enter the boiler and the water inside the boiler has to be chemically adjusted or balanced to prevent scale formation, corrosion and imbrittlement.

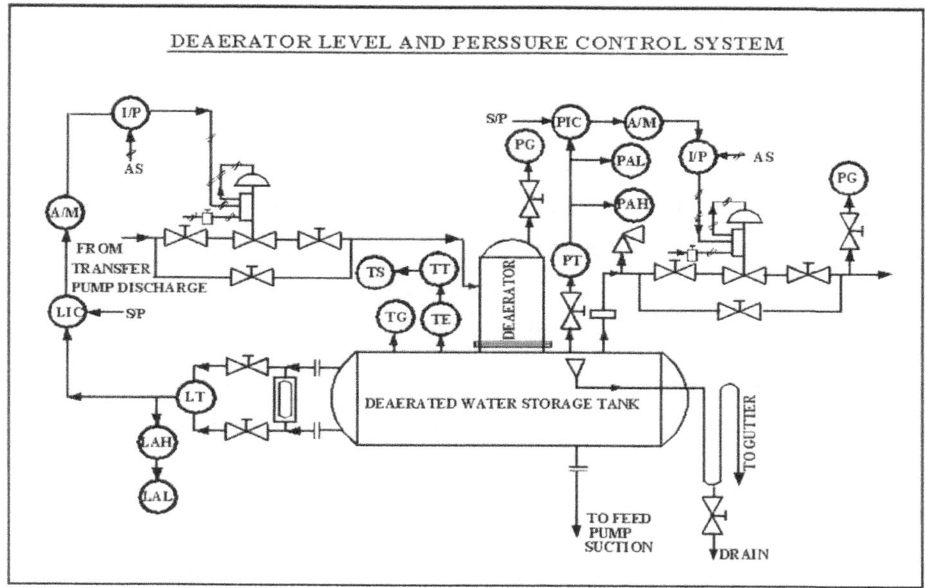

Fig.74.3 boiler feed water deaeration system

Conventional treatment:

Conventional treatment normally involves use of caustic soda, soda ash, and phosphate to precipitate calcium & magnesium. Tannin and lignin are used for sludge conditioning. Sodium sulphite or aqueous solution of hydrazine is used in chemical deaeration. The dosing of chemicals should be done through metering pumps.

Addition of caustic soda neutralises any free carbon dioxide in feed water. Alkalinity and pH is decided by correct dosing of caustic soda, pH should be in the range of 10.0 – 11.5. When value falls down below 9.5 the magnetite film may be attacked. When pH is high, due to presence of free hydroxide alkalinity caustic soda gets concentrated and converts magnetite layer to soluble Ferrates and Ferrite. This exposes metal surface to attack. It is therefore essential to keep the pH of boiler water within limits.

Under normal conditions the boiler feed water contain a combination of carbonates and hydroxides of sodium. Determining of pH of boiler water and controlling the treatment based on it cannot be so accurate because boiler water pH only does not give accurate indication of the amount of either the carbonates or the hydroxides present. Alkalinity analysis measures the total contribution of carbonates, bicarbonates, hydroxide and other alkaline compounds. Boiler water is tested for two different alkalinities

- Methyl orange alkalinity or total alkalinity (M)
- Phenolphthalein alkalinity (P)

To prevent corrosion particularly from carbon dioxide a minimum caustic alkalinity need to be maintained in the boiler water. Caustic alkalinity is calculated from total alkalinity (M) and phenolphthalein alkalinity (P). When phenolphthalein alkalinity is less caustic soda can be used to increase the alkalinity. If phenolphthalein alkalinity is less and methyl orange alkalinity is more, then washing soda is used.

Deviation in pH, which can occur due to any sugar contamination in the feed water have detrimental effect on magnetite film and rebuilding of the film must take place immediately so as to prevent pitting. The magnetite film also protects the caustic corrosion and hydrazine corrosion.

By use of chemicals the alkalinity of boiler water should be maintained in specified limits and should be 15-20% of total dissolved solids. Excess alkalinity results in foaming, carry over and caustic imbrittlement. Alkali corrosion and caustic imbrittlement are caused by excess concentration of sodium hydroxide.

Any hardness (Ca Mg salts) entering the boiler is likely to form scale inside the boiler tubes. The addition of phosphate prevents the formation of scale as the phosphate reacts with calcium and magnesium salts to form a sludge that does not form scale but remain in suspension and is removed by blow downs. One of the derivatives of sodium phosphate is added to the boiler water to precipitate the soluble calcium and magnesium salts and prevents deposition of scale. The phosphate has the advantage that they are stable and do not break down even at high temperatures. Another advantage of using phosphate is that no critical ratio are involved. The various phosphate compounds used for boiler water treatment act differently. Alkalinity can be controlled to some extent by the choice of phosphate salts. Trisodium phosphate increases the alkalinity. Disodium, monosodium phosphates, Mexa Meta phosphate reduce the alkalinity. Hence as per requirement of the alkalinity specific sodium phosphate should be used. Phosphate content should be maintained to 20-30 ppm in boiler water.

Chemical oxygen scavengers:

Sodium sulphite and hydrazine are the most widely used chemical oxygen scavengers. Both these chemicals can be catalysed, which greatly accelerates their rate of oxygen removal. Sodium sulphite is usually less expensive and is easy to handle, however it adds unnecessarily quantity of dissolved solids in boiler water. Hence additional blow-downs may be required. . It begins to decompose at temp above 280° C. Therefore it is recommended for low and medium pressure boilers. The activity of the sodium sulphite may be increased by addition of 5 % cobalt chloride in sodium sulphite. For low-pressure boilers an excess of 20-40 ppm sodium sulphite is generally maintained in the boiler water while the boiler water is in service. The excess sulphite takes cares of any fluctuations in the boiler feed water. Sodium sulphite reacts with oxygen to produce sodium sulphate

$$2Na_2SO_3 + O_2 = 2Na_2SO_4$$

The sodium sulphate formed remains in dissolved state in boiler water and passes in blow down. Hydrazine reacts with dissolved oxygen to form nitrogen and water

$$O_2 + N_2H_4 = 2H_2O + N_2$$

Hydrazine is slightly more expensive. It is toxic and should be handled carefully. It provides an effective and protective film of magnetite (Fe_3O_4) and does not add much-dissolved solids, hence blow down requirement is not in comparison with sodium sulphite. Hydrazine may be used in boilers at different pressures. In order to achieve good corrosion protection a continuous magnetite film (Fe_3O_4) must be maintained. Hence it is necessary to maintain a reserve concentration of chemical oxygen scavenger (COS). Therefore 30-40 ppm sodium sulphite or 5-10 ppm hydrazine should be kept in reserve in the boiler drum. Hydrazine also reacts with ferric oxide to form a protective layer of magnetite.

$$6Fe_2O_3 + N_2H_4 = 4Fe_3O_4 + 2H_2O + N_2$$

It also reduces cupric oxide to passive cuprous oxide.

74.13 Antifoaming Agents:

The addition of certain organic compounds as antifoaming agents in very small amount affects the surface condition of the small bubbles and causes them to combine again. Therefore use of antifoaming agents help in controlling the foaming and to some extent priming in low and medium pressure boilers but with high-pressure boilers those are not quite effective.

74.14 All organic conditioning of Boiler water:

Traditionally sodium sulphite and hydrazine are used as oxygen scavenger along with phosphate and caustic for corrosion control. Occasionally, morpholine and cyclohexyl amine are also used for corrosion control. This conventional treatment gives good corrosion protection if monitored strictly. However in practice such close control is difficult and any upset will lead to irreversible damage.

Introduction of polymerization technology has now made it possible to prepare specific polymers for boiler water treatment. Combination of such polymers with amines can now provide cost effective and reliable substitute for conventional phosphate treatment.

The magnetite layer formed by phosphate treatment may crack during operation providing access to corrodants. Under strict operating conditions, these cracks seal itself. However during upsets the corrosion goes unchecked. Ingress of oxygen and absence of hydrazine cause pitting corrosion during such upsets. Further the under deposits continue unchecked. Thus occasional upsets in phosphate programme cause corrosion and fouling in boiler.

Neutralising Amines

Amines are commonly used for film formation and corrosion protection. Ammonia and cyclohyxyl amines are being used for condensate corrosion control. Such neutralising amines react with carbonic acid in condensate to form neutral non-corrosive salt. However condensation of amines depends on partial pressure in system and dose control is required on condensate pH and dose. These amines do not prevent oxygen attack in condensate system.

Polyamines

The disadvantage of neutralising amines is overcome by using polyamines. Polyamines are versatile, thermally stable, multipurpose polymers, designed for boiler water treatment. The polyamines are synthesised to keep advantage of filming and neutralising amine and overcome disadvantage of basic chemicals.

Polyamines form a monomolecular elastic but stable film all over metal surface.

Due to its polybasic nature, same molecule of amine surface increasing film stability. Thus disadvantage of aliphatic amine is overcome.

Polyamine forms a close-knit monomolecular layer, which does not allow diffusion of oxygen. Thus oxygen cannot come in contact with metal, preventing oxygen corrosion and eliminating use of chemical scavenger like hydrazine. The polymeric nature of amine helps to disperse iron oxides by selective adsorption. Such dispersed iron oxides, then can be removed through blowdowns keeping system free from deposition. The polyvalent nature of polyamine gives buffering action in boiler. Thus pH is kept uniform throughout the system. Unlike trisodium phosphate polyamine is all organic in nature and thus do not add to boiler water TDS. Thus boiler blowdown can be reduced.

Polyamines ensure equal concentration of amine throughout the boiler system. The polyamine in steam phase neutralise carbonic acid condensate, reducing acid attack. The formation of polyamine film further protects condensate lines from corrosion and oxygen attack.

Elimination of caustic reduce foaming tendency of boiling water reducing carry over. The deposition of phosphate is completely eliminated. Further the elastic film formed over inside turbine and turbine blades protects it from other salt deposition.

74.15 Total dissolved solids:

Total dissolved solids in boiler water should be restricted as per specification. The control is important to prevent carry-over of solids. Electrical conductivity method is quick, simple and quite accurate for determination of TDS. The total concentration of soluble and suspended solids in boiler water depends on solids entering through feed water and chemicals added in boiler water for treatment and frequency as well as quantity of blow downs. Because of evaporation TDS concentration goes on increasing and because of chemical treatment the concentrated dissolved solids get precipitated in sludge form. The concentration of suspended and dissolved solids is controlled by blow downs of boiler water. Frequent and short blow downs control the suspended solids, sludge and dissolved impurities in boiler water more effectively in acceptable limits.

Generally there is a common feed water tank for all boilers. It is a normal practice to add chemicals for boiler water treatment to this tank. It does not allow for a proper control on the quantity of chemicals dosed in any one boiler. If one or two boilers take in feed at the time chemicals are added to the tank, these boilers would suck in a major portion of the treatment whereas the other boilers would receive only the remaining chemical charge. This can result in excessive caustic alkalinity in some of the boilers.

74.16 Prevention of corrosion in boiler during off-season:

Severe corrosion can take place in boilers if proper precautions are not taken when the boiler remains idle during off-season. Hence after stopping the boiler for off-season, after cleaning the boiler thoroughly the boiler is filled with treated water to normal operating level. Sufficient caustic soda is added to the boiler water for a hydrate reading of 1000 ppm and a sulphite reading of 100-200 ppm (Hydrazine reducer to 5-10 ppm). Then the boiler is operated for 2-3 hours at a steam pressure of 80-100 psig in order to remove the dissolved gases. Afterwards the boiler is cooled down so that the pressure returns to zero, and then the boiler is filled with additional quantity of water until it reaches the highest level. At this point all the valves are closed tightly so that air should not come in contact with the heating surfaces. The boiler water is to be tested regularly for alkalinity, pH and the sulphite or hydrazine content. In event of shortage additional quantity of chemicals are to be added for better protection from corrosion. Chemical conditioning of boiler during off-season prevents corrosion and pitting and improves considerably life of boiler tubes and accessories.

-OOOOO-

75 STEAM TEURBINES

A machine that converts some natural force into mechanical motion is called as prime mover. All heat engines are prime movers. A steam turbine, is also is a prime mover. It converts heat energy from high pressure steam into mechanical by motion by driving the turbine rotor.

The turbines are designed to suit the steam pressure and temperature conditions of the boiler. The turbine is coupled to an alternator through closed reduction gear. The alternator develops 3 phase alternating current of 50 cycles/Sec. and 400-440 volts.

In U.S. A. the standard frequency is 60 cycles whereas in India, Europe and some other parts of the world 50 cycles/sec. is standard frequency. The alternator speed has to be submultiple of 3000 for this frequency.

The steam nozzle is a passage of varying cross-section by means of which a part of the enthalpy of steam is converted into kinetic energy as the steam expands from a higher pressure to a lower pressure. Therefore, a nozzle is a device designed to increase the velocity of steam.

Fig. 75.1 power turbine

Nozzles are the elements of a turbine in which steam jets are formed. The steam discharges from the nozzle as a high velocity jet. Numbers of small blades are fixed to the rim of a revolving wheel or rotor. Jets of steam of high velocity are obtained by the expansion of steam through the nozzles and are directed on to the blades. The effective force of these jets acting on the blades rotates the turbine shaft.

As that steam flows and pass through the turbine's spinning blades, the steam expands gets cooled giving its thermal energy to the turbine blades and the rotor rotates at high speed. Thus thermal energy is converted into kinetic energy.

Fig. 75.2 multistage turbine

Low pressure steam coming out from turbine called as exhaust steam is used for process work or may be condensed to water in the form of condensate in condenser. This condensate is again fed into the boiler.

The steam turbine works on Rankine cycle in which work done is the difference between the initial and final total heat of the steam, which are measured at constant entropy.

Steam turbine is a most versatile prime mover used for variety of applications. The steam turbine runs at constant speed and typically designed to rotate at 3,000–15,000 rpm. It can be built from 10 hp to 100,000 hp. It is relatively quiet and smooth in operation.

In sugar mills the steam turbines are used for

1. Power turbine to generate electricity,
2. Prime movers for the mills, boiler water feed pumps, forced or induced-draft fans, blowers, compressors etc.

Multistage steam turbines - used for power generation, have thermodynamic efficiencies that vary from 65 % to 90 % depending upon design and capacity of the turbine.

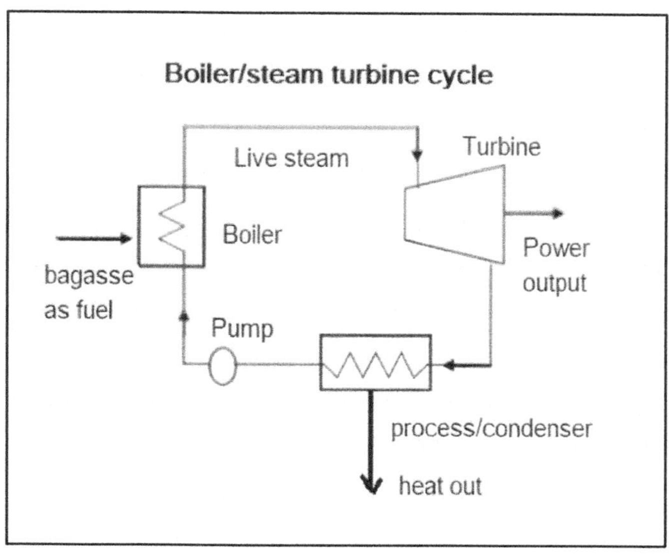

Fig. 75.3 steam- condensate cycle

The steam turbines used at mills and boiler feed water pumps, I.D. F.D. fans etc. are typically, single stage impulse type turbines having about 25-30 % efficiency. In sugar factories the efficiency of steam turbine was given less importance till that the exhaust production was less than process requirements. However, with profuse vapour bleeding in modern plants steam consumption has brought down now up to 30-35% on cane. Therefore and now with cogeneration trend, the efficiency of the turbine is being given due importance. Therefore now in modern sugar factories the mills are driven by electric motors.

In sugar mills steam leaving the boiler has pressure of 32kg/cm2 to 124 kg/cm2 and temperature 350 to 550°C. The efficiency of the turbine increases with increase in steam pressure and temperature. The highest temperature, which may be allowed in the steam turbine, is governed by the quality of material used for manufacture turbine.

Mechanical drive turbines are rated in horse power and turbine generator unit in Megawatts.

The electricity generated at power turbines is used to run the factory and surplus electricity generated can be exported to state grid.

Principal parts of turbine:

Rotor: This is main moving element of the turbine. In impulse turbine it is a shaft on which wheels are mounted that are fitted with blades. The rotor of a reaction turbine is a drum it is stepped or tapered so as to increase diameter towards the lower pressure end.

Casing: It surrounds the rotor and holds it internally. The bearings, auxiliaries and steam lines are part of the casing.

Bearing: The bearings are provided with pressure oiling system.

Shaft Seals: Where the shaft emerges from the casing it needs sealing to prevent steam leakage at both ends. In case of backpressure turbine at both ends the steam may leak to atmosphere. In case of condensing turbines there is vacuum at exhaust side and sealing is needed to prevent inflow of air.

Labyrinth glands with steam leak-off at high pressure end and steam or water sealing at the condenser end are employed.

The multistage impulse turbines must also be internally sealed between the shaft and diaphragms.

Steam inlet valves: In a typical installation the steam line leading to the turbine will contain one or more gate valves that isolate the turbine from the steam line. These are fully opened when preparing to start the turbine.

Next in sequence is the turbine throttle valve. This is manually operated and used to regulate the rate of starting the turbine. When the turbine is warmed and reached to its normal speed this valve is left fully open.

An emergency trip valve may either be built into throttle valve or become a separate unit following it. This is tripped shut if the turbine over speeds if governor does not operate properly.

It can also be connected to act on other emergencies - water in extraction lines, generator under voltage, loss of oil pressure.

Following the emergency valve the steam flows through the governor valve and into the first stage nozzles. The steam flow is regulated to maintain constant speed of the shaft.

Further the steam flows expansively through the turbine without control.

In case of constant pressure extraction cum condensing turbines there is automatically regulated valve at the extraction point.

Lubricating system: Oil is required for lubricating the bearings. In most of the turbines, same oil pressure system is used for both bearing lubrication and governor servomechanism operation. An integral oil pump, oil reservoir, oil filter and oil cooler are included in the system.

The turbines may vary in designs several ways. There are impulse turbines, reaction turbines, but we are not going in details about these differentiations. There are varied product range steam turbines from Single Stage Steam to Large Multi Stage Steam Turbines for generating power up to a single installed capacity up to 50 MW in sugar factories. From cogeneration point of view broadly there are four type of multistage turbine that are used for power generation.

1. **Back pressure turbines:** Back pressure turbines are available in single stage or multi-stage options. These types of turbine were being installed till 1990 when mostly cogeneration and export of electricity was not followed. Back pressure turbines are commonly used in sugar industry as exhaust steam is fully utilized for process.
2. **Condensing turbines:** Now in many sugar factories a back pressure turbine is used for generation of power for captive assumption and a separate condensing turbine is installed for export of power.
3. **Extraction backpressure turbines:** If there is a need for steam extraction when there is medium pressure demand within tolerable limit, then extraction cum back-pressure steam turbines are installed.

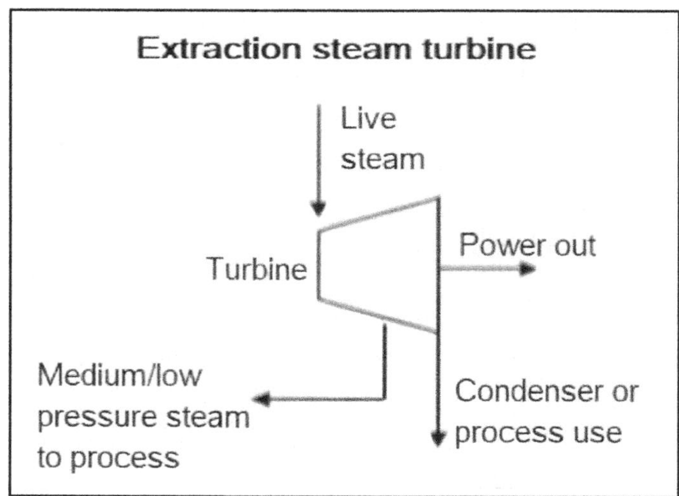

Fig. 75.4 Extraction condensing turbine

4. **Extraction condensing turbines:** single or double extraction condensing multistage turbines of large capacity up to 50 MW now are used in sugar Co-Gen plants. The steam requirement for process and ethanol production is

available by extraction from the turbine. The remaining steam passes to the condenser.

Reduction gear box for power turbines:

To increase the torque produced by the input from the turbine while decreasing the input speed, a reduction gearbox is employed. The steam turbine directly drives a shaft with pinion of a small number of teeth. This pinion drives the big pinion on main shaft that is extended to the alternator.

Generally Helical Gears Box are provided for Steam Turbine. The gear box is of compact design with high stability and safe and reliable operation. The input end and the output end of the gear box are all provided with sealing devices, thereby effectively avoiding the oil leakage problem.

Alternator:

The steam turbine is coupled to an alternator through reduction gear box. An alternator is an electrical generator. Alternators work on the principle of electromagnetic induction. According to this law, for producing the electricity we need mechanical energy, magnetic field and a conductor. The alternator converts mechanical energy into electrical energy in the form of alternating current. The electrical output from the alternator is delivered to the bus bars through transformer, circuit breakers and isolators.

Prime requirements of the steam turbine are reliability since failure of turbines can make great difficulties in factory operation. The reliability in control and instrumentation systems of the turbine is most important. Now electronic governing systems are being used instead of mechanical governor.

Diesel Generating Sets: generally two generating sets one of 500KW and second of 250 KW are provided which can be used at the time of starting of the season and in emergency. A high capacity diesel engine is coupled with is coupled with alternator capable of developing continuously the rated power at three phase, 4 wire, 50 Hz at normal voltage of 420 VAC.

-OOOOO-

76 COGENERATION AND EXPORT OF ELECTRICITY

76.1 General:

Some 30-35 years back in most of the factories the steam consumption was around 44-50 % on cane and bagasse consumption about 20-24%. The boiler were of pressure 21 kg/cm^2 g or $32kg/cm^2g$. All mills were driven by single stage low efficiency turbines because the higher exhaust produced was being utilized in the boiling house. The electric power was being produced for captive consumption only. The efficiency of steam turbine was given less importance as demand of exhaust for was higher. Most of the factories were not able to dispose-off the bagasse in profit. Therefore factory managements were not much thinking of saving of bagasse.

Some factories with better steam economy were saving higher percentage of bagasse and installed small paper plant to use bagasse as raw material. The paper plants were also not much profitable and most of the paper plants closed down.

In the last three decades the concept has been changed. The potential of bagasse as a renewable source of energy has gained wide acceptance. The bagasse is now being used for generating surplus electric power that is sold to the state grids. Bagasse used as cogeneration fuel is environmental friendly because we save natural resources like coal.

76.2 Major developments in cogeneration:

1 **High pressure and high temperature steam:** The total electric power that can be produced depends on pressure and temperature of live steam. In the following table steam required kilograms per KWh is given for various pressures and temperatures for exhaust gauge pressure 1.0 kg/cm^2. It can be seen from the table that as live steam pressure increases the steam requirement per KWH generation of electric power decreases.

Table76.1 steam required kg/kwh at turbine efficiency 65%

Sr. No.	Live steam inlet pressure at turbine	Live steam temperature °C	Steam required in kg/Kwh at turbine efficiency 65%
1	20	320	11.55
2	30	375	10.26
3	42	440	7.95
4	62	480	6.98
5	86	510	6.36

Thus as the steam pressure and increases the electric power generated per MT of steam is increased. Therefore steam pressure and temperature have to be higher. Pressure 66-160 Kg/cm^2 and temperature 450-550°C respectively. The boiler design must be suitable for highly efficient operation.

2 **Efficient use of power in factory**: by using hydraulic drives / DC motors at mill the efficiency is increased. This increases cogeneration power to operate at high efficiency (65-70%). The single stage power turbines used in old days are much inefficient producing high exhaust steam. Now in many old sugar factories, - that have gone for cogeneration under modernization programme, the mill turbines are replaced by DC drive/ hydraulic drive.

3 **Reduction in steam consumption:** in modern sugar factories the steam consumption, - with profuse vapour bleeding and steam economy, has been brought down to 32-35% on cane. Even in some factories it has been brought down to 27% on cane. This saved steam either saves the bagasse or produces extra power by condensation route.

4 **Bringing down bagasse moisture:** In modern efficient mills the moisture percent in bagasse is about 48%. In some factories bagasse first is send out. Then it is conveyed to boiler by return bagasse carrier. This helps to reduce the moisture up to 1%. This increases calorific value of bagasse and boiler efficiency increases. Bagasse moisture also can be brought down by installing dryer, however its design working investment and return are doubtful and therefore commonly not being followed.

76.3 Turbines for cogeneration:

The turbine installed is designed to suit the steam pressure and temperature conditions of the boiler. With cogeneration trend, the efficiency of the turbine is being given due importance. In cogeneration most of the factories select two turbo-alternator sets to operate in parallel, one conventional back pressure set and the other extraction /condensing set.

1 **Back pressure steam turbines (BPST):** These gives thermal efficiency in the range of 80 – 90 % as the heat in the exhaust steam is recovered in the boiling house.

2 **Extraction condensing steam turbines (CEST)**: In these type of turbines the remaining low pressure steam required in boiling house is extracted at automated controlled rate from the extraction system. The remaining steam passes to the condenser. Extraction-cum-condensing/back-pressure steam turbines achieve thermal efficiency in the range of 55 – 75%. This depends on live steam pressure

and temperature at the inlet of turbine. The quantity of exhaust extracted its pressure and temperature. The quantity of exhaust going to condenser its pressure and temperature.

3 Condensing steam turbines (CST): Condensing turbines are specifically used electric power to be exported generated from the excess steam. These works at low thermal efficiency between 20 – 35% due to wastage of substantial useful heat in condensing of the steam. Let us study cogeneration for a 5000TCD /230 TCH plant.

Table 76.2 power generation

Sr. no.	Particulars	Value
1	Crushing rate	230 TCH
2	Bagasse % cane	28.5
3	Bagasse burning % cane	27.0
4	Bagasse burning MT /hr.	61.2
5	Steam / Fuel ratio	2.2
6	Steam generation MT/hr.	136.62
7	Live steam pressure	86 kg/cm^{2}g
8	Live steam temperature	510°C
9	Exhaust pressure	1.0 kg/cm^{2}g
10	Condensing pressure	0.1 (a)
11	Turbine efficiency	70 %
12	Live steam required kg/kwh at 1.0 kg/cm^{2}g back pressure	6.00
13	Live steam required kg/kwh at 0.1(a) pressure	4.00
14	Exhaust requirement for process % on cane	30.00
15	Exhaust requirement MT/hr.	69.00
16	Live steam that goes to backpressure turbine 1MT bleeding at 7 kg/cm2g pressure for miscellaneous use.	67.00
17	Elec. Power generated from 69 MT live steam at exhaust of 1.0 kg/cm2g back pressure	11.16 MW
17	The remaining steam going to condensing turbine	69.62
18	Elect. Power generated at condensate turbine	17.40 MW
19	Total power generated	28.56 MW
20	Captive power consumption	7.00 MW
21	Export of power	21.56 MW

Table 76.3 capacity of power turbines recommended

Sr. No.	Turbine required	Capacity
1	**A bleeding backpressure steam turbine** working at live steam 86kg/cm^2 and temperature 510°C, bleeding at 7 kg/cm^2 g, Exhaust back pressure 1.0 kg/cm^{2}g	12.00 MW
2	**A condensing steam turbine** working at live steam 86kg/cm^2 and temperature 510°C, exhaust pressure 0.1 (a)	20.00 MW

The boilers recommended are two boilers of 75 MT/hr. capacity. At pressure 87kg/cm^{2}g and temperature 510°C.

The following table gives steam rating (live steam Kg/KWh) for power generation at various live steam pressure, temperatures and back pressures.

-OOOOO-

77 SYRUP CLARIFICATION

77.1 General:

The colour of plantation white sugar is normally in the range of 100-150 International Unit (I U). The International standard given by 'Codex Alimentarius Commission' for colour of white sugar no.2 (plantation white sugar) is 150 I U maximum. However in international market plantation white sugar of colour below 100 I U is easily accepted and fetches more prices.

Sugar factories which get high purity juices and maintain proper control over the process can produce plantation white sugar of colour below 100 I U. However this is also not possible consistently throughout the season.

Many factories in Gujarat and in Uttar Pradesh produce low colour bolder grain sugar with heavy sulphitation of syrup and melt and by following three and half or four massecuite boiling schemes. In such case the massecuite percentage and consequently steam consumption increases. With recirculation of sugar the inversion loss also increases.

To improve quality of raw sugar or plantation white sugar, the syrup quality has to be improved. To improve syrup quality some industrial chemical manufacturers have developed process for syrup clarification and supplied the technical knowhow of the process to some sugar factories.

77.2 Syrup clarification process:

Syrup clarification process is originated from the phosphatation process used in sugar refineries for clarification of refinery melt. Syrup from evaporator is taken into a receiving tank where colour precipitant (80-120 ppm dose on solids) is continuously added. Then the syrup is send to heat exchanger to raise its temperature to 80-85° C. At higher temperature the reactions are faster and viscosity of the syrup also less. The heated syrup is then send to aeration tank. When the syrup passes from heater to aeration tank, in pipeline addition of phosphoric acid (160 to 220 ppm on weight of solids in syrup) and of lime sucrate is made. The pH of the syrup is maintained to about 6.8. The syrup is aerated in aeration tank and is send to flotation clarifier. Flocculent (10-15 ppm on solids in syrup) is added in the syrup automatically when it passes from aeration tank to clarifier.

The colour precipitant reacts with colouring matters such as polyphenols, amino acids and other nonsugars and precipitates them. Addition of phosphoric acid and lime sucrate forms a tricalcium phosphate precipitate that occludes precipitated impurities. Flocculent coagulates the precipitated impurities.

Fig. 77.1 Flotation Clarifier

In floatation clarifier the floc entrapped in the air bubbles comes up in the form of scum. The scum is then removed by rotating scrappers. The clarified syrup is withdrawn through a weir-box and is send for filtration to a multi-bed or deep-bed filter. The residence time in the clarifier is about 25-35 minutes.

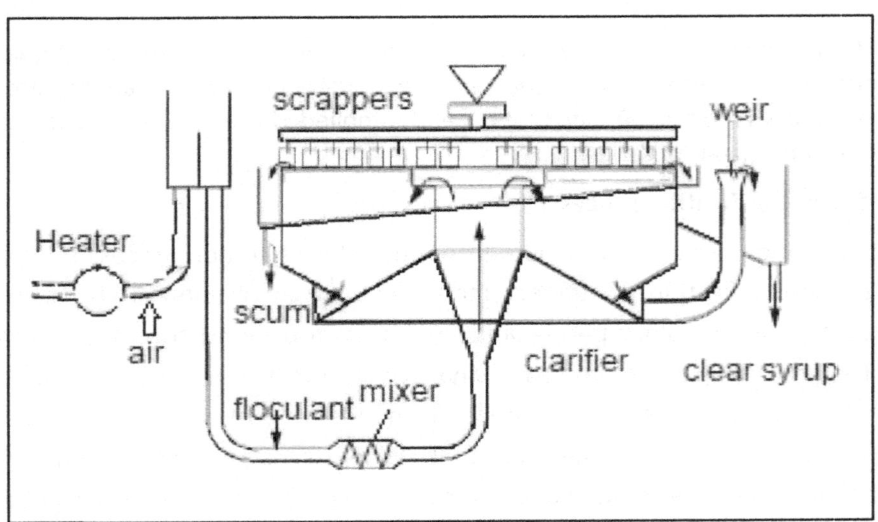

Fig. 77.2 Floatation Clarifier

The clarified syrup under flow from clarifier is also not much clear and some suspended solids pass along with it. Therefore filtration of the clarified syrup is necessary. In deep bed filters the syrup is passed through a special filtering medium that entraps all the suspended impurities that are passed along with the syrup through the flotation clarifier.

The pressure drop across the filter medium is monitored continuously and when it reaches to a preset value, back-washing of the filter is started automatically. The filter inlet and outlet valves are closed. The suspended impurities entrapped in the filtering medium are released by air and backwash and are recycled to inlet of syrup clarifier. The filter returns back automatically to its normal filtration cycle.

The more difficult task in syrup clarification is to remove the precipitated solids and its treatment. The scum separated from syrup is send to juice clarification station and mixed with muddy juice for filtration. Here some precipitated solids pass with filter cake but some are again redissolve and are recycled in the process. The redissolution and reprecipitation is not desirable.

Some technologists suggest the use of powdered activated carbon. The research in China found that polyaluminium chloride (trade name G 409) can be used as a colour precipitant. They found 55-57 % colour removal in syrup clarification and 40-48% reduction in colour of sugar.

77.3 Results of syrup clarification:

The suppliers of the syrup clarification system claim the following benefits of the system.

KEY BENEFITS

- **80% reduction in clarified juice turbidity**
- **10-20% reduction in raw sugar color**
- **Substantial reduction in insoluble solids e.g. bagacillo**
- **Substantial reduction in ash**
- **Substantial improvement in the filterability of the raw sugar**
- **Reduction in pan boiling cycle times**
- **Reduced starch levels in raw sugar**
- **Standard sizes up to 1,000 TCH capacity in a single system.**

Nigam[1] has given the following results of syrup clarification.

Table 77.1 results after clarification of syrup

Particulars	**Normal value**
Purity rise	0.6 to o.9 units
Percentage ash removal	1.0 -.3.0 %
Percentage colour removal	12 –21%
Increase in RS % 100 Bx	2.7 to 3.1 %
Percent ash in sugar	0.04 to 0.13 %
Sugar colour	70 –90 I U

With syrup clarification system, about 60 to 80 percent turbidity is removed.

The viscosity is reduced by 5-7 percent. The sugar loss in final molasses can be reduced and hence gain in recovery may be about 0.02 to 0.04 units.

In India some 40 sugar factories have installed syrup clarification. However as compare to investment and recurring expenses little gain in sugar prices are observed. Therefore some factories have now stopped the process of syrup clarification.

References:

- Nigam R B - STAI 2002 p M 53-59

-OOOOO-

78 DIVERSIFICATION OF INTERMEDIATE PRODUCTS

Diversification of intermediate products of sugar processing is not new thing. It is being followed occasionally in many countries depending upon sugar and ethanol demand in that country. In Brazil it is being followed regularly for more than fifty years and the percentage of diversion is decided by the government according to the international sugar prices.

Government of India has decided to reduce Sugar production, by considering diversion of process intermediates – juice, syrup or B heavy, to Ethanol production. This is fulfilling the ethanol demand for blending in petrol reducing the excess sugar production in the country. At present 10% of ethanol mixing in petrol is being implemented. However it is to be increased up to 20%. This is increasing demand of ethanol in the country.

The new policy of Govt. of India allows Ethanol production from BH molasses, sugar cane juice, syrup, sugar Beet, sweet sorghum, corn, cassava, rotten potatoes, and rotten food grain etc. which is unfit for human consumption.

Further Government of India has announced revised prices for yield of Ethanol from B heavy and syrup. The prices are Rs 57.61/litre in case of ethanol produced from BH and Rs 62.65 /litre in case of Ethanol produced from Syrup. (The prices may change time to time).

In response to this new government policy many sugar factories are now following diversifications of clarified juice, or unsulphited syrup and /or B heavy molasses for production of ethanol.

When two massecuite system is followed And B heavy is diverted to ethanol production the massecuite % cane is reduced by about 10-11% on cane and consequently the steam consumption at pan station is reduced by about 3 to 3.5% on cane.

Many factories are diverting syrup of 35-40 brix, 15 to 20 % on solids for ethanol production. This also decreases steam consumption at pan station. The reduction in steam consumption at pan floor reduces vapour bleeding from evaporator. This gives difficulties in operation of evaporator by extra load on the last bodies. However in some factories the C pan that remains idle is frequently used for syrup concentration.

Deterioration or spontaneous combustion of syrup / molasses in storage:

In the season the syrup and B heavy molasses is diverted and without much delay used for ethanol production. However to use these in the off season that are to be stored for long time in tanks for 3 to 6 months. This case does not arise in case of factories that are near equator and work for almost round the year.

However in prolonged storage the syrup / B heavy molasses get deteriorated. Therefore preservation of syrup / molasses is very important. With deterioration the TRS (total reducing sugars that give ethanol are destroyed) It affects ethanol yield which may be a direct loss to the factory.

Sometime however if in summer days proper care of syrup / B heavy molasses is not taken by cooling and circulation, spontaneous combustion may start. In spontaneous combustion the reducing sugars react with nitrogenous compounds like amino acids and amides. This is called as Maillards reaction. Spontaneous combustion is also called as froth fermentation as high level froth is formed during these reactions. The froth fermentation once started cannot be stopped by any mean. The molasses gets charred completely.

Experiments and research are being conducted to restrict the deterioration of the syrup /molasses. The experiments are going by two ways –

1 Thermal treatment of syrup: The syrup is heated and then cooled down and send to tanks for storage.
2 Enzymatic inversion of syrup: In this process the sucrose is inverted in 24 hours by enzymatic reactions and then it is concentrated up to 82° brix and after cooling send to tanks for storage.

-OOOOO-

79 COMPUTERIZATION IN SUGAR INDUSTRY

There is no doubt that automation and process control has an important role in sugar industry. Integrated instrumentation and control systems (PLC with SCADA) can be integrated with the help of process control software and networking of computers.

There are different types of computer control systems like Supervisory control and Data Acquisition (SCADA), Distributed Control System (DCS), or MIS (Management Information System) etc. The control systems are selected based on the number of process variables. Some companies like ABB, Siemens have developed such systems.

Fig. 79.1 Computer control system

Generally separate computerization is done at each station. The input data at each station is feed to the computers. Parameters are measured calculated and indicated on the computer screen. Upper and lower limits are defined and figures exceeding upper or lower limit are shown in colour change.

Various gauges, indicators and different instruments are used to acquire the data such as temperature, pressure, vacuum, pH values or polarization etc. On line measured values from the process are recorded automatically at preset intervals. The calculations such as purities are assigned with the help of formulae. In addition to standard mathematical functions, all formulae specific to sugar process are permanently stored in the system. Laboratory data is either acquired automatically or entered by persons working in laboratory.

The acquired data is used in calculating material and energy balances. The data is evaluated and archived in different levels of data format.

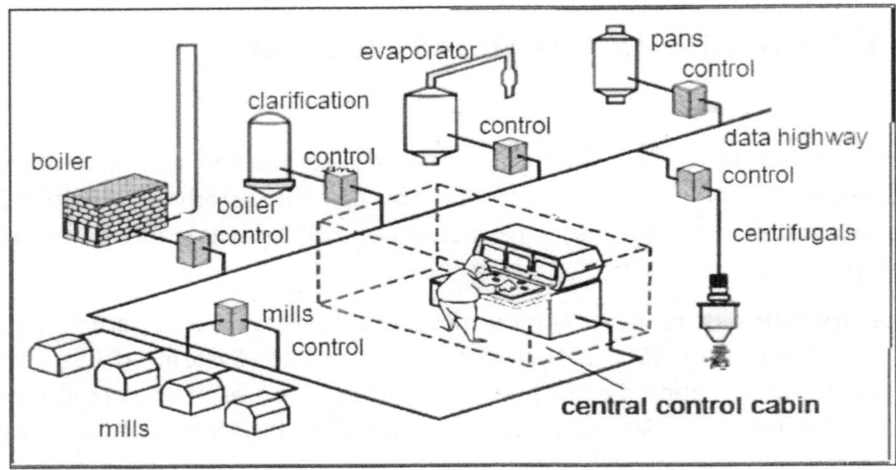

Fig. 79.2 Central Computer control system

Process data, laboratory data and calculated values are available at glance for comparison and evaluation. The information systems are capable of computing the data in different formats. Daily reviews of laboratory and process data are saved.

The systems display actual on line values / hourly values and also average values. The system can display daily manufacturing report, weekly, monthly and run reports at the finger of tips.

The data collected and computed Log sheets of each station can also be made from the computed data. The system helps to monitor accurately the control parameters without human error.

The data can also be viewed and controlled from central control room which may be adjacent to works manager's and process manager's office. It is possible to access all important process parameters at any place.

The operating staff and factory managers receive information of the current factory status continuously so that after analysing and evaluating the data they can intervene in the process whenever necessary. This helps in

- Quick and easy control over the process,
- Optimum capacity utilization,
- Reduction in down time,
- Conservation of energy,
- Reduction in maintenance work.

-OOOOO-

80 EFFLUENT TREATMENT PLANT

80.1 Effluents of Sugar Factory:

During manufacture of sugar effluents that contain high pollution load are produced. The effluent from cane sugar factories is non-toxic and contains organic compounds like carbohydrates. It has to be treated to bring down the BOD and COD values before disposal. The indiscriminate disposal of the effluent in soils and / or surface water causes environmental problems. With the enforcement of water pollution acts in all countries, it has become necessary to treat effluents from the sugar industry.

The quantity of effluent discharged from factory varies considerably. It varies from 100 to 300 l/t of cane from factory to factory. The variations are due to different reasons. Main reasons are housekeeping conditions of the plant and machineries, mode of operation and water use. Thus the total quantity of effluent for a 2500 TCD plant can be 500 -600 m^3/day.

Effluent is primarily treated by physio-chemical process and secondary treatment by biological process. It is treated and generally used for land irrigation or dispose of into nearby streams. In some factories the treated and chlorinated filtered waste water is reused in the factory. The effluent treatment plant is to be selected upon local conditions.

80.2 Waste water generated in sugar factory:

The waste water generated in sugar factory can be broadly classified as follows

Mill house - Oils and greases and cooling water are used at mill bearings. These may leak and pass into gutters. Juice may leak into gutters. Water used for daily washing and cleaning of mill section also pass to gutters. Thus effluent is generates at mills.

Boiling house – water used for washing and cleaning of clarification and mud filter station does to gutters and makes an effluent. Periodical chemical boiling and tube cleaning of juice heaters evaporator set and pans generate effluent. Pump leakages and excess condensate water increase the waste water and added to effluent flow. Sometime juice /syrup /molasses may leak at pumps, pipes and gutters and increase the load of effluent.

1. Spray pond overflow or blow downs are waste/effluent water added to the effluent flow. Sometimes entrainment may occur at pans or last body of evaporator and spray pond water gets polluted. The entrainment from the rest of the bodies of evaporator passes along with the condensate to the overhead tank.
2. Boiler blow down and power plant cooling tower blow down are waste water but generally does not content any effluent than dissolved solid and does not give much load to the effluent treatment.
3. Laboratory and domestic use of water also waste / effluent water but it is not added to main stream of waste water/ effluent.

80.3 Waste Prevention at Source:

The pollution load in sugar mills can be reduced with a better water and material management in the plant. Judicious use of water and its recycle, wherever practicable, can reduce the volume of effluent.

Optimum use of water can reduce the effluent quantity. Floor washing with water can be avoided and sweeping with dry bagasse and burning the same can reduce pollution load.

Well-designed save-alls and external catch-alls can control entrainment in the condenser water. Efficient operation of evaporators can stop or reduce entrainment to the minimum level.

Proper handling and storage of molasses to eliminate spill overs. Molasses has an extremely high biochemical oxygen demand (BOD) of the order of 900,000 mg/l, hence proper care should be taken to avoid spillage of molasses and it into waste water. If any leakages, spillage or overflow occurred, it should be removed by dry cleaning with bagasse or bagacillo.

80.4 Characteristics of Effluent:

The effluent from cane sugar factories is of non-toxic and contains organic compounds like carbohydrates. This effluent has to be treated to bring down the BOD and COD values before disposal. The effluent contains high percentage of suspended solids which deposit and cause blockage in drainage and gutters. The indiscriminate disposal of the effluent in soils and / or surface water causes environmental problems.

Table 80.1 characteristics of effluent

Sr. No.	Characteristic	value
1	pH value	4.0 to 6.7
2	oil and grease content	10-50 mg/lit
3	Total dissolved solids	3000 to 6000
4	Suspended	400 to 800
5	Biochemical Oxygen Demand (BOD) 5 days at 20°C mg/l	600 to 2000

6	Chemical Oxygen Demand (COD) mg/l	1000 to 4500
7	Total nitrogen mg/l	10 to 40

When in periodical cleaning boiling house vessels are cleaned, the effluent generated are acid and/or alkaline and may damage aquatic life if not treated carefully and allowed to pass into surface water steams.

The effluent from leakages and overflow contains high concentration of sugar and other carbohydrates. During anaerobic conditions intolerable odour is developed. Hydrogen sulphide gas is produced. The effluent imparts black colour because of precipitation of iron by hydrogen sulphide. The average characteristics of sugar factory effluents are given here.

Table 80.2 Standard norm for treated effluent for sugar industry:

Sr. No.	Parameters	Standards given by state pollution control Board
Limiting concentration in mg/l, except for pH		
1.	pH	5.5-9.0
2.	Oil & Grease	10
3.	BOD (3 days 27°C)	100
4.	Sulphate	1000
5.	Suspended Solids	100
6.	COD	250
7.	Chloride	600
8.	Total Dissolved Solids	2100

Note: In some standard COD of treated effluent to be maintained is below 150 ppm and BOD below 30 ppm.

80.5 Effluent treatment and disposal:

Various methods of treatment and disposal can be grouped under as under

- Treatment in lagoons or stabilization ponds,
- Effluent treatment plant,
- Biochemical treatment - Anaerobic / aerobic digestion of effluent.

Treatment in lagoons or stabilization ponds:

Treatment of sugar factory effluents by lagooning (or in oxidation pond) can be followed where there is sufficient land available for lagooning and winter condition are not severe. Hence the retention of effluent in lagoons can be short with reasonable degree of purification.

The factory for which sufficient land is available, treatment of effluent in stabilization ponds is simple and economical. The effluent may be treated in two stages, the first stage being primarily anaerobic digestion open deep pond and the second stage being aerobic oxidation in open shallow pond. The anaerobic pond may have liquid depth of not less than 2.5m and detention period of 6 days. The reduction in BOD is about 60 percent.

The effluent from this pond can be purified further in an aerobic oxidation pond having liquid depth less than 1.2m and detention period varying from 10 to 12 days. The BOD removal by oxidation pond is around 70 %.

The overall reduction in BOD in the treatment plant may be of the above 90%. The final effluent will have a BOD between 60 – 100 mg/l. The effluent can be used for irrigation purposes.

If this treated effluent is to be discharged into inland surface water, it is necessary to dilute it to bring down the BOD level below 30 mg/l.

Effluent treatment plant:

The current waste water treatment technologies adopted by sugar industries is conventional Activated Sludge Process (ASP) to treat effluent. ASP works on the principle of growth activity of aerobic bacteria to break down the organic waste by coagulation and flocculation process.

The regulatory agency like the CPCB has directed to each sugar industry that maximum effluent discharge limit of 200 liters per tonne of cane crushed.

It is also mandatory to install online continuous emission monitoring systems (OCEMS) at the outlet of the ETP for continuous evaluation of treated effluent quality and to store the treated effluent into the lagoon and apply time to time to agricultural land.

The chemicals such as lime (CaO), commercial poly aluminum chloride (PAC) and magna flocculent are used as common coagulants to remove suspended solids as well as dissolved solids. The doses of coagulants are determined by the jar test.

The following flowchart gives the operations carried out in the effluent treatment plant. In this flowchart an operation can either be eliminated or added as per requirement of the factory. The operation of ETP should be such that it will give treated effluent of standards given by state pollution board.

1) Screen chamber:

The screen chamber (Bar Screen) is used to remove the large floating objects. The untreated effluent contains large suspended solids like bagasse paper etc. These are to be removed to avoid choking pipe system and clogging the pumps impellers and aberration to equipment.

The bar screen is of 10 mm wide and 50 mm deep bars arranged with spacing of 20 mm between 2 adjacent bars. Solid material retained over the bars is to be removed frequently and the bar screen is to be kept clean,

All the bigger suspended solids and floatable material retained on the bars and more clear effluent is passed for further treatment. The retained matter is to be removed frequently and the bar screen is to be kept clean.

The screened effluent is then taken to the oil and grease removal tank by gravity.

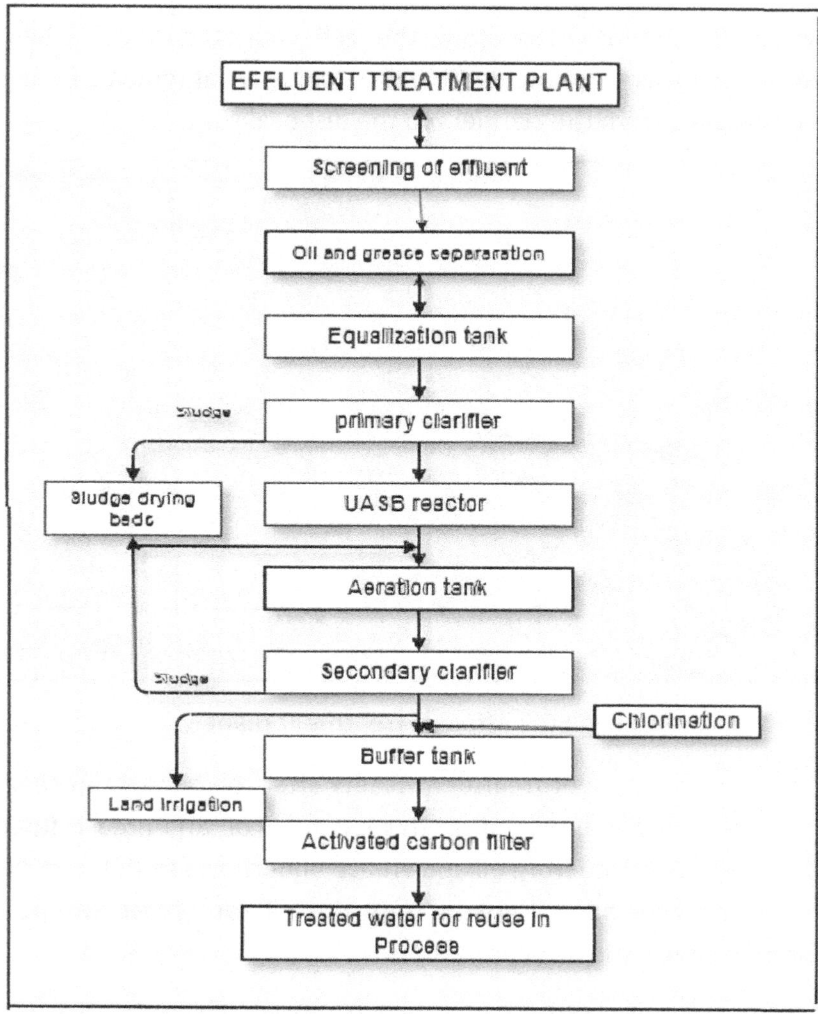

Fig, 80.1 Flowchart of effluent treatment plant

2) Oil and grease removal:

Oil and grease from the effluent may cause damage to pumping unit, obstruction to further biological treatments. Oil is lighter than water. This property is used to separate out oil and grease from the effluent. However if the boiler blow-downs and excess condensate is mixed with effluent and effluent flow is turbulent with more travel distance, the oil gets emulsified and does not float easily. Therefore the boiler blow-downs and condensate should be kept separate and should be mixed after oil and grease separation. The effluent flow should steady.

If the oil and grease escapes the trap it disturbs the further process of treatment. In aeration tank the oil and grease gets churned and the oil may cover the cell walls of bacterial. Therefore the cell cannot consume their food – carbohydrates and other compounds. Then BOD value is not reduced to an expected level. The oil escapes from the clarifiers to the disposal site.

Fig, 80.2 effluent treatment plant

In most of the effluent treatment plants after the bar screen chamber, the oil and grease chamber is provided. In this chamber oil and grease floats on the surface and the effluent free from oil and grease underflows to the next tank. There are various patterns for oil and grease trap. The oil and grease will be collected manually or mechanically in a separate sump.

Therefore removal of oil and grease is very much essential and the traps should be situated as far as close to the source. As oil and grease passes to the effluent mainly from mills, it is advantageous to provide oil and grease trap near mill section.

3) Equalization Tank:

Equalization basins or tank is used for temporary storage and dilution of effluent. The equalization basin has variable discharge control. The equalization tank should not work as settling tank. The solids should be kept in suspension, for this the water must be in motion. The mixing is carried out with the help of mechanical stirrers. Now in many factories aerator is also provided to the basin. Bypass arrangement is also made for cleaning purpose.

The raw effluent is passed into the equalization tank where it is mixed continuously with floating aerator or air diffusion system to homogenize the effluent. Hence characteristics of the effluent are made uniform.

Generally the effluent is having low pH value. So the equalization tank are thoroughly mixed by dosing lime or caustic soda or soda ash in the equalization tank to neutralize the pH of the raw effluent before passing to the clarifier.

From outlet of equalization tank a steady flow of the composite effluent is maintained to the primary clarifier.

4) Primary Clarifier:

Purpose of the primary clarifiers is to settle down the suspended solids of higher specific gravity and remove them from the effluent. The neutralized effluent from Equalization Tank is pumped to the primary clarifier. Different types of primary clarifiers (like conventional Or lamella type clarifier) are used to removing suspended solids.

When the effluent is detained in clarifier without disturbance for about one to two hours, the heavier suspended matter settles down. The supernatant waste water passes to next treatment unit. The lower specific gravity suspended matter of that is floating, pass along with waste water. Whereas the settled sludge is taken out from the bottom of the clarifier and is sent to sludge drying beds.

5) **Aeration Tank:**

Aeration is the process by which air is circulated through and mixed with or dissolved in a liquid. An aeration tank is provided to aerate the effluent.

The clarified effluent from primary clarifier is subjected to aeration tank. The biological treatment of effluent by aeration process with sludge culture is very sensitive. The efficiency depends on pH, temperature, suspended solids, air in contact, culture growth.

The biological treatment efficiency also depends on mixed liquor suspended solids (MLSS). MLSS is the concentration of suspended solids, in an aeration tank. The units MLSS is primarily measured in milligrams per litter (mg/L) or kilograms per cubic metre [kg/m^3]. Mixed liquor is a mixture of wastewater and activated sludge in an aeration tank. MLSS is an important for activated sludge process to ensure that there is a sufficient quantity of active biomass (quantity of organic pollutant) available to consume for microorganisms. This is can noted as the food to microorganism ratio (F/M ratio). MLSS removes COD and BOD and purifies waste water. The microbial culture concentration is to be maintained in the range of 1500-4000 mg/lit.

6) Secondary clarifier:

It is cylindrical concrete tank with conical bottom. There is centre well to which the effluent water is fed. The centre stirrer is rotated at 2-3 RPH. There is circumferential overflow from which treated effluent is collected and is send for agriculture use. Sludge is collected at the bottom from where it is partly recirculated to aeration tank and rest is send to sludge drying beds by pumping.

7) Hypo dosing system:

Alternatively the clarified water is then feed to hypo-dosing system where it comes in contact with hypo solution which helps in disinfection of treated water. The chemical name of hypo is sodium thiosulfate. Chemical formula for hypo is $Na_2S_2O_3$. Hypo is also called as hyposulphite of soda. It is a crystalline solid. Sodium thiosulfate is readily soluble in water.

When sodium hypochlorite is released in water, it produces hypochlorous acid. This acid then reacts with pathogens in the water, like bacteria, viruses and protozoa, and deactivates them, preventing them from being able to reproduce or pose a risk to human health.

8) Dual media filter:

The treated water is pumped through dual media filter which polishes the water before active carbon filter. Basically it removes any residual suspended matter or any kind of smell /colour present in the water.

9) Activated carbon filter:

Activated carbon has much higher adsorption capacity than other materials. The activated carbon is mixed with treated waste water and allowed to remain in contact for one hour for adsorption of impurities from the water. Then the activated carbon is separated by filtration. The resulting water is free from dissolved organic substances. About 80% of the dissolved organic matter can be removed by this method. Powdered or granulated carbon available in market can be used.

10) Sludge Drying Bed:

Sludge drying beds are used for dewatering of the sludge. Sludge from the clarifier is discharged to sludge drying beds at intervals. These are the sand beds of 250 mm of sand over about equally thick well-graded gravel layer, underlain by perforated drainage lines spaced 2.5 to 6 m apart. The bed has a slope towards the discharge end.

Sludge drying beds are used for dewatering of the sludge. These are the sand beds of 250 mm of sand over about equally thick well-graded gravel layer, underlain by perforated drainage lines spaced 2.5 to 6 m apart. The bed has a slope towards the discharge end. Sludge from the clarifier is discharged to sludge drying beds.

After the primary treatment, the effluent is passed in a specially designed bioreactor where the organic matter is utilized by microorganisms such as bacteria, algae, or fungi for either anaerobic or aerobic wastewater treatment.

Anaerobic treatment is generally used to treat higher organic loading wastewater, whereas aerobic treatment is used to treat lower organic loading wastewater.

Anaerobic digestion is a series of biological processes in which microorganisms break down biodegradable material in absence of oxygen and produce methane, carbon dioxide and other biomass. There are four stages; Hydrolysis, Acidogenesis, Acetogenesis and Methanogenesis in the process.

Hydrolysis: this is the first step in the conversion of organic material to biogas. In this stage, certain bacteria break down organic polymers like carbohydrates into simple sugars so that the next group of bacteria can further process the material.

Acidogenesis: this is the second step in the conversion of organic materials to biogas. In this stage, certain bacteria called acidogenic bacteria convert the simple sugars and amino acids into carbon dioxide, hydrogen ammonia and organic acids.

Acetogenesis: this is third step in the conversion of organic materials to biogas. In this stage, certain bacteria called acetogenic bacteria convert the organic acids ito acetic acid, carbon dioxide and hydrogen.

Methanogenesis: this is final step in the conversion of organic materials to biogas. In this stage, certain microbes called methanogens convert the intermediate products produced in the preceding stages into biogas (primarily methane and carbon dioxide). The solid and liquid left over from this process i.e. digestate consists of dead bacteria and material that the microbes cannot use.

In anaerobic process mass flow meters is used to monitor fast, accurate and stable flows of methane and carbon dioxide. The biogas produced during an anaerobic treatment process can be used as a source of renewable energy (natural gas/methane).

In aerobic treatment oxygen is required. Microorganisms digest the organic matter and other pollutants like nitrogen and phosphorus and produce carbon dioxide, water, and other biomass. Control of mass flow and stable flows of air and oxygen is very important to achieve fast and complete process.

In effluent treatment plants generally a both anaerobic and anaerobic systems are used. The anaerobic process is followed by aerobic system.

Air pollution:

Air pollution by the boiler flue gases is regulated as per state and central government rule. The emissions of oxides of sulphur (SO_x) and oxides of nitrogen (NO_x) and fly ash (fine particulate) are important in controlling air pollution.

Atmospheric emissions like SO_x, NO_x, fly ash, volatile organic compounds etc. arise from combustion of bagasse in boilers. Bagasse handling equipment also forms bagasse dust that flies with air.

-OOOOO-

81 CHEMICALS USED IN SUGAR INDUSTRY

Many chemical as regular chemicals and some chemicals as auxiliary special are used in process in sugar factories.

81.1 Biocides:

The objective of these chemicals is to control bacterial decomposition of sucrose from cane juice at the mills. The biocide to be selected for use of mill sanitation should have the following features.

- it should confirm to the food regulation rules
- it should not be excessive volatile
- it should withstand in in presence of organic material for long time
- It should exhibit broad antimicrobial spectrum at pH 5.00 to 5.5
- It should have low mammalian toxicity.
- It should have no adverse effect on the final product.
- Detectable residue should not appear in sugar or molasses.
- It should be safe and noncorrosive.
- Easy to store and handle
- It should be economically viable

Most of the biocides in the market have following base

- halogen-derivatives
- quaternary ammonium salts
- Dithiocarbamates.
- formaldehyde
- polymer based compounds

Halogen derivatives:

i) Chlorine: Solutions of sodium hypochlorite, NaOCl, containing 8 to 10% available chlorine are commonly used as biocides for mill sanitation. The antimicrobial action of chlorine comes through the hypochlorous acid formed when free chlorine is added to water.

$$Cl_2 + H_2O \text{ ----- } HCl + HClO$$

The hypochlorous acid is further decomposes under

$$HClO \text{ ----- } HCl + O$$

The nascent oxygen liberated is a strong oxidising agent and through its action microorganisms are killed. Chlorine also directly combines with proteins of cell membrane and enzymes. The hypochlorites are used at 20-100 ppm dose.

However now this chemicals are not much used in the Industry because chlorine is unstable to light, heat and activity. The lower the pH, the more bactericidal, but it is less stable is the solution. Moreover antimicrobial activity of hypochlorites is reduced in presence of organic matter, inorganic reducing agents such as sulphides nitrites and compounds of Fe and Mn. Hypochlorites are corrosive to metals.

ii) Iodine: Iodine is used for mill sanitation in the form of Iodophors, which are combination of Iodine with non-ionic surfactants. They contain about 1.5 to 2.0% available Iodine which is slowly released from the complex form. Iodine is a strong oxidising agent, which oxidises proteins. Iodophors are used at about 5-7 PPM dose. However Proteins and other organic compounds diminish activity of iodine, but less than chlorine.

iii) Fluorine: Fluorine is used as biocide in the form of ammonium bifluoride, NH_4 HF_2, which is a white transparent solid with a slightly acid odour and acid reaction. Fluorine reacts vigorously and oxidises cellular constituents. Ammonium bifluoride is used at 5-10 ppm dose. Fluorine reacts with practically all-organic and inorganic compounds; therefore bactericidal activity in cane juice is reduced. Fluorine reacts vigorously at ordinary temperature with most metals to form fluoride. Thus it is highly corrosive.

Quaternary ammonium compounds:

These may be regarded as long chain alkyl relatives of ammonium salts are used for mill sanitation. It penetrates the cell and precipitate the cell contents causing death of cell. The dose varies from 2 to 10 ppm. It got some additional features like toxicity, high solubility in water, stability in solution and noncorrossiveness. But these compounds are most effective on alkaline side, pH 9-10 and are less effective as pH decreases below 7. However, with these limitations also the unique property of these compounds is the ability to produce bacteriostatic effect in very dilute solution.

Dithiocarbamates:

They are derivatives of dithiocarbamic acid, which is unstable in Free State but when reacted with another molecule of the acid or with some metal, stable and highly bactericidal compounds are obtained. Chemicals used for mill sanitation are of 2 types:

I. dialkyl dithiocaromates
II. monoalkyl dithiocaromates

The dithiocaromates inhibit certain enzymes. This arrest metabolic activity of the microorganisms causing its death. The dose used is about 10-20 ppm. Dithiocaromates are most satisfactory chemicals for addition in the juice. In addition as they inactivate the enzymes dextranase and invertase they reduce formation of dextran and invert sugars. As per our experience quaternary ammonium based biocides and Dithiocarbamates are the most effective biocides.

Formaldehyde:

Formaldehyde: formaldehyde is a traditional biocide available in aqueous solutions as a formalin, which contains 37 to 40% formaldehyde. Formaldehyde combines with organic nitrogen compounds such as proteins and nucleic acids of microorganisms. Formaldehyde has a serious disadvantage of releasing noxious fumes, which are irritating to tissues and eyes when used at the mills. Therefore nowadays it is generally not used for mill sanitation.

81.2 Lime:

Lime is added in the raw juice as milk of lime, a white milky liquid containing slaked lime in suspension and solution. The lime attracts moisture and carbon dioxide from the atmosphere. In order to reduce this absorption compact stacking is essential. Stock at mills should be kept just sufficient for one week and fresh supply of lime should arrive to factory every week.

Quality of lime

A complete examination of lime sample consist of

1. determination of percentage of available CaO
2. determination of percentage of certain impurities
3. A slackability test.

Lime may contain high percentage of unburnt lime or dead burned lime and also presence of clay or sand. These reduce the quality of lime. Therefore the most important criterion for the lime quality is available CaO.

But the percentage of available CaO and of impurities is not sufficient to classify lime completely. In many sugar factories a gradual increase in the pH of limed juice is observed due to delayed dissolution of unslaked lime particles. This is because of

- The use of over-burnt lime. The dead burn particles of which hydrate very slowly
- Inadequate straining by using a sieve with too large openings.

The complete slacking of such lime takes long time. This property is checked in laboratory and is called as slackability test of lime. Lime is slaked under standard conditions. Slakability test is equally important to confirm the lime quality.

Specification of lime:

1. The percentage of available CaO of high grade lime should be above 80 %
2. Slakability test
    - **Residue after 10 minutes 4-10 % and after**
    - One hour less than 2 %.

Table 81.1 maximum limit for percentage of impurities in lime

Sr. no.	Impurity	Max. permissible Percentage
1	moisture	2%
2	solids insoluble in HCl	2%
3	silica (SiO_2)	2%
4	Iron and Aluminium	2%
5	Magnesium oxide	2%
6	sulphates	0.2%
7	carbonates	2%

81.3 Sulphur:

Yellow sulphur or octahedral α form of sulphur used for generation of SO_2 gas in sulphitation process. Elemental sulphur mostly occurs in USA, Italy and Japan. The atomic weight of sulphur is 32.06. The melting point is 119°C. the boiling point of sulphur is 445°C. The sulphur should be very pure with sulphur content more than 99%. The maximum permissible limits for impurities are

- Moisture 0.5 %
- Ash 0.1 %
- Bituminous matter 0.1 %
- Arsenic 0.05 %

81.4 Phosphoric Acid:

Phosphoric acid is generally used in juice clarification, clear juice and in some factories in syrup and massecuites as supplementary chemical. It is used in all the process of sugar manufacture as also in sugar refinery which follow phosphatation process. High quality clear transparent colourless phosphoric acid of about 75-85 % pure contains about 55-60% P_2O_5 that contains only traces of impurities such as iron lead arsenic fluorine etc.

81.5 Polymer flocculants:

The flocculants are normally derived from polyacryl amides-polymers with very high molecular weight ($2*10^7$). These days, it has been observed that many a times the product comes in adulterated form causing much decrease efficiency. It is therefore suggested that irrespective of price, genuine products should be procured.

The use of polyelectrolytes as flocculent in treated juice becomes essential when juices do not settle well by normal treatment. This is observed when colloidal contents in the juice considerably increase and/or their nature drastically changes due to effect of drought, disease and staling. Under such condition addition of suitable flocculent in the form of diluted solution in the treated juice facilitates the flocculation of the precipitates and coagulates the colloids of the juice. The normal dose of these flocculent vary in the range 0.5ppm – 2.0 ppm. Excessive usage cause retardation in filtration of mud, increase in massecuite viscosity. In short retention clarifier the continuous regular use of polymer based flocculent is essential.

Factors, which affect the performance of polymer flocculent in juice clarification are- time of agitation, degree of Hydrolysis (minimum30%) solvent ionic strength, concentration of polymer solution, pH, influence of P_2O_5 precipitate and concentration of calcium in clear juice.

Most commonly used brands of flocculants are as under

Sedipur-TF-2, A F 200	Magnofloc-LT-26	Seperan A.P.-30
Nalco-7415-SC	Indfloc-236 and 237	D.K.-8020
Chemifloc-SA-2	Zuclar 110	Pristal 2935
Midland PCS 3000	Olin 4790	Talosep A3 and A5

81.6 Antiscalents: It is a synthetic-polymer with sequestrates and anionic surfactant based on sodium polymethacrylate.

pH 5 to7
Sp. Gr. 1.22

Some of antiscalent being used in Indian Sugar Industry are-

- Indion-8102
- Sweetchem-100 etc.

81.7 Surface active chemicals or viscosity reducers:

Use of appropriate surfactant enables to reduce viscosity of low-grade massecuite shortening of boiling time and improves purgeability of massecuite at centrifugals. This facilitates proper separation of sugar crystals and final molasses. Some of the good surfactants used in Hawaii, Australia, Philippines, and South- Africa etc. are Hodeg CB-6, Fabcon-viscaid, Midland PCS-5002, and Mazu 400. The surfactants commonly used in India are - Metacrystallon-C, pan-aid, white all, Tensolite etc. Generally these chemicals are called pan additives.

Sodium hydrosulphite decreases viscosity of low-grade massecuites, increases C massecuite brix and reduces boiling time by about half an hour. When used for A massecuite colour is improved. A dose of 3-4 kg/100 MT of massecuite is recommended. A viscosity reducer containing 75% calcium hypochlorite, known as Maxehlor is effective is reducing massecuite viscosity during boiling. The optimum dosage according to him was 2.0 to 2.5 Lt. Of maxchlor per 25-30 tonnes of C-massecuite. 0.5 Ltr.to be added during graining and remaining quality was added ½ hr. before dropping. Some factories use soda ash (Na_2Co_3) as viscosity reducer in case of C-massecuite, but it is generally added in the solution form during crystalliser treatment of C-massecuite.

Whichever may be the surfactant, it should be used in a controlled manner as these being molassigenic compounds tend to increase molasses purity. Some viscosity reducers increase colour density of the mother liquor.

81.8 Other various chemicals:

Magnesia: Magnesia is much expensive than lime. However to reduce severity of scale at evaporator systematic use of magnesia or magnesium oxide by mixing it with lime has been tried. The clear juice obtained is highly turbid that affects plantation white sugar quality in sulphitation factories. Moreover now effective antiscalents that reduce scale formation at evaporator are available.

Bentonite: (magnesium aluminium silicate): Bentonite is a form of clay available in Argentina gives good results in clarification. The advantages claimed are lower lime and sulphur consumption with greater removal of organic non-sugars. Less scale in evaporators and low viscosity of syrup and molasses. Moreover the stale cane can be well clarified.

Kieselguhr (diatomaceous earth): Kieselguhr of different grades is used as filter aid or filtering medium in sugar refineries. It is available in market under different trade names such as Celite, Dicalite etc.

Perlite: Perlite ore is crushed and then shattered at 800-1100°C. The perlite so obtained is graded the same way as diatomaceous earth and are sold under different trade names.

Blankit: Sodium-hydro-sulphite commonly called as Blankit has been used mostly in A massecuite boiling to improve the sugar colour and to reduce the viscosity of mother liquor. However keeping quality of sugar obtained from Blankit bleached massecuite gets considerably reduced. In case of C-strike Blanket acts as viscosity reducer.

Hyposulphite of soda ($Na_2HS_2O_4$): Commonly known in India as **hydrous.** This is sold in market under different trade names. It is added to vacuum pans during boiling of A massecuite in plantation white sugar manufacture. This release SO_2 in the boiling massecuite that acts on colouring matter, lime salts and iron salts.

Hydrogen-peroxide: Hydrogen Peroxide is an efficient bleaching agent. Hydrogen peroxide (H_2O_2) is used sometimes in few factories as a secondary bleaching agent. On decomposition yields only oxygen and water. Moreover it use in the process does not need any plant modification and can be used as processing aid. . It can be used in place of Blankit in A-pans.

Sodium carbonate (soda ash): When stale cane arrived at the factory or when the juice is to be stored in clarifier for longer time during general cleaning lime produces soluble salts that are passed along with clear juice. This gives trouble in further process. At such time use of lime is made limited and soda ash can be used to neutralise the excess acidity. This is common practice in many countries. Soda ash is more expensive than lime also it gives high colour to the juice therefore use of soda ash is to be done only when it is necessary.

Antifoaming agents: Foaming or frothing in cane of mixed juice and sulphited syrup is a common phenomenon in any sugar factory. It is caused mainly due to occlusion of air in the liquids with high viscosity and surface tension. Higher the surface tension of liquor the foam is more stable. At this time antifoaming agents are used.

Turkey-red Oil: Turkey-Red-oil is sulfonated castor oil. It is most commonly used as an antifoaming agent defoamers to subside the frothing in sulphited syrup. One part of T.R. oil is dispersed in eight to ten parts of water and it is spread over the foam surface. There is a wide variation in average consumption figures.

Enzymes: nowadays in enzymes amylase and dextranase are used for disintegration of starch and dextran respectively in juice either at mill or in syrup at pan station. This reduces viscosity of the juices and syrup.

Table 81.2 Standard norms for consumption of chemicals and lubricants In double sulphitation sugar factories

Sr. No.	Chemical	Standard Norm
1	Lime	0.12-0.22 % on cane for sulphitation factories
		0.07-0.13 % on cane for defecation factories
2	Sulphur	0.04-0.06 % on cane
3	Phosphoric acid	Depends upon original P_2O_5 content in mixed juice hence no fixed norm
4	Biocide	5-15 ppm on cane depending upon the efficacy of the chemical
5	Flocculating agent	With conventional clarifier seldom used in case of mud settling trouble. With short retention clarifier regularly 2-4 ppm dose is used.
6	Viscosity reducing agents	Occasionally used about 1-3 kg per 100tonnes of C massecuite depending upon the efficacy of the chemical.
7	caustic soda	No fixed norm but generally used 1.5-2.5 kg per 100 tonnes of cane for evaporator cleaning
8	Washing soda	Generally 10 % of caustic soda
9	Common salt	Generally 10 % of caustic soda
10	Hydrochloric acid	Sparingly used for cleaning of evaporator and pans
11	Boiler chemicals	The boiler water chemicals generally used are Caustic soda, washing soda, sodium phosphate, catalysed sodium sulphite etc. no fixed norms depends on quality of boiler feed water.
12	Lubricants	Approximately 10 kg per 100 tonnes of cane.

-OOOOO-

82 SUGAR GODOWNS

82.1 General:

Sugar is hygroscopic in nature and absorbs moisture when relative humidity of the atmosphere is high, particularly above 60-65. Therefore sugar deteriorates on long term storage. Hence to keep or maintain good quality of sugar in storage the godown should be well constructed. Maintenance of the godowns is also equally important. Superior quality sugar deteriorates slowly. On the other hand inferior quality sugar deteriorates faster. In India it is found that about 1.0-2.0 % of the total sugar bags stored in the godowns get damaged or gets deteriorated to varying degree depending upon sugar quality and storage conditions.

Fig. 82.1 Sugar Godown

There are two main reasons for deterioration of sugar in godowns.

i) High humidity in godowns
ii) Higher temperature in godowns

Therefore to prevent deterioration of sugar low humidity has to be maintained in the godown. The entry of moisture into godowns could be from outside atmosphere and subsoil water.

During rainy season the humidity in the outside atmosphere is about 85 %. In such weather humid air can enter in the godown through leak in the doors, ventilators, roofs etc.

The ingress of moisture from subsoil water into the godown through the floor and walls of the godowns results in increasing humidity inside the godown. Such ingress of moisture is appreciable in low level areas particularly in the rainy season when subsoil water goes up. Therefore the floor and walls of the godowns should be waterproof / humidity proof.

Fig. 82.2 sugar bag conveyor

82.2 Considerations in Deciding Sugar Godowns Areas:

- The godowns should be away from spray pond, molasses storage tanks, water storage tanks effluent treatment plant, steam blow downs and in opposite direction of wind from the above places.
- The godowns should be near to sugar house / sugar bagging section.
- The godowns should be on well-raised platform with proper drainage to all sides.
- Doors of the godowns should be in the opposite direction to monsoon wind.

82.3 Stacking of sugar bags:

- The sugar should be dry and its temperature should be below 42°C at the time of bagging. If it is above 45° C, the sugar bags may be hard in the stacks. The moisture % in sugar at the time of bagging should be less than 0.08%.
- The relative humidity in the godowns should be less than 60 preferably less than 55. Doors, windows should be tightly closed during monsoon. In hot sunny days when the relative humidity is less the godowns doors and windows should be opened.
- The temperature in the godown should be about 37°C and should be maintained steady.
- The sugar bags should be stacked 0.7-1.0 m away from the walls of the godown in order to prevent these bags from the heat of the wall. The stack should be continuous and compact to reduce contact with the atmosphere.
- The bags are piled up to the height of about 8-m. (25 bags) with a margin of about 3-m. from the roof. The pile can go higher up in the interest of keeping quality of sugar, but there is danger that the bottom layer may be crushed.

- The lowest layer of sugar bags should not rest on concrete floor but should be insulated from it by means of a layer of plastic woven bags or tarpaulins.
- The uppermost layer of sugar bags should be covered with tarpaulins in order to prevent the bags from heat of the roof and hold up any moisture dripping from the roof by condensation or leakage.
- Lanes 0.7-1.0 m. should also be provided after every 8 m long stack for facility of proper handling of sugar bags and ventilation.

Fig. 82.3 sugar bags stalked in godown

When humidity in the godowns is higher the sugar deteriorates faster. This can be seen in the following table. The observation were made in two different factories.

Table 82.1 deterioration of sugar due to high humidity

Sr. No.	Month	Relative humidity	Pol % sugar	Moisture % sugar	Sugar colour IU
Factory A - sugar grade S-30					
1	May	52	99.80	0.050	91
2	July	67	99.70	0.070	125
3	September	65	99.65	0.072	130
4	November	59	99.62	0.076	170
Factory B - sugar grade S-30					
1	May	48	99.80	0.040	70
2	July	56	99.75	0.050	77
3	September	54	99.72	0.055	78
4	November	54	99.71	0.056	81

It can be seen from the above table that in factory A where the humidity in the gown is above 60 for long duration the colour development is very fast. The temperature is also important.

- During rainy season the godown doors should be kept open only at the time when truck is being taken inside or taken outside. Wet trucks should not be allowed inside the godown.
- To keep the relative humidity at minimum level in the godowns quick lime can be kept inside the godowns at 10-12 places and by changing it after every 2-3 days.
- A thermometer and hygrometer should be kept in the godown for regular observation of temperature and humidity inside the godown.

-OOOOO-

83 MOLASSES STORAGE TANKS

83.1 General:

The mother liquor- separated from low-grade massecuite in centrifugals, called as final molasses or waste molasses. In a general way sugar cannot be further recovered from this molasses hence called as final molasses. It is sent out to molasses storage tank. The molasses production is generally 3.5 to 4.3% on cane. It is thick highly viscous semi liquid with 80 to 90% dissolved solids having 1.4 -1.45 density. It is of dark brown colour due to presence of caramels and melanoidins that are formed in the process. Composition of molasses is as follows:

Table 83.1 Composition of molasses

Sr. No.	Particular	Percentage
1	Sucrose	28 - 35%
2	Reducing Sugar	10-20%
3	Ash	9-13%
4	Water	12-18%
5	Organic Nonsugars	15-20%

Government of India has given standard grades as per total reducing sugars content in the molasses

Table 83.2 Indian standard grades for cane molasses:

Sr. No.	Characteristic	molasses grade		
		Grade-1	Grade-2	Grade-3
1	Density, in degrees Brix at 27.5° C Min.	85	80	80
2.	Ash (Sulphited), percent by mass, (Calculated for 100° Brix), Max.	14.0	17.5	17.5
3.	Total reducing matter as invert sugars percent by mass, Min.	50.0	44.0	40.0

In transfer from the centrifugals to molasses storage tank air gets entrapped in the molasses. Molasses with its total sugar of 50-55% is an ideal raw material for production of ethyl alcohol.

Proper storage of molasses without loss, dilution or deterioration is very much essential for high alcohol yield from it. Therefore storage of final molasses in steel tanks becomes necessary. Bureau of Indian standards gives specification of steel tank for storage of molasses.

However, during prolonged storage the molasses deteriorates slowly. This deterioration is not due to any microorganisms. The deterioration or decomposition is caused by chemical reactions taking place in the molasses.

83.2 Spontaneous destruction of molasses:

When the molasses is stored hot and temperate in storage is above 43° C and no circulation of molasses is followed, the molasses may go under rapid spontaneous destruction. In spontaneous destruction of molasses, sugars as well as other organic compounds are decomposed completely. This results in charring of the entire mass with rise in temperature. The phenomenon is also called as froth fermentation as voluminous froth is formed in these reactions. Although the reaction is called as froth fermentation, it is due to chemical reaction and not due to microbial action. The chemical reactions are not known but are exothermic and CO_2 is liberated. The spontaneous destruction / froth fermentation once started cannot be stopped or controlled by any mean.

83.3 Calculation of volumes of tanks:

The volume of the steel tank for storing molasses for a factory of 2500 TCD capacity is calculated as follows:

Table 83.3 calculations for number of tanks for a factory

Sr. No.	Particulars	Quantity
1	Average duration of the season in days	160
2	Total cane crushing during the season tonnes	4,00,000
3	Average molasses production % cane	4.0
4	Total molasses production in tonnes for the season	16000
5	Volume of molasses in m^3/ t at 90 °Bx	0.708 m^3
6	Total volume of molasses produced in the season	11328 m^3
7	Add 10 % in volume for foam in m^3	1132 m^3
8	Say	12500m^3
9	Number of recommended tank of 3000m^3 capacity (effective capacity 2850m^3) (70 % of the total production)	3 nos.

83.4 Steel tank:

In many countries the storage of the molasses in steel tanks has been made compulsory. The molasses storage tanks are adjacent to the factory to facilitate easy transport of molasses to the tanks. The steel tanks as far as possible should be away from sugar godowns, spray ponds, and effluent drains. It is necessary that tanks are built on good cement concrete / masonry platform. The diameter and height of the molasses storage tanks usually depends upon the size of the ground area available and volume of molasses required being stored.

Fig. 83.1 Molasses storage tank

The tank is built of mild steel plate as recommended in IS 226-1975. It is designed with due consideration of wind load. The tank is painted with anticorrosive paint from inside, outside and top. It is fitted with following accessories and fittings.

1. molasses Inlet
2. molasses outlet
3. Drainage / washout valve
4. Manholes
5. Safety railing around the roof of the tank
6. Vent pipe covered with cap, with wire netting
7. Sampling cock – a 25-mm drip cock
8. Water spraying coil
9. Ladder,
10. Level indicator.
11. Indicator thermometer
12. Molasses circulation and lifting pump.

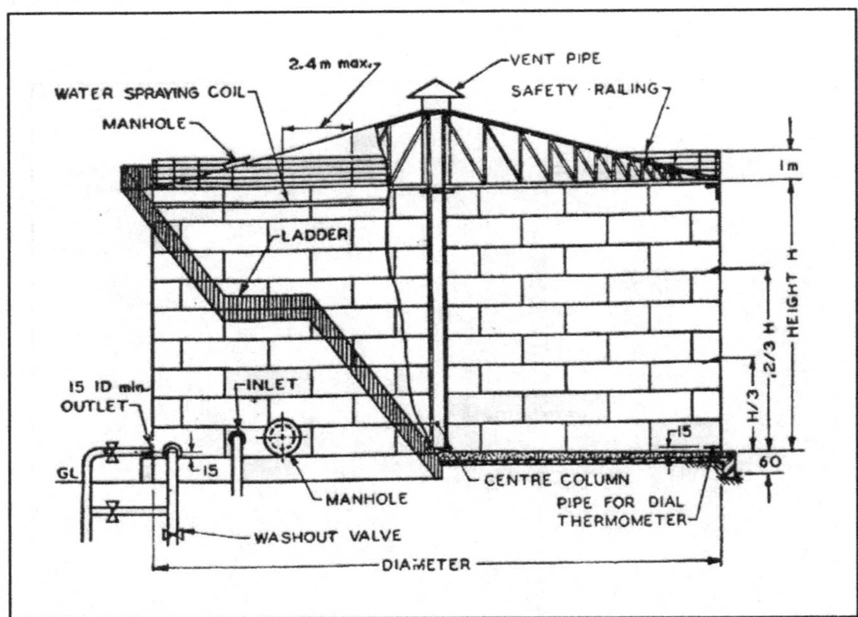

83.2 Molasses tank details

83.5 Preventive measures in storing the molasses:

While storing the molasses all the necessary preventive measures are to be taken to minimize the degradation of the molasses and avoid spontaneous destruction. The following measures are to be taken in the process.

1. Vacuum in the pans during boiling should not be less than 640 mm. Low grade massecuite should not be boiled at temperature more than 69°C.
2. The temperature of water / diluted molasses used as lubricant at C massecuite centrifugals should not be above 55°C.
3. Steaming at molasses pumps should be avoided. Instead molasses may be diluted for easy flow.
4. Avoid mixing of fresh molasses with old molasses
5. Molasses stored in the tank should be regularly circulated.
6. The steel tank should be kept cool by running cold water spray around exterior surface of the tank. Temperature of molasses should not exceed more than 42°C.
7. An arrangement for diluting the molasses has to be provided.
8. Analysis of molasses at regular intervals is to be followed.

-OOOOO-

84 SUGAR SEED SLURRY PREPARATION

84.1 General:

We have seen in the chapter "Graining Technique" grain or fine crystals are made in syrup / molasses of desired purity in the vacuum pan for intermediate and low grade (B and C) massecuites. When grain is made by true seeding method the required numbers of microscopic sugar crystals (of size 3-5 microns) in slurry form are introduced into the boiling graining material. The required sugar slurry is regularly prepared in sugar factories. Now in India readymade sugar slurry is available in market.

An organic solvent is used as medium for seed slurry preparation. When an organic solvent which has lower boiling point than water is used, the solvent explode when the slurry is fed into the pan and the microscopic sugar crystals get distributed in the supersaturated molasses. The following solvents are generally used.

- Denatured ethyl alcohol /rectified spirit
- Isopropyl alcohol
- Mixture of ethyl alcohol and glycerine
- Coconut oil

84.2 Grinding method:

600 grams of best quality super small grain size sugar and 3 litters of rectified spirit are to be taken for sugar slurry preparation in a ball mill. The mill is run for 32 hours continuously with 65 RPM. Sugar slurry of 4-5 micron size will be ready. The size of the slurry is then measured.

Now in south India instead of rectified spirit coconut oil is used. One kg of sugar is taken in four litters of coconut oil. The ball mill is rotated for 24 hours.

84.3 Precipitation method:

A known quantity of sugar is dissolved in a specific volume of water so as to get a saturated solution at a particular temperature. The saturated solution is poured in a ball mill containing rectified sprit. An inclined round ball mill having smooth inner surface is generally employed.

560 g of good quality sugar is to be dissolved in 200 ml of water at 70°C so as to get a supersaturated solution upon cooling. This hot solution free from any turbidity is slowly poured into the running ball mill having 2 litres of rectified spirit. A long stem funnel dipped in the spirit is used to pour the sugar solution. The sugar solution immediately gets distributed in the spirit.

Fig.84.1 inclined ball mill

The ball mill is run for about 30 minutes so that the sugar comes out in the form of precipitate to give uniform size of slurry. After 30 minutes the whole mass is transferred from ball mill to a cylinder and allowed to settle for 10-15 minutes after which supernatant alcoholic layer is replaced by fresh volume of alcohol.

This slurry is again run in the ball mill for another 15-20 minutes to obtain the final slurry for use in the pan. The size of slurry obtained by this method will be in between 3 to 5 micron. Slurry prepared should be used within 36-48 hours.

84.4 Ball mill:

Ball mills of two different designs are available in market.

(1) Ball mills with cylindrical porcelain jar of size 10" x 10" mounted horizontally on steel frame.

(2) Round porcelain ball mill with inclined steel frame.

There are about 130-150 balls of different sizes. The frame is attached to reduction gearbox and is rotated by an electric motor. The speed of rotation of ball mill is kept at 60-65 rpm.

Fig. 84.2 Inclined ball mills

Table 84.1 Details of the ball mill - Jar Particulars:

Sr. No.	Particular	Size in mm
1	Height	300
2	Diameter	315
3	Mouth inner Dia	115
4	Mouth outer Dia	160
5	Holding capacity	10 litres
6	Bottom dia	170
7	Thickness of the cell	20
	Number of Balls	**Size mm dia**
1	5	48
2	35	27.5
3	55	21.7
4	60	18.5
total	155	--

The average size of slurry depends on,

- Number of balls in the mill
- Size of the ball mill
- The rpm of the ball mill
- Number of hours of rotation.

84.5 Weight of sugar in slurry required for graining:

Microscopic sugar crystals of size 4 to 5 micron in the form of slurry required

in grams for each tonne massecuite can be calculated as follows:
Let massecuite purity be 56.00, Crystal content % massecuite - 34.00 %
Hence 1MT of massecuite will contain 340 kg of sugar in crystal form.

Let

W_1 - Weight of fine microscopic sugar in the slurry required for one MT of massecuite

W_2 - Weight of sugar crystals in one M.T. of masse i.e. 340 kg or 340000 Gms

N - Number of crystals of W_1 Gms weight of microscopic sugar in slurry.
Since the numbers of crystals are not changed, there are same numbers of crystals in one tonne of the final massecuite.

L_1 - Average size of fines in slurry i.e. 5 micron or 0.005 mm.

L_2 - Average size of sugar crystals in the final massecuite i.e. 200 micron or 0.2 mm.

Then

$$\frac{W_1}{W_2} = \frac{1.11 \text{ X } L13 \text{ X } 1.58 \text{ X } N}{1.11 \text{ X } L23 \text{ X } 1.58 \text{ X } N}$$

$$\frac{W_1}{W_2} = \frac{L_1^3}{L_2^3}$$

$$W_1 = \frac{L_1^3}{L_2^3} \text{ X } W2$$

$$W_1 = \frac{(0.005)^3}{(0.200)^3} \text{ X } W2$$

$= (0.025)^3$ X 340 X 1000

= 0.000015625 X 340 X 1000

= 0.015625 X 340

W_1 = 5.3125 g

Thus weight of fine microscopic sugar crystals in slurry required for one M.T. of C massecuite is 5.315 g.

-OOOOO-

85 SUGARS OF DIFERENT VARIETIES

There are special sugars of different varieties in most of the countries. There are no specific properties and no strict definitions are given for these types of sugars. Most of these are categorized under natural, organic, chemical free or vegan brown sugars. These sugars are originally manufactured from cane.

Khandsari sugar: in Asian and African khandsari sugar is largely manufactured. In India and Pakistan there are thousands of small khandsari units of capacity from 50 MT/day to 500 MT/day capacity. The process is open pan boiling process. The juice is screened, boiled. Some milk of lime vegetable extract of ladies figure or guar etc. is used in clarification. The scum that comes up while concentrating is removed by spade. The highly concentrated syrup called rab. Powdered sugar is mixed in the rab and it is transferred to crystallizers. It is stored for many days and when crystals appear the massecuite is cured in small centrifugal. In recent years there has been improvement in this industry. The machinery machineries are being used. The sugar produced from some factories is quite competitive with plantation white sugar. This sugar has no taxes as this production comes under village industry.

Fig. 85.1 khandsari sugar

Demerara sugar: this sugar is originally manufactured in the Demerara area of Guyana, South America. It is a light brown sugar with large, slightly sticky crystals. It is natural and unrefined cane sugar. Now this sugar is manufactured and supplied to U. S. and European countries from many other countries like Mauritius. Originally assumed to be a product of a sugar cane mill, but nowadays it is also produced in England and Canada in refineries. It is popular for tea and coffee in England, Australia and Canada, but not very well known in the U.S.

Fig. 85.2 Demerara sugar

Muscovado sugar: it is very dark brown sugar with a strong molasses flavour. It was originally described as an unrefined sugar crystallized from the first boiling in the mill. The crystals are larger than brown sugar and sticky with strong flavour. Now light colour muscovado sugar is also available in the market. The original term of "muscovado" referred to the low quality dark brown sugar crystallized from cane juice. This muscovado sugar was then refined in Europe. The word has its origin in old Spanish and Portuguese words for "unrefined." it was produced in the European colonies of the Americas.

Turbinado Sugar: A "partially refined raw sugar" from which some of the surface molasses film has been removed by steam or water. It is a light golden to brown colour with large crystals and mild cane flavour. The crystals are dry and free-flowing. It is the same as washed raw sugar.

Barbados Sugar: this is also one special sugar available in U S A and other advanced countries.

Bura sugar: this is homemade sugar manufactured by dissolving white sugar in little water and again the concentrating the thick syrup till crystals reappears. Some cow ghee is added while concentrating the syrup. This sugar is used in manufacturing sweets like pedha, barfi etc.

All of the above sugars are considered as unrefined and hence more natural amongst the health caring people. It is assumed that these sugar contain more iron and other nutrients. They are sold on many health food web sites. In marketing strategy these sugars described "vegan and organic." Therefore these sugars are more expensive.

-OOOOO-

86 TECHNICAL CONTROL GENERAL

86.1 Introduction:

Chemical control section of a sugar factory is an important branch of the entire factory organization. It provides technical guidance to the staff working in the factory it provides all the efficiency figures to the management of the factory.

A factory laboratory plays a key role in analysis of various samples and technical accounting of the factory working. Accuracy in sampling, analysis and reporting is very much essential for reliable technical reports.

The chemical control and gives accurate account of sugar entered in the factory, losses occurred and sugar recovered.

It also gives time account i. e. the time for which mills operated, the time lost for various reasons such as general cleaning, cane shortage, stoppages due to machinery break downs and other miscellaneous reasons such as stoppages due to heavy rains etc.

The chemical control provides accounts for capacity utilization including and excluding stoppages. It also provides information about the steam used by sugar factory and bagasse consumed, bagasse saved and extra fuel if any used (in terms of bagasse). It also provides information about the chemicals used for process as well for boilers and for general cleaning purposes. It gives information about the oil and greases used in the factory.

For this purpose following data is essential.

- Weighing & measuring
- Sampling
- Routine analysis
- Calculations
- Record keeping

A part from routine analysis, analysis of raw material such as lime, sulphur, phosphoric acid that are used in the process as well as analysis of sugar and molasses is the essential.

The account of various inputs in process, which affects cost of production like chemicals, extra fuel etc. for this, records of steam, fuel & stores material are required to be maintained. Consumptions of these materials are entered in various manufacturing reports.

Various technical reports contain information about working results of the factory, sugar account, molasses account, time account, stores material account and efficiency figures besides quantities of various material handled.

In short, chemical control of a factory depends upon

a) Laboratory analysis
b) Weights of material input as well as output.
c) Standard methods of reporting.

Chemical control also helps to maintain process parameters at specified levels. Optimization of process parameters helps in increasing the capacity utilization and also in minimizing losses and improving quality of sugar

86.2 method of sugar estimation in cane and intermediate products

Basic concepts of brix, pol and purity have to be understood to know the concentrations of sugar at various stages and the sugar balance in the factory.

Brix: The percentage of total solid (w/w) in the solution is called as brix. Measurement of brix is made by brix hydrometer. Brix hydrometer is a glass spindle that floats in juice. Brix hydrometer (spindle), are made for the standard temperature, 20° / 27.5°C and are graduated in 0.1° Bx. inside the brix spindle there is a thermometer to measure the temperature of the juice.

The level up to which the brix spindle is dipped in juice depend on density of juice which ultimately depends on temperature and percentage of dissolved solids and their average density. The brix spindle is made and calibrated based on sucrose concentration. Therefore if 20 Gms of sugar is dissolved in 80 Gms of water to get 100 Gms sugar solution, the brix of the solution would be 20 deg. It is because the brix hydrometer or brix spindles are calibrated for pure sucrose solution.

But if we take 20 Gms sugar and 5 Gms common salt and 75 Gms water to get a sugar salt solution of 100 Gms, the brix of solution would be about 25°.

In a sugar factory we always have to analyse juices and other intermediate products for brix percentage which are impure and contain nonsugars / impurities in dissolved state.

In factory working brix is considered as a substance. Hence the term tons of brix is commonly used for tones of solids in intermediate materials.

There is one more method for measurement of brix that is refractometric brix. In some countries refractometric brix is used for routine analysis and record.

If we evaporate such as solution to dryness some solid matter will result. The percentage by weight of this solid matter is represented by the term 'Brix' or gravity solid.

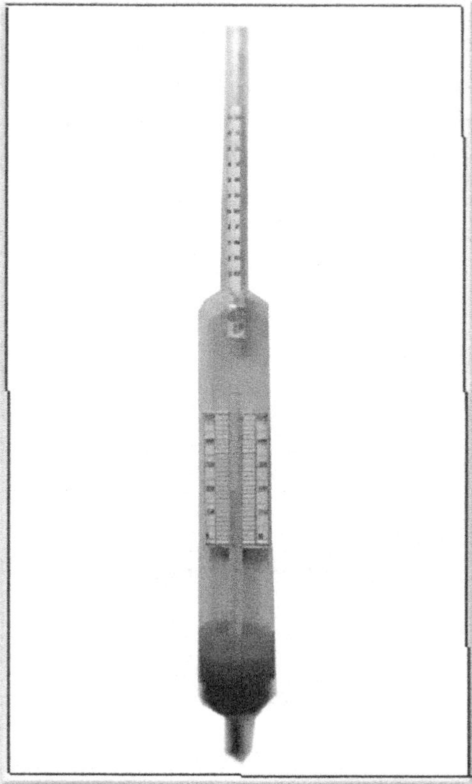

Fig. 86.1 Brix spindle

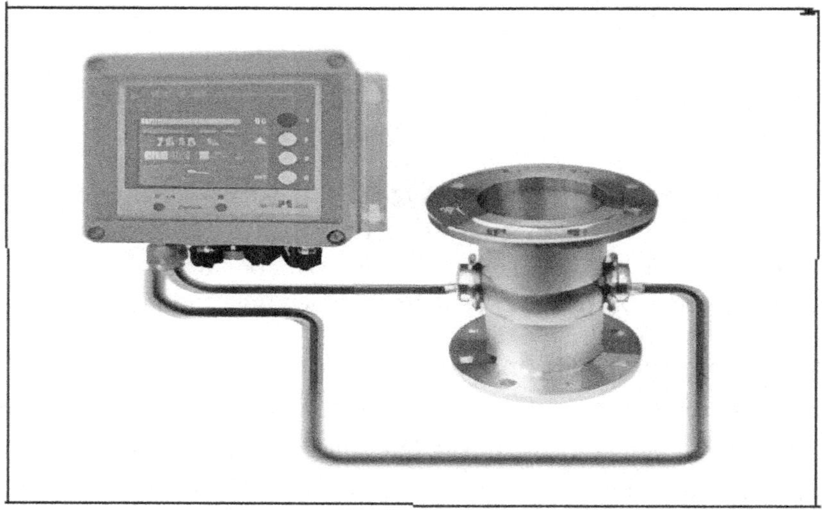

Fig. 86.2 microwave on line brix measurement

Pol or polarization:

Property of sugar is that if "polarised light is passed through a solution of sucrose; the plane of polarization is turned towards right by certain number of degrees. The number of degrees by which the plane gets rotated is directly proportional to the percentage of sugar (sucrose) present in solution.

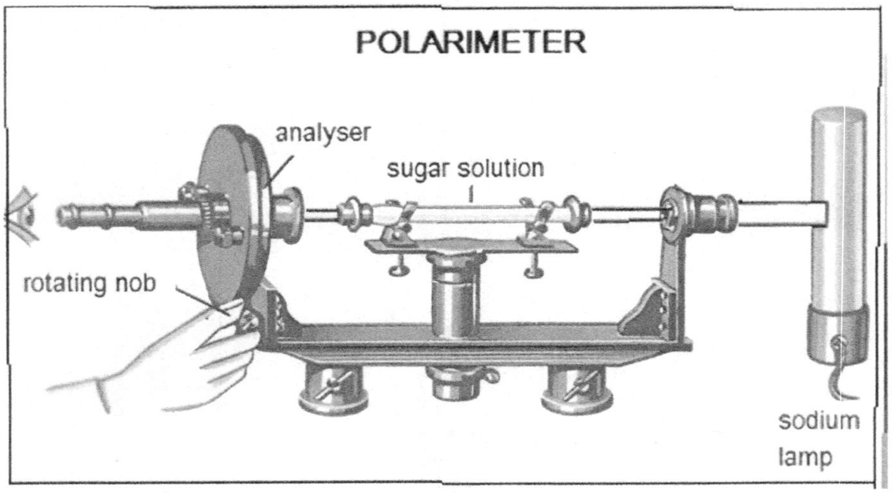

Fig.86.3 Polarimeter

Pol of a solution is the concentration (on a w/v basis) of a solution of pure sucrose in water having the same optical rotation as the sample at the same temperature. The quantity of 26.000 g of pure sucrose, weighed in air under standard conditions, dissolved in water to produce 100.00 ml of solution, will have a pol of 100°Z. Hence pol is defined as the 'Value determined by the direct polarization of normal or half normal solution in a polarimeter".

Glucose, fructose etc. are also sugars. But sugar which we recover from sugarcane in crystalline form in a sugar factory is most pure organic compound industrially recovered is sucrose. Therefore the term 'pol' derived from polarization denotes for all practical purposes only sucrose which we called it as sugar.

For solutions containing only pure sucrose in water, pol is a measure of the concentration of sucrose present. However for solutions containing sucrose and other optically active substances, pol is the algebraic sum of the rotations of the constituents present. In factory pol percent is an approximate measure of the sugar percentage in the juices / syrup / massecuite / molasses.

To measure pol reading of a sample of the juice, it is clarified by adding basic lead acetate powder (which coagulates colloidal impurities and removes some colorant) followed by filtration. The pol of the clarified solution is read in a polarimeter tube of standard length in a sugar polarimeter (self-indicating polarimetric saccharimeter). The pol of a sugar solution (a w/v measure) can be converted to pol % juice (a w/w measure) using the brix value of the solution prior to temperature correction by using Schmitz's Table. In case of syrup / massecuites and molasses the material is diluted and then clarified for measurement of pol.

Fig. 86.4 Digital automatic Polarimeter

If the juice from deteriorated cane cannot be clarified by basic lead acetate powder, a solution of lead nitrate/sodium hydroxide (Herles' Reagent) may be used as the clarifying agent.

Purity: - The purity denotes percentage by weight of sugar in solid matter or,

$$\text{Purity} = \frac{\text{Pol \%}}{\text{Brix \%}} \times 100$$

The term purity will indicate in a sugar factory, the proportion of sucrose present in total dissolved solids by weight.

Factory control:

Mainly factory control is divided into two parts

1. Mill House control or milling control.
2. Boiling House Control

We further are going to study the above topics in the next chapters. Before that we hereunder will go through some common definitions of terms that are used in technical control of the factory. The fundamental equation for mill house control is

Cane + Added Water = Mixed juice + Bagasse

From analysis and weight of mixed juice, and analysis & weight of bagasse the pol % cane is calculated. Direct analysis of cane is not taken for factory control.

Pol in cane = Pol in Mixed Juice + Pol in Bagasse

For factory control the cane, mixed juice, imbibition water, filter cake, sugar and final molasses are weighed.

86.3 Losses in sugar manufacture:

All losses in the process fall into two general classes Determined and Undetermined (Known and unknown) Chemical loss occurs either as inversion or decomposition of sucrose.

A) Determined losses:

i) **Bagasse Loss:** Sugar loss in bagasse is a part of the mill house control. This loss is usually the well-controlled in regular working factories.

ii) **Filter Cake Loss:** With almost universal adoption of rotary vacuum filters to replace the press cake, the loss of sugar in filter cake has considerably reduced.

iii) **Final Molasses Loss:** This loss depends on cane quality, Massecuite boiling system followed Treatment of low grade massecuite and curing system

B) Undermined Loss:

i) Apparent Losses: These are not real losses but play a large role in the reported losses of the factory. They are caused by some error in cane, mixed juice, imbibition water, sugar or final molasses weighment. These losses also can occur due to improper composite sampling and error in analysis. Apparent loss may also occur due to higher apparent pol reading. The apparent losses can be controlled by careful supervision of weights and analysis.

ii) Mechanical losses: Overflow leakages of juices, syrup or molasses are mechanical losses Entrainment is also included in mechanical loss. If care is taken the modern vessels and equipment can reduce this loss considerably

iii) Inversion and decomposition losses: Importance of pH control in the prevention of inversion loss is very important. Both inversion and decomposition losses can be reduced by a strict pH control at clarification station. However in sulphitation process the syrup is sulphited to a pH Of 5.0-5.2 and the material remains acidic in all the sugar boiling process till the final molasses passes out and hence the acidic inversion in the pan boiling stage cannot be avoided.

86.4 Routine sampling and Analysis:

Cane is directly measured for pol and fibre percentage directly for checking. Primary juice, last mill juice, mixed juice, bagasse, clear juice, and syrup samples are collected every 15 minutes and analysed after every 2 / 4 hours for efficient chemical control.

Composite samples of A, B and C massecuite and AH, AL, BH, CL molasses, melt and final molasses are collected frequently and are analysed at the end of every 8 hrs. shift.

Every day composite sugar samples from bags are to be collected and analysed for pol % sugar and moisture contents of sugar.

86.5 Definitions:

1) **Cane:** "The raw material delivered to the factory including clean cane, trash tops, leaves etc." The cane received to the factory is ***Gross cane***. When extraneous matter is deducted from it ***net cane*** is obtain.
2) **Absolute Juice:** "All the dissolved solids in the cane plus all the water" The ISSCT decided to abandon entirely the term 'normal juice or undiluted juice" because it has several different meanings. Because of presence of undetermined water or colloidal water, it becomes difficult to define the juice as it exists in the cane.
3) **Primary Juice:** "All the juice expressed undiluted."
4) **Secondary Juice (ISSCT):** The diluted juice that joins the primary juice to form mixed juice.
5) **Mixed Juice (ISSCT):** The mixture of primary and secondary juices, which enter the boiling house.
6) **Last mill juice (ISSCT):** The juice expressed by the last mill of a tandem.
7) **Last expressed juice (ISSCT):** The juice extracted by the last two rollers of a tandem.
8) **Residual juice:** The juice left in the bagasse minus fibre. The last expressed juice is to be used in calculation of fibre.
9) **Bagasse:** The residue of cane after crushing in one mill or a train of mills. Bagasse are named successively as first mill bagasse, second mill bagasse and so onto last mill bagasse or simply bagasse.
10) **Fibre (ISSCT):** "The dry water in soluble matter in the cane" Note that the definition includes all the insoluble including soil and stones. The true fibre is not determined in the factory control.
11) **Imbibition (ISSCT): "**The process in which water or juice is applied to the bagasse to enhance the extraction of juice at the next mill. The term is applied to the fluid used for the purpose"

12) **Maceration (ISSCT):** "A form of imbibition in which the bagasse is steeped in an excess of fluid. The term is also applied to the fluid used and loosely as alternative to the term imbibition" I. e. maceration is a special kind of imbibition. Maceration is also called as bath maceration as employed in Australia and Fiji. In many regions maceration is commonly used for imbibition.
13) **Extraction (ISSCT):** The portion (usually percentage) of a component of cane, which is removed by milling. Familiar components in this connection are juice, Brix, Pol and sucrose, and the 'extraction' is qualified accordingly. Extraction alone normally signifies Pol extraction. The term juice extraction needs an accompanying specification of the reference juice and the basis. e. g. absolute juice. Brix basis.
14) **Pol Extraction:** Pol in mixed juice percent pol in cane.
15) **Sucrose Extraction:** Sucrose in mixed juice percent sucrose in cane.
16) **Java Ratio:** This arbitrary milling ratio first used in Java. It expresses the relationship between sucrose (Pol) in the first expressed juice and sucrose (Pol) in the cane.

$$\text{Java Ratio} = \frac{\text{Sucrose (Pol) \% Cane}}{\text{Sucrose (Pol) \% first expressed juice}} \times 100$$

It was previously believed that irrespective of any mill equipment and usual variations in the fibre content of the cane, the ratio remain reasonably constant. But many technologists do not agree with it. The ISSCT defines the ratio without recommending it.

17) **Filtrate (ISSCT):** The liquid that has passed through the screen or cloths, of the filters.
18) **Filter Cake (ISSCT):** The material retained on the screen or cloths of the filters.
19) **Clarified Juice (ISSCT):** The finished product of the clarification process.
20) **Syrup:** The concentrated juice from the evaporators before crystallisation. However this meaning is for sugar factories only, in refineries the syrup has opposite meaning, where it is used to solutions from which sugar has been removed by crystallisation.
21) **Massecuite:** A mixer of sugar crystals and mother liquor when syrup or molasses is concentrated. Massecuites are distinguished by letters and /or numbers to indicate their relative purity and position in massecuite boiling scheme.
22) **Magma:** A mixture of sugar from low grade massecuite and molasses or syrup.
23) **Molasses:** The mother liquor separated from the sugar crystals when a massecuite is spun in a centrifugal machine. Letters and/or numbers corresponding to those of the massecuite further indicate the molasses. The discharge before the washing is termed as heavy molasses and diluted with wash called as light molasses. The final molasses is the molasses from which no more sugar can be removed economically.

24) Brix (Spindle Brix): The percentage by weight of the solid in a pure sucrose solution, as determined by brix hydrometer. The apparent percentage by weight of dissolved solids in juices or impure sugar solutions as determined by brix hydrometer. Brix is considered as substance and accounted as Kg of brix or tonnes of brix or brix extraction. Spindle brix solids and gravity solids are same.

25) Refractometer Brix, Refractometer solids (ISSCT): Percentage by weight of solids as determined by the refractometer (either by direct sugar scale or by the reference tables of refractive indices and percent sucrose.)

26) Solids by drying, dry substances total solids (ISSCT): The material remaining after drying the product examined.

27) Normal weight (ISSCT): "The weight of sucrose which when dissolved in water to a volume of 100 ml at 20°C and tested at 20°C in a sugar polarimeter under conditions specified for the instrument, gives a reading 100 degree of scale (26.000 g.)"

28) Sucrose, (Pol): Sucrose defined by ISSCT as "the chemical compound so named also known as saccharose or cane sugar" ISSCT has not specified any method for determination of sucrose. However ICUMSA gives method (no. GS 3/5/7/8-3, 1994) of gas chromatograph for quantitative analysis of sucrose.

Pol is the value determined by direct or single polarisation of the normal weight solution in a saccharimeter. The term is used in calculations as the real substance. Because of its simplicity and convenience, pol is used in sugar industry of all parts of the world Similar to Brix Pol is also taken as real substance and expressed as 'Kg pol' 'Tonnes of pol' or 'Pol extraction'.

29) Purity: Purity of a sugar material is the sugar present in percentage expression of the solid matter. Since sugar can be expressed as pol or sucrose and the solids as brix (spindle) refractometer solids or solids on drying it follows that that the purity can be expressed in several forms.

$$\text{i) Apparent purity} = \frac{\%\ \text{Pol}}{\text{Spindle Brix}}$$

$$\text{ii) Ref. Purity} = \frac{\%\ \text{Pol}}{\text{Ref. Brix}}$$

$$\text{iv) True Purity} = \frac{\text{(True) Sucrose}}{\text{Dry substances}}$$

iii) Gravity Purity = $\dfrac{\text{(True) Sucrose}}{\text{Spindle Brix}}$

v) Ref. Sucrose Purity = $\dfrac{\text{(True) Sucrose}}{\text{Brix}}$

When purity is not specifically mentioned it is understood as apparent purity. The apparent purity is used in routine analyses and reports. ISSCT has adopted gravity as the standard for comparative control. The ISSCT permits the use of refractometer brix instead of hydrometer brix with compromise that the method of determination must be stated.

30) **Reducing Sugars (ISSCT):** Reducing substances in cane and its products interpreted as invert sugars.

31) **Ash (ISSCT):** The residue remaining after incinerating the product under specified conditions.

32) **Recovery:** The percentage of the pol in cane that passes into mixed juice is termed as mill extraction and the percentage of the pol in mixed juice that passes into the sugar produced is termed as boiling house recovery. The product of these two is known as overall recovery

33) **Boiling house recovery or extraction** (B.H.R.): The value of the ratio of sucrose (Pol) in commercial sugar produced to the sucrose (Pol) in mixed juice.

34) **Overall extraction:** It is the values of the ratio of the sucrose (Pol) in commercial sugar to the sucrose (Pol) in cane.

-OOOOO-

87 MILL HOUSE CONTROL

87.1 General:

Mill house control means assessment of the performance of mills. The mill house control deals with

- Weight of cane crushed,
- Determination of the amount of sugar extracted in mixed juice,
- Determination of sugar loss in bagasse,
- Quantity of imbibition water used at mills.

The material input and output at the mills gives fundamental or basic formula of mill house control.

Input of materials at mills = Output of the materials at mills

Or Cane + Added water = Mixed juice + Bagasse

The cane and mixed juice are weighed. The added water is either weighed or measured volumetrically or by magnetic flow meter and weight is calculated. Weight of bagasse is calculated from the above basic formula.

87.2 Sugar Balance at mills:

Sugar (pol) in cane = Sugar (pol) in mixed juice + Sugar (Pol) in Bagasse

Or

Sugar (Pol) % cane = Sugar (Pol) in M.J. % cane + Sugar (Pol) in Bagasse % cane

Where,

$$\text{Sugar (Pol) in M.J. \% cane} = \frac{\text{Sugar (Pol) \% M.J. X M.J. \% cane}}{100}$$

And,

$$\text{Sugar in Bagasse \% cane} = \frac{\text{Sugar (Pol) \% Bagasse X Bag. \% cane}}{100}$$

Mixed juice is analysed for Bx. % & Pol % and bagasse is analysed for Pol% and Moisture%.

87.3 Mill Extraction: The quantity of sugar extracted in the mixed juice per 100 sugar in cane i.e. percentage of sugar extracted in the mixed juice out of total sugar in cane.

$$\text{Mill Extraction} = \frac{\text{Sugar (Pol) in M.J. \% cane}}{\text{Sugar (Pol) \% cane}} \times 100$$

$$= \frac{\text{Pol \% M.J. x M.J. \% Cane}}{\text{Pol \% M.J. X M.J. \% Cane + Pol \% Bag. X Bag. \% Cane}} \times 100$$

Example:

Given particulars	tonnes
Cane crushed	2560
Added Water	880
Mixed juice	2705

Analysis Particulars	Brix%	Pol%	Purity	Moisture%
M J	15.40	12.70	82.46	
Bagasse		2.2		

Calculation:

M.J. % Cane = 105.66
A. W. % Cane = 34.36
Bag. % Cane = 28.70

Then

$$\text{Mill extraction} = \frac{12.70 \times 105.66}{12.70 \times 105.66 + 2.2 \times 28.70} \times 100$$

$\therefore$ Mill extraction = 95.48

87.4 Composition of bagasse:

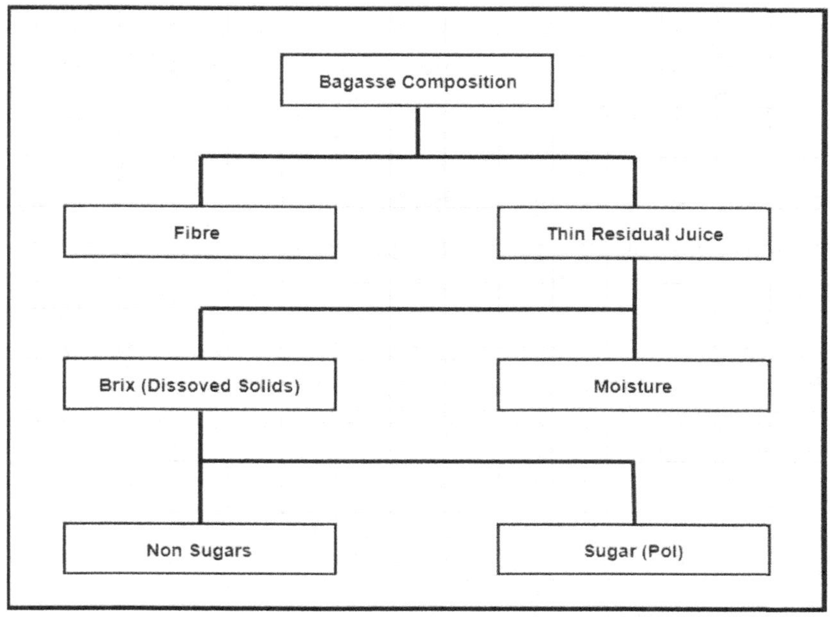

Fig. 87.1 composition of bagasse

Fibre % Bagasse:

Bagasse = Fibre + Moisture + Brix
Fibre = Bagasse – Moisture – Brix

Now if bagasse = 100 then,

$$\text{Fibre \% Bag.} = 100 - \text{Moist. \% Bag.} - \text{Brix \% Bag.}$$

The bagasse is analysed in factory laboratory for pol and moisture percentage and the brix percentage of bagasse is calculated

Brix % Bagasse:

While determining brix % bagasse, it is assumed that purity of thin residual juice in bagasse is same as that of last expressed juice (LEJ)

Thus

$$\therefore \text{Brix \% Bagasse} = \frac{\text{Pol \% Bagasse}}{\text{Purity (LEJ)}} \times 100$$

Fibre % cane:

$$\text{Fibre \% cane} = \frac{\text{Fibre \% Bag X Bag \% cane}}{100}$$

Laboratory method for Fibre % cane: by Rapi-pol Extractor

The sugar cane sticks are cut into five pieces. 250 Gms of this sample along with 2000 ml water is transferred to rapi-pol vessel. Run the Rapi-pol extractor for 15 minutes. Decant the whole liquid from vessel. No particle of fibre should be lost. Then again 2 to 3 litres of water is added to remaining material and run the extractor for 15 minutes. Repeat the same above procedure for 2 to 3 times. Then take out the fibrous mass from vessel after well squeezing with hands and dry the fibrous mass completely in drier at 105°C and weigh it. The fibre % cane will be calculated as,

$$\text{Fibre \% cane} = \frac{\text{Weight of the fibrous mass}}{250} \times 100$$

87.5 Reduced Mill extraction by Noel Deerr:

Apart from the cane preparatory devices, number of rollers in milling tandem and operating conditions such as roller RPM hydraulic pressure at rollers etc., the mill extraction is greatly influenced by fibre percentage in the cane. The fibre percentage in cane depends on variety of cane, agricultural practices, water availability, maturity, and summer etc. Extraction will also vary according fibre percentage in cane. Therefore assessment of mill extraction by comparison with mill extraction of the same factory for different periods or that with another factory cannot be done.

The fibre takes along with it certain quantity of thin juice and the juice lost in bagasse per unit fibre depends on working performance of milling tandem. When the mill working remains steady, the juice lost per unit fibre remains fairly constant. Hence juice lost in bagasse varies in proportion with fibre % cane and consequently mill extraction changes with change in fibre % cane. Therefore mill extraction with different fibre % cane cannot be compared.

To eliminate the effect of variation in fibre % cane Noel Deerr proposed formula to reduce the extraction to common basis of 12.5 % fibre, that is called as Reduced Mill Extraction ($RME_{\text{Noel Deerr}}$).

Suppose,
For unit juice in cane e is mill extraction.
Then juice lost in bagasse will be (1 – e)
(Here for simplicity the pol extraction is considered as juice extraction).
But unit cane contains (1 – f) juice.

Therefore juice lost per unit cane will be (1 – e) (1 – f)
The unit cane contains f fibre. Therefore juice lost in bagasse per unit fibre will be

$$\frac{(1 - e)(1 - f)}{f}$$

Now suppose for a milling tandem working different period with the same condition but cane of different fibre percentage.

Case I E is mill extraction when F is fibre per unit cane which is equal to 0.125.
Case II e is mill extraction when f is fibre per unit cane,

Now as mill is same and working conditions are also same, the juice lost per unit fibre will be same in both the cases.

Hence,

$$\frac{(1 - E)(1 - F)}{F} = \frac{(1 - e)(1 - f)}{f}$$

$$\therefore \frac{(1 - E)(1 - 0.125)}{0.125} = \frac{(1 - e)(1 - f)}{f}$$

$$\text{Assume} \quad \frac{(1 - e)(1 - f)}{f} = V$$

$$\text{Then} \quad V = 7(1 - E)$$

$$\text{Or} \quad RME_{(Noel\ Deerr)}\ E = \frac{7 - V}{7}$$

E is the mill extraction at 12.5 % fibre or called as reduced mill extraction ($RME_{Noel\ Deerr}$). The term that is denoted by the symbol **V** is the "Lost juice per unit fibre".

Now with the reduced mill extraction figures comparison of the same factory for different periods is possible. Also comparison of different factories is possible.

With the figures of RME we can come to conclusion that in which period the same mill is working better or for different factories which mill is working better. The factory of which RME is higher is working better than other factory.

87.6 Reduced Mill Extraction by B L Mittal

In the reduced mill extraction formula derived by Noel Deerr the value of 'V' (juice retention factor), the absolute juice in bagasse % fibre, in the formula is calculated from the following equation.

$$V = \frac{(100 - e) \times (100 - f)}{f}$$

Now as per definitions given by ISSCT

100 – f = absolute juice % cane

And 100 – e = sucrose or pol in bagasse % sucrose or pol in cane.

The dissolved solids in the juice are not dispersed uniformly. However the definition of absolute juice does take in to consideration the uneven dispersion of the dissolved solids in the cane. This is also true for sucrose in the juice. It is clear from the evident that there is a considerable drop in juice purity from the first mill to the last mill.

Hence the extraction, which is calculated on the basis of pol or sucrose, is not same as the absolute juice extraction. It is therefore wrong to presume that the expression (100-e) x (100-f) is the value of lost absolute juice or the absolute juice in bagasse % cane.

Deerr's formula for reduced mill extraction also fails in its main purpose to reduce the extraction values to a common basis of 12.5% fibre in cane. In other words, the influence of variations in fibre % cane is not completely eliminated. This will be clear from the following example.

Let us take example, suppose pol% cane, pol% bagasse & fibre % bagasse (m) all being independent of fibre % cane remain constant, their values being 13.0, 2.5 and 48.00 respectively. Then the reduced extraction for different values of fibre % cane is determined. The results are given in the following table.

Reduced Extraction by Deerr's formula for different values of fibre % cane, when pol% cane, pol% bagasse and fibre % bagasse are constant.

Table 87.1 RME by N D for different values of fibre % cane

Sample	Fibre % cane	Bag.% cane (100 f/m)	Pol% Bag.	Pol in bag % cane pbc	Pol% cane Pc	100-e	100-f	V	RME = (7-v) x 100 / 7
1.	11.0	22.92	2.5	0.573	13.0	4.41	89.00	35.68	94.90
2.	12.0	25.00	2.5	0.625	13.0	4.81	88.0	35.27	94.96
3.	12.5	26.04	2.5	0.651	13.0	5.01	87.5	35.06	94.99
4.	13.0	27.08	2.5	0.677	13.0	5.21	87.0	34.86	95.02
5.	14.0	29.17	2.5	0.729	13.0	5.61	86.0	34.46	95.08
6.	15.0	31.25	2.5	0.781	13.0	6.01	85.0	34.06	95.13
7.	16.0	33.34	2.5	0.833	13.0	6.41	84.0	33.66	95.19
8.	17.0	35.42	2.5	0.885	13.0	6.81	83.0	33.25	95.25

It can be seen from the above table that the values of reduced mill extraction (RME) go on increasing with the increase in values of fibre % cane or in other words, the influence of fibre content of cane has not been completely eliminated.

To overcome this inaccuracy or error B L Mittal has derived another formula for RME. Derivation of B L Mittal's RME formula is given below.

Derivation of Reduced Mill Extraction by Mittal:

Let e = actual mill extraction or sucrose or pol basis

Pc = Pol % cane

Pbc = Pol in bagasse % cane

fc = fibre % cane

$$\text{Pol in bagasse percent pol in cane} = \frac{Pbc}{Pc} \times 100$$

And also pol in bag percent pol in cane = 100 – e

$$\therefore \quad 100 - e = \frac{Pbc}{Pc} \times 100$$

$$\therefore \quad Pbc = \frac{(100 - e)\, Pc}{100}$$

$$\text{i.e. Pol lost in bagasse \% cane} = \frac{(100 - e)\quad Pc}{100}$$

$$\text{Now pol lost in bag \% fibre} = \frac{\text{Pol in bag \% cane}}{\text{fibre \% cane}} \times 100$$

$$= \frac{(100 - e)\, Pc}{fc}$$

$$= \frac{(100 - e)\, Pc}{fc} = M \text{ say}$$

$$\therefore \quad M \times fc = (100 - e)\,Pc$$

$$= \frac{M \times fc}{Pc} = (100 - e)$$

Or

$$e = 100 - M \times \frac{fc}{Pc}$$

Now let fc = 12.5, then the extraction is reduced to,

$$E = 100 - M\,\frac{12.5}{Pc}$$

Substituting the value of M in above equation,

$$\text{Then } E = 100 - \frac{(100 - e)\,Pc}{fc} \times \frac{12.5}{Pc}$$

or

$$\text{Then } E = 100 - \frac{12.5\,(100 - e)}{fc}$$

Now we know that 100 – e = Pbc x 100

Substituting the value of (100 –e) in above equation becomes

$$E = 100 - \frac{100 \times 12.5 \times Pbc}{Pc \times fc}$$

Thus RME by Mittal is given by formula

$$e' = 100\left(1 - \frac{12.5 \times Pbc}{Pc \times fc}\right)$$

Simple method of derivation of reduced mill extraction by Mittal:

Since the loss at the mills is (other things being equal) proportional to the fibre f of cane, it is more logical in order to eliminate the influence of fibre to relate the pol loss to the fibre content.

Therefore

$$\frac{1 - E}{F} = \frac{1 - e}{f}$$

Hence

$$RME_{Mittal} = E = 1 - \frac{(1 - e)\,F}{f}$$

Putting the value of F as 0.125 the equation becomes

$$RME_{Mittal)}\ E_{12.5} = 1 - \frac{0.125\,(1 - e)}{f}$$

Where f is the actual fibre per unit cane e is mill extraction.

$$\text{Or}\quad RME_{Mittal} = 100\left(1 - \frac{12.5\ Pbc}{Pc \times fc}\right)$$

The term Pbc & Pc in above formula can be accurately determined while 'f' is inferentially calculated. To check whether this new formula eliminates the influence of varying fibre values, reduced extractions in case of example No.1 are calculated; The results are given below reduced extraction by Mittal's formula for different values of fibre % cane and, pol % cane, pol% bagasse and fibre % bagasse remaining constant at 13.0, 2.5 & 48.0 respectively.

Table 87.2 RME by Mittal for different values of fibre % cane

Sample No.	Fibre % cane (f)	100 Pbc / Pc	Reduced Mill Extraction
1.	11.0	4.41	94.99
2.	12.0	4.81	94.99
3.	12.5	5.01	94.99
4.	13.0	5.21	94.99
5.	14.0	5.61	94.99
6.	15.0	6.01	94.99
7.	16.0	6.41	94.99
8.	17.0	6.81	94.99

It is seen from the results that reduced RME by Mittal's formula remains the same for all values of fibre % cane.

87.7 Undiluted juice lost in bagasse % fibre:

For the purpose of calculation, it is assumed that bagasse has the same brix as that of primary juice and purity as that of last expressed juice (last mill juice).

Suppose
1) Brix % primary juice = A
2) Brix % bagasse = B
3) Juice = 100
4) Fibre % bagasse = M

For 'A' Brix undiluted juice = 100

$$\text{For 'B' Brix undiluted juice} = \frac{B}{A} \times 100$$

But now 'B' juice in 100 bagasse which contains M fibre
Hence

M fibre takes 100B / A juice in the bagasse

$$\therefore \text{ 100 fibre takes } \frac{100\,B}{A} \times \frac{100}{M} \text{ juice in bagasse.}$$

$$\text{Juice lost \% fibre} = \frac{\text{Brix of bagasse X 10,000}}{\text{Brix of primary juice X fibre \% bagasse}}$$

This is simple formula to calculate juice lost % fibre for which no weighment quantities are required.

87.8 Primary Extraction:

When maximum juice is extracted at the first mill the task of extracting the juice by secondary mills is reduced. Further the overall extraction of the milling tandem is increased. It is confirmed that when the mill working remains steady, a gain of 1% in primary extraction gives a gain in total extraction of 0.12 % in 12-roller milling tandem and 0.10 % in 15 roller tandem and 0.09 % in 18 roller tandem.

It is therefore, necessary to observe first mill extraction carefully two or three times a week. A tandem of 3-roller unit should attain a minimum of 60-63 % primary extraction. It may reach easily to 70-72 % and as high as 75 to 78 % according to fibre % cane, preparatory index, feeding at the mill, mill setting, hydraulic pressure maintained, proper juice drainage and optimum mill roller speed.

With a view to obtain high primary extraction, the first mill should be provided with Donnelly chute, under feed roller and a lotus roller.

Primary extraction is the percentage of pol (sugar) extracted of the pol (sugar) in cane by the primary / first mill.

Thus we can say,

$$\text{Primary extraction} = \frac{\text{Pol \% Primary Juice X Primary Juice \% Cane}}{\text{100 X Pol \% Cane}}$$

Primary juice and primary bagasse samples are collected for analysis. While collecting the sample of primary bagasse care is to be taken to stop the secondary juice being imbibed over the bagasse emerging out from the first mill.

- Primary juice is analysed brix, pol and purity,
- Primary bagasse is analysed for pol % primary bagasse and fibre % primary bagasse in rapi-pol extractor.

The primary juice % cane and primary bagasse are calculated as follows:

$$\text{Primary bagasse \% cane} = \frac{\text{Fibre \% cane}}{\text{Fibre \% primary bagasse}} \times 100$$

- The fibre % cane is taken from the previous day daily manufacturing report,
- The fibre % primary bagasse is found out in laboratory by Rapi pol extractor,

Primary juice % cane = 100 – primary bagasse % cane

Example

Let us take	Pol % P J	=	15.6
	Pol % P B	=	8.7
	Fibre % P B	=	35.5
	Fibre % cane	=	13.7

$$\text{Then P B \% cane} = \frac{13.70}{35.50} \times 100 = 38.59$$

And P J % cane = 61.41

Putting the value in the equation we get,

$$\text{Primary extraction (P E)} = \frac{(15.6 \text{ X } 61.41/100)}{(15.6 \text{ x } 61.41/100) + (8.7 \text{ X } 38.59/100)} \text{ X } 100$$

Primary extraction (P E) = 68.70

87.9 Equivalent Ratio Quotient Value (ERQV):

ERQV figures indicate mill sanitation condition. More purity drop indicate poor mill sanitation. The normal difference between crusher juice purity (1st expressed juice) and mixed juice purity should be of the order of 1.2 to 1.4. Purity drop greater than 1.4 indicates increasing sucrose loss due to inversion mainly bacterial. In chemical control, ERQV is a measure of mill sanitation efficiency and is given by the equation.

$$\text{ERQV MJ/PJ} = \frac{1.4 \text{ X M. J. Pty} - 40}{1.4 \text{ X P. J. Pty} - 40}$$

$$\text{ERQV LMJ/PJ} = \frac{1.4 \text{ X LMJ Pty} - 40}{1.4 \text{ X PJ Pty} - 40}$$

An ERQV MJ/PJ value of about 98 indicates good mill sanitation performance and ERQV = LMJ / PJ, a value of about 85 is considered satisfactory. This figure is also termed as "Proportionate Purity Ratio".

87.10 Added water extracted % Added water:

$$\text{Added Water \% Fibre} = \frac{\text{Added Water \% Cane}}{\text{Fibre \% cane}}$$

$$\text{Added Water extracted \% M.J.} = \frac{\text{Bx. \% P.J.} - \text{Bx. \% M.J.}}{\text{Bx. \% P.J.}} \text{ X } 100$$

$$\text{A.W. extracted in M.J. \% Cane} = \frac{\text{A.W. Extracted \% M.J. X M.J. \% Cane}}{100}$$

$$\text{A.W. Extracted \% A. W.} = \frac{\text{A.W. extracted in M.J. \% Cane}}{\text{A.W. \% Cane}} \text{ X } 100$$

87.11 Brix Free Cane Water:

Brix free cane water is the portion of water the moisture associated with fibre and cannot be separated from it mechanically. It does not contain any sugar or brix, it is also called as colloidal water. It is closely associated with protoplasmic constituents of living cells and fibre in cane. The general value of B.F.C.W. % fibre varies from 16 to 30. It helps in checking the chemical control figures.

Therefore we can assume,

Cane + Fibre + undiluted juice + BFCW

Or BFCW = Cane – undiluted juice – Fibre

Or BFCW % Fibre = Cane % Fibre – undiluted juice % Fibre – 100

I) Suppose Fibre % Cane is 'F'

Then For F Fibre content ≡ 100 Cane

$$\therefore \text{ 100 Fibre content} \equiv \frac{100 \text{ X } 100}{F} \text{ ----- cane}$$

$$\therefore \text{ Cane \% Fibre} = \frac{100 \times 100}{\text{Fibre \% Cane}}$$

II) $$\text{Undiluted juice \% Bag} = \frac{\text{Bx. \% Bag.}}{\text{Bx. \% P.J.}} \text{ X } 100$$

Similarlly

$$\text{Undil. juice \% Cane} = \frac{\text{Bx. \% Cane}}{\text{Bx. \% P.J.}} \text{ X } 100$$

Therefore,

$$\text{Undil. Juice \% Fibre} = \frac{\text{Undil, juice \% Cane}}{\text{Fibre \% Cane}} \text{ X } 100$$

$$\therefore \quad \text{Until. Juice \% Fibre} = \frac{\text{Bx. \% Cane X 100 X100}}{\text{Bx \% P.J. X Fibre \% Cane}}$$

Putting the values we get,

$$\text{BFCW \% Fibre} = \frac{100 \times 100}{\text{Fibre \% Cane}} - \frac{\text{Bx, \% Cane X 100 x 100}}{\text{Bx.\% P. J. X Fibre \% Cane}} - 100$$

$$\text{Or BFCW \% Fibre} = \frac{100 \text{ X } 100}{\text{Fibre \% Cane}} \left(1 - \frac{\text{Bx. \% Cane}}{\text{Bx. \% P.J.}} \right) - 100$$

-OOOOO-

88 BOILING HOUSE CONTROL

88.1 Terms used in boiling house control

1) **Commercial Sugar:** The product of the sugar factory as sent to the market. It is not 100 % pure, but contains very little quantities of impurities (non-sugar).

2) **Standard Granulated:** The term used to indicate the finest commercial product and considered as equivalent to pure sugar.

3) **Equivalent Standard Granulated:** The commercial sugar is expressed in terms of standard granulate.

4) **Boiling house recovery or extraction** (B.H.R.): The value of the ratio of sucrose (Pol) in commercial sugar produced to the sucrose (Pol) in mixed juice.

5) **Overall extraction: It** is the values of the ratio of the sucrose (Pol) in commercial sugar to the sucrose (Pol) in cane.

6) **Basic boiling house recovery:** The theoretical boiling house recovery as calculated by S-J-M formula in which J in gravity purity of M.J. and S=100 and m=28.57.

88.2 S-J-M Formula

This formula is used for calculating the available sugar from the raw material or from the other material. Suppose,

S = Sugar purity or final products.
J = Purity of Initial material or purity of raw material.
M = Purity of molasses or by products.
1 = Solids in initial material i.e. solids in raw material.
C + D = Total solids in final products (C = Sugar & D = Non-sugar)
1 – (C + D) = Solids in the co product (molasses)

Sugar in initial material = Sugar in final product + Sugar in by-product

$$1 \times J = (C + D)\,S + (l - (C + D))\,M$$
$$J = (C + D)\,S + M - M\,(C+D)$$
$$= (C + D)\,S - (C+D)\,M + M$$
$$J\text{-}M = (C + D)\,(S - M)$$

Or $$(C + D) = \frac{(J - M)}{(S - M)}$$

Multiplying both the sides by S / J

$$\frac{S(C + D)}{J} = \frac{S\,(J - M)}{J\,(S - M)}$$

But S (C + D) = Sugar in final product
And 1 x J = Sugar in initial product

Therefore

$$\frac{\text{Sugar in final product}}{\text{Sugar in initial product}} = \frac{S\,(J - M)}{J\,(S - M)}$$

Calculating available sugar by SJM formula

In factory working generally stock of material is taken volumetrically then its sample is analysed for Brix, Pol, and Purity From reference table density of the material is found out from the brix. Then the sugar in the original material is calculated and to that figure the SJM formula is applied.

Suppose,

- Volume of syrup V m^3
- Brix of syrup B
- Syrup purity J
- Density of the syrup D

The density of syrup can be found out from the reference table. Then sugar in the syrup will be

$$\frac{V \times D \times B \times J}{100 \times 100}$$

We can apply SJM formula to the sugar in syrup (Original Material) and can found out available sugar from the syrup.

$$\text{Available Sugar} = \frac{V\,D\,B\,J}{100 \times 100} \times \frac{S\,(J - M)}{J(S - M)}$$

Putting the value of sugar purity as 100 the formula becomes

$$\frac{V\,D\,B\,(J - M)}{100\,(100 - M)}$$

Example: Given

Syrup volume	50 m^3
Syrup Brix	60°
Purity of syrup	80.00
Purity of final molasses	30.00
Sugar purity	100
Syrup density	1.28

Putting the values in the above equation we get

$$\text{Available sugar} = \frac{50 \times 1.28 \times 60\,(80 - 30)}{100(100 - 30)} = 27.428 \text{ tonnes}$$

Therefore available sugar = 27.428 tonnes

Calculating available molasses by SJM formula:

[Solids in the original material] – [solids in the final product (sugar)]
= [solids in the coproduct i.e. Final Molasses]

But solids in the final product is the Available Sugar

Therefore,

$$\text{Available Molasses} = \frac{[\text{Solids in final Molasses}] \times 100}{\text{FM Bx.}}$$

$$= \frac{[(\text{Solids in the original Material}) - (\text{Available Sugar})] \times 100}{\text{FM Bx.}}$$

$$= \left(\frac{V\,D\,B}{100} - \frac{V\,D\,B\,(J - M)}{100\,(100 - M)}\right)\frac{100}{\text{FM Bx.}}$$

$$\text{Available Molasses} = \frac{V\,D\,B}{FM\ Bx.}\left(1 - \frac{(J - M)}{(100 - M)}\right)$$

The example given for available sugar is continued for available molasses
Let the final molasses brix be 90, then putting the values

$$\text{Available Molasses} = \frac{50 \times 1.28 \times 60}{90}\left(1 - \frac{(80 - 30)}{(100 - 30)}\right)$$

Available Molasses = 12.190 t

88.3 Winter's formula for available sugar:

Let

- S = Total sugar present in the original material.
- B = Total solid (brix) present in the original material.
- B – S = Total non-sugar in the original material.
- M = purity of the final molasses i.e. co product.

When purity of FM is M means its Brix is 100 and sugar is M.
I.e. FM contains M sugar when brix is 100 and
100 - M = non-sugar in final molasses
That means 100 – M non-sugars carry M sugar
Hence unit non sugar carry along with it

$$\frac{M}{100 - M}$$

Therefore total non-sugar in original material are (B – S), that will carry with them

$$\frac{(B - S)\,M}{(100 - M)} \text{ Sugars}$$

Thus $(B - S)\dfrac{M}{(100 - M)}$ is non available sugar

But total sugar present in original material = S

Therefore
Available sugar = Total sugar – Non-available sugar.

$$\therefore \text{Available sugar} = S - (B - S)\frac{M}{100 - M}$$

Technologist observed that generally when the molasses is well exhausted unit non-sugars carry with it 0.4 sugar and final molasses purity is 28.57. Winter's formula is generally used to calculate Basic boiling house recovery when the final molasses purity is assumed to be 28.57.

Then the above formula becomes

$$\text{Available sugar} = S - 0.4\,(B - S)$$

Then Basic Boiling House Recovery

$$= \frac{\text{Available sugar at 28.57 FM purity}}{\text{Total sugar in the original material}}$$

$$\text{BBHR} = \frac{S - 0.4\,(B - S)}{S}$$

88.4 Clarification Factor:

The purpose of clarification of juice in sugar factory is to remove as much as non-sugar from it as possible with the least loss of sugar rendering the juice desirably sparkling bright and transparent. The colour of sugar its keeping quality and sugar losses in final molasses all have got that bearing on the clarification efficiency.

Good clarification means better sugar less losses and unsatisfactory clarification means (results in) production of inferior quality of sugar with poor keeping quality and higher manufacturing losses.

$$\text{Clarification factor} = \frac{\text{N. S. in removed in filter cake \% Cane}}{\text{N.S. in M. J. \% Cane}}$$

$$= \frac{\text{N. S. in M.J. \% Cane} - \text{N.S. in C.J. \% Cane}}{\text{N. S. in M.J. \% Cane}}$$

I) N. S. in M.J. % Cane = Bx % M.J.x M.J. % Cane – Pol % M.J.x M.J. % Cane

II) Pol in C.J. % Cane = Pol in M.J. % Cane – Pol in F. C. % Cane

$$\text{III) Bx in C. J. \% Cane} = \frac{\text{Pol in C.J. \% Cane X C. J. Pty}}{100}$$

IV) N. S. In C J. % Cane = (III – II)

$$\text{Clarification Factor} = \frac{(\text{I} - \text{IV})}{\text{I}}$$

Higher the value of clarification factor better is considered the efficiency of clarification. The clarification factor may be very high showing high efficiency figure of clarification though loosing much sugar in F.C. for i. e. If the filter cake is not washed thoroughly. This will result in higher removal of non-sugars as well as higher sugar.

Normally a clarification factor figure of 20-25 for carbonation factories and 2-10 for sulphitation factories is to be indicated as good working efficiency of clarification process.

88.5 Losses in sugar manufacture:

All losses in the process fall into two general classes Determined and Undetermined (Known and unknown) Chemical loss occurs either as inversion or decomposition of sucrose.

A) Determined losses:

Bagasse Loss: Sugar loss in bagasse is a part of the mill house control. This loss is usually the well-controlled in regular working factories.

Filter Cake Loss: With almost universal adoption of rotary vacuum filters to replace the press cake, the loss of sugar in filter cake has considerably reduced.

Final Molasses Loss: This loss depends on cane quality, Massecuite boiling system followed Treatment of low grade massecuite and curing system

B) Undermined Loss:

Apparent Losses: These are not real losses but play a large roll in the reported losses of the factory. They are caused by some error in cane, mixed juice, imbibition water, sugar or final molasses weighment. These losses also can occur due to improper composite sampling and error in analysis. Apparent loss may also occur due to higher apparent pol reading. The apparent losses can be controlled by careful supervision of weights and analysis.

Mechanical losses: Overflow leakages of juices, syrup or molasses are mechanical losses Entrainment is also included in mechanical loss. If care is taken the modern vessels and equipment can reduce this loss considerably

Inversion and decomposition losses: Importance of pH control in the prevention of inversion loss is very important. Both inversion and decomposition losses can be reduced by a strict pH control at clarification station. However in sulphitation process the syrup is sulphited to a pH Of 5.0-5.2 and the material remains acidic in all the sugar boiling process till the final molasses passes out and hence the acidic inversion in the pan boiling stage cannot be avoided.

88.8 Pol Balance:

1) Pol % Cane = Pol in M.J. % Cane + Pol In Bag. % Cane

2) $$\text{Pol in M.J. \% Cane} = \frac{\text{Pol \% M.J. X M.J. \% Cane}}{100}$$

3) $$\text{Pol in Bag. \% Cane} = \frac{\text{Pol \% Bag. X Bag \% Cane}}{100}$$

4) $$\text{Pol in F.C. \% Cane} = \frac{\text{Pol \% Filter Cake X F.C. \% Cane}}{100}$$

5) $$\text{Pol in FM \% Cane} = \frac{\text{Pol \% FM X FM \% Cane}}{100}$$

6) $$\text{Pol in Sugar \% Cane} = \frac{\text{Pol \% Sugar X Sugar \% Cane}}{100}$$

7) Total losses = Pol % Cane – Pol in Sugar % Cane
= Loss in (Bag. + F.C. + F. M. + Unknown)

8) Unknown loss =Total losses – Determined loss (Losses in Bag.+ F.C.+ FM)

88.9 Boiling house recovery

The efficiency of the boiling house is defined in a manner analogous to that of the mill plant. The term "boiling-house recovery" indicates the ratio of the sucrose obtained in the sugar manufactured to that entering in the mixed juice.

$$\text{Boiling house recovery} = r = \frac{\text{Sucrose obtained in sugar manufactured \% cane}}{\text{Sucrose in mixed juice \% cane}}$$

It is seen that the recovery takes into account the following losses:

- Loss of sugar in filter cake
- Loss of sugar in molasses
- Undetermined losses, (leakages, entrainment, inversion, etc.).

88.10 Reduced boiling house recovery by Noel Deer's formula:

$$R_{85\ N.D.} = \frac{100(85 - M_V)}{85(100 - M_V)}$$

Where M_V is the virtual final molasses purity and is calculated by the formula

$$M_V = \frac{100\ J\ (1 - R)}{(100 - RJ)}$$

Reduced boiling house recovery by Gundu Rao's formula:

$$R_{85\ G.R.} \quad r + k \times \frac{M}{1 - M} \left(\frac{17 - 20J}{17J} \right)$$

Where r = B. H. R.

$$k = \frac{\text{N.S. in clear juice \% Cane}}{\text{N.S. in mixed juice \% Cane}}$$

M = Final Molasses purity

J = Mixed juice purity

-OOOOO-

List of reference books

Sr. no.	Book	Author
1	Cane Sugar Handbook	Spencer
2	Handbook of Cane Sugar Engineering	E. Hugot and G.H. Jenkins 3rd Ed. 1986
3	Manufacture and Refining of Raw Sugar	V.E. Baikow, 1967.
4	Principles of Sugar Technology vol. I and II	Edited by P. Honig
5	Manufacture of Sugar from Sugarcane	C.G.M. Perk 1973.
6	Machinery and Equipment of the Cane Sugar Factory (eBook)	L.A. Tromp. 1946.
7	Cane Sugar (eBook)	Noel Deerr,1921
8	Introduction to cane sugar technology	Jenkins 1967
9	Handbook of cane sugar technology	R B L Mathur 1982
10	Cane sugar manufacture in India (eBook)	D P Kulkarni 1992
11	System of cane sugar factory control (eBook)	ISSCT publication 1955
12	On cane sugar and the process of its manufacture in java (eBook)	H. C. Prinsen Geerligs
13	Chemical Control (eBook N S I Kanpur)	Jahar Singh

-OOOOO-

Made in United States
Orlando, FL
12 August 2024

50301179R00339